Conservation of Faunal Diversity in Forested Landscapes

JOIN US ON THE INTERNET VIA WWW, GOPHER, FTP OR EMAIL:

WWW: http://www.thomson.com
GOPHER: gopher.thomson.com
FTP: ftp.thomson.com
EMAIL: findit@kiosk.thomson.com

A service of I(T)P®

Conservation of Faunal Diversity in Forested Landscapes

Edited by

Richard M. DeGraaf
Northeastern Forest Experimental Station, University of Massachusetts, USA

and

Ronald I. Miller
Senior Associate, Pioneer Geographic Designs, Massachusetts, USA

CHAPMAN & HALL
London · Weinheim · New York · Tokyo · Melbourne · Madras

Published by Chapman & Hall, 2–6 Boundary Row, London SE1 8HN, UK

Chapman & Hall, 2–6 Boundary Row, London SE1 8HN, UK

Chapman & Hall GmbH, Pappelallee 3, 69469 Weinheim, Germany

Chapman & Hall USA, 115 Fifth Avenue, New York, NY 10003, USA

Chapman & Hall Japan, ITP-Japan, Kyowa Building, 3F, 2-2-1 Hirakawacho, Chiyoda-ku, Tokyo 102, Japan

Chapman & Hall Australia, 102 Dodds Street, South Melbourne, Victoria 3205, Australia

Chapman & Hall India, R. Seshadri, 32 Second Main Road, CIT East, Madras 600 035, India

First edition 1996

© 1996 Chapman & Hall

Typeset in 10/12 Sabon by Acorn Bookwork
Printed in Great Britain by St Edmundsbury Press Ltd, Bury St Edmunds, Suffolk

ISBN 0 412 61890 7

A catalogue record for this book is available from the British Library

Library of Congress Catalog Card Number: 95–83390

To Arnold D. Rhodes and Jack Ward Thomas and to Larry D. Harris and Richard G. Wiegart, with gratitude for their early support.

Contents

Contributors

Per Angelstam
Grimsö Wildlife Research Station
Department of Wildlife Ecology, Forest Faculty
Swedish University of Agricultural Sciences
S-730 91 Riddarhyttan
Sweden

Roger Bergström
Swedish Hunters' Association, Research Unit
Box 7002
S-750 07 Uppsala
Sweden

Duncan Cameron
Department of Biology
York University
4700 Keele Street
North York
Ontario
Canada M3J IP3

Göran Cederlund
Grimsö Wildlife Research Station
S-73091 Riddarhyttan
Sweden

Sharon K. Collinge
Department of Environmental Design
University of California
Davis, California 95616
USA

Richard M. DeGraaf
USDA Forest Service
Northeastern Forest Experiment Station
Amherst, Massachusetts 01003-4210
USA

R. Michael Erwin
National Biological Survey
Patuxent Wildlife Research Center
Laurel, Maryland 20708-4015
USA

Richard T.T. Forman
Graduate School of Design
Harvard University
Cambridge, Massachusetts 02138
USA

Mark R. Fuller
Raptor Research and Technical Assistance Center
3948 Development Avenue
Boise, Idaho 83705
USA

Todd K. Fuller
Department of Forestry and Wildlife Management
Holdsworth Hall
University of Massachusetts
Amherst, Massachusetts 01003-4210
USA

Paul A. Gray
Terrestrial Ecosystems Branch
Ministry of Natural Resources
PO Box 7000
Peterborough, Ontario
Canada K9J 8M5

Jay B. Hestbeck
USDI National Biological Survey
Massachusetts Cooperative Fish and Wildlife Research Unit
Holdsworth Hall
University of Massachusetts
Amherst, Massachusetts 01003
USA

Pekka Helle
Meltaus Game Research Station
Finnish Game and Fisheries Research Institute
SF-97340 Meltaus
Finland

Ian Kirkham
Zimbabwe Natural Resources Management Programme
Research and Technical Branch
Department of Natural Resources
6 Deary Avenue
Belgravia
Harare
Zimbabwe

David B. Kittredge, Jr
Department of Forestry and Wildlife Management
Holdsworth Hall
University of Massachusetts
Amherst, Massachusetts 01003-4210
USA

Naoki Maruyama
Department of Wildlife Management
Tokyo University of Agriculture and Technology
Saiwai cho 3-5-8
Fuchu
Tokyo 183
Japan

Ronald I. Miller
Senior Associate
Pioneer Geographic Designs
PO Box 805
Northampton, Massachusetts 01060
USA

Gerald J. Niemi
Natural Resources Research Institute
University of Minnesota
5013 Miller Trunk Highway
Duluth, Minnesota 55811-1442
USA

John H. Rappole
National Zoological Park
Conservation and Research Center
1500 Remount Road
Front Royal, Virginia 22630
USA

Harry F. Recher
Department of Ecosystem Management
University of New England
Armidale
New South Wales 2351
Australia

Henry L. Short
USDI Fish and Wildlife Service
300 Westgate Center Drive
Hadley, Massachusetts 01035-9589
USA

Kjell Sjöberg
Department of Animal Ecology
Swedish University of Agricultural Sciences
S-901 83 Umeå
Sweden

Ralph W. Tiner
USDI Fish and Wildlife Service
300 Westgate Center Drive
Hadley, Massachusetts 01035-9589
USA

Kunihiko Tokida
Japan Wildlife Research Center
Yushima Ohta Building
Yushima 2-29-3
Bunkyoku
Tokyo 113
Japan

James W. Wiley
Grambling Cooperative Wildlife Project
PO Box 4290
Grambling State University
Grambling, Louisiana 71245
USA

Joseph M. Wunderle, Jr
International Institute of Tropical Forestry
USDA Forest Service
PO Box B
Palmer
Puerto Rico 00928-2500
USA

Preface

The human population is.expected to reach six billion around the year 2000. Forest ecosystems will be dramatically influenced by the encroachment of civilization and the increasing intensity of management for wood products from the world's forests, open woodlands and shrublands. Wildlife populations will be affected as never before as land-use patterns, habitat quality and cultural values change. Of course, the magnitude and rate of wildlife population change will vary by region or taxonomic group. Some changes will be fairly obvious or predictable. Some species will become rare. Population increases of ungulates, often producing plant composition changes at stand and landscape level, will occur as hunting declines, other factors being equal. Grassland species will decline with the reversion of open land to forest as agriculture declines as a way of life in Western countries. Other changes will be more subtle, such as long-term declines of certain migratory New World warblers in boreal and temperate forests as land uses change on their wintering grounds in the Neotropics. Still other changes, although extensive, will go largely unseen, for example in populations of small mammals on the forest floor in response to changes in mast availability and predation.

Many citizens of affluent societies, notably in North America and western Europe, are outraged by obvious practices such as the ivory trade, rhinoceros poaching and logging in the tropics. In all forested regions of the world, however, people are still extracting fiber or harvesting wildlife resources; humans have shaped the landscape ever since wood was first used for fuel. Wood is still the primary fuel for the third of the earth's population that lives in the developing world. Flooding and soil loss are catastrophic in many deforested areas of the world. The effects of exploitive logging on wildlife communities are often poorly reported.

In other areas, land has reverted to forest cover after almost complete clearing. New England, in the northeastern United States, is one such area. In the 1840s, the region was primarily cleared for crops and pasture; the degree of forest fragmentation was very high. By 1960 the region was primarily forested – the result of land abandonment brought about largely by the opening of the Midwest via the Erie Canal and hastened by the California gold rush and the Civil War. Periodic, small-scale harvests of merchantable hardwoods and softwoods still occur in southern New England but the region is generally characterized by mixed stands of variable quality. Thousands of miles of stone fences grid the forest, testimony to an agrarian past. High quality softwood was available from the Pacific Northwest, so there was little incentive to manage New England forests. All presettlement tree species still occur in the region, although some are plagued by introduced pests.

Northern Europe has undergone a similar land-use history. In Sweden, in the 1860s, one-fourth of the rural population left the land; many emigrated to the midwestern United States. In central Sweden today, as in New England, one can still see the signs of an agricultural past written in the landscape. A major difference in Scandinavia has been the intense management of valuable softwoods to the virtual exclusion of deciduous trees. Scandinavia is again heavily forested, but with greatly simplified, homogeneous stands compared with the original forest.

Wildlife changes have accompanied these changes in the land. In New England, extensive clearing and hunting in the eighteenth and early nineteenth centuries had extirpated many forest game species. The effect on species of little economic importance went unrecognized. Wolves, mountain lions, beaver, moose and elk were among the first to be eliminated. White-tailed deer and wild turkey soon followed. Reversion of agricultural land to hardwood forest after the clearing of white pine produced high populations of ruffed grouse in sapling forests. Today the mature forest supports wild turkey, fisher and black bear, species that had been absent since the land was cleared for settlement. Now, early successional forest species – ruffed grouse, cottontails, and many grassland and shrubland birds – are declining. The native prairies that existed in eastern North America were among the first areas to be settled, and their wildlife communities probably the first to be extirpated. Grassland species that had later extended their ranges eastward with land clearing are now retreating back toward the Great Plains. While it will be a challenge to maintain grassland species in New England, the opportunities for maintaining or creating diverse forest wildlife communities are abundant.

In Scandinavia, timber values, land-use values, laws and cultural values are different from those in the US; the resulting highly ordered, simple and efficiently managed forests permit fewer immediate opportunities to restore wildlife communities to a condition that resembles the original forest. Most top predators such as brown bear, lynx and wolf have been rare or absent for a long time, and species associated with old deciduous forest, such as the white-backed woodpecker, are reduced to small isolated patches of suitable habitat. Recent concerns about biodiversity are now mandating change in Scandinavia but it will take a fairly long time to diversify Scandinavian forests and wildlife communities. This process is compounded by the fact that the simplified forest so completely occupies the landscape and the range of possible habitats is so restricted that it may be difficult to determine what is good habitat for some uncommon species.

Australia represents an even more extreme situation. Sparsely forested to begin with, most of the continent's original forest had been cleared within two centuries after settlement. Today, it is fragmented by logging, grazing, fire and introduced species. Rates of local and regional wildlife extinctions are very high and proportions of the threatened forest vertebrates are among the highest in the world.

Trends in forest cover, intensity of management and composition vary greatly throughout the world. The potential for managing wildlife communities varies accordingly. Success will entail efforts that are large scale and long term. Pre-settlement landscapes or forest conditions are probably unattainable. Climates today are different from those of the Little Ice Age when trees in pre-settlement forests germinated. Indeed, climates and forests distribution have changed dramatically throughout the Holocene.

In many areas the natural disturbance patterns and the effects of early humans that produced the forest landscapes we seek to maintain are just beginning to be understood. The dynamics of changes in plant and animal populations are complicated by the multitude of factors that influence biogeography. It is still only possible today to understand the influences of local and perhaps regional patterns of change on forest wildlife from consideration of habitats and movements.

We now know that forests changed greatly throughout pre-settlement times. A useful course of action is to allow (or emulate) natural disturbance regimes to exert their influences over as large a landscape as is practical so that as many native forest vertebrates as possible can maintain viable populations. Such a naturalistic approach will require active management because people will always require wood products from the forest.

This book is composed of a series of original contributions, grouped in three sections: current variations in the status of forest wildlife diversity, wildlife population and habitat change in various landscapes, and examples of effective conservation tools and strategies. The first section presents global views of forest and wildlife habitat trends as well as trends for sensitive taxa: forest wetland birds, an old-forest species, top predators, forest raptors and large forest carnivores. The next section presents wildlife responses to landscape change, whether from forestry, agriculture or natural disturbance. Examples range from Japan to Europe to the Caribbean. The final section presents some tools and strategies for the conservation of forest wildlife diversity, with the use of reference landscapes in northern Europe, forest practices in Australia and the use of remote sensing to map key species distributions or identify unprotected areas that have high wildlife diversity. Throughout, the perspective is on the large scale: landscape, region or continent. The larger the scale, the more effective will be efforts to conserve forest wildlife diversity.

Acknowledgements

We thank the following for their careful reviews of the various chapters: Robert A. Askins, Connecticut College, New London; Victor Van Ballenberghe, US Forest Service, Pacific Northwest Research Station, Anchorage, Alaska; Boguslaw Bobek, Jagiellonian University, Krakow, Poland; Frank W. Davis, University of California, Santa Barbara; Francis C. Golet, University of Rhode Island, Kingston; Lennart Hansson, Swedish University of Agricultural Sciences, Uppsala; William M. Healy, US Forest Service, Northeastern Forest Experiment Station, Amherst, Massachusetts; Richard T. Holmes, Dartmouth College, Hanover, New Hampshire; Malcolm L. Hunter, Jr, University of Maine, Orono; Harto Lindén, Finnish Game and Fisheries Research Institute, Helsinki, Finland; H. Gyde Lund, US Forest Service, Washington, DC; Douglas M. Muchoney, Conservation and Research Center, Smithsonian Institution, Front Royal, Virginia; Michael E. Nelson, National Biological Survey, Ely, Minnesota; William A. Patterson III, University of Massachusetts, Amherst; and Daniel A. Welsh, Canadian Wildlife Service, Neppean, Ontario.

We also thank Mary A. Sheremeta for manuscript typing and editorial assistance.

Part One

Prevailing Contrasts in the Status of Forested Biodiversity

In the twentieth century, transformations of the environment on the surface of the earth seem to occur at an ever more rapid pace due to the expanding development of human civilization. The forests of the earth endure as some of the last storehouses of biodiversity.

In this first section we attempt to establish a context for the typical changes that take place in forest habitats and wildlife and to portray their current status. The historical patterns of land-use change and wildlife response to this change in New England over the past several millennia are presented in Chapter 1.

Forest responses to glacial retreat, the activities of agricultural Indians and periodic hurricanes are discussed as background agents of change. The wave of clearing that followed settlement by Europeans produced widespread declines of forest fauna, which were largely reversed soon after land abandonment. Landcover and faunal changes are evident as successional habitats give way to mature forest. The New England landscape, like all landscapes, continues to respond to past disturbance.

A broad overview of forest habitat and wildlife changes and the remaining extent of the world's forests is presented in Chapter 2. Trends in the extent and conditions of forests are based upon a variety of published and unpublished sources in this chapter and sources of information for an in-depth treatment of the status of the forests are also identified. This provides a broad geographic perspective for the status of the world's forests.

Chapter 3 presents an assessment of the general status of forested wetlands and some associated waterbird species in North and Central America. The principal ecological requirements of waterbirds in forested wetlands are discussed. The forested wetlands and waterbird species for each country in North and Central America for which adequate information is available are identified and the known data sources are discussed.

Management recommendations are presented to mitigate the effects of human activities on wetlands and their avifauna.

Chapter 4 examines the status of the large Eurasian woodland grouse, capercaillie (*Tetrao urogallus*), across northern Europe. The status of this species is considered in relation to factors such as social behavior, food species and the influence of forestry activities. Habitat changes caused by forestry practices are primary reasons for decreases in populations of this species. The means by which levels of landscape scale influence seasonal natural history patterns in capercaillie such as colonization and food resources are discussed. Finally, the current changes in the forest landscape most influencing the status of capercaillie are considered in this chapter.

Chapter 5 considers how changes in forest habitats due to increasing human pressures seriously impact large carnivore populations in a variety of ways. This chapter outlines the status of large forest carnivores worldwide. It identifies the factors impacting these species and it recommends solutions for some persistent conservation problems. Evidence is also presented in regard to the intensity of studies that have been executed for each carnivore species that is considered. Historical and present-day vistas of adverse human impacts on carnivore populations are presented and the issues related to the conservation of carnivores are considered. Six categories are introduced that describe what needs to be done for successful carnivore conservation and what we need to know to insure the consequences of conservation actions.

1

The importance of disturbance and land-use history in New England: implications for forested landscapes and wildlife conservation

Richard M. DeGraaf and Ronald I. Miller

1.1 INTRODUCTION

Forested landscapes are dynamic – they are always in transition and have always been in a state of constant change. Though the rates and agents of change vary temporally and spatially, change underlies the natural state of the world's forests. Land-use patterns and disturbance regimes are two of the dynamic agents that can have profound effects on the abundance, distribution and diversity of terrestrial vertebrates (Burgess and Sharpe, 1981; Saunders *et al.*, 1987; Wilson 1988). The need for conservation and management of forest and wildlife resources requires humans to influence the uses of land and the sizes and distributions of wildlife populations to maintain specific habitat conditions. Identification of the primary agents of change and their extent and magnitude are necessary to maintain forested ecosystems and their faunal components. An understanding of the background rates and causes of change in forested landscapes can help to guide conservation efforts on many scales.

In this first chapter we synthesize information about forested landscapes in New England. This information, gathered from historical records, is used to demonstrate the continuing patterns of change influencing conditions for the region's forest fauna. Many studies are being conducted today that utilize new technologies that can be used to monitor forest habitats in New England (Miller and Griffin, 1994) and around the globe (Kattan *et al.*,1994). With respect to satellite imagery, however, most studies to date

Conservation of Faunal Diversity in Forested Landscapes. Edited by
R.M. DeGraaf and R.I. Miller. Published in 1996 by Chapman & Hall. ISBN 0 412 61890 7.

have assessed the effectiveness of particular technologies rather than the development of tools for planning and detecting change. We therefore believe that a synthesis of information remains the primary focus in our efforts to understand patterns of change over time in forests in New England and elsewhere.

New England has undergone dramatic land-use changes since European settlement. In the mid eighteenth century the rate of land clearing for agriculture increased substantially and spread westward with increased population, immigration and more advanced technology. Native prairies, woodlands and extensive forests – which had largely obliterated pre-settlement Indian clearings – were converted to subsistence farms by the early nineteenth century. Virtually all of the arable land in southern New England was in crops and pasture by 1820–1840 (Raup, 1966; Irland, 1982). These changes constitute an extensive and intensive habitat change compared with, and in addition to, the long-term changes resulting from glaciation, cyclonic storms and Indian activities in the region.

Today New England is again heavily forested, almost to the maximum extent achieved over the last few millennia. Many long- and short-term changes in the landscape have occurred since the end of the Holocene. These include extinctions, local extirpations and recolonizations by vertebrates. The region is subject to constant natural disturbance. But in New England the superimposition of dramatic change – clearing for settlement and reforestation after land abandonment – has resulted in few vertebrate extinctions over the last two centuries and high rates of recolonization. How does this situation compare to other regions in the world today?

1.2   LONG-TERM CHANGES IN FOREST AND FAUNA

1.2.1   Eastern North America

Natural rates of change, e.g. from fire or wind, in boreal and temperate forests are slow in terms of human history but are rather fast in relation to biological and geological change. Until recently, vegetation displacement by the Wisconsin glaciation was thought to be in belts or zones of vegetation similar to latitudinal distributions today, i.e. ice sheets bordered by arctic tundra and associated with zones of intact boreal forest and temperate (deciduous) forest communities in sequence southward. The distance of displacement between these forest communities still remains a question for scientific research (Brown, 1970; Deevey, 1949) but recent investigations of vegetation displacement have focused upon pollen records which reveal that boreal tree species were dispersed far to the south and west of New England (Watts, 1970, 1980; Delcourt, 1979).

During this period the trees were not associated in predictable communities (although this may have always been true on time scales important to trees). The evidence indicates that, when the ice sheet melted, particular species of both coniferous and deciduous tree species migrated northward independently from one another during the Holocene. Throughout the Holocene the data indicate that forest communities have been, in all likelihood, rather chance combinations of species that were rarely stable for more than 2000 or 3000 years (see Davis, 1981, for review). Species that comprised what are now mapped as single forest types (e.g. Braun, 1950) have been shown to have followed quite different historical migration routes. For example, beech (*Fagus*) lagged 4000 years behind hickory (*Carya*) when it migrated into Ohio deciduous forests. In Connecticut, 700–800 km to the northeast, beech preceded hickory, which followed about 3000 years later (Davis, 1976). Therefore, plant communities should be viewed as ensembles of distinct species rather than as integrated biological units. Similar differences are reported from studies of the interglacial floras of Great Britain before rising sea levels isolated Great Britain from Europe (West, 1970). In that region forest types have all been shown to be different from one interglacial period to the next, depending upon the order of arrival of tree species.

Davis (1981) summarized the migration rates of trees during the Holocene in eastern North America and concluded that the speed with which forests adjust to climate is exceeded by the speed of climate change, even though the rate at which trees have extended their ranges is quite rapid (i.e. 300 m/ yr on average). Rates for different species vary: 100 m/yr for chestnut (*Castanea dentata*); 200 m/yr for balsam fir (*Abies balsamea*), maple (*Acer*), and beech; 350 m/yr for oak (*Quercus*); and up to 400 m/yr for jack pine (*Pinus banksiana*) and red pine (*P. rigida*). Considering the differences in seed dispersal rates between conifers and hardwoods, between heavy-seeded and light-seeded species and between wind-dispersed and animal-dispersed species, clear patterns of migration rates are not apparent. Full vegetational response must, of necessity, lag behind climate change; given that climate varies continuously on all time scales, vegetation responses can be viewed as being in relative equilibrium (see review in Webb, 1986).

At the onset of the Holocene, fairly abrupt changes occurred in the composition of forests and woodlands south and east of the Wisconsin ice sheets. Spruce, especially white spruce (*Picea glauca*), the most abundant species at that time, suddenly and synchronously became rare over a wide region from Minnesota to Nova Scotia. This species was replaced by a more diverse forest of conifers that included jack pine (*Pinus banksiana*), red pine (*P. rigida*), white pine (*P. strobus*) and balsam fir (*Abies balsamea*) and hardwoods that included paper birch (*Betula papyrifera*), elm (*Ulmus*) and oak (*Quercus*). The decline of spruce may have been rapid (Watts, 1983) and paleo-Indians, probably hunters of woodland caribou (*Rangifer caribou*), may have witnessed the forest change (Pielou, 1991).

The course of forest change is fairly well understood in New England.

The pollen record shows that spruce forests thrived in southern New England as the climate warmed, beginning approximately 12 000 years ago. Spruces, especially *Picea glauca*, were among the first trees to appear after melting of the ice sheets, although other species including balsam poplar (*Populus balsamifera*), black ash (*Fraxinus nigra*), and hophornbeam (*Ostrya virginiana*) probably co-occurred with *P. glauca*. There is evidence that forests in eastern North America are still responding to the last glacial cycle (see Davis, 1976, for a review) but it is likely that they are now responding to climate change within the Holocene. Several tree species have moved southward in the last 2000–3000 years (Webb, 1988).

1.2.2   Regional disturbance regimes

The eastern deciduous forest biome is influenced by large-scale disturbances throughout its range, but especially around the edges of the biome (Figure 1.1). Fire is an important source of disturbance at this scale and its importance varies in different areas of the region (Runkle, 1990). Fire frequency is high in the south and east on the sandy soils of the Atlantic coastal plain (Nelson and Zillgitt, 1969); in the western grasslands where precipitation is low; and in the north, due to the increased importance of flammable conifers such as *Pinus* and *Picea* (Whitney, 1986). Other important sources of climatic disturbance include blowdowns and hurricanes, which are most common on the Atlantic Coast and in New England (Nelson and Zillgitt, 1969; Foster, 1988a,b). In addition, tornadoes are common along the western edge of the biome though they are infrequent elsewhere in this biome (Fujita, 1976).

The center of the eastern deciduous biome was relatively stable, influenced predominantly by small-scale disturbances such as gaps formed by the fall of individual trees (Brokaw, 1982; Runkle, 1981, 1984) and small fires on dry ridges (Harmon *et al.*, 1983). Geographic factors (e.g. inland or coastal position, latitude, topography) are moderated, to a considerable degree, by climatic factors (e.g. mean temperatures, frost-free periods) and by weather factors (e.g. winds, humidity, storm frequencies). Climatic factors interact with physiography to produce disturbance regimes that in turn influence the presence of certain tree species (Runkle, 1990) and successional patterns (Braun, 1950) that occur in the eastern deciduous forest.

1.2.3   An example: Indian clearings and primeval openings

Humans have inhabited eastern North America for about 10 000 years. The native Indian societies of New England were mobile, and the size and location of villages changed in response to social and ecological needs (Cronon, 1983:38). Dense temporary settlements were established at key coastal and river fishing sites in the spring, but in the fall, when it was time

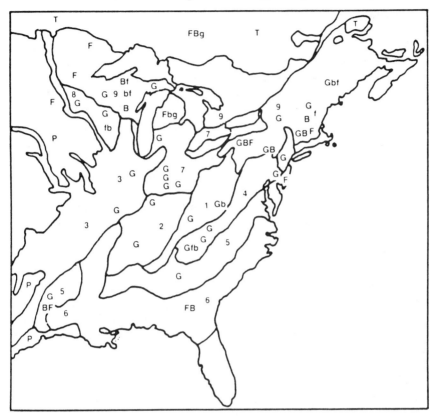

Figure 1.1 Geography of disturbance for eastern deciduous forest. F and f locations indicate where fire was of major or minor importance, respectively; B and b, where big blowdowns were of major or minor importance, respectively; G and g, where gaps were of major or minor importance, respectively; T = taiga. Literature on which the figure is based, and names of forest regions numbered on the figure, are given in Braun (1950) (after Runkle, 1990). (Reproduced, by permission, from *Canadian Journal of Forest Research*, 20.)

to hunt, the same people would scatter over a wide area. Houses and tools were readily moved or abandoned (Morton, 1632). In the period AD 500–1000, Indians south of the Kennebec River in Maine shifted from food gathering to food production and storage (Likens, 1972). Indians raised corn, beans, squash, pumpkins and tobacco as part of their annual subsistence cycles; except for tobacco, the crops were commonly interplanted (Mood, 1937; Russell, 1961). These fields were on average used for eight to ten years (Wood [1634] 1977:35) until the soil fertility declined and new fields were needed. The period of cultivation varied by site. Fields were fallowed after only two years on sandy soils of Rhode Island (Winthrop,

1863) and Cape Cod (Champlain [1605] 1968:88). On finer-textured, more fertile inland soils, fields were used for up to 30 years (Pratt, 1976:15–16).

Agriculture strongly affected the size of New England Indian populations. For example, non-agricultural Indians in Maine sustained population densities of about 19 per 100 km^2 while agricultural Indians in southern coastal New England maintained about 40 people on a similar land area (Whitney, 1994:101). The total Indian population in New England in 1600 is estimated to have been 70 000–100 000 (Jennings, 1975:15–31; Snow, 1980:31–42; Cook, 1976) within an area of about 83 000 km^2, although the population was concentrated, to a great extent, along the coasts and major rivers.

Most early estimates of post-contact Indian populations in New England and in the Northeast in general are calculated by multiplying the number of warriors per tribe (from historical records) by the average family size (Cook, 1976; Whitney, 1994). But estimating pre-contact populations is difficult because many, if not most, coastal Indian populations were exposed to the Europeans' diseases such as smallpox by the early 1600s and many populations experienced catastrophic declines soon after contact (Whitney, 1994:97–100). The epidemics of 1616–1619 and 1633, for example, resulted in massive die-offs in coastal areas (Cook, 1973). Estimates of mortality vary greatly, ranging up to 95% of the total population (Whitney, 1994:100). Snow (1980:34) indicates that early estimates of mortality were conservative and should be revised upward.

American Indian population sizes have been subject to debate. The hypothesis that unrecorded exogenous pandemics greatly reduced Indian populations after contact with Europeans has been challenged recently. The absence of children in parties of Europeans making contact with Indians, at least outside the sphere of Spanish colonization, calls into question the role of epidemics as decimating factors shortly after contact. Smallpox and other epidemics that devastated American Indian populations were childhood diseases in Europe at the time, and nearly all adult Europeans who survived had lifelong immunity and were not contagious to Indians they might encounter (Snow, 1995). Recent evidence suggests that patterns of regional or even local epidemics affected some parts of the continent decades or even centuries before others (Ubelaker, 1988).

Indian population densities on the offshore New England islands of Nantucket, Martha's Vineyard and Block Island were about 10 times higher than on the mainland coasts, likely due to the rich marine resources there and the intensive cultivation of corn (Cook, 1976:45). Densities were considerably lower among the non-agricultural Indians that lived north of the 120–130-day frost-free, corn cultivation zone (Bennett, 1955).

Indian settlements and agriculture produced a substantially open landscape in some. areas. Tribal villages in southern New England varied in size and permanence but all societies cleared land for dwellings, crop fields and firewood (Day, 1953).

Whitney (1994:100–107) reviewed the impact of Indian agricultural practices on the southern New England landscape. Corn cultivation resulted in major alterations of the landscape. Indians did not lack the technology or means of clearing large areas of a heavily wooded landscape: girdling and fire together provided an efficient mechanism for clearing the forest. Large trees were girdled with stone hatchets in spring and felled a year later by firing brush piled around the bases; the fallen trunks and branches were used for fuelwood (Loskiel, 1794:1–55). It is clear that New England Indians were knowledgeable about the use of fire (Patterson and Sassaman, 1988) but how extensively Indians used fire to prepare land for cultivation or to drive game is a subject of debate. Bromley's (1935:64) review of early accounts indicates that the forest was open and park-like in large areas. Russell (1983) disputes the openness of the landscape due to Indian burning on the grounds that accounts were second-hand and designed to encourage colonists from the Old World to settle lands that were already partly cleared. It is clear that Indian burning resulted in a mosaic of fields and forests in all stages of succession in the vicinities of villages and old village sites.

The Indians' use of firewood was especially great and was not restricted to dead or fallen trees; some areas, such as the site of Boston, were almost entirely cleared by Indians (Bromley, 1935). Verrazzano in 1524 encountered stretches of 25–30 leagues of treeless land near Narragansett Bay in present-day Rhode Island (Wroth, 1970). One of the main reasons that Indians moved to new winter camps was that the trees had been cleared for corn cultivation and fuel.

Determinations of the total amount of land under Indian cultivation in New England have been based upon the estimated population sizes, the yields of corn and rates of consumption (Whitney, 1994:102). Most scholars and early observers report that an acre of corn supported one to five people or about one family (Willoughby, 1906; Kroeber, 1939:146; Bennett, 1955; Pring [1625] 1906). Kroeber (1939) used a low estimate of Indian populations and assumed 1 acre of corn was raised per person and calculated that the total area under cultivation was less than 1% of the landscape at any one time (see Whitney, 1994, for a review).

The use of fallowing rendered the area impacted larger than that under cultivation. Whitney (1994:102) applies a factor of 2 to 5 to the area under cultivation, based on the assumption that a patch was cultivated for an average of 10 years and that it required 20–50 years to restore soil fertility (Thomas, 1976). The area around the villages was cleared for fuelwood, buildings and stockades. Agricultural Indians probably returned to arable sites periodically. If so, they would encounter stands of poletimber-size trees which would have been relatively easy to clear again. Such areas would be devoid of large old trees and would have been readily identifiable by settlers long after the Indians were gone.

Indians of southern New England generally burned the forests in spring or fall to facilitate travel and drive game (Wood, 1634:25; Morton, 1637:52,54). This annual burning produced open, park-like woodlands with rich herbaceous layers. In addition, the forest was burned to open land for cultivation (Trumbull, 1797; Parker, 1910), to improve the forage for game (Dwight, 1823, Vol. 4:50) and to enhance the production of strawberries, blueberries and other fruiting shrubs (Bromley, 1945). Periodic low-intensity ground fires tended to eliminate white pine (*Pinus strobus*) (Patterson and Backman, 1988) and enhance mast-producing species such as oaks (Brown, 1960). An abundance of deer, turkey, rabbit, grouse and heath hens were noted by the earliest English colonists (Wood [1634] 1977); Josselyn, 1675; Higgeson, 1630). This abundance of game was produced by Indian agriculture, fuelwood cutting and periodic intentional burning of woodlands which created a patchy landscape of fields and forest in various stages of succession.

Prior to European settlement, much of the northeastern coastal forest of the United States from Maine to Virginia had a considerable amount and variety of open habitats (Plate A). The dominant habitat of the region was forest but there is ample evidence that extensive grasslands and oak openings were common in eastern North America before European settlement (e.g. Askins, 1993; Niering and Dreyer, 1989).

The heath hen (*Tympanuchus c. cupido*) occupied scattered grasslands, native prairies and blueberry barrens of the east from Maine to Virginia and on the larger offshore islands. The openness of the landscape is revealed by the abundance of the heath hen in the mid seventeenth century: it was then 'so common in the area around Boston that laborers and servants stipulated that employers not serve it more than a few times in a week' (Winthrop [1636] in Nuttall, 1834:67–8).

Indian activities in southern New England created a mosaic landscape of fields and forests once corn cultivation became established, probably 1000–1500 years ago. It is not known how far north into the region's interior Indians modified the landscape but their impact was dramatic where they practised agriculture. For example, the town of Wilbraham in south-central Massachusetts was named Minnechaug ('Berry Land') by the Indians; the openness of the 'whole territory' is described as 'so devasted by fires, that in many places there were no forest trees, in other places hardly any shrubbery grew . . . The tradition is handed down to us that the country was so bare that a deer could be seen from mountain to mountain' (Stebbins, 1864).

1.3   SHORT-TERM HABITAT CHANGES IN NEW ENGLAND

In southern New England, modern forest composition differs, depending upon site and land use history, from that encountered by the first white settlers in the early seventeenth century. Where agricultural Indians had lived, forests in the eighteenth century had reclaimed old Indian clearings and probably did not contain large old trees. Indian populations dwindled soon after contact with Europeans due to disease and conflict; early settlers who penetrated interior southern New England likely encountered a landscape that was more heavily forested than that which had existed a century or two earlier.

Our knowledge of forest response to land use is detailed only since European settlement. The effects of several centuries of land use in southern New England are different depending upon scale. The distribution of modern and pre-settlement forest types are similar at the regional scale (despite structural changes and the loss of some species) but at the (smaller) landscape scale, modern forest characteristics are strongly related to land-use history (Foster, 1992). Most of southern and central New England experienced the same land-use history. Land was cleared slowly for agriculture and small industry between 1700 and 1750, after which time the rate of clearing increased dramatically. By 1820 more than 75% of the arable landscape was in farm crops and pasture (Raup, 1966). The height of land clearing in central Massachusetts for commercial agriculture was achieved between 1820 and 1840, after which there began a long period of land abandonment that resulted largely from the opening of the fertile Midwest via the Erie Canal in 1825. After land abandonment, white pine (*Pinus strobus*) rapidly invaded former (impoverished) tilled land and dry pasture and produced nearly pure stands that were cut in the last major land clearing in New England around 1910 (Fisher, 1918). About half of the tilled fields and one-fifth of pastures were still open in 1908, and were invaded by *P. strobus* or planted to conifers (Plate A).

A severe hurricane passed through the region in 1938 and more than 7 million cubic meters of timber were blown down (Curtis, 1943). Human land use and the consequent invasion of *P. strobus* on agricultural sites enhanced the damage caused by the storm. *P. strobus* and other conifers are particularly susceptible to wind damage and the high proportions of these species on the landscape at that time due to past land use was largely responsible for the great extent of damage to the central New England landscape (Rowland, 1941). In terms of natural processes, the effect of the 1938 hurricane was greatly influenced by the structure, composition and spatial distribution of forest vegetation that resulted from 200 years of occupancy by European settlers (Foster and Boose, 1992). Storms of similar magnitude occurred in 1635 and 1815 (Channing, 1939) and are estimated to occur, generally, at 150-year intervals.

Today the landscape of New England is again mostly forested. About 65% of southern New England and more than 90% of northern New England is completely forested. Each year, except on the industrial forest-lands in Maine, the age and extent of the forest increases (Brooks and Birch, 1988). Original forest species, especially *Picea* and *Tsuga*, persist in woodlots that are located in swamps and other poorly drained sites (Foster, 1992). Several species present in the original forest are now greatly reduced by pathogens. Oaks and other hardwoods have been stressed in many areas by the gypsy moth (*Lymantria dispar*). The chestnut blight (*Endothea parasitica*) has rendered *Castanea* an understory tree or shrub. Dutch elm disease has essentially eliminated *Ulmus* from the forest and the beech bark disease (*Nectria*) has greatly reduced numbers of large *Fagus*. Currently the hemlock woolly adelgid (*Adelges tsugae*) is causing widespread mortality of *Tsuga* in Connecticut. These infestations are all helping to change dramatically the species composition of the New England forested landscape.

1.4   FAUNAL RESPONSES TO FOREST HABITAT CHANGES IN NEW ENGLAND

1.4.1   Vegetation and faunal changes during the last interglacial period

A most dramatic occurrence between 9000 and 12 000 years ago was the abrupt extinction of many species of North American mammal. Before the onset of the Holocene about 10 000 years ago, over 50 species of large Ice Age mammals became extinct at the end of the Wisconsin glaciation, which had covered the region for the previous 100 000 years. Most of these mammalian extinctions occurred in the brief period 9000–12 000 years ago, and no mammals have become extinct on the North American continent during the last 10 000 years (Meltzer and Mead, 1983).

During this 3000-year period, between 35 and 40 species of large mammal disappeared (Kurtén and Anderson, 1980; Webb, 1984). This wave of extinctions has never been satisfactorily explained (Pielou, 1991:251). Most of these extinct mammals were large and several closely resembled species that are present today. They included mastodons, mammoths, sabertooths, bison, woodland muskoxen, camels, horses, short-faced bears, giant ground sloths and dire wolves. The dire wolf, which became extinct about 10 000 years ago (Pielou, 1991), was a little larger than the present timber wolf and it had bigger teeth (Kurtén and Anderson, 1980). The dire wolf and the timber wolf shared much of the North American continent although dire wolves were much more numerous. Similarly, pleistocene horses closely resembled modern horses (which arrived with early Spanish explorers). However, for most of the extinct species, extant counterparts do not exist. For example, the giant beaver and Shasta ground sloth were unique forms that are not complemented by any present-day North American mammal species.

Atmospheric or astronomical causes for this North American wave of extinctions are doubtful because these events are not synchronous with similar waves of extinction in South America, Europe, Africa and Australia (Pielou, 1991; Horton, 1984). Two major hypotheses have been advanced: one proposes that the extinctions resulted from human overkill, and the other holds that they were due to changing environments.

The human overkill hypothesis is supported by some strong circumstantial evidence. Extinctions were most numerous and sudden in North America, South America and Australia (albeit earlier – Aboriginal people arrived there about 40 000 years ago). These continents were first invaded by humans who had developed their hunting skills elsewhere. The mammals were naïve prey that had not developed wariness toward armed hunters (Martin, 1984; West, 1983; Horton, 1984; see also review by Pielou, 1991).

In North America, it has been suggested that these extinctions were caused by the exploitation of mammals by human hunters, especially by the Clovis culture. (This is the name given the humans who became established in North America 11 500 years ago. Widespread from Nova Scotia on the Atlantic coast to California on the Pacific coast, all groups apparently made fluted spearheads of the same design; they were the earliest widespread human occupants of the continent.) Most extinctions on this continent occurred among species that descended from ancestral forms that lived there for at least one million years. Pleistocene species that survived human exploitation to the present are essentially all (the pronghorn (*Antilocapra americana*) is the sole known exception) immigrants from Asia and probably possessed the adaptation to learn from encounters with human predators (Pielou, 1991:257).

Opponents of the overkill hypothesis cite the small size of the human population, especially of the Clovis people in North America, and suggest it is improbable that they directly caused extinctions (Whittington and Dyke, 1984). The archeological record also shows only moderate association between extinct mammal fossils and human remains at 'kill sites' (Pielou, 1991:258). For example, mammoth bones have only been found at about a third of known Clovis sites (West, 1983).

Many environmental change theories are proposed for North America. Each theory suggests an explanation for change that is usually tied to a specific site or species. In New England, accelerated warming about 10 000 years ago led to the almost simultaneous disappearance of the spruce forest of eastern North America and the tundra in Maine. Both of these habitats were replaced by birch, then by white pine, and then by hemlock/hardwood forest (see Pielou, 1991, for review). In Beringia (the region exposed during the Wisconsin glaciation and now largely flooded by the Bering Sea, including adjacent Siberia and Alaska), if extensive arctic steppe existed at the height of the Wisconsin glaciation, it disappeared because of climate change or because of marine incursion, during glacial melting, along with herds of grazers (Pielou, 1991:262). As spruce forests were replaced in the northern latitudes between 10 000 and 12 000 years ago (King and Saunders, 1984), the hypothesis is that mastodons declined. Habitat loss is also cited as the causal factor in the extinction of giant beavers (Kurtén and Anderson, 1980) and of stag-moose (Pielou, 1991).

For mammals, one environmental change hypothesis links declines to changes in climatic periodicity. Rapidly increased seasonality of Holocene climates has been proposed as a cause of mammalian declines related to the lack of synchrony between the reproductive cycle and the seasonal cycle (Kiltie, 1984).

Environmental change occurs during each interglacial. It seems unlikely that different species of large mammal in various parts of the continent each went extinct at approximately this same time (i.e. between 9000 and 12000 years ago) (Pielou, 1991). However, vegetation change was time-transgressive – similar changes occurred at different times depending on location. Great changes in vegetation composition did occur during this period, but the direct causes of faunal extinction remain unclear.

1.4.2   Changes in more recent times

The variations of climatic patterns have significantly influenced the distribution of forests and their constituent fauna across New England. A noticeable warming known as the hypsithermal period occurred between 5000 and 9000 years ago. The distribution of forests and individual tree species shifted hundreds of kilometers to the north during this period. Evidence today of forests and forest fires that lie far to the north of the present tree line (Bjorck, 1985; Bryson *et al.*, 1965) indicates that cooling since the hypsithermal period resulted in withdrawal of forests southward and their replacement by tundra (Ritchie *et al.*, 1983). More recently, European settlers experienced bitter winters in North America in the seventeenth and eighteenth centuries, part of what is now known as the Little Ice Age of the fifteenth to nineteenth centuries (Lamb, 1982; Bradley, 1985:239). Dendrochronology identifies this period of climatic cooling between 1350 and 1870 (Luckman and Kearney, 1986).

(a)   Changes in forest fauna abundances

Since the onset of the hypsithermal, and before the European settlement of eastern North America, faunal abundance was great and there were few extinctions (see Pielou, 1991; Martin and Klein, 1984). Most of the faunal changes in the past 200–300 years were due to the vast scale and intensity of habitat change that swept across eastern North America during European settlement, and to the reafforestation that has occurred since the mid 1800s. A great many shifts in the abundances and ranges of vertebrate species occurred (e.g. De Vos, 1964) as a result of these changes. Nevertheless very few of these species were rendered extinct solely from the effects of habitat change.

No forest mammal became extinct during the settlement period in New England, although the gray wolf, mountain lion and caribou, among others, were extirpated. Therefore change in the distribution and abundance of mammals is the best indicator of faunal change. Both white-tailed deer (*Odocoileus virginianus*) and black bear (*Ursus americanus*) have remained among the most utilized large mammal species in New England forests, and their abundances serve as excellent indicators of change during the past 5000 years. White-tailed deer and black bear were the most common food animal recorded in Indian middens approximately 4300 years ago in southern Massachusetts and Rhode Island (Loskiel, 1794:65). White-tailed deer were the predominant vertebrate found in these middens but black bear also comprised a recognizable portion of the harvest (Waters, 1962). Deer were taken in autumn and bear in winter during hibernation (Godman, 1826:124).

White-tailed deer abundance in pre-settlement North America has been conservatively estimated at 40 million (7.7/km^2) (Seton, 1929 [3-1]:244; McCabe and McCabe, 1984). White-tailed deer densities in disturbed forests of the northeastern United States immediately prior to European settlement were estimated at between 3.8 and 5.8/km^2 (Mattfeld, 1984); similar estimates were made for Wisconsin (Dahberg and Guettinger, 1956). In contrast, undisturbed northern hardwood forests were estimated to have supported 1 deer/km^2 (Cooperrider, 1974).

Deer were exploited for food and hides more or less continuously as settlement progressed. Hides were shipped to England and Europe from the period of earliest settlement; for example, Quebec exported 132 271 deerskins in 1786 alone (Schoen *et al.*, 1981). Development of the repeating rifle and the refrigerated railroad car after the Civil War coincided with the peak of logging and settlement of the upper Midwest. Between 50 000 and 100 000 hides were shipped from the lower peninsula of Michigan in 1880 (Jenkins and Bartlett, 1959; Mershon, 1923) and market hunting was intense throughout the occupied deer range (see McCabe and McCabe, 1984, for an exhaustive review). Hunting declined following periods of deer scarcity. The Lacey Act in 1900 in the United States prohibited interstate traffic of wild game taken in violation of state law and allowed deer to recolonize their former range. From a low population of 300 000 in the United States (Trefethen, 1970) and between 350 000 (Trefethen, 1970) and 500 000 (Seton 1909) in North America, modern white-tailed deer populations probably approach pre-settlement levels (McCabe and McCabe, 1984). The mosaic pre-settlement forest landscape is now known to have supported high densities of white-tailed deer (Schoen *et al.*, 1981).

Pre-settlement bear density in the primeval deciduous forest in the east has been estimated at between 19/100 km^2 (Shelford, 1963) and 8/100 km^2 (Seton, 1929, vol. 2, pt 1:129). Black bears were apparently never extirpated completely from Massachusetts and they reached their lowest population between 1860 and 1880 (Cardoza, 1976). Bear density in western Massachusetts is currently about 30/100 km^2 (Fuller, 1993).

(b)   Species extinctions and extirpations

The historical evidence associated with the well-known bird extinctions in eastern North America illustrates the composite factors that generated these events. For example, human persecution figured prominently in the extinctions of the coastal great auk (*Pinguinis impennis*) and Labrador duck (*Camptorhynchus labradorius*). Simultaneous taking in great numbers and rapid habitat alteration significantly contributed to the extinction of both the passenger pigeon (*Ectopistes migratorius*) and the heath hen (*Tympanuchus c. cupido*).

Extirpated on the mainland by 1869, the heath hen persisted on Martha's Vineyard (off the coast of Massachusetts) well into the twentieth century (Gross, 1928). The small remaining population was subject to extreme predation pressure by goshawks and to low productivity. A wildfire reduced the spring population by more than 80% in 1916. A subsequent ban on fire allowed forests to replace the open heaths, and the loss of open habitat sealed the bird's fate. The heath hen was last seen in 1932 (Gross, 1932).

By the mid nineteenth century, the few remaining large breeding colonies of passenger pigeons were located in the upper Midwest (Schorger, 1955). For example, a colony that covered about 850 square miles and contained a population of 136 million nesting pigeons was located in central Wisconsin in 1871. At least 1.2 million birds were taken there by 600 professional hunters (Schorger, 1936) but the kill represented less than 1% of the adult population (Bucher, 1992). Pigeon populations were initially reduced due to clearing and fragmentation of the mature midwestern forests of beech (*Fagus grandifolia*), oak (*Quercus* spp.) and chestnut (*Castanea dentata*), upon which the passenger pigeon, a nomadic mast specialist, depended. Detection of (patchy) areas of mast abundance was highly dependent upon social facilitation. The combination of loss of critical breeding habitat and lack of social facilitation at low densities was enough to lead the passenger pigeon to extinction, even in the absence of human persecution (Bucher, 1992). The Carolina parakeet (*Conuropsis carolinensis*) was a bird of mature floodplain forests that foraged in fields. This species may have became extinct due to the clearing of riverine forests, especially large hollow roost trees and mature deciduous woodlands, but it was also an agricultural pest and it therefore came under added pressure due to the extermination efforts of farmers.

Local and regional extirpations of game and large predators are well documented as the landscape was settled. Wolves (*Canis lupus*) were gone from New England by about 1850 (Goodwin, 1935). Records of John Pynchon, a fur trader in Springfield, Massachusetts, show that he shipped 8992 pelts of beaver (*Castor canadensis*) from the Connecticut River drainage between 1652 and 1657 (Judd, 1857). Beaver create a mosaic of successional stages as they colonize and abandon flowages. With the decline of Indian populations soon after European settlement in the early seventeenth century and heavy trapping of beaver, forests in southern New England reverted to fairly continuous forests.

The heavily forested landscape encountered and recorded by settlers penetrating interior southern and central New England (and the Blue Ridge and Cumberland Mountains in the southern highlands from western Virginia to Tennessee and Kentucky) came to be viewed as the primeval forest condition but the landscape may have been fairly open in many areas prior to the sixteenth century. Furthermore, colonists from largely pastoral/ agricultural, densely populated England, where forests had long since been reduced to woodlots, likely viewed the patchy forests of the New England and mid-Atlantic coasts as 'wild and savage' deep forests (Bradford [1856] 1981:70).

1.4.3 Modern transformations

Reversion of the New England landscape to forest, except near urban centers, began soon after 1840 and continues to the present time. This afforestation process has brought about faunal changes that are fairly well understood.

Grassland and shrubland birds are the most rapidly declining species in eastern North America (Askins, 1993). Mature forest birds in eastern North America have been shown to be quite tolerant of patchy disturbance within extensive areas of forest (Webb *et al.*, 1977; Maurer *et al.*, 1981; DeGraaf, 1991) but grassland and shrubland birds are specialists that quickly disappear from a site as succession proceeds (Figure 1.2). Some early-successional birds extended their ranges eastward from the Great Plains as land clearing progressed, e.g. brown-headed cowbird (*Molothrus ater*). These species are area-sensitive or are habitat specialists (Askins, 1993). Bobolinks, for example, desert grassy hayfields when proportions of alfalfa increase (Kantrud, 1981) or when field area drops below 10 ha (Bollinger and Gavin, 1992). Grasshopper sparrows (*Ammodramus savannarum*) occur only in grassland that is interspersed with bare ground (Smith, 1963; Whitmore, 1981). Henslow's sparrows (*A. henslowii*) occur only in fields with a deep litter layer, standing dead forbs and tall, dense grass (Zimmerman, 1988). Yellow-breasted chats (*Icteria virens*) inhabit brushy old fields until they begin to be invaded by overtopping trees (Shugart and James, 1973; Thompson, 1977). The decline of agriculture and reversion of forest have essentially eliminated grassland birds from the New England landscape, although they were common 30–40 years ago (Askins, 1993; Bagg and Eliot, 1937). Resident forest birds have increased in abundance in extensive woodland; pileated woodpecker (*Dryocopus pileatus*), which requires large (> 50 cm in diameter) trees for nest and roost cavities, has significantly increased in abundance over the last 25 years (Bull, 1974; Norse, 1985). Wild turkey (*Meleagris gallopavo*) have been restored by transplants from

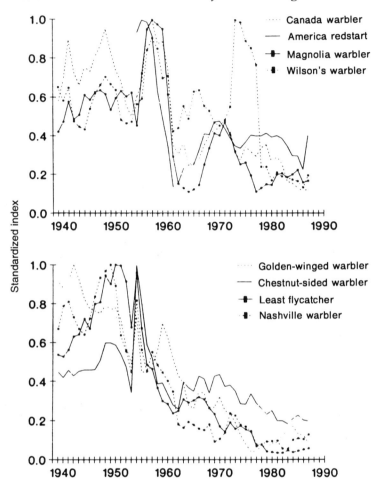

Figure 1.2 Annual indices of some migrating land birds that nest in the forests of New England. Samples (number observed/hr) were obtained in eastern Massachusetts during spring migrations (Hill and Hagan, 1991). Indices were standardized by dividing the highest annual observation rate for a particular species into each annual observation rate for that species. (After Litvaitis, 1993. Reprinted by permission of Blackwell Scientific Publications, Inc.)

wild stock throughout (and exceeding) their former range (Kennamer and Kennamer, 1990). Some Neotropical migratory birds, especially species that are primary-forest specialists on the wintering grounds, e.g. wood thrush (*Hylocichla mustelina*) and cerulean warbler (*Dendroica cerulea*), are declining in some parts of their breeding ranges (Rappole *et al.*, 1992; Robbins *et al.*, 1992) due largely to events on the wintering grounds or during migration (DeGraaf and Rappole, 1995; Terborgh, 1989; Hagan and Johnston, 1992).

Among terrestrial mammals, only the elk (*Cervus elaphus*), caribou (*Rangifer caribou*), wolf and mountain lion (*Felis concolor*) have not recolonized New England (or northeastern North America) since extirpation (although individuals of the latter two species, probably released from captivity, are occasionally reported). Human-caused mortality – as indexed by road density and thus human access – seems to limit wolf distribution and numbers: wolf survival is low when road densities exceed 0.6–0.7 km/km^2 (Thiel, 1985; Mech *et al.*, 1988; Mech, 1989; Fuller *et al.*, 1992). Mountain lions in good habitat in the western United States occur where road densities are less than 0.4 km/km^2 (Van Dyke, 1983; Van Dyke *et al.*, 1986). Road densities exceed these thresholds throughout most of the northeastern United States. Both species may be replaced ecologically, to a large extent, by eastern coyotes (*Canis latrans*), which now occur widely throughout northeastern North America (Moore and Parker, 1992).

1.5 THE HISTORICAL PATTERN OF DISTURBANCE IN NEW ENGLAND

Since the onset of European settlement, the forest landscape has been largely cleared and reafforested. Indeed, much of eastern North America was cleared and reafforested – largely through neglect; vast timber reserves existed farther west during this period. This settlement period represents a time of great change for the New England landscape but there were few extinctions and virtually all extirpated wildlife species later recolonized the area. For comparative purposes, the relatively constant climate associated with this period does not provide a 'baseline' pre-settlement landscape condition for New England.

Initially, land cleaning spread westward and northward from New England more or less as a wave (Figure 1.3). Land clearing began in earnest in interior southern New England around 1750 and its peak occurred in the period 1820–1840 (Raup, 1966). Forest clearing peaked in southern Vermont in 1860 when it was only beginning farther west in the Lake States. By the time extensive land clearing occurred in the upper Midwest, land abandonment in New England was fairly well advanced and forests were re-established on former farmland (Raup, 1966). Vertebrate species that had large distributions and were relatively mobile and those that survived at low numbers in remnant woodlots and inaccessible woodlands were able to recolonize former ranges in New England as forests became re-established.

Land clearing in New England was never complete. Up to 90% of the *arable* land was cleared for crops and pasture in central Massachusetts in 1820–1840 and in southern Vermont by 1860. Nevertheless, wood was the primary fuel for cooking and heating and supplies were permanently main-

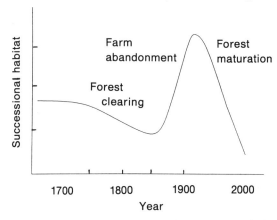

Figure 1.3   Suggested pattern of events that influenced the abundance of early successional habitat in northern New England from 1650 to the present. (After Litvaitis, 1993. Reprinted by permission of Blackwell Scientific Publications, Inc.)

tained near areas of settlement and on sites not suited to agriculture. At the peak of clearing, forests still covered 27% of Connecticut, 35% of Vermont, 40% of Massachusetts, 50% of New Hampshire and 79% of Maine (Harper, 1918).

Forest cutting for fuelwood alone in New England was far greater than today's harvest for all uses (Irland, 1982). In woodlots maintained for fuel, large old trees, especially oaks and hickories, were selectively removed. The sprouting of these trees after cutting perpetuated the farm 'sprout lot' in which trees rarely grew older than 30 years (Irland, 1982:59). As a result, woodland trees 100 years old or more became scarce (Bromley, 1935:65). For example, 95% of the hardwood forest in Lichfield County, Connecticut, was less than 41 years old by 1909 (Irland, 1982:59).

Fuelwood use declined in the 1820s when cast-iron stoves became widely used; such stoves generated the same amount of heat as a fireplace on one-quarter to one-fifth of the fuel (Bull, 1830; Fletcher, 1950). The introduction of coal from northeastern Pennsylvania, once canals were built to transport it to eastern cities (Powell, 1978:3), further reduced the demand for fuelwood. By 1840, coal stoves were common in New York and Philadelphia (Hoglund, 1962), while wood was still the dominant fuel in rural northern New England.

The replacement of fireplaces by stoves and the introduction of coal produced a dramatic decrease in the use of fuelwood in the nineteenth century (Reynolds and Pierson, 1942) and hastened the reversion of the land to forest. Whitney (1994:216) traces the trend: a house in colonial southern New England used 30+ cords per year; by 1846, use had declined to 13–14 cords (Emerson, 1846:15) and was less than 7 cords in many towns by 1888

(Blake, 1888:77). Population growth offset fuelwood use but total fuelwood consumption was halved between the 1870s and 1910 (Reynolds and Pierson, 1942).

Land was cleared by hand and wood was hauled by draft animals. These methods were inefficient by modern standards; mountains, steep ravines, rocky hills and deep swamps were difficult for humans to clear or farm, and likely served as refugia for many species which probably persisted at low densities. As forests invaded old fields, scattered woodlots ceased to be 'islands' of forest and the landscape assumed a more completely forested character fairly soon after land abandonment. Even the agriculturally impoverished soils of New England were capable of producing trees soon after abandonment.

These are the historical patterns of change in New England forests. They provide a backdrop to the present-day ecological perspective for forest disturbances around the globe.

1.5.1 The impact of recent landscape change

In the United States methods using satellite imagery are being developed to produce countrywide maps of forest cover (Zhu and Evans, 1994). For coniferous forests, similar methods are being developed to measure both forest cover and fragmentation (Ripple, 1994). Most importantly, satellite imagery from different dates can be used for the assessment of land-cover change in forested areas (Green *et al.*, 1994; Bauer *et al.*, 1994, Vogelmann, 1995). These applications of satellite imagery produce useful monitoring tools (e.g. land-cover maps) for both the forests and the fauna that inhabit them. Currently, however, it must be noted that the resolution of the satellite imagery limits detailed applications (Miller, 1994a). For example, deriving an association between land-cover change and land-use change remains difficult using satellite imagery (Green *et al.*, 1994).

The general patterns of landscape change that have produced faunal changes in the forests throughout New England also occur in many other parts of the world. The influence of these changes on the fauna is relatively moderate in heterogeneous temperate forests, in comparison with the effects in tropical and semitropical regions of the world – though the resilience of some tropical forest ecosystems must also be noted (Wilson, 1988).

1.5.2 Disturbance regimes, biogeographic patterns and faunal distributions

Two well-known approaches for analyzing factors influencing species diversity patterns are:

- the modeling of the landscape as a pattern of isolated and/or interconnected habitat fragments, patches and segments (e.g. MacArthur and Wilson, 1963, 1967);

- the modeling of the landscape as a composite of diverse environmental gradients (e.g. the work of R.H. Whittaker).

In the first case, species within habitats result from a balance between immigration and extinction rates. In the second case, the convergence of gradients results in distinct habitats with characteristic species ensembles. These two perspectives have proven useful in testing hypotheses about species diversity patterns over the past 30 years. In addition, habitat features can most accurately and realistically be modeled from a synthesis of these two perspectives. However, current mapping and modeling methodologies still require development and refinement to achieve this goal (Miller, 1994b).

In other parts of the world, the consequences of human encroachment on species distribution patterns within forests dispersed across the landscape differ considerably from species to species (Miller, 1994). For example, in the forests of Gabon, elephant (*Loxodonta africana*) densities are found to increase with distance from roads and major rivers (Michelmore, 1994). This reflects the influence of human disturbance on elephant distribution since these are the same areas where human settlements are concentrated. The transformation of broadleaf forests from Mexico to northern South America is producing a significant reduction in habitat for many migratory bird species and this is producing serious population reductions (e.g. wood thrush – Rappole *et al.*, 1994). Similarly, intensive logging and agriculture in the Andean foothills are causing severe declines in the cerulean warbler, a migratory bird that winters only in primary humid evergreen forest along an extremely narrow elevational zone between 620 and 1300 meters – this zone is among the most intensively disturbed regions in the Neotropics (Robbins *et al.*, 1992). A balance is needed today between the need to preserve these species and the continuing pressures of sustained timber harvest. It has recently been recommended that forest conservation today requires a balance between the intense use of selected timber stands with the absolute protection of primary forests (Lugo, 1994).

The status of a forest biome is significantly governed by the disturbance regimes that dominate in the region under consideration. For example, the dominant climatic patterns in a region regulate the height and density of the tree canopy in many parts of the world. At the scale of individual forests, the occurrence and rate of factors such as tree fall gaps may, in fact, significantly influence the availability of habitats (Runkle, 1981, 1984). Nevertheless, there have been relatively few extirpations of vertebrates in New England, indeed in northeastern forests, despite the major changes that have occurred on the land. These temperate forest communities appear to be relatively resilient in contrast to semitropical and tropical forests.

1.5.3 Forest landscape change worldwide and its possible relation to faunal distribution changes

All facets of biogeography across all landscape scales need to be taken into account when we consider the relationship between landscape change and faunal distribution changes. In recent years, much discussion has focused upon the negative impacts of global warming, but for some mammals in some areas of the United States it has recently been shown that future climatic changes may enhance the availability of refuge sites (Stokes and Slade, 1994).

The effects of human activities upon species distributions are complicated because of the interwoven patterns between biogeography and species niches. An illustration is provided by the great debate generated regarding the severity of the effects of clearcutting on the forest fauna of North America. Clearcutting in heavily forested northern New England provides habitat for many early successional birds that are adapted to disturbances from hurricanes and fire (DeGraaf 1987, 1992). In contrast, clearcutting is shown to have a significant negative impact on populations of California red-backed voles (*Clethrionomys californicus*) in forest remnants in south-western Oregon (Mills, 1995). While clearcutting benefits species and communities that are adapted to disturbance by fire, for example, at least in boreal and temperate forests, the potential exists to fragment the remaining forest and reduce habitat suitability for mature-forest species. For example, when 30% or less of original forest habitat remains, spotted owl (*Strix occidentalis*) and capercaillie (*Tetrao urogallus*) decline faster than expected, i.e. faster than habitat losses would indicate (Thomas *et al.*, 1990; Rolstad and Wegge, 1987, 1989). For many species, however, the effects of clearcutting are unknown. In some cases, the certain effects of this practice are not known even for entire taxonomic groups. For example, we do not currently know whether clearcutting significantly reduces the populations of salamanders in the southern Appalachians (Ash and Bruce, 1994). In New England northern hardwood forests, redback salamander (*Plethodon cinereus*) populations return to pre-harvest levels about 50 years after clearcutting (DeGraaf and Yamasaki, 1992). The impacts of change on forest fauna still require extensive study before these phenomena will be truly understood.

Sustainable forestry is a new concept that proposes a more integrated approach to the use and conservation of forested ecosystem; however, it is not currently known how sustainable forest ecosystem management will evolve from conventional sustained yield timber practices (Aplet *et al.*, 1993).

A recent Finnish study (Kuusipalo and Kangas, 1994) attempts to weigh the balance between the needs and benefits of sustainable forestry and those of biological conservation. However, this study concludes that the criteria for the evaluation of biodiversity alternatives need to be carefully considered based upon local conditions of forests and fauna. The criteria for biodiversity conservation in forested landscapes vary significantly according to local conditions. It is not usually appropriate to use a standardized set of criteria that will weigh the importance of biodiversity across different regions and scales.

The importance of forest corridors for interconnecting habitat patches across the fragmented landscape has been underscored in the past decade (e.g. Harris, 1984). Forested corridors have recently been proposed as possible key tools for maintaining avian forest communities in Chile (Willson *et al.*, 1994). The importance of corridors should be considered within the context of the prevailing local conditions.

(a)   Distributions in the temperate zone

In the temperate zone, forests are dynamic and most vertebrates have broad geographic ranges. For example, many vertebrates such as black bear, white-tailed deer, wild turkey, ruffed grouse, microtime rodents, numerous woodpeckers, thrushes, most wood warblers, parids and raptors have broad, even continental ranges (American Ornithologists' Union, 1983; Hall and Kelsorn, 1959; Hamilton and Whitaker, 1979). Birds exhibit habitat specificity to a greater degree than other vertebrate taxa (for review, see Cody, 1985).

Therefore, recolonization from distant source populations is possible for most temperate zone species. In North America most temperate bird species occur in many forest types (DeGraaf *et al.*, 1991) and they have wide distributions. All contemporary native forest mammals except the elk, caribou, wolf and mountain lion have recolonized New England.

(b)   Distributions outside the temperate zone

One generally accepted biogeographic pattern is that a high proportion of tropical plant and animal species exhibit restricted geographic distributions in contrast to species in the temperate regions. A comprehensive recent review of endemism as it occurs around the globe points out the clear distinction between the distribution patterns of restricted species as they occur in the temperate zone in contrast to the tropics (Thirgood and Heath, 1994). In contrast to eastern North America, the New World tropics contain a large number of endemic plant species with confined distributions (Gentry, 1986). In the tropics frugivorous birds and arboreal mammals are particularly characterized by habitat specializations that restrict their ranges.

A consideration of the threatened vertebrate species around the globe provides one perspective on the distributions of forest fauna. Approximately 9% of the total number of threatened mammals worldwide are found in seasonal woodlands (a category defined by the World Conservation Monitoring Center as including temperate forest) while 43% are found in tropical forests (World Conservation Monitoring Centre, 1992: Figure 17.6). For birds, approximately 22% of the threatened species are found in temperate forests and 43% are found in tropical forests. (Significantly, about 40% of threatened birds are located on oceanic islands – World Conservation Monitoring Centre, 1992.) A total of 219 threatened bird species occur in North and Central America, while more than double that number, 535 bird species, occur in South America (ibid Table 17.3). Based upon a general estimate founded upon the species–area relationship (Reid and Miller, 1989), the World Resources Institute estimates that if current trends of worldwide forest loss continue, species loss from closed tropical forest in Asia, Africa and South America will reach between 3 and 17% during the next 30 years.

(c)   The New England landscape
The New England landscape is comprised of a mosaic of habitats. Many of these habitats are created by the geologic, climatic and human agents that implement environmental change. These are environmental changes that exert a variable influence on forest species composition through time. The forest vegetation changes produced by harvesting, fire and soil disturbance are dependent upon the variable colonization rates of plant species (Walker, 1994). Therefore, the vegetative composition of disturbed forest communities varies considerably through time, depending upon the species inhabiting the region. The diversity of both plant and animal species inhabiting different expanses of disturbed forest communities will ultimately depend on the species colonization rates in surrounding areas.

The normal state of communities and ecosystems is not the result of an equilibrium but rather a response to the last disturbance event, i.e. a relatively discrete event in time that disrupts ecosystem, community or population structure and changes resources or the physical environment (Pickett and White, 1985). Non-equilibrium explanations of community structure, such as patch dynamics, explain high species diversity and the coexistence of similar species in terms of disturbance regimes and recruitment (Reice, 1994). Disturbance creates opportunities for colonization of vacated spaces by new species and this enhances biodiversity (ibid). Forest management policy needs to recognize the role of disturbance in maintaining biodiversity. Limiting or eliminating the effects of disturbance are not constructive practices if they preclude the occurrence of species that would naturally recolonize the area of concern and reduce plant and animal diversity over time.

1.6 SUMMARY

Disturbance appears to be relentless across the New England landscape. New England is subject to repeated long-term changes that result from climatic change and to shorter-term changes from periodic hurricanes. In addition, New England and much of the mid-Atlantic coast were subject to intensive slash-and-burn agriculture for at least the past 1000 years.

The wave of intensive forest clearing that commenced in the 1750s wrought profound changes in the landscape. Much of New England was cleared for subsistence farming by 1840, abandoned for the fertile Midwest, and reclaimed by forest by 1910. While some faunal extirpations occurred, few extinctions resulted from dramatic changes in vegetation that took place in less than two centuries. Recolonization by most species was possible because of their wide distributions in temperate North America, the patchiness of the region forests even at the height of clearing and the rapid reversion of the land to forest cover after abandonment.

Such a response to landscape change is not likely where many species are highly habitat specific or have very limited distributions, as is the case in the New World tropics. Monitoring of landcover change and faunal distributions at regional scales is critical to forest wildlife conservation efforts today.

1.7   ACKNOWLEDGEMENTS

We thank Robert A. Askins (Connecticut College, New London), William M. Healy, (US Forest Service, Northeastern Forest Experiment Station, Amherst, Massachusetts) and William A. Patterson III (University of Massachusetts, Amherst) for their critical reviews of the manuscript. We thank Nancy Haver for preparing the illustration of land-use history in southern New England and Mary A. Sheremeta for typing the manuscript.

1.8   REFERENCES

Ablet, G., Johnson, N., Olson, J. and Sample, V. (eds) (1993) *Defining Sustainable Forestry*. Island Press, Washington DC.

American Ornithologists' Union (1983) *Check-list of North American Birds*, 6th edn. American Ornithologists' Union, Lawrence, Kansas.

Ash, A.N. and Bruce, R.C. (1994) Impacts of timber harvesting on salamanders. *Conserv. Biol.* 8(1):900–901.

Askins, R.A. (1993) Population trends in grassland, shrubland, and forest birds in eastern North America. *Current Ornithology* 11:1–34.

Bagg, A.C. and Eliot, S.A. (1937) *Birds of the Connecticut Valley in Massachusetts*. The Hampshire Bookshop, Northampton, Massachusetts.

Bauer, M.E., Burk, T.E., Ek, A.R. *et al.* (1994) Satellite inventory of Minnesota forest resources. *Photogrammetric Engineering and Remote Sensing* 60(3):287–298.

Bennett, M.K (1955) The food economy of the New England Indians, 1605–1675. *J. Political Economy* 63:369–397.

Bjorck, S. (1985) Deglaciation chronology and revegetation in northwestern Ontario. *Canadian J. Earth Sci.* 22:850–871.

Blake, W. (1888) *History of the Town of Hamden, Connecticut, with an Account of the Centennial Celebration, June 15th, 1886.* Price, Lee and Company, New Haven, Conn.

Bollinger, E.K. and Gavin, T.A. (1992) Eastern bobolink populations: Ecology and conservation in an agricultural landscape, in *Ecology and conservation of Neotropical migrant landbirds* (eds J.M. Hagan and D.W. Johnston), pp. 497–506. Smithsonian Institution Press, Washington, D.C.

Bradford [1856] (1981) *Of Plymouth Plantation, 1620–1647.* Reprint: The Modern Library, New York.

Bradley, R.S. (1985) *Quaternary Paleoclimatology.* Allen and Unwin, Boston.

Braun, E.L. (1950) *Deciduous forests of eastern North America.* Hafner Press, New York. 596 pp.

Brokaw, N.V.L. (1982) The definition of treefall gap and its effect on measures of forest dynamics. *Biotropica* **14**(2):158–160.

Bromley, S.W. (1935) The original forest types of southern New England. *Ecol. Monogr.* **5**:61–89.

Bromley, S.W. (1945) An Indian relict area. *Sci. Monthly* **60**:153–154.

Brooks, R.T. and Birch, T.W. (1988) Changes in New England forests and forest owners: implications for wildlife habitat resources and management. *Trans. N. Am. Wildl. and Nat. Resour. Conf.* **53**:78–87.

Brown, J.H., Jr (1960) The role of fire in altering the species composition of forests in Rhode Island. *Ecology* **41**:310–316.

Brown, R.J.E. (1970) *Permafrost in Canada: Its influence on northern development.* University of Toronto Press, Toronto. 234 pp.

Bryson, R.A., Irving, W.M. and Larsen, J.A. (1965) Radiocarbon and soil evidence of former forests in the southern Canadian tundra. *Science* **147**:46–48.

Bucher, E.H. (1992) The causes of extinction of the passenger pigeon. *Current Ornithology* **9**:1–36.

Bull, J. (1974) *Birds of New York State.* Doubleday/Natural History Press, Garden City, N.J.

Bull, M. (1830) Experiments to determine the comparative quantities of heat evolved in the combustion of the principal varieties of wood and coal used in the United States. *Transactions of the American Philosophical Society* **3**:1–63.

Burgess, R.L. and Sharpe, D.M. (1981) *Forest island dynamics in man-dominated landscapes.* Springer Verlag, New York.

Cardoza, J.E. (1976) The history and status of the black bear in Massachusetts and adjacent New England states. Research Bull. 18, Mass. Div. of Fish and Wildl., Westborough, Massachusetts. 113 pp.

Champlain, S. de [1605] (1968) Discovery of the coast of the Almouchiquois as far as the 42nd degree of latitude and details of this voyage, in *Sailors' narratives of voyages along the New England Coast 1524–1624* (ed. G.P. Winship), pp. 64–97. Reprint: Burt Franklin, New York.

Channing, W. (1939) *New England hurricanes 1635–1815–1938.* Walter Channing, Inc., Boston, Massachusetts.

Cody, M.L. (ed.) (1985) *Habitat selection in birds.* Academic Press, Orlando, Florida.

Cook, S.F. (1973) The significance of disease in the extinction of New England Indians. *Human Biology* **45**:485–505.

Cook, S.F. (1976) The Indian populations of New England in the seventeenth century. *Publication in Anthropology* 12:1–91. Univ. of California, Berkeley.

Cooperrider, A.Y. (1974) Computer simulation of the interaction of a deer population with northern forest vegetation. Ph.D. Thesis. SUNY College of Environmental Science and Forestry, Syracuse, New York. 220 pp.

Cronon, W. (1983) *Changes in the Land: Indians, colonists, and the ecology of New England*. Hill and Wang, New York. 241 pp.

Curtis, J.D. (1943) Some observations on wind damage. *J. Forestry* 41:877–882.

Dahberg, B.L. and Guettinger, R.C. (1956) *The white-tailed deer in Wisconsin*. Tech. Wildl. Bull. 14. Wisconsin Conservation Department, Madison. 282 pp.

Davis, M. B. (1976) Pleistocene biogeography of temperate deciduous forests. *Geosci. Man.* 13:13–26.

Davis, M.B. (1981) Quaternary history and the stability of forest communities, in *Forest Succession* (eds D.C. West, H.H. Shugart and D.B. Botkin), pp. 132–153. Springer-Verlag, New York. 517 pp.

Day, G.M. (1953) The Indian as an ecological factor in the northeastern forest. *Ecology* 32:329–346.

Deevey, E.S. (1949) Biogeography of the Pleistocene. Part I. Europe and North America. *Geol. Soc. Am. Bull.* 60:1315–1416.

DeGraaf, R.M. (1987) Managing northern hardwoods for breeding birds, in *Managing Northern Hardwoods*. Faculty of Forestry Misc. Pub. 13. State University of New York, Syracuse, pp. 348–362.

DeGraaf, R.M. (1991) Breeding bird assemblages in managed northern hardwood forests in New England, in *Wildlife and Habitats in Managed Landscapes* (eds J.E. Rodiek and E.G. Bolen), pp. 153–171. Island Press, Washington, DC.

DeGraaf, R.M. (1992) Effects of even-aged management on forest birds at northern hardwood stand interfaces. *Forest Ecology and Management* 46:95–110.

DeGraaf, R.M. and Rappole, J.H. (1995) *Neotropical Migratory Birds: Natural history, distribution, and population change*. Cornell University Press, Ithaca, New York.

DeGraaf, R.M. and Yamasaki, M. (1992) A nondestructive technique to monitor the relative abundance of terrestrial salamanders. *Wildlife Soc. Bull.* 20:260–264.

DeGraaf, R.M., Scott, V.E., Hamre, R.H. *et al.* (1991) *Forest and Rangeland Birds of the United States*. Agric. Handbook 688, US Department of Agriculture, Washington, DC. 625 pp.

Delcourt, P.A. (1979) Late quaternary vegetation history of the eastern Highland Rim and adjacent Cumberland Plateau of Tennessee. *Ecol. Monogr.* 49:255–280.

De Vos, A. (1964) Range changes of mammals in the Great Lakes region. *Am. Midland Nat.* 71:210–231.

Dwight, T. (1823) *Travels in New England and New York*. Printed for W. Baynes and Son and Ogle, Duncan and Co., London. 4 vols.

Emerson, G.B. (1864) *Report on the Trees and Shrubs Growing Naturally in the Forests of Massachusetts*. Dutton and Wentworth, State Printers, Boston.

Fisher, R.T. (1918) Second-growth white pine as related to the former uses of the land. *J. Forestry* 16:493–506.

Fletcher, S.W. (1950) *Pennsylvania Agriculture and Country Life 1640–1840*. Pennsylvania Historical and Museum Commission, Harrisburg.

Foster, D.R. (1988a) Disturbance history, community organization and vegetation dynamics of the old-growth Pisgah Forest, southwestern New Hampshire, USA. *J. Ecol.* 76:105–134.

Foster, D.R. (1988b) Species and stand response to catastropic wind in central New England, USA. *J. Ecol.* 76:135–151.

Foster, D.R. (1992) Land-use history (1730–1990) and vegetation dynamics in central New England, USA. *J. Ecol.* 80:753–772.

Foster, D.R. and Boose, E.R. (1992) Patterns of forest damage resulting from catastrophic wind in central New England, USA. *J. Ecol.* 80:79–98.

Fujita, T.T. (1976) *Tornado map.* University of Chicago, Chicago.

Fuller, D.P. (1993) Black bear population dynamics in western Massachusetts. MS thesis, University of Massachusetts, Amherst. 136 pp.

Fuller, T.K., Berg, W.E., Radde, G.L. *et al.* (1992) A history and current estimate of wolf distribution and numbers in Minnesota. *Wildl. Soc. Bull.* 20:42–55.

Gentry, A.H. (1986) Endemism in tropical versus temperate plant communities, in *Conservation Biology: The science of scarcity and diversity* (ed. M.E. Soulé), Sinauer Associates, Sunderland, USA, pp. 153–181.

Godman, J.D. (1826) *American Natural History. 1.* H. Carey and I. Lea, Philadelphia. 362 pp.

Goodwin, G.G. (1935) *The mammals of Connecticut.* Conn. State Geol. Nat. Hist. Survey, Bull. 53. Hartford. 221 pp.

Goodwin, G.G. (1936) Big game animals in the northeastern United States. *J. Mammal.* 17:48–50.

Green, K., Kempka, D. and Lackey, L. (1994) Using remote sensing to detect and monitor land-cover and land-use change. *Photogrammetric Engineering and Remote Sensing* 60(3):331–337.

Gross, A.O. (1928) The heath hen. *Mem. Boston Soc. Nat. Hist.* 6:491–588.

Gross, A.O. (1932) Heath hen, in *Life Histories of North American Gallinaceous Birds* (ed. A.C. Bent), pp. 264–280. US Natl. Mus. Bull. 162, Washington, DC.

Hagan, J.M., III, and Johnston, D.W. (eds) (1992) *Ecology and Conservation of Neotropical Migrant Landbirds.* Smithsonian Inst. Press, Washington, DC. 609 pp.

Hall, E.R. and Kelson, K.R. (1959) *The mammals of North America* (2 vols). Ronald Press, New York.

Hamilton, W.J., Jr. and Whitaker, J.O., Jr. (1979) *Mammals of the Eastern United States.* Cornell University Press, Ithaca, New York.

Harmon, M.E., Bratton, S.P. and White, P.S. (1983) Disturbance and vegetation response in relation to environmental gradients in the Great Smoky Mountains. *Vegetatio* 55:129–139.

Harper, R.M. (1918) Changes in the forest area of New England in three centuries. *Journal of Forestry* 16:442–452.

Harris, L.D. (1984) *The Fragmented Forest.* University of Chicago Press, Chicago. 211 pages.

Higgeson, FH. [1630] (1929) New England's plantation, in *Massachusetts Historical Society Collections*, 1st series, 1:117–124.

Hoglund, A.W. (1962) Forest conservation and stove inventors – 1789–1850. *Forest History* 5:2–8.

Horton, D.R. (1984) Red kangaroos: last of the Australian megafauna, in *Qua-*

ternary Extinctions, (eds P.S. Martin and R.G. Klein), pp. 639–680. University of Arizona Press, Tucson.

Irland, L.C. (1982) *Wildlands and Woodlots.* University Press of New England, Hanover, NH. 217 pp.

Jenkins, D.H. and Bartlett, I.H. (1959) *Michigan White-tails.* Michigan Dept. Conservation, Game Div., Lansing. 8 pp.

Jennings, F. (1975) *The Invasion of America: Indians, colonialism, and the cant of conquest.* University of North Carolina, Chapel Hill.

Josselyn, J. (1675) An account of two voyages to New England, in *Massachusetts Historical Society Collection,* 3rd series, 3(1833). 273 pp.

Judd, S. (1857) *The fur trade on Connecticut River in the seventeenth century,* New England Hist. General Reg. N.S.1:217–219.

Kantrud, H.A. (1981) Grazing intensity effects on the breeding avifauna of North Dakota native grasslands. *Can. Field. Nat.* 95:404–417.

Kattan, G.H., Alvarez-Lopez, H. and Giraldo, M. (1994) Forest fragmentation and bird extinctions: San Antonio eighty years later. *Conserv. Biol.* 8(1):138–146.

Kennamer, J.E. and Kennamer, M.L. (1990) Current status and distribution of the wild turkey, 1989, in *Proceedings of the Sixth Wild Turkey Symposium,* (eds W.M. Healy and G.B. Healy), pp. 1–12. Nat. Wildl. Turkey Federation, Edgefield, So. Carolina. 228 pp.

Kiltie, R.A. (1984) Seasonality, gestation time, and large mammal extinctions, in *Quaternary Extinction,* (eds P.S. Martin and R.G. Klein), pp. 299–374. University of Arizona Press, Tucson.

King, J.E. and Saunders, J.J. (1984) Environmental insularity and the extinction of the American mastodont, in *Quaternary Extinction* (eds P.S. Martin and R.G. Klein), pp. 315–339. University of Arizona Press, Tucson.

Kroeber, A.L. (1939) *Cultural and natural areas of native North America.* Publications in American Archaeology and Ethology 38. Berkeley, University of California.

Kurtén, B. and Anderson, E. (1980) *Pleistocene Mammals of North America.* Columbia University Press, New York.

Kuusipalo, J. and Kangas, J. (1994) Managing biodiversity in a forestry environment. *Conserv. Biol.* 8(2):450–460.

Lamb, H.H. (1982) *Climate, History and the Modern World.* Methuen, London.

Likens, G.E. (1972) Mirror Lake: Its past, present, and future? *Appalachia* 39:23–41.

Litvaitis, J.A. (1993) Response of early successional vertebrates to historic changes in land use. *Conserv. Biol.* 7:866–873.

Loskiel, G.H. (1794) *History of the Mission of the United Brethren among the Indians of North America.* Printed for the Brethren's Society for the furtherance of the Gospel, London. 233 pp.

Lugo, A.E. (1994) Preservation of primary forest in the Luquillo Mountains, Puerto Rico. *Conservation Biology* 8(4):1121–1131.

Luckman, B.H. and Kearney, M.S. (1986) Reconstruction of Holocene changes in alpine vegetation and climate in the Maligne Range, Jasper National Park, Alberta. *Quaternary Res.* 26:244–261.

MacArthur, R.H. and Wilson, E.O. (1963) An equilibrium theory of insular zoogeography. *Evolution* 17:373–387.

MacArthur, R.H. and Wilson, E.O. (1967) *The Theory of Island Biogeography.* Princeton University Press, Princeton, New Jersey. 203 pp.

Martin, P.S. (1984) Prehistoric overkill: the global model. Pages 354–403 in *Quaternary Extinction,* (eds P.S. Martin and R.G. Klein), pp. 354–403. University of Arizona Press, Tucson.

Martin, P.S. and Klein, R.G. (eds) (1984) *Quaternary Extinction.* University of Arizona Press, Tucson.

Mattfeld, G.F. (1984) Northeastern hardwood and spruce-fir forests, in *White-tailed Deer, Ecology and Management* (ed. L.K. Hall), pp. 305–330. Stackpole, Harrisburg, PA. 870 pp.

Maurer, B.A., McArthur, L.B., and Whitmore, R.C. (1981) Effects of logging on guild structure of a forest bird community in West Virginia. *Am. Birds* 35:11–13.

McCabe, R.E. and McCabe, T.R. (1984) Of slings and arrows: an historical retrospection, in *White-tailed Deer, Ecology and Management* (ed. L.K. Hall), pp. 19–72. Stackpole, Harrisburg, PA. 870 pp.

Mech, L.D. (1989) Wolf population survival in an area of high road density. *Am. Midl. Nat.* 121:387–389.

Mech, L.D., Fritts, S.H., Radde, G.L., and Paul, W.J. (1988) Wolf distribution and road density in Minnesota. *Wildl. Soc. Bull.* 16:85–87.

Meltzer, D.J. and Mead, J.I. (1983) The timing of the late Pleistocene mammalian extinctions in North America. *Quaternary Res.* 19:130–135.

Mershon, W.B. (1923) *Recollections of my Fifty Years Hunting and Fishing.* The Stratford Co., Boston. 259 pp.

Michelmore, F. (1994) Keeping elephants on the map: Case studies of the application of GIS for conservation, in *Mapping the Diversity of Nature,* (ed. R.I. Miller). Chapman & Hall, London, pp. 107–126.

Miller, R.I. (1994a) Setting the scene, in *Mapping the Diversity of Nature,* (ed. R.I. Miller). Chapman & Hall, London, pp. 3–18.

Miller, R.I. (ed.) (1994b) *Mapping the Diversity of Nature.* Chapman & Hall, London. 218 pp.

Miller, R.I. and Griffin, C.R. (1994) A New England conservation project: Gap Analysis. *Global Biodiversity* 4(1):9–10.

Mood, F. (1937) John Winthrop, Jr, on Indian corn. *New England Quarterly* 10:128–9.

Moore, G.C. and Parker, G.R. (1992) Colonization by the eastern coyote (*Canis latrans*), in *Ecology and Management of the Eastern Coyote,* (ed. A.H. Boer), pp. 23–27. Wildlife Research Unit, University of New Brunswick, Fredericton. 194 pp.

Morton, T. (1632) New English Canaan, *Publications of the Prince Society, XIV,* (ed. L.F. Admans) (1883) p. 177. Boston.

Morton, T. (1637) *New English Canaan or New Cannan.* J.F. Stam, Amsterdam. 188 pp.

Nelson, T.C. and Zillgitt, W.M. (1969) *A Forest Atlas of the South.* USDA For. Serv., Southern For. Exp. Sta., New Orleans, Louisiana.

Mills, L.S. (1995) Edge effects and isolation: red-backed voles in forest remnants. *Conservation Biology* 9(2):395–403.

Niering, W.A. and Dreyer, G.D. (1989) Effects of prescribed burning on *Andropogon scoparius* in postagricultural grasslands in Connecticut. *Am. Midl. Nat.* 122:88–102.

Norse, W.J. (1985) Pileated woodpecker, in *The Atlas of Breeding Birds of Vermont* (eds S.B. Laughlin and D.P. Kibbe), pp. 168–169. University Press of New England, Hanover, New Hampshire.

Nuttall, T. (1834) *Manual of the Ornithology of the United States and Canada,* Vol. 1. Boston.

Parker, A.C. (1910) *Iroquois uses of maize and other food plants.* NY State Mus. Bull. 144. Albany. 119 pp.

Patterson, W.A. III and Backman, A.E. (1988) Fire and disease history of forests, in *Vegetation History,* (eds R. Huntley and T. Webb III), pp. 603–632. Kluwer Academic Publishers, Dordrecht.

Patterson, W.A. III and Sassaman, K.E. (1988) Indian fires in the prehistory of New England, in *Holocene Human Ecology in Northeastern North America,* (ed. G.P. Nicholas), pp. 107–135. Plenum Publ. Corp., New York.

Pielou, E.C. (1991) *After the Ice Age: the Return of Life to Glaciated North America.* University of Chicago Press, Chicago.

Powell, H.B. (1978) *Philadelphia's First Fuel Crisis: Jacob Cist and the Developing Market for Pennsylvania Antracite.* Pennsylvania State University Press, University Park.

Pratt, P.P. (1976) *Archaelogy of the Oneida Indians. Man in the Northeast.* Occasional Publications in Northeastern Anthropology No. 1, George's Mills, New Hampshire.

Pring [1625] (1906) *A voyage set out from the citie of Bristol, 1603.* Reprinted in H.S. Burrage (ed.), *Early English and French Voyages, 1524–1608,* (ed. H.S. Burrage), pp. 341–352. Charles Scribners' Sons, New York.

Rappole, J.H., Morton, E.S., and Ramos, M.A. (1992) Density, philopatry, and population estimates for songbird migrants wintering in Veracruz, in *Ecology and Conservation of Neotropical Migrant Landbirds* (eds J.M. Hagan III and D.W. Johnston), pp. 337–344. Smithsonian Inst. Press, Washington, DC. 609 pp.

Rappole, J.H., Powell, G.V.N. and Sader, S.A. (1994) Remote-sensing assessment of tropical habitat availability for a nearctic migrant: The wood thrush, in *Mapping the Diversity of Nature* (ed. R.I. Miller), pp. 91–103. Chapman & Hall, London.

Raup, H.M. (1966) The view from John Sanderson's farm: A perspective for the use of the land. *Forest History* 10:2–11.

Reice, S.R. (1994) Nonequilibrium determinants of biological community structure. *American Scientist* 82:424–435.

Reid, W.V. and Miller, K.R. (1989) *Keeping Options Alive: The Scientific Basis for Conserving Biodiversity.* World Resources Institute, October 1989.

Reynolds, R.V. and Pierson, A.H. (1942) *Fuelwood used in the United States 1630–1930.* US Dept Agric. Circular 641, Washington, DC.

Ripple, W.J. (1994) Determining coniferous forest cover and forest fragmentation with NOAA-9 Advanced Very High Resolution Radiometer data. *Photogrammetric Engineering and Remote Sensing* 60(5):533–540.

Ritchie, J.C., Cwynar, L.C., and Spear, R.W. (1983) Evidence from northwest Canada for an early Holocene Milankovitch thermal maximum. *Nature* 305:126–128.

Robbins, C.S., Fitzpatrick, J.W., and Hamel, P.B. (1992) A warbler in trouble: *Dendroica cerulea,* in *Ecology and Conservation of Neotropical Migrant Landbirds* (eds J.M. Hagan III and D.W. Johnston), pp. 549–562. Smithsonian Inst. Press, Washington, DC. 609 pp.

Rolstad, J. and Wegge, P. (1987) Distribution and size of capercaillie leks in relation to old forest fragmentation. *Oecologia (Berl.)* 72:389–394.

Rolstad, J. and Wegge, P. (1989) Capercaillie *Tetrao urogallus* populations and modern forestry – a case for landscape ecological studies. *Finnish Game Research* 46:43–52.

Rowlands, W.P. (1941) Damage of even-aged stands in Petersham, Massachusetts by the 1938 hurricane as influenced by stand condition. MFS thesis, Harvard University, Cambridge, Massachusetts.

Runkle, J.R. (1981) Gap regeneration in some old-growth forests of the eastern United States. *Ecology* 62:1041–1051.

Runkle, J.R. (1984) Development of woody vegetation in treefall gaps in a beech-sugar maple forest. *Holart. Ecol.* 7:157–164.

Runkle, J.R. (1990) Gap dynamics in an Ohio *Acer–Fagus* forest and speculations on the geography of disturbance. *Can. J. For. Res.* 20:632–641.

Russell, E.W.B. (1983) Indian-set fires in the forests of the northeastern United States. *Ecology* 64:78–88.

Russell, H.S. (1961) New England Indian agriculture. *Bull. Massachusetts Archaeological Society* 22:58–61.

Saunders, D.A., Arnold, G.W., Burbidge, A.A., and Hopkins, A.J.M. (eds) (1987) *Nature Conservation: the Role of Remnants of Native Vegetation.* Surrey Beatty and Sons, Chipping Norton, NSW.

Schoen, J. W., Wallmo, O.C. and Kirchhoff, M.D. (1981) Wildlife–forest relationships: Is a re-evaluation of old growth necessary? *Trans. N. Am. Wildl. Nat. Res. Conf.* 46:531–544.

Schorger, A.W. (1936) The great Wisconsin nesting of 1871. *Proc. Linnaean Soc.* 48:1–26.

Schorger, A.W. (1955) *The Passenger Pigeon, Its Natural History and Extinction.* University of Oklahoma Press, Norman.

Seton, E.T. (1909) *Life Histories of Northern Mammals.* Vol. 1. Charles Scribner's Sons, New York. 673 pp.

Seton, E.T. (1929) *Lives of Game Animals.* Doubleday, Doran and Co., Garden City, NY. 4 vols.

Shelford, V.E. (1963) *The Ecology of North America.* University of Illinois Press, Urbana. 610 pp.

Shugart, H.H., Jr. and James, D. (1973) Ecological succession of breeding bird populations in northwestern Arkansas. *Auk* 90:62–77.

Smith, R.L. (1963) Some ecological notes on the grasshopper sparrow. *Wilson Bull.* 75:159–165.

Snow, D.R. (1980) *The Archaeology of New England.* Academic Press, New York.

Snow, D.R. (1995) Microchronology and demographic evidence relating to the size of pre-Columbian North American Indian populations. *Science* 268:1601–1604.

Stebbins, R.P. (1864) *An Historical Address delivered at the Centennial Celebration of the Town of Wilbraham, June 15, 1863.* George C. Rand and Avery, Boston.

Stokes, M.K. and Slade, M.A. (1994) Drought induced cracks in the soil as refuges for small mammals: An unforseen consequence of climatic change. *Conserv. Biol.* 8(2):577–580.

Terborgh, J.W. (1989) *Where Have All the Birds Gone?* Princeton Univ. Press, Princeton, New Jersey.

Thiel, R.P. (1985) The relationship between road densities and wolf suitability in Wisconsin. *Am. Midl. Nat.* 113:404–407.

34   Disturbance and land-use history in New England

Thirgood, S.J. and Heath, M.F. (1994) Global patterns of endemism and the conservation of biodiversity, in *Systematics and Conservation Evaluation,* (eds P.L. Forey, C.J. Humphries and R.J. Vane-Wright), pp. 207–228. Oxford University Press, Oxford, England.

Thomas, J.W., Forsman, E.D., Lint, J.B. *et al.* (1990) *A Conservation Strategy for the Northern Spotted Owl.* USDA Forest Service. 427 pp.

Thomas, P.A. (1976) Contrastive subsistence strategies and land use as factors for understanding Indian–White relations in New England. *Ethnohistory* 23:1–18.

Thompson, C.F. (1977) Experimental removal and replacement of territorial male yellow-breasted chats. *Auk* 94:107–113.

Trefethen, J.B. (1970) The return of the white-tailed deer. *American Heritage* 21:97–103.

Trumbull, B. (1797) *A Complete History of Connecticut.* Hudson and Goodwin, Hartford. 587 pp.

Ubelaker, D.H. (1988) North American Indian population size, AD 1500 to 1985. *Am. J. Phys. Anthro.* 77:289–294.

Van Dyke, F.G. (1983) A western study of cougar track surveys and environmental disturbances affecting cougars related to the status of the eastern cougar *Felis concolor cougar.* Ph.D. Thesis, State Univ. New York, Syracuse. 244 pp.

Van Dyke, F.G., Brocke, R.H., and Shaw, H.G. (1986) Use of road track counts as indices of mountain lion presence. *J. Wildl. Manage.* 50:102–109.

Vogelmann, J.E. (1995) Assessment of fragmentation in southern New England using remote sensing and Geographic Information System technology. *Conservation Biology* 9(2):439–449.

Walker, J. (1994) Impacts of forest management on plant diversity. *Bulletin of the Ecological Society of America* 75(3):171–172.

Waters, J.H. (1962) Some animals used as food by successive cultural groups in New England. *Bull. Archaeological Soc. Conn.* 31:32–45.

Watts, W.A. (1970) The full-glacial vegetation of northwestern Georgia. *Ecology* 51:19–33.

Watts, W.A. (1983) Vegetation history of the eastern United States, in *Late Quaternary Environments of the United States, Vol. 1, The Late Pleistocene,* (ed. S.C. Porter), pp. 294–310. Univ. of Minnesota Press, Minneapolis.

Webb, S.D. (1984) Ten million years of mammal extinctions in North America, in *Quaternary Extinction* (eds P.S. Martin and R.G. Klein), pp. 189–210. University of Arizona Press, Tucson.

Webb, T. III. (1986) Is vegetation in equilibrium with climate? How to interpret late-Quaternary pollen data. *Vegetatio* 67:75–91.

Webb, T. III. (1988) Eastern North America, in *Vegetation History* (eds. B. Huntley and T. Webb III), pp. 385–414. Kluwer Academic Publishers, Dordrecht, The Netherlands.

Webb, W.L., Behrend, D.F. and Saisorn, B. (1977) Effect of logging on songbird populations in a northern hardwood forest. *Wildl. Monogr.* 55:6–36.

West, F.H. (1983) The antiquity of man, in *Late Quaternary Environments of the United States, Vol. 1, The Late Pleistocene* (ed. S.C. Porter), pp. 364–382. Univ. of Minnesota Press, Minneapolis.

West, R.G. (1970) Pleistocene history of the British flora, in *Studies in the Vegeta-*

tional History of the British Isles (eds D. Walker and R.G. West), pp. 1–11. Cambridge University Press, London.

Whitmore, R.C. (1981) Structural characteristics of grasshopper sparrow habitat. *J. Wildl. Manage.* 45:811–814.

Whitney, G.G. (1986) Relation of Michigan's presettlement pine forests to substrate and disturbance history. *Ecology* 67:1548–1559.

Whitney, G.G. (1994) *From Coastal Wilderness to Fruited Plain: Temperate North America 1500 to the Present.* Cambridge University Press, Cambridge, UK.

Whittington, S.L. and Dyke, B. (1984) Simulating overkill: experiments with the Mosimann and Martin model, in *Quaternary Extinction* (eds P.S. Martin and R.G. Klein), pp. 451–465. University of Arizona Press, Tucson.

Willoughby, C.C. (1906) Houses and gardens of the New England Indians. *American Anthropologist* 8:115–132.

Willson, M.F., De Santo, T.L., Sabag, C., and Armesto, J.J. (1994) Avian communities of fragmented south-temperate rainforests in Chile. *Conserv. Biol.* 8(2):508–520.

Wilson, E.O. (ed.) (1988) *Biodiversity.* National Academy Press, Washington, DC.

Winthrop, J. Jr (1863) Letter to John Winthrop, Sr (April 7, 1636). *Massachusetts Historical Society Collections, 4th Ser.,* 6:514.

Wood, W. [1634] (1977) *New England's Prospect* (ed. A.T. Vaughan), University of Massachusetts Press, Amherst.

World Conservation Monitoring Centre (1992) *Global Biodiversity: Status of the Earth's Living Resources.* Chapman & Hall, London. 585 pp.

Wroth, L.C. (ed.) (1970) *The Voyages of Giovanni de Verrazzano, 1524–1528.* Yale University, New Haven.

Zhu, Z. and Evans, D.L. (1994) US forest types and predicted percent forest cover from AVHRR data. *Photogrammetric Engineering and Remote Sensing* 60(5):525–531.

Zimmerman, J.L. (1988) Breeding season habitat selection by the Henslow's sparrow (*Ammodramus henslowii*) in Kansas. *Wilson Bull.* 11:17–24.

—

–2

Changes in global forest distribution

David B. Kittredge, Jr

2.1 INTRODUCTION

A tremendous amount has been written recently about the worldwide destruction of forests. Forests are valued as an important source of wood for industrial and residential use, fiber and fuel, and as providers of clean water, clean air and outdoor recreation. Forests store tons of carbon, which otherwise might be in the atmosphere in the form of carbon dioxide contributing to the build-up of greenhouse gases and so increasing the potential for global warming (Dixon *et al.*, 1994). In addition, some of the planet's greatest biological diversity is found in forests. Thousands of plant and animal species depend on forested habitat for their existence. In fact, there is a great diversity of forests themselves, ranging from the boreal forests of the northern hemisphere to the moist and dry forests of tropical latitudes. Such a variety of ecosystems, that collectively cover roughly 39% of the world's land surface (WCMC, 1992), is the key to the global diversity of life.

It is the goal of this chapter to provide a broad overview of the extent of and changes in the world's forests, and to provide a broad geographical perspective for subsequent chapters on forest wildlife communities and species. Estimates of forested area can vary considerably, and it is not the intention to provide an authoritative estimate of the forested state of the planet. Mather (1990), for example, maintained that a comprehensive review of global forests is virtually impossible, due to the 'confusing combination of a plethora of statistical information and a paucity of data. Information is scattered and diverse, and attempts at compilation and synthesis are all too rare'. Several books on the subject have been published recently and the reader is referred to these for a more in-depth treatment (Laarman and Sedjo, 1992; Mather, 1990).

Conservation of Faunal Diversity in Forested Landscapes. Edited by
R.M. DeGraaf and R.I. Miller. Published in 1996 by Chapman & Hall. ISBN 0 412 61890 7.

For the purposes of this overview, trends in the existence of forests are based on past, present and future estimates from a variety of sources. Since the discussion deals with forests on a global scale, the finest resolution here is at the continental level.

2.2 TOTAL FOREST AREA

2.2.1 Sources of information

Reporting the estimated area of forests worldwide in the past, present and future is complicated and inexact. Geopolitical boundaries fluctuate over time, making it difficult to report consistent changes. For example, forest area in Latvia was counted separately from Russia in a global inventory of forest resources in 1923 (Zon and Sparhawk, 1923), yet it was reported as part of the Soviet Union in later years. With the dissolution of the Soviet Union, information is reported again on the basis of republics that were formerly a part of that union. Tracking the extent of and changes in global forests over decades is made complicated by changes in political boundaries and names.

Another major reason for the difficulty in estimating the global extent of forests over time is one of definition. Terms such as 'forest', 'woodland', 'forested land', 'interrupted woods', 'major forests', 'forestland', 'timberland', 'dry forests', 'moist forests', 'open forest', 'closed forest' and 'unproductive forest' have been variously used over time to describe land on which some number of trees is growing (Figure 2.1). In some cases, trees do not actually have to be growing on the land. Production yearbooks published by the Food and Agriculture Organization of the United Nations (UN FAO), for example, include in their definition of 'forest land' areas that have been cleared but are intended for reforestation (Mather, 1990). Some definitions have an economic basis: 'forest area' was defined by Zon and Sparhawk (two economists who worked for the US Forest Service) in 1923 as land covered with woody growth of economic importance. Other definitions have an ecological basis. A recent report on global biodiversity (WCMC, 1992), for example, used 58 different classifications to describe the extent of forest worldwide. So ecologically sensitive were these designations that they included four specific categories of rainforest: eastern oceanic constantly humid forests; humid forests with short dry season; mixed forests with short dry season; and constantly humid evergreen forests.

The World Resources Institute (WRI, 1990) reported three published independent estimates of the extent of the world's forested ecosystems. They ranged from a low of 34 million km^2 to a high of 57 million km^2. Needless to say, this is probably attributable to differences in definition. It may also be due, in part, to differences in methods of estimation. Ground surveys are conducted in some cases but in others remote sensing data are

Figure 2.1 'Natural' forest in Costa Rica. (Photo: H. Gyde Lund, USDA Forest Service.)

used to estimate the extent of forests in a country. In yet other cases, governments are asked to respond to a survey conducted by an international agency. Responses and methods vary from country to country and in their reliability. In India, for example, remote sensing indicated an annual deforestation rate for the early 1980s of approximately 1.5 million hectares, which was roughly 10 times greater than earlier estimates reported by the UN FAO. In fact, the satellite imagery indicated that large areas that were thought to be forested actually had virtually no trees (Repetto, 1990).

In short, there is a great variety of definitions and methods of estimation used to describe the amount of forests in a country or region, and thus worldwide. As a result, the estimates provided in this chapter need to be interpreted cautiously. They are useful to indicate relative trends but cannot be considered absolute. Likewise, it is difficult to interpret anything but broad trends over time, since geopolitical boundaries change, as do methods of estimation and definitions. It is no wonder that a Swedish forest scientist, attempting to assemble information on the world's forests, remarked that modern science has accumulated more definitive information on the craters of the moon than of the forests of the earth (R. Persson, 1977, in Laarman and Sedjo, 1992).

2.2.2   Trends

Over time, there has been a worldwide loss of forest. One global estimate put the decline at roughly 18% between 1700 and 1980 (J.F. Richards, 1990, in WCMS, 1992). Mather (1990) estimated that there has been roughly a 2% decline in the amount of forest worldwide between 1975 and 1985, based on data from FAO Production Yearbooks. For that period, Mather breaks it down to a 0.7% decline in the 'developed' world and a 3.0% decline in the 'developing' world. Regionally, he reports roughly a 5% decline in Latin America and a 4% decline in Africa. He further reports that Central America, parts of southeast Asia and west Africa experienced a 10% decline between 1975 and 1985.

Not all areas declined, however. Mather (1990) reported an increase of a little over 1% in Europe and a 4% increase in the former Soviet Union. He attributes the slight European increase to agricultural surpluses in the 1980s which resulted in national programs to convert land use from agriculture to forest. Increased forest area in the former Soviet Union can be attributed to plantation establishment. It is estimated that there are 21.9 million hectares of plantations in the former Soviet Union, and additional plantations are being established at a rate of approximately 1.3 million hectares annually (Mather, 1990).

Actual estimates of forest area (Figure 2.2) are reported here on the basis of either continents or broad geographic areas (e.g. Scandinavia, Pacific region, former USSR). Based on similarity of forested conditions, forest areas have been further combined into two regional categories: those located primarily in the northern hemisphere (i.e. those regions dominated primarily by boreal and temperate forests) and those located primarily in the southern hemisphere (i.e. those regions dominated by moist or dry tropical forests). Estimates from 1963 are from a world forest inventory conducted by the UN FAO (FAO, 1963). Estimates from 1990 are by the World Resources Institute (WRI, 1994).

With the exception of North America, which experienced a 36% decrease in forest area between 1963 and 1990, forest area in the northern hemisphere (Figure 2.2a) remained relatively stable. In fact, the former Soviet Union reported an increase in forest area. In the southern hemisphere (Figure 2.2b), several regions experienced significant declines. Africa lost an estimated 24% during this period. Declines of this magnitude are reflected in other estimates (FAO, 1988; WCMC, 1992). Closed tropical moist forests in Africa were deforested at a rate of 61% between 1976 and 1985. Those in South and Central America were deforested at a rate of 60% between 1976 and 1980, and by 63% between 1981 and 1985. The apparent increase in forest land in Central and South America (Figure 2.2b) is most likely attributable to differences in definition between the two sources. This illustrates the difficulty in monitoring forestland change on the global scale.

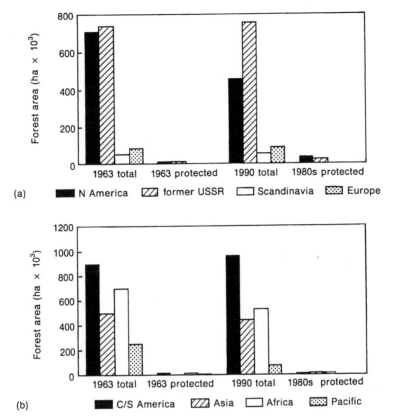

Figure 2.2 Estimated forest area by region: (a) northern and (b) southern hemispheres. (Sources: FAO, 1963; WRI, 1994.)

2.2.3 Causes of change

Factors contributing to deforestation are many and they vary in importance from region to region. The practice of shifting cultivation is the principal cause in the tropics (Lanley, 1982, in WCMC, 1992). Apparently, 70% of the deforestation in Africa, 50% in Asia and 35% in Central and South America were caused by transient agriculture. Permanent conversion to pasture is another prominent cause of deforestation. Prior to 1980, approximately 72% of the deforestation in Brazil was attributed to conversion to pastures, as opposed to shifting cultivation (Mather, 1990). During times of drought, especially in subSaharan regions, trees are sometimes cut to provide forage for livestock (Lund, 1994) – yet another agricultural activity that can cause deforestation.

In some cases, commercial timber harvesting may result in deforestation.

This is not necessarily related to internal population pressures for land or wood, as much as to external or international markets for timber products. Often, harvesting for timber utilization is not the simple cause of the loss of forest. Roads constructed to access timber provide means for farmers to move into an area and either clear land for agriculture or further and more completely exploit the area for fuelwood (Figure 2.3).

Other causes are population pressures and exploitation for fuelwood and charcoal (WCMC, 1992). Most fuelwood exploitation occurs in woodlands or farmed areas, rather than in closed forest (Mather, 1990) (Figure 2.4). Fuelwood use varies regionally, ranging from a low of about 16% of total energy consumption in Chile to over 95% of total energy consumption in both Chad and Uganda (Mather, 1990). Likewise, consumption per capita ranges from 0.5 cubic meters in Chile to over 1.6 cubic meters in Chad and Uganda (Mather, 1990). While deforestation due to a reliance on fuelwood may be important in some areas, it is less so in others.

Government policy can stimulate deforestation. For example, the Malaysian government has promoted agricultural development of rubber and palm oil plantations (Figure 2.5); and in Brazil, tax credits encourage further development of cattle ranches (Johnson and Cabarle, 1993). Elsewhere, land settlement and forest conversion are officially promoted to alleviate urban crowding (Johnson and Cabarle, 1993). The granting of timber concessions often seriously damages forests and makes them vulnerable to subsequent deforestation. Such concessions are often concerned only with removing the most valuable species in the shortest amount of time and extraction can result in considerable damage to residual stands. For example, studies of tropical logging have shown that even where as little as 5–20% of the timber volume is removed, between 20 and 50% of the residual stand can be damaged (Johnson and Cabarle, 1993). Such logging often pays little heed to the ecological processes of the tropical forest, such as pollinating mechanisms (Johnson and Cabarle, 1993).

Not all causes of change are artificially induced. There may be some changes in forest cover caused by catastrophic wind or fire events. Some recent estimates of tropical deforestation are based on remotely-sensed satellite data (Skole and Tucker, 1993; Lund, 1993) and it is sometimes impossible to distinguish between forest change caused by humans and that caused by natural events (Malingreau, 1993). The former can be a permanent change, whereas the latter can represent merely a change in successional condition of the forest. In terms of wildlife habitat, human-induced deforestation results in a loss of forest habitat whereas natural catastrophic disturbance might result in a change in habitat.

All in all, the importance of these different deforestation factors depends on the influence of human population pressure. No single cause has a more significant worldwide effect. All factors are related to the increase in human population and the demands that humans place on resources.

Figure 2.3 (a) Logging road system through forested landscape in Papua New Guinea. (Photo: Ray Allison, retired, USDA Forest Service.) (b) Logging road in Papua New Guinea showing extent of disturbance. (Photo: H. Gyde Lund, USDA Forest Service.)

Figure 2.4   Fuelwood resulting from agricultural clearing in Costa Rica. (Photo: H. Gyde Lund, USDA Forest Service.)

Figure 2.5   Plantations of sego palm or palm oil intermingled with natural forests in Papua New Guinea. (Photo: H. Gyde Lund, USDA Forest Service.)

In the face of forest loss, some forest has been set aside with a protected status (i.e. declared by government to be withdrawn from any use other than the protection of a natural condition; Figure 2.2). This clearly represents a very small proportion of forest land in either the northern or southern hemispheres. Of note is the fact that protected forest area in North America increased three-fold during this time period, though it still represents a small fraction (approximately 8%) of the total. Likewise, protected forest area roughly doubled in the former Soviet Union during this period but still comprises only 3% of total forest area.

2.3   FOREST AREA PER CAPITA

Estimates of absolute numbers of forested hectares can be misleading, especially when it comes to the ability of the forest to provide suitable habitat for wildlife. Since the principal overall reason for deforestation is the influence of humans, it is worth considering the amount of forested area per capita in the various regions.

The regions of the northern hemisphere have experienced a moderate to small decline in the number of forested hectares per capita (Figure 2.6a). This is attributable to the absolute decline in area, as well as the increase in population over time. Most notably, between 1963 and 1990, Scandinavia and Europe have remained relatively stable by this measure, with the former Soviet Union and North America declining somewhat (40% and 61%, respectively).

The shifts have been more dramatic in the regions of the southern hemisphere (Figure 2.6b). Of particular note in the recent past (between 1963 and 1990) are the decline in the Pacific region (78%) and the small proportion of forest land per capita in both Asia and Africa.

2.4   HUMAN POPULATION

Forest area per capita has declined not only because of a loss of the absolute number of forested hectares but also because the human population has increased. Human population trends reported here are based on information reported in WRI (1994). Some areas of the world have relatively stable human populations, such as Europe and Scandinavia, where increases between 1950 and 1995 were of 32% and 27%, respectively (Figure 2.7a). These regions are expected to be stable in the future: between 1995 and 2025, for example, their populations are predicted to increase by 5% and 8%, respectively.

Regions in the southern hemisphere have undergone greater population growth and are predicted to increase more dramatically in the future (Figure 2.7b). In particular, Asia and Africa experienced considerable population growth between 1950 and 1995 (147% and 234%, respectively), and are expected to continue this growth by 2025 (44% and 113%, respectively). In

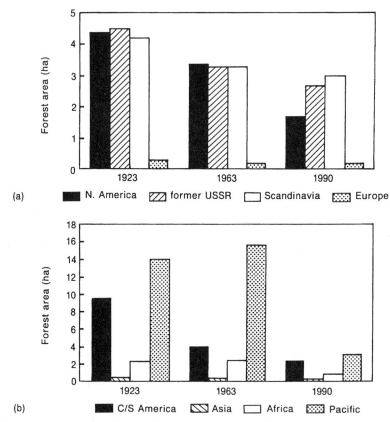

Figure 2.6   Forest area per capita, by region: (a) northern and (b) southern hemispheres. (Sources: Zon and Sparhawk, 1923; FAO, 1963; WRI, 1994.)

South America the human population grew by 201% between 1950 and 1995, and the small Pacific region grew by 129%. Both are expected to undergo large population increases in the future (46% and 43%, respectively).

Mather (1990) discussed the relationship between forest area and population trends on a national or global scale: 'In countries with rapidly expanding populations the forest land area is rapidly contracting. Conversely, where the population is stagnating or growing only slowly, forest land is expanding.' He attributed this to the often-cited reasons of agricultural expansion and fuelwood collection. He even cites circumstances in which the two can act synergistically, resulting in more forest clearance. Declines in nearby fuelwood availability, for example, result in increased use of dung and crop residues as fuel. Thus, less is available for fertilizer and agricultural yields decline, resulting in more land being cleared for crop production. Clearly, the loss of forest land is a complex issue, inextricably

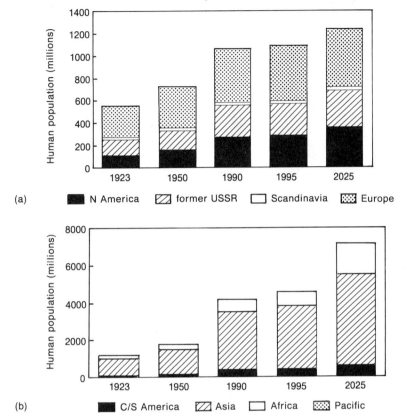

Figure 2.7 Estimated human population, by region: (a) northern and (b) southern hemispheres. (Sources: Zon and Sparhawk, 1923; WRI, 1994.)

tied to population growth. It is important to distinguish however, between the agents of deforestation, such as peasant farmers, shifting cultivators and rural landless people, and the root causes of deforestation (Mather, 1990). Often 'social and political structures, and the mode of "development" pursued in many developing countries, are therefore the driving-force behind deforestation, rather than population pressure per se' (Mather, 1990). Government policies resulting from debt-burdened economic situations and the practices of national and international financial institutions can create conditions which favor tropical deforestation (Repetto, 1990).

2.5 WOOD PRODUCTION AND CONSUMPTION

Humans have relied on wood from forests for centuries (Perlin, 1991) and the increasing global population will look to forests more and more to

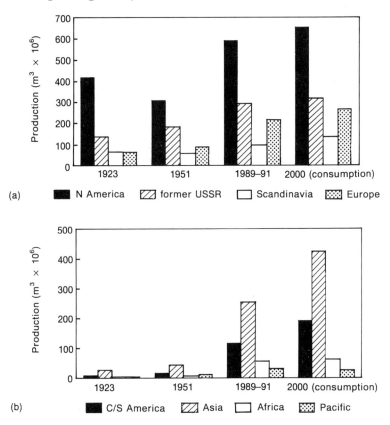

Figure 2.8 Estimated production of industrial wood, by region: (a) northern and (b) southern hemispheres. (Sources: Zon and Sparhawk, 1923; FAO, 1951; WRI, 1994; FAO, 1988.)

meet societal needs. Prominent among these needs are wood for industrial use, such as construction, paper and similar applications. The other principal use of wood worldwide is for fuel and more wood is consumed globally for fuel than for all industrial uses combined (WRI, 1990). While the use of wood from forests is not necessarily incompatible with a wildlife habitat function, it can potentially have a serious impact on the quality of habitat, as well as the absolute amount of forest available to serve this function.

Regions in the northern hemisphere have traditionally been active producers of wood for industrial uses (Figure 2.8a). North American forests have yielded the greatest amount of wood in this category and this region is predicted to be the greatest consumer of industrial wood by the year 2000 (WRI, 1990). Even the small and highly populated region of Europe (per capita forest area of 0.20 hectare in 1990) has been an active producer of industrial wood, due no doubt to its excellent infrastructure and technological advances.

In contrast, regions in the southern hemisphere have not historically been large producers of industrial wood (Figure 2.8b), though this is starting to change. Asian forests, for example, increased their output of industrial wood by over 464% between 1951 and 1989–91. Central and South American forests increased production by a similar order of magnitude.

Fuelwood use has generally declined from forests in the northern hemisphere (Figure 2.9a) and is projected to continue to decline in the future. This decline stands in stark contrast to the situation in the forests of the southern hemisphere (Figure 2.9b), where fuelwood use has risen dramatically since 1951 and is predicted to continue in this manner.

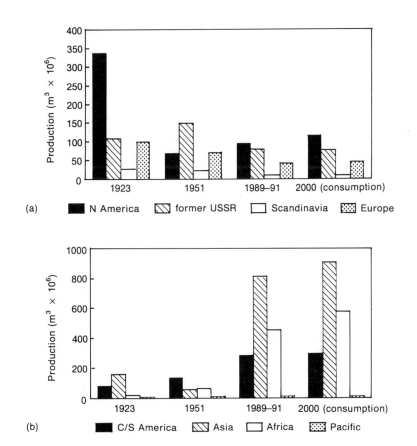

Figure 2.9 Estimated production of fuelwood, by region: (a) northern and (b) southern hemispheres. (Sources: Zon and Sparhawk, 1923; FAO, 1951; WRI, 1994; FAO, 1988.)

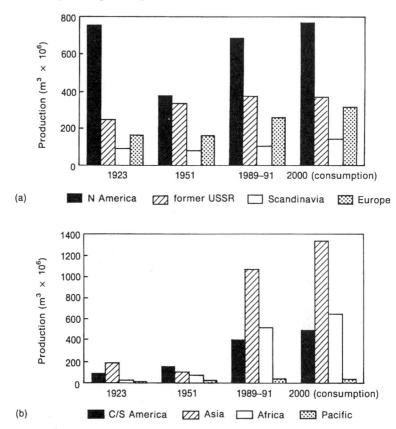

Figure 2.10 Estimated total production of wood, by region: (a) northern and (b) southern hemispheres. (Sources: Zon and Sparhawk, 1923; FAO, 1951; WRI, 1994; FAO, 1988.)

In general, the total production or estimated future consumption of wood by hemispheres is quite different (Figure 2.10). In the north, production is primarily based on increased industrial wood output. Total future consumption is estimated to grow moderately (e.g. North America, with a total increase of 12% between 1989–91 and 2000) or hardly at all (e.g. former USSR, with an estimated decrease for the same period of 1%). In the south, output from forests or consumption of wood will grow more significantly between 1989–91 and 2000. This is primarily driven by the dramatic demand for fuelwood in Asia, Africa and Central and South America: fuelwood consumption is expected to increase in these regions by 11%, 27% and 5%, respectively.

2.6 OVERALL IMPLICATIONS

2.6.1 Northern hemisphere

In general, the forest land-base is relatively stable in the regions of the northern hemisphere. There has not been a recent dramatic loss of boreal or temperate forestland, and the amount of forest per capita has declined somewhat but not significantly. Human population growth is low in these regions.

Total wood consumption in this hemisphere is projected to increase but this is based on an increase in industrial wood output, not fuelwood. The production of industrial wood from northern forests does not necessarily represent a potential cause of deforestation. In many parts of Canada, Scandinavia and the former Soviet Union, wood production takes place in rural or remote areas. Older forests are generally replaced, either naturally or through planting, by young forests. The species composition of these new forests may be very different from the forests they replaced. For example, if artificial reafforestation is successful, natural boreal conifer forests may be replaced by less diverse coniferous monocultures. In some cases, these monocultures might be composed of exotic or even genetically improved species. If artificial regeneration efforts fail, coniferous boreal forests may be replaced by early successional deciduous boreal species such as aspen (*Populus*) or birch (*Betula*) species. Alternatively, depending on the harvesting method and resulting environmental conditions, the site may be occupied by earlier successional herbaceous vegetation such as raspberries (*Rubus* spp.). In any event, since such harvesting in rural or remote areas will probably not be followed by human incursion, there will not necessarily be a loss of forestland (deforestation) but, rather, a change in the composition and structure of the forest, from older and coniferous in nature to younger and coniferous, deciduous or herbaceous (Kuusela, 1992). Such potential changes have serious implications for wildlife species such as moose (*Alces alces*), hazel grouse (*Bonasa bonasia*) and many other species.

Probably the most significant potential for change in the forested regions of the northern hemisphere is in the former Soviet Union. The forest area, economy and political climate in Scandinavia, North America and Europe are all relatively stable but this is not necessarily the case in the former Soviet Union (Rosencranz and Scott, 1992). The Russian Federation alone occupies one-seventh of the world's land area; it possesses nearly one-fourth of the world's timber volume and more than half of all the world's boreal forests (Krankina and Dixon, 1992). Even in 1923 (Zon and Sparhawk, 1923), Gifford Pinchot, first Chief of the Forest Service in the United States, recognized the potential of this vast forested region: 'Lumbermen of this and other countries are thinking seriously of getting timber from the great but little-known forests of Asiatic Russia. But can Russia meet the demand? Without accurate knowledge of the world's timber resources there can be no intelligent policy of forest consumption.' Changes in forest policy and use in this politically volatile region could have a profound effect on these forest ecosystems and the wildlife species that depend on them.

Much of the forest in the former Soviet Union has remained untouched due to a lack of roads and other infrastructure necessary to access timber supplies. This is especially the case east of the Ural Mountains, where roughly 75% of the growing stock is located yet where only 11% of the population of the former Soviet Union resides (Barr and Braden, 1988). If the criterion of economic cost-effectiveness was applied to the production of industrial wood from these Siberian and far-eastern forests most would remain unexploited in this century (Barr and Braden, 1988).

In the past, central economic planners wanted to develop this eastern region and establish wood-using industries but the extreme distance to any market made this unfeasible. It is estimated that the currently accessible European/Uralian forests in the former Soviet Union could alone sustain current national annual harvesting levels of 400 million m^3 if efficient silviculture and forest management practices were applied (Barr and Braden, 1988).

An important variable that potentially affects the future of these vast boreal forests in the former Soviet Union is their economic value now in an open, international market. A desire to generate hard currency may provide the stimulus to increase harvest rates and exceed an allowable cut or biologically sustainable level of productivity (Krankina and Dixon, 1992). Joint ventures with firms from Japan, South Korea, Sweden, Norway and the United States may provide the economic ability to access previously isolated areas. They could also provide modern milling and harvesting technology, which would minimize waste of timber. Standard milling technologies in the former Soviet Union reportedly use up to three times the amount of wood to produce an equivalent amount of finished product, due to tremendous waste (Rosencranz and Scott, 1992).

There is concern, however, that increased economic activity in the forest will result in more clearcutting (the most cost-effective short-term means of harvesting timber) and potential environmental damage. Forests may not regenerate themselves easily in the harsh northern climate (Rosencranz and Scott, 1992) and artificial regeneration efforts have often failed in the past (Krankina and Dixon, 1992). The harsh northern climate and thin soils prone to erosion have resulted in an estimate of as much as 32% of the forestland of the former Soviet Union being 'unexploitable' (Kuusela, 1992).

There also is uncertainty about the potential for economic exploitation in the vast stretches of the Russian forest. Backman and Waggener (1990) report that the near-term potential for joint ventures conducting business in the former Soviet Union will be strained, due to political uncertainty, unclear regulations, a muddled bureaucracy and a lack of understanding of basic business concepts. In addition, the quality and value of the timber may be sufficiently low to discourage economic exploitation (Cardellichio *et al.*, 1990).

Forests in the former Soviet Union are not only potentially at risk from economic exploitation by the unsustainable harvest of wood. An estimated one million hectares of forest has been killed in the recent past by industrial pollution (Krankina and Dixon, 1992). This trend may continue in the future, in the absence of sufficient air quality monitoring and controls. Similarly, boreal forests in the former Soviet Union and elsewhere throughout the northern hemisphere may be subjected to change based on a warming climate, if some predictions of global warming are correct (Shugart and Smith, 1992).

By virtue of its sheer size and the huge proportion of the northern forest in the former Soviet Union, future actions there may have significant impacts on wildlife populations. This may especially be the case for species that require extensive forested areas, such as lynx (*Felis lynx*), bears (*Ursus* spp.), sable (*Martes zibellina*) and tiger (*Panthera tigris*), among others. It is important to remember, however, that in most cases future activities will not necessarily result in the absolute loss of forest or an increase in human population.

2.6.2 Southern hemisphere

There has been a recent decline in the amount of forest in regions of the southern hemisphere. This is concurrent with a dramatic increase in population and the growth of the consumption of fuelwood in many regions. Shifting cultivation (removal of the forest for agricultural use and subsequent abandonment several years later due to exhausted soils) and poor farming practices are responsible for much of the deforestation (WCMC, 1992). Fuelwood exploitation removes almost all wood of any size and leaves little forest to regenerate itself. The burgeoning population is interested in agricultural uses of the land which are not compatible with the maintenance of forested habitat for wildlife, especially species such as jaguar (*Panthera onca*), anaconda (*Eunectes murinus*) and many primates. Even if land is not actually deforested, the tropical forest can be fragmented, resulting in isolated habitats as well as the creation of edge. Such edge, the effects of which extend into the residual forest, represents seriously modified habitat, due to greater exposure to wind, micrometeorological differences, improved access for livestock, hunters and others, and consequent biological and physical effects (Skole and Tucker, 1993).

Forests in this region are primarily moist or dry tropical or subtropical in nature, and are very diverse. Roughly half of the world's species depend on tropical forests (Repetto, 1990). Collectively, tropical forest types (wet evergreen; moist deciduous; dry and subtropical dry; and other subtropical) represent 60% of the planet's forest area (Mather, 1990). Their conversion in the future threatens the biodiversity that depends on these ecosystems.

A wide variety of data suggests that deforestation is a serious threat to forested regions in the southern hemisphere. It is less clear, however, if this general trend is leading in an irreversible direction towards the elimination of tropical forests, or if it is part of a trend of contraction and subsequent stability of forest area that has been experienced historically in Europe and North America (Mather, 1990).

Because there are differences in definition, estimates vary as to the amount of tropical forest loss. Some run from between 10 and 40 hectares per minute (Mather, 1990). Based on these, straight-line estimates indicate that the total tropical forest biome could disappear between 2020 and 2120 (Mather, 1990). Such global projections do not accurately reflect regional trends, however. For example, tropical moist forest in insular and peninsular situations is the most threatened by deforestation (Mather, 1990). Forests in west Africa, Central America and Southeast Asia are likely to disappear soon (perhaps in the next 10–20 years; Mather, 1990). Human population growth in these areas is rapid and forested areas are accessible to these growing populations.

Annual deforestation rates in the tropics also reportedly vary by ecotype. Between 1981 and 1990, annual deforestation advanced at rates of 0.6% in tropical rain forest, 0.9% in dry deciduous and very dry tropical forests, 1.0% in moist deciduous tropical forests and 1.1% in hill and mountain tropical forests (Aldhous, 1993). This represents an overall annual estimated global tropical deforestation rate of 0.8% (Aldhous, 1993).

There are some areas in the tropics that are considered not to be immediately threatened by total deforestation. Areas such as the Congo basin, the western Amazon in Brazil and northern South America (e.g. French Guiana, Guyana and Surinam) have large unbroken tracts of forest and low human population densities (Mather, 1990). Because of these remote areas, Mather (1990) believes that tropical forests will not disappear completely in the next century. Losses of tropical forests on islands and in peninsular situations could seriously jeopardize the existence of many wildlife species, however. This may be especially serious for the neotropical migrant bird species, which breed wholly or in part north of the Tropic of Cancer and winter south of that latitude (DeGraaf and Rappole, 1995).

To date, forest management has not been very sustainable in the humid tropics. The consensus of experts seems to be that it is technically possible but that significant obstacles exist, in the form of social influences that stimulate deforestation, as well as official government policies. Sustainable natural forest management (Johnson and Cabarle, 1993) must focus more on the social context in which a forest exists, as well as its vitally important and highly complex ecological functions. To succeed, management must not focus solely on the extraction or exploitation of timber. Its goal instead must to be to manage for a sustainable condition in which timber and other products can be removed periodically (Johnson and Cabarle, 1993).

It is important to locate spatially ecologically sensitive natural forest management activities and areas of natural forest reserves in which no management would take place. Reserves should be located in areas that are now remote, with low human populations and where deforestation is not a threat. Natural forest management activities should be located in areas of high human population, along the development/agricultural fringe, where there exists good access to markets. In general, the 'use it or lose it' philosophy can be applied, whereby unless natural forest management is implemented and multiple benefits can be derived from a standing and sustainable forest, its value will not be recognized and it will be converted to other uses (Johnson and Cabarle, 1993). Natural forest management is not a panacea, however. It requires investment to monitor its progress, and likewise maintain reserves. Financial benefits should not be directly deposited into general government coffers but reinvested in the forest (Johnson and Cabarle, 1993). Without such control, natural forest management could introduce agricultural incursions and the beginnings of deforestation.

2.7 CONCLUSION

There is no doubt that the amount of forest area in the world is decreasing. Regardless of standards, definitions, means of measurement or estimation, and changes in geographic names or boundaries, there are now fewer hectares of forest than there were 10, 20 or 100 years ago. There is some debate about the rate at which forests are being converted to other uses, and this varies regionally. Likewise, there is differing opinion regarding the implications of the decreasing number of hectares of forest. Some believe the statistics indicate an irreversible trend towards the elimination of some types of forest (e.g. tropical forests), while others put the decrease into a historical context comparable to the earlier decrease of forest in more developed parts of the world, such as Europe and parts of North America. Following these historic decreases, the area of forest stabilized or even increased.

One of the most important factors associated with the decrease in forested area is the influence of human populations. There is a high positive correlation between human population growth and deforestation. It is important to remember, however, that local human populations are not necessarily the cause of the deforestation, but the agents of that process. The actual causes are often poorly designed or implemented development or governmental policies created in distant capitals.

In general, forested regions in the northern hemisphere have not changed greatly in area. Human populations are increasing at a low rate, and the per capita amount of forest is relatively stable. Fuelwood use is declining but the production of wood for industrial uses is expected to increase. An increase in wood production from these northern forests is not likely to result in a decrease in forested hectares but it will result in areas of younger forest, with a possible shift in species composition.

In contrast, some regions in the southern hemisphere have experienced a greater decrease in forested area, especially where human population growth has been dramatic. Production or consumption of wood for fuel is expected to increase greatly in this region in the future. This, coupled with conversion to pastureland, shifting agricultural uses and exploitation for timber resources, will result in more deforestation.

Studies of scarcity of renewable resources such as croplands, water and forests suggest that it can result in social unrest, the displacement of populations, violence and armed conflict in the developing world (Homer-Dixon *et al.*, 1993). This pattern of resource scarcity, social instability and unrest has been documented in India, Bangladesh, Senegal, the Philippines, Indonesia, Brazil and Central America. Social unrest and violence are not conditions that are conducive to the development and implementation of policies designed to protect and judiciously manage natural resources in a sustained manner.

The solution to the global decrease in forest area is inextricably tied to human population growth. Forests will continue to be converted to other uses such as food production and other means of earning a livelihood unless human population growth and distribution can be influenced. In a larger sense, emphasis in the future needs to be on making non-forest lands more sustainably productive (G.H. Lund, personal communication).

Sustainable productivity does not apply only to the agricultural side of the equation. Currently, less than one-tenth of 1% of the world's tropical forests are managed for sustained productivity (Repetto, 1990). Even in boreal and temperate forests of the northern hemisphere, a relatively small proportion of forested area is managed (WRI, 1990; e.g. 49% in the United States, 59% in Sweden, 15% in Norway; statistics unavailable for the former Soviet Union and Canada). Forests that are actively managed benefit from investments of time and money to enhance potential yields (Winjum *et al.*, 1993). Based on this definition, roughly 12.5% of the forests of the former Soviet Union are actively managed. Furthermore, perhaps as few as 9% of the world's boreal forests and 25% of the temperate forests are actively managed (Winjum *et al.*, 1993).

From a historical standpoint, the world may now be shifting to a new phase of management of forested resources on a sustained basis, following a long period of extraction and exploitation of natural forests (Laarman and Sedjo, 1992). This pattern of human use is somewhat analogous to the development of agricultural systems, whereby civilization learned to practise agriculture on the same areas repeatedly, rather than hunt and gather food on a widespread basis (Laarman and Sedjo, 1992). International efforts are underway to realize a 'global forest management' goal, whereby active forest management would be enhanced, loss of forests would be reduced and forested area would be expanded (Winjum *et al.*, 1993).

Intensive cropping of wood in some forests in the future on a sustainable basis will leave other forested areas free from exploitation. It has been estimated that as little 5–10% of the world's forests could be used to support society's needs for wood products (Laarman and Sedjo, 1990). Tropical plantations have been shown to produce extraordinarily high yields. Likewise, the application of agroforestry systems can be used to stem the tide of shifting cultivation and leave original forests in an undisturbed state. The challenge is to locate:

- significant areas of forest reserves to preserve unique communities, ecosystems and biota;
- areas of natural forest management that are sensitive to ecological patterns and processes that produce benefits for both local populations and government treasuries;
- areas of intensive forest production (probably on land that has already been converted from primary forest), where high outputs can offset further conversion of natural, primary forests and meet increasing global demands for wood.

This 'triad' approach to land allocated to forestry is eloquently described by Seymour and Hunter (1992) and has potential application around the world.

The ability to allocate forest land uses wisely will be facilitated by excellent information on the status of the world's forests. The variety of definitions, methodologies and policies literally makes knowing about the extent of forest and its rate of change difficult at best (Wardle, 1993; Janz, 1993). A unified or more comprehensive approach to monitoring the rate of change would also enable more sophisticated analysis of the underlying complex of causes (Turner *et al.*, 1994). The use of technology such as remote sensing will greatly improve the global inventory of forest resources, but cooperation between interested agencies on timing, definitions, accuracy and standards is also needed (Lund, 1993).

It is difficult to forecast the future of forests globally. Care must be used in interpreting information, and to void mistaken assumptions based on data from different parts of the world. It is safe to say that temperate and boreal forests in the northern hemisphere will not change much and will continue to support viable wildlife populations in the future. The outlook is less optimistic in tropical forests of the southern hemisphere. The potential exists, however, to alter the trend of rapid deforestation through sustained management and sound government and development programs.

2.8 SUMMARY

The forests of the world are being converted to other uses. This decline in forested area has serious consequences for wildlife populations, and regional and global biodiversity.

There is a lack of consensus on the rate at which forest conversion is occurring. This is in part attributable to a combination of factors such as differences in standards, definitions, means of measurement or estimation, and geographic variation. The broadest estimates indicate that forest conversion varies regionally. One important factor related to forest conversion is the influence of human populations. There is a high positive correlation between human population growth and deforestation.

In the northern hemisphere, forested regions have not changed greatly in area and human populations are increasing at a slow rate. The per capita amount of forest is relatively stable. Fuelwood use is declining, but the production of industrial wood is expected to increase. This increase in industrial wood is not expected to result in conversion to non-forest use, but to areas of younger forest, with a possible shift in species composition.

Human population growth has been much greater in the southern hemisphere. Resulting increases in fuelwood consumption, timber exploitation, shifting agricultural use and conversion to pastureland have all combined to have a negative effect on the amount of forest. The causes vary by region and often act in synergy, making them difficult to assess or address.

Opinions vary concerning the implications of forest conversion. Some believe that the statistics indicate an irreversible trend towards the elimination of certain types of forest. Others put the decreases into a historical context comparable to earlier declines in more developed parts of the world, such as Europe and parts of North America. Following historical decreases, the area of forest stabilized or even increased.

Only broad assumptions can be made about the future of the world's forests, due to difficulties in estimation and monitoring. Temperate and boreal forests in the northern hemisphere will probably not change dramatically and will continue to support viable wildlife populations. The outlook is less optimistic for tropical forests of the southern hemisphere. The trend of conversion can be slowed or reversed by sustainable forest management, inspired by enlightened government and development programs. In the future, sustainable forest management may be located in optimal areas for production, leaving other areas in a natural state. Decisions such as these need to be made at the landscape scale, and will require intergovernmental cooperation, excellent resource information and ongoing monitoring, and knowledge of species and habitat requirements.

2.9   REFERENCES

Aldhous, P. (1993) Tropical deforestation: not just a problem in Amazonia. *Science* **259**:1390.

Backman, C.A. and Waggener, T.R. (1990) *Soviet Timber Resources and Utilization: an interpretation of the 1988 national inventory.* Center for International Trade in Forest Products, Working Paper no. 35. University of Washington, Seattle. 296 pp.

Barr, B.M. and Braden, K.E. (1988) *The Disappearing Russian Forest: a dilemma in Soviet resource management*. Rowman and Littlefield, London. 252 pp.

Cardellichio, P.A., Binkley, C.S. and Zausaev, V.K. (1990) Sawlog exports from the Soviet Far East. *Journal of Forestry* 88(6):12–17, 36.

DeGraaf, R.M. and Rappole, J.H. (1995) *Neotropical Migratory Birds: natural history, distribution, and population change*. Cornell University Press, Ithaca, New York.

Dixon, R.K., Brown, S., Houghton, R.A. *et al.* (1994) Carbon pools and flux of global forest ecosystems. *Science* 263:185–1990.

FAO (1951) *Yearbook of Forest Product Statistics*. Food and Agriculture Organization of the United Nations, Rome, Italy.

FAO (1963) *World Forest Inventory*. Food and Agriculture Organization of the United Nations, Rome, Italy.

FAO (1988) *Forest Products: world outlook projections*. FAO Forestry Paper no. 84. Food and Agriculture Organization of the United Nations, Rome, Italy. 350 pp.

Homer-Dixon, T.F., Boutwell, J.H. and Rathgens, G.W. (1993) Environmental change and violent conflict. *Scientific American* 268(2):38–45.

Janz, K. (1993) World forest resources assessment 1990: an overview. *Unasylva* 44:3–9.

Johnson, N. and Cabarle, B. (1993) *Surviving the cut: natural forest management in the humid tropics*. World Resources Institute, Washington, DC. 72 pp.

Krankina, O.N. and Dixon, R.K. (1992) Forest management in Russia. *Journal of Forestry* 90(6):29–34.

Kuusela, K. (1992) The boreal forests: an overview. *Unasylva* 170(43): 3–13.

Laarman, J.G. and Sedjo, R.A. (1992) *Global Forests: Issues for Six Billion People*. McGraw Hill, New York. 337 pp.

Lund, H.G. (1993) Politically correct global mapping and monitoring, in *Proceedings, 'Mapping Tomorrow's Resources'*. Utah State University, Logan, Utah. 23–24 April 1992. pp. 47–54.

Malingreau, J.P. (1993) Satellite monitoring of the world's forests: a review. *Unasylva* 174(44):31–38.

Mather, A.S. (1990) *Global Forest Resources*. Timber Press, Portland, Oregon. 341 pp.

Perlin, J. (1991) *A Forest Journey: The Role of Wood in the Development of Civilization*. Harvard University Press, Cambridge, Massachusetts.

Repetto, R. (1990) Deforestation in the tropics. *Scientific American* 262(4):36–42.

Rosencranz, A. and Scott, A. (1992) Siberia's threatened forests. *Nature* 355:293–294.

Seymour, R.S. and Hunter, M.L. Jr (1992) *New forestry in eastern spruce–fir forests: principles and applications to Maine*. Maine Agricultural Experiment Station Misc. Publication 716. 36 pp.

Shugart, H.H. and Smith, T.M. (1992) Modelling boreal forest dynamics in response to environmental change. *Unasylva* 170(43):30–38.

Skole, D. and Tucker, C. (1993) Tropical deforestation and habitat fragmentation in the Amazon: satellite data from 1978 to 1988. *Science* 260:1905–1910.

Turner II, B.L., Meyer, W.B. and Skole, D.L. (1994) Global land-use/land-cover change: towards an integrated study. *Ambio* 23(1):91–95.

Wardle, P. (1993) Forestry statistics in the global partnership for environment and development. *Unasylva* **44**:51–54.

Winjum, J.K., Meganck, R.A. and Dixon, R.K. (1993) Expanding global forest management: an 'easy first' proposal. *Journal of Forestry* **91**(4):38–42.

WCMC (World Conservation Monitoring Center) (1992) *Global Biodiversity: status of the earth's living resources.* Chapman & Hall, London. 585 pp.

WRI (World Resources Institute) (1990) *World Resources 1990–1991.* Oxford University Press, Oxford, UK. 383 pp.

WRI (World Resources Institute) (1994) *World Resources 1994–95.* Oxford University Press, Oxford, UK. 400 pp.

Zon, R. and Sparhawk, W.N. (1923) *Forest Resources of the World*, volumes I and II. McGraw Hill, New York. 493 pp.

3

The status of forested wetlands and waterbird conservation in North and Central America

R. Michael Erwin

3.1 INTRODUCTION

Forested wetlands represent one of the more prominent yet least understood ecosystems in the world (Lugo *et al.*, 1990). These wetlands have been studied less than most other systems primarily because of their inaccessibility and aversion (e.g. noxious insects and poisonous reptiles) to humans (Mitsch and Gosselink, 1986; Bacon, 1990). Nonetheless, forested wetlands provide habitat for an exceedingly diverse array of fish and wildlife resources throughout the world, especially in the tropics (Lugo and Snedaker, 1974; UNESCO, 1978).

This chapter focuses on the status and conservation of various forested wetlands and one of the major wildlife components in these systems: the waterbirds. The geographic focus is North America; Central America is included as well although the information from that region is much more limited. First, terms are defined; then follows an outline of the general status of forested wetlands and the waterbirds often associated with them. Next is a summary of the principal ecological requirements of waterbirds that are met by forested wetlands, then a list of regional and national status reports for forested wetlands and selected waterbird species where the information is adequate. Lastly, the chapter presents management recommendations that attempt to mitigate the effects of human activities on these wetland systems and the wildlife they support. While it may be tempting to relate changes in forested wetlands causally with changes in specific waterbird populations, these correlations are usually weak. Conditions on the breeding, migration and wintering grounds may all affect populations in

Conservation of Faunal Diversity in Forested Landscapes. Edited by
R.M. DeGraaf and R.I. Miller. Published in 1996 by Chapman & Hall. ISBN 0 412 61890 7.

complex ways. Instead, the chapter tries to indicate what the most likely limiting factors are.

3.1.1 Definitions

The nomenclature of Lugo *et al.* (1990) is followed for wetlands in the North and Central America and the Caribbean. In this system, the term 'forested wetlands' refers to 'any wetland with a significant component of woody vegetation, regardless of the height of the plants'. In common terms, these are mangroves and other halophytic (i.e. salt-tolerant) shrubs, bogs, forested swamps (or palustrine forested wetlands), shrub wetlands and riparian (river) systems, respectively (Dahl and Johnson, 1991). The list of common names for wetlands included under these categories in North and Central America is shown in Table 3.1 (modified from Lugo,

Table 3.1 Nomenclatures for different forested wetlands in North and Central America and the Caribbean (modified from Lugo, 1990, Table 1.2)

Type	Location
North America	
Bog forests, moors, birch bog	Alberta, Canada
Riverside swampland	Canada
Tamarack-dominated swamp; prairie swamp	Northern Lake Michigan to Illinois
Alder swamp	New York State
Bog forests	Wisconsin–Michigan
Fan palm oases	California
Deepwater swamp, shallow-water swamp, peaty freshwater swamp	Southern US
Tupelo gum swamp, cypress–gum	Alabama
Bottomland forest	North-central Oklahoma
Cypress–gum, cedar, maple–gum, mixed-hardwood swamp	Virginia
Cypress heads, domes; white cedar swamp	Florida
Low floodplain	Oklahoma
Central America	
Mangrove forest, *Mora prioria,* mixed flooded forests	Panama
Mangrove forests, freshwater tidal swamp forests, bamboo thickets, *Bactris* swamps	Honduras
Mangrove forests, riverine, swamp forests	Nicaragua
Raphia swamps	Costa Rica

1990). This list is not exhaustive but represents some of the diversity of types encountered in different regions (Lugo, 1990). This definition of forested wetlands is broader than the one used by Cowardin *et al.* (1979) who consider palustrine forested, scrub–shrub and riverine wetlands to be exclusive.

Nomenclature for birds also can be confusing. The term 'waterbirds' may include one or a combination of the following: loons (Gaviiformes), grebes (Podicipediformes), waterfowl (Anseriformes), seabirds (Charadriiformes, Procellariiformes and Pelecaniformes), wading birds (Ciconiiformes) and shorebirds (e.g. Charadriiformes). Here, the focus is restricted to wading birds and selected members of the Pelecaniformes (pelicans, cormorants) and Anseriformes (waterfowl) because the species in these groups are most often associated with forested wetlands for nesting, roosting and feeding (Tables 3.2 and 3.3). Common names for species are used in the following text; scientific names are listed in Tables 3.2 and 3.3.

3.2   SOURCES OF INFORMATION

The published literature, government reports, proceedings of meetings, symposia and theses have been canvassed for this chapter, and key individuals in several foreign countries have been contacted. To determine current trends for hunted waterbird species, assistance was requested from the Office of Migratory Bird Management (OMBM), US Fish and Wildlife Service (USFWS). In addition, National Biological Service (NBS) biologists conduct statistical trend analysis of Breeding Bird Survey data for many North American species (Droege and Sauer, 1990). Such data analysis was required for a number of colonial wading bird species, but some of these species cannot be adequately monitored by the roadside survey methods used by the Breeding Bird Survey (Erwin *et al.*, 1993).

For North American waterfowl, reliance was placed on Smith (1995) for 20 years of aerial survey data from the North American waterfowl breeding ground surveys. Except for mallards, these aerial surveys are not efficient at detecting populations of most forested wetland species (G. Smith, USFWS, and K. Reinecke, NBS, personal communications). The years from 1973 to 1992 were used in determining gross estimates of annual percentage population changes. A regression model was used to analyze log transforms of the annual counts. The slope term was then transformed back to the exponential scale, after incorporating the variance term as well, to yield the estimate of annual percent change (APC). The APC assumes that the counts have a lognormal distribution with no autocorrelation from year to year. If autocorrelation is present (which it probably is), the true APC is smaller; thus the APC reported here is probably an upper bound on the estimate of yearly percentage change (J. Hatfield and J. Sauer, NBS, personal communications).

Table 3.2 Nongame waterbird species[a] and their uses of different types of forested wetlands in North and Central America

Species	Use[b]			US status[c]	Source[d]
	Nesting	Roosting	Feeding		
Brown pelican (Eastern) (*Pelecanus occidentalis*)	2	2	N	+	BBS; USFWS, 1985
Double-crested cormorant (*Phalacrocorax auritus*)	2	2	N	+	BBS
Olivaceous (neotropical) cormorant (*P. olivaceus*)	1	1	2	?	
Anhinga (*Anhinga anhinga*)	1	1	1	?	
Magnificent frigatebird (*Fregata magnificens*)	1	1	N	NA	
Great frigatebird[e] (*Fregata minor*)	1	1	N	NA	
Great blue heron (*Ardea herodias*)	1	1	2	+	BBS
Great white heron (*A. h. occidentalis*)	1	1	2	0	
Great egret (*Casmerodius albus*)	1	1	2	0	

Species				
Snowy egret (*Egretta thula*)	1	N	0	
Little blue heron (*E. caerulea*)	1	2	?	
Reddish egret (*E. rufescens*)	1	2	Unstable	OMBM, 1987
Cattle egret (*Bubulcus ibis*)	1	N	+	BBS
Chestnut-bellied heron[e] (*Agamia agamia*)	1	1	NA	
Green-backed heron (*Butorides striatus*)	1	1	?	
Black-crowned night heron (*Nycticorax nycticorax*)	1	2	?	
Yellow-crowned night heron (*N. violaceus*)	1	1	?	
Rufescent tiger heron[e] (*Tigrisoma lineatum*)	1	1	NA	
Bare-throated tiger heron[e] (*T. mexicanum*)	1	1	NA	
Boat-billed heron[e] (*Cochlearius cochlearius*)	1	1	NA	
Green ibis[e] (*Mesembrinibis cayennensis*)	1	1	NA	
White ibis (*Eudocimus albus*)	1	N	0	P. Frederick, unpub. data
Scarlet ibis[e] (*Eudocimus ruber*)	1	2	NA	

Table 3.2 continued

Species	Use[b]			US status[c]	Source[d]
	Nesting	Roosting	Feeding		
Glossy ibis (*Plegadis falcinellus*)	1	1	N	0	
White-faced ibis (*P. chihi*)	2	2	N	Unstable	OMBM, 1987
Roseate spoonbill (*Ajaia ajaja*)	1	1	1	?	
Wood stork (*Mycteria americana*)	1	1	2	Endangered	Recent increases (J. Ogden, pers. comm.)
Jabiru[e] (*Jabiru mycteria*)	1	1	N	Endangered in Mexico	

[a]'Waterbird' here includes only members of the orders Pelecaniformes and Ciconiiformes (see Table 3.3 for waterfowl).

[b]1 = primary or exclusive habitat use; 2 = secondary, minor use; N = little or no use.

[c]+ = increasing population (in most of its range); − = decreasing (in most of its range); 0 = apparently fairly stable; ? = insufficient information or conflicting data; NA = not applicable to US.

[d]BBS = trend analysis program developed by the Breeding Bird Survey, Laurel, MD (B. Peterjohn, unpublished); OMBM, 1987 = report on *Species of Management Concern* developed by the Office of Migratory Bird Management, USFWS.

[e]Species restricted to Mexico or Central America.

Information about wetlands and waterbirds in Mexico and Latin America is fragmented. The chapter has relied heavily on Scott and Carbonell's (1986) directory to identify major forested wetlands and associated water-birds, and upon data from numerous aerial surveys of waterfowl conducted by the USFWS since the 1930s in cooperation with Mexico (Saunders and Saunders 1981; Spencer, 1984).

3.3 SCOPE

3.3.1 Wetland status – general

The condition and trends of wetlands have been increasingly important both to humans (Conservation Foundation, 1988) and wildlife (Scott and Carbonell, 1986; Sharitz and Gibbons, 1989; Finlayson and Moser, 1992) throughout the world. The prospects for global climate change causing sig-nificant sea level rise affecting coastal wetlands in the next 25 years is daunting (Titus *et al.*, 1991). Rainfall patterns in continental regions are also expected to alter freshwater wetland processes (Titus *et al.*, 1991). A number of international meetings have been held in the past two years focusing on wetland processes and the implications for biological diversity (see Dugan, 1991).

The United States has one of the most active wetland inventory programs in the world. Since 1975, the USFWS has supported a National Wetlands Inventory (NWI) (Tiner, 1984). In addition to conducting a complete wetland inventory, the NWI also sponsors a Status and Trends program that evaluates changes at selected sampling locations every decade (Dahl and Johnson, 1991). The resulting reports have indicated that, in the past 200 years, the United States has experienced loss or alteration of more than half of its wetlands (Dahl and Johnson 1991). Of 10 categories of wetlands listed in Dahl and Johnson (1991: Table 3), the second highest loss (6.2%) from the 1970s to the 1980s was of palustrine forested wetland (i.e. the areas generally referred to as 'bottomland hardwood forests' or 'wooded wetlands'; see Anon., undated). Riparian wetlands and shrub wetlands experienced very little change.

Outside of the United States, there are few comprehensive quantitative data on wetland status and trends. In Canada, no extensive wetland inventories have been completed on a national scale (National Wetlands Working Group, 1988; V. Glooschenko, NBS, personal communication), although the populated areas of some provinces are being mapped in some detail (J. Jeglum, Forestry Canada, personal communication), but it is believed that the rate of loss of Canadian wetlands of all types was much less than that of the United States until the 1960s (Lynch *et al.*, 1963, in Bellrose, 1980). However, since then, intensive agricultural expansion in the prairie provinces has reduced wetlands significantly, including both prairie

Table 3.3 Waterfowl species associated (to varying degrees) with forested wetlands in North and Central America

Species	Use[a]			US status[b]	Source
	Nesting	Roosting	Feeding		
Black-bellied whistling duck (*Dendrocygna autumnalis*)	1	1	2	?	
Muscovy duck[c] (*Cairina moschata*)	1	1	2	–	Saunders and Saunders, 1981
Canada goose (*Branta canadensis*)	N	N	2	+	Smith, 1995
Wood duck (*Aix sponsa*)	1	1	1	+	Sauer and Droege, 1990
Green-winged teal (*Anas crecca*)	N	N	2	0	Smith, 1995
American black duck (*A. rubripes*)	1	2	2	–	Smith, 1995
Mallard (*A. platyrhynchos*)	N	N	2	0	Smith, 1995
Northern pintail (*A. acuta*)	N	N	2	–	Smith, 1995

Gadwall (A. strepera)	N	N	2	0	Smith, 1995
Common goldeneye (Bucephala clangula)	1	1	2	0	Smith, 1995
Barrow's goldeneye (B. islandica)	1	1	2	0	Smith, 1995
Bufflehead (B. albeola)	1	1	2	0	Smith, 1995
Hooded merganser (Lophodytes cucullatus)	1	1	1	+	Smith, 1995
Common merganser (Mergus merganser)	1	1	2	+	Smith, 1995

[a] 1 = primary or exclusive habitat use; 2 = secondary or minor use; N = little or no use.
[b] + = increasing population (in most of its range); – = decreasing (in most of its range); 0 = apparently fairly stable; ? = insufficient information, or conflicting data.
[c] Native species in Mexico and Central America.

wetlands and aspen parkland (i.e. forested) wetlands (Aus, 1969, in Bellrose, 1980).

In Mexico, there has been little inventory work on forested wetlands in most of the country and little attempt has been made to classify wetlands (Olmsted, 1992). Recently, remote sensing has been used in the Yucutan Peninsula (Southworth, 1985; Greene *et al.*, 1988) but has not focused on wetlands. Some high quality vegetation community maps exist for the Usu-macinta River region but have not been published (J. Correa and J. C. Ogden, personal communications). In Mexico, more research has been done on coastal wetland systems than on palustrine or riverine systems (Olmsted, 1992). Limited reports were located summarizing national status and trends of forested wetlands in Central American countries (Seeliger, 1992).

3.3.2   Waterbird populations – general trends

In North America, declines in most waterfowl populations have been well documented in recent decades (Bellrose, 1980; USDI, 1986; Chandler, 1989). These declines seem to be due to some combination of:

- temporary loss of breeding habitat resulting from drought;
- permanent loss of breeding habitat for some species resulting from agricultural development;
- permanent loss of wintering habitat resulting from drainage and agricultural development (e.g. bottomland hardwoods in the southern US);
- overharvesting of certain waterfowl species (Bellrose, 1980, USDI, 1986).

Declines seem to be most severe in Canada and the United States for American black ducks (*Anas rubripes*) and northern pintail (*A. acuta*) (Table 3.4). The mid-continental population of mallard (*A. platyrhynchos*) that breeds in the Canadian–US prairies and winters in the lower Mississippi River valley has probably also declined at least in part because of bottomland habitat loss (Fredrickson and Heitmeyer, 1988; K. Reinecke, USFWS, personal communication). Some population changes of wintering waterfowl have been reported for the early years of the Mexico surveys (Saunders and Saunders, 1981). In contrast, few comprehensive population data are available for resident waterfowl species in Mexico or Central America.

Population declines have also been documented for nongame wetland birds; half of the species listed in 1987 as being of national (US) 'concern' were aquatic species (OMBM, 1987). The wood stork (*Mycteria americana*) and both eastern and California brown pelicans (*Pelecanus occidentalis*) have been listed as Federally Endangered or Threatened, although the south Atlantic population of the eastern brown pelican has been delisted (USFWS, 1985). The most severe declines of wading birds have occurred in south Florida, due largely to changes in water management in the freshwater

Table 3.4   Estimates of annual changes in waterfowl populations based upon USFWS aerial breeding ground surveys in the United States and Canada, during the period 1973–1992 (from Smith, 1995)

Species	*Annual change[a] (%)* *Type 2*
Black duck	−6.03
Bufflehead	1.26
Canada goose	6.97
Gadwall	−0.67
Goldeneye(s)	−0.50
Green-winged teal	−0.03
Mallard	1.27
Merganser(s)	2.81
Northern pintail	−6.13

[a] Assumes autocorrelation; incorporates variance of the slope estimate.

regions of the Everglades (Kushlan and Frohring, 1986; Kushlan, 1989; Davis and Ogden, 1994). Some wading birds have declined elsewhere in the US but not as dramatically as in Florida (Ogden, 1978).

In Canada, nongame waterbird species are not being monitored nationally but there are some provincial monitoring efforts for the great blue heron (*Ardea herodias*) in Quebec and Ontario (DesGranges, 1980; Des-Granges and Laporte, 1983; Dunn *et al.*, 1985). Numbers of breeding birds appear to be increasing markedly in Ontario (Anon., 1992) and possibly in Quebec (J.-L. DesGranges, Canadian Wildlife Service, personal communication) over the past decade, but the census effort in Quebec has not been consistent over time.

In Mexico and in Central America, there appear to be no comprehensive national inventories for resident waterbirds. However, there are a number of regional inventories for some species of waterbirds, such as in the Usumacinta-Grijalva River delta (states of Tabasco and Campeche) in eastern Mexico (Sprunt and Knoder, 1980; Ogden *et al.*, 1988; Correa, 1992; Hartásanchez, 1993). Occasional surveys for Jabirus and other waders have been conducted in Belize (C. Luthin, personal communication) and along the Miskito Coast region of Nicaragua (P. Frederick, personal communication).

3.4   WATERBIRDS AND FORESTED WETLAND HABITAT USE

Ecological associations of waterbird species and their forested wetland habitats vary markedly by species, season and geographical region. At one

Table 3.5   Types of waterbird foods produced in forested wetlands (modified from Fredrickson and Heitmeyer, 1988)[a]

Wetland type	Plant[b]	Animal
Scrub–shrub wetland	Seeds	Insects
		Crayfish
		Snails
		Fish
Forested wetlands		
Baldcypress–water tupelo	Tupelo drupes	Crayfish
		Fish
Overcup oak–red maple	Samaras	Small crustaceans
		Crayfish
		Spiders
Pin oak–Nuttall oak	Acorns	Small crustaceans
	Samaras	Crayfish
		Spiders
Willow oak–Cherrybark oak	Acorns	Small crustaceans
	Samaras	Crayfish
		Spiders

[a]Wetland types in the Upper Mississippi River valley of the United States.
[b]Important waterfowl foods.

extreme, the wood duck (*Aix sponsa*) may spend most or all of its life within forested wetlands, depending on these habitats for food, nest sites and roost sites throughout the year (Haramis, 1990; Kirby, 1990; Fredrickson *et al.*, 1990). At the other extreme, the magnificent frigatebird (*Fregata magnificens*) may depend on mangrove vegetation only for a nesting or roosting structure and not for food resources at all.

For waterfowl, eight of the 14 species in Table 3.3 depend on forested wetlands because of their hole-nesting requirements; feeding in forested wetlands may be secondary (Bellrose, 1980). Hole-nesters also depend on these habitats for roosting. Forested wetlands may be important feeding habitats for wintering mallards and northern pintails in regions such as the southern United States (Fredrickson and Heitmeyer, 1988), where a number of plant and animal foods are available to waterfowl in mixed bottomland hardwood stands (Table 3.5). As forested bottomlands have been converted to agricultural fields (MacDonald *et al.*, 1979), species such as mallard and Canada goose have opportunistically shifted to feeding in flooded fallow fields (Fredrickson and Heitmeyer, 1988).

As with waterfowl, nongame waterbird use of forested wetlands is based primarily on nesting and roosting (Table 3.2). Woody vegetation provides support for the stick nests built by most wading birds. Shrubs and trees permit the birds to elevate their nests to reduce predation by mammals and reptiles (Palmer, 1962). Of all the species listed in Table 3.2, the only ones that often nest in habitats other than woody vegetation (Spendelow and Patton, 1988) are:

- brown pelican (bare ground);
- double-crested cormorant (rock, bare ground, artificial platforms);
- green-backed heron (duck blinds);
- all three ibis species (occasionally in *Phragmites australis*, *Scirpus* spp., or on the ground);
- black-crowned night-heron (*Phragmites australis*).

In general, waterbirds roost primarily in woody vegetation, often in the same vicinity as the nesting colonies (Erwin, personal observation; H. Kale, Florida Audubon Society, personal communication). However, little research has been done on roosting habitat use.

Feeding sites for the majority of waterbirds are in non-forested wetlands, especially persistent emergent marshes in tidal or freshwater systems (Kushlan, 1978); however, in some regions, forested wetlands are important feeding habitats. For example, in the tropics and subtropics, mangrove communities provide abundant fish and crustaceans (Odum, 1971; Lugo and Snedaker, 1974; Odum *et al.*, 1982) that, in turn, support anhingas, cormorants, pelicans, wood stork, roseate spoonbill, white ibis and yellow-crowned night-heron. Swamps and riparian wooded habitats support great blue heron, black-crowned night-heron and green-backed heron throughout most of the interior of Canada, the United States and Mexico (Palmer, 1962). For both wading birds and waterfowl, there is a critical point: if a habitat is to provide all the requirements for nesting, feeding, roosting and molting, a complex of forested and emergent wetlands needs to be included. In some cases (e.g. Mississippi valley mallards), agricultural fields can also be included as an important component (Heitmeyer and Frederickson, 1981).

3.5 STATUS REPORTS BY COUNTRY AND REGION

3.5.1 Canada

(a) Wetlands
The general status of wetlands in Canada has been summarized by the National Wetlands Working Group (NWWG, 1988) of the Canada Committee on Ecological Land Classification. The summary concludes that very little of Canada's wetlands has been inventoried and trends therefore cannot be quantified over regions, provinces or nationally. Less than 5% of southern Canada has had any study of changes in land use (Lynch-Stewart, 1983); the recent emphasis has been on southern Ontario and along the St

Lawrence River (NWWG, 1988). In Ontario, as elsewhere, agriculture has resulted in the loss of more than half of the pre-settlement wetlands (Aus, 1969; Snell, 1987; NWWG, 1988). In the prairie provinces of Canada, as in the US, much of the pothole wetland landscape has been lost to grain-crop expansion (Aus, 1969, in Bellrose, 1980; Turner *et al.*, 1987; Chandler, 1989).

Of the forested wetlands in Canada, much of the wetland loss has resulted from urbanization along the St Lawrence Seaway. The rapid wetland loss between 1950 and 1978 has been mapped (Le Groupe Dryade, 1981, in NWWG, 1988) and evaluated (Environment Canada, 1986). In 1985, recognizing the importance of the natural resources of the region, the Canadian Wildlife Service and Ontario Ministry of Natural Resources initiated a cooperative study of the wetlands of the St Lawrence to assess the significance to waterbirds of these wetlands and to update an inventory of wetlands (NWWG, 1988).

(b) Wildlife

At least in part as a result of wetland breeding habitat loss, mid-continental populations of mallard, northern pintail and green-winged teal have declined over the past decade (Bellrose, 1980; USFWS, 1986; Stewart *et al.*, 1988; Smith, 1995). The decline has no doubt been exacerbated by the long-term drought in the prairie regions resulting in very poor reproduction (Chandler, 1989). How much of this decline is due to wintering habitat loss, especially in the bottomland hardwoods of the lower Mississippi River region of the United States, is subject to conjecture (Stewart *et al.*, 1988). Determining where the bottlenecks are in the annual life cycle of a particular species (e.g. the mallard) has proven to be resistant to conventional analyses (Johnson *et al.*, 1988).

The long-term decline of a facultative forested wetland species, the black duck (*Anas rubripes*) (Table 3.4), has resulted in debate over whether the problems are related to habitat or hybridization (USFWS, 1986; Rusch *et al.*, 1989). In spite of efforts to reduce harvest in both Canada and the United States, excessive harvesting may still be a factor in the decline (USFWS, 1986; Chandler, 1989; J. Longcore, USFWS, unpublished data). Concern for the status of the species has resulted in the formation of habitat joint ventures in Canada and the United States (USDI, 1986). The focus of these efforts is to ensure the protection and enhancement of breeding, migration and wintering habitat in eastern Canada and the United States. Because forest area in maritime Canada and New England has either remained stable or increased since the late 1880s, any habitat-related problems are probably unrelated to forested wetlands.

Of the nongame species depending upon forested wetlands in Canada, few population data exist. Increasing populations of the great blue heron in eastern Canada have already been noted (section 3.3.2). A recent (1991) survey of wading birds along the coastline of the Canadian Great Lakes was initiated to establish a population monitoring baseline (H. Blokpoel, CWS, unpublished data). In British Columbia in western Canada, great blue herons have been censused irregularly (Butler 1987, 1992). Some population decline may have occurred in coastal British Columbia in the Straits of Georgia region since some long-established colonies have been abandoned in recent years (Butler, 1987). Also, in one of the largest watersheds in British Columbia, the Fraser River, large-scale timbering and industrial pollution have stimulated concern for waterbirds and other wildlife (Dorcey, 1991; Savard, 1991).

In general, species diversity and abundance of wading birds is much lower in Canada than in the United States and Latin America (Sprunt *et al.*, 1978). The tropical origin of the Ciconiiformes and the shorter seasonal production of prey at higher latitudes probably account for this pattern.

In summary, the predominant wetland loss in Canada has been in the non-forested wetlands, especially in the prairie provinces. These changes, due largely to agriculture, have affected nesting waterfowl. Forested wetlands, by contrast, have experienced some local and regional losses (especially the upper St Lawrence River and Fraser River) but waterbird populations do not appear to have been strongly affected.

3.5.2   United States

(a)   Wetlands

The classification, evaluation and assessment of trends in wetlands have progressed faster in the United States than in any other country. The National Wetland Inventory was initiated by the US Fish and Wildlife Service in 1975 to map completely all wetlands in the lower 48 states (Tiner, 1984). More recently, a Status and Trends Program has been added to estimate changes by wetland type based on an intensive sampling program (Dahl and Johnson, 1991).

Results have shown that, of the 90 million hectares of wetlands that existed in the conterminous United States at the time of European settlement, only about 41 million remain (46% of the original area) (Tiner, 1984). From the mid 1950s to the mid 1970s, annual wetland losses averaged 200 000 ha, with 183 000 ha of palustrine (freshwater) wetland accounting for the vast majority. Of the forested wetlands, most of the loss occurred in the lower Mississippi River valley as bottomland hardwoods were converted largely to soybean fields (MacDonald *et al.*, 1979; Clark and Benforado, 1981; Tiner, 1984). Shrub wetlands were affected mostly in North Carolina as pocosin wetlands were converted to crops or pine plantations, or were mined for peat (Tiner, 1984; Cashin *et al.*, 1992).

Trends from the mid 1970s to mid 1980s indicated that, on a national basis, palustrine forested wetlands were suffering the second highest loss rate (6.2% per decade) (Dahl and Johnson, 1991). Riverine and palustrine shrub wetlands showed little change, while estuarine vegetated wetlands had a 1.5% decadal loss rate (Dahl and Johnson, 1991). The southeast region had the highest loss rate of forested wetlands, including the Carolinas, Georgia, Florida, Mississippi, Louisiana and Arkansas.

Concerns over losses of palustrine wetlands especially have resulted in intensive investigations in particular states. For instance, in the states surrounding the Chesapeake Bay, Tiner (1989) summarized changes by state (Table 3.6). Scrub–shrub wetland changes have been minimal in the five states, but losses of forested wetlands (palustrine forested in this case) have been marked during the period from 1950 to the late 1970s. Virginia and Delaware experienced the highest losses of forested wetlands (Table 3.6; Tiner, 1989).

In Florida, the pattern is similar, with high losses of palustrine forested wetlands but little net change in scrub–shrub wetlands (Frayer and Hefner, 1991). From the 1970s to the 1980s, about 3% of palustrine forested wetlands were lost (from more than 2.3 million ha, a loss of 77 000 ha) (Table 3.6).

Farther west, in Louisiana, Turner and Craig (1980) estimated loss rates of forested wetlands in that state to be about 35 300 acres (*ca.* 15 000 ha) per year during the 1970s. In the coastal region, waterway alteration for oil extraction and shipping have affected estuarine shrub habitats in many regions (Chabreck, 1988).

Table 3.6   Changes in area (ha) in palustrine vegetated wetlands in the Middle Atlantic and southeastern United States

State	Wetland type		Estimated remaining area[a]	Source
	scrub–shrub	forested		
Delaware	−2850	−11,442	56,302	Tiner, 1989
Maryland	−2249	+811	94,454	Tiner, 1989
Pennyslvania	+863	+5049	145,973	Tiner, 1989
Virginia	+3102	−24,829	279,143	Tiner, 1989
West Virginia	+2182	+2250	26,412	Tiner, 1989
North Carolina[b]	−21	−10,750	16,400	Cashin *et al.*, 1992
Florida	+1800	−77,000	2,830,000	Frayer and Hefner, 1991

[a]For Delaware to West Virginia, change from 1950s to late 1970s (estimate); for Florida, change from 1970s to late 1980s; for North Carolina, change form 1950s to late 1980s.
[b]Coastal plain region only.

(b) Wildlife

Among waterfowl species, the general pattern has already been summarized (Tables 3.3 and 3.4 and section 3.3). For forest-using species, the pattern is complex (Table 3.3). Wood duck populations have generally increased in North America (Sauer and Droege, 1990). This increase has been attributed at least in part to the intensive nesting box management that has been implemented on state and federal refuge lands (Bellrose, 1980; Haramis, 1990; Nichols and Johnson, 1990); however, a recent interpretation of wood duck trends suggests that too few boxes were erected to have been a significant factor in increases seen during the early decades of the recovery (Bellrose, 1990). Surprisingly, even though the wood duck now comprises a larger proportion of the hunter's bag in the eastern United States, its populations have still increased (Nichols and Johnson, 1990).

The Canada goose, in some areas, feeds occasionally in forested wetlands during the fall and winter (Fredrickson and Heitmeyer, 1988) and its opportunistic feeding has resulted in expansion into fallow agricultural fields in much of its range (Bellrose, 1980). Populations of the Canada goose, like several other goose species, have increased in recent years in the United States (Weller, 1988; Smith, 1995; Table 3.3).

Other species that are increasing in the United States and also depend on forested wetlands include common and hooded mergansers (Smith, review; Table 3.3). These species are predominantly cavity nesters in the northern tier of states and in the parklands of Canada, but feed in a wide variety of wetlands during migration and winter (Bellrose, 1980).

As indicated earlier, several species of waterfowl are experiencing declines in at least part of their range in the United States (Table 3.3). The most dramatic declines have been the black duck, the northern pintail and Mississippi Flyway populations of mallard (Smith, 1995; USDI, 1986). Some evidence has also been found for regional (flyway) declines of blue-winged teal and green-winged teal in recent years (USFWS, Office of Migratory Bird Management, unpublished reports). Declines of these species are probably unrelated to loss of forested wetlands, except for mallards in the Mississippi River valley.

Among nongame waterbirds, some strong population trends have been reported (Table 3.2). Habitat changes in forested wetlands have probably had little to do with these trends, the strongest of which have been associated with changes in non-forested wetland conditions (Sharitz and Gibbons, 1989) or in agricultural pesticide practices, specifically DDT and its derivatives (Blus *et al.*, 1971; Stickel, 1973; Fleming and Clark, 1983; USFWS, 1985). The widespread contamination of agricultural soils during the 1950s and 1960s with DDT probably resulted in impaired reproduction of many species including pelicans and double-crested cormorants (Blus *et al.*, 1971; Fleming and Clark, 1983). In general, the concern for waterbird populations, except for brown pelicans (USFWS, 1985), was relatively small compared with the alarming declines in raptor populations (Stickel, 1973).

An organized national program to inventory colonial waterbirds was attempted in the 1970s; it was discontinued in the mid 1980s but is now being re-evaluated (Erwin *et al.*, 1993). Using Breeding Bird Survey data to assess national and regional trends is only useful for a few of the wading birds species that are widely distributed across coastal and interior wetlands. Analyses indicate that the eastern brown pelican, double-crested cormorant, great blue heron and cattle egret are increasing in most areas (Table 3.2). In fact, cormorant populations have increased across their range so rapidly that they are now considered a nuisance to aquaculturists and commercial fishermen in Canada and the United States (Nettleship and Duffy, 1995). In many states, control permits have been issued to reduce fish depredation (Nettleship and Duffy, 1995). The eastern brown pelican seems to be another 'success story' in the Endangered Species program and has been delisted (USFWS, 1985).

In contrast, several species appear to be unstable or declining over large regions. The white-faced ibis and wood stork are species of, respectively, regional or national concern (Table 3.2; OMBM, 1987). Wood stork declines in Florida may be partially offset by increases in Georgia and South Carolina (Ogden *et al.*, 1987). Reddish egrets had been listed by the USFWS as a declining species (OMBM, 1987); however, more recent evidence suggests that the species may be slowly expanding its range along both Florida coasts (J. Rodgers, Jr and J. Ogden, personal communications). Also, because many wading birds nest in a limited number of mixed-species colonies, many individual states list all of the Ciconiiformes that breed in their states in a 'special concern' category (for example, see Parnell, 1977).

Other attempts to summarize wading bird inventories, habitat use and distribution over large regions or nationally have been published (Portnoy, 1977; Ogden, 1978; Scharf, 1978; Osborn and Custer, 1978; Sprunt *et al.*, 1978; Erwin and Korschgen, 1979; Sowls *et al.*, 1980; all summarized in Spendelow and Patton, 1988). Several of these attempted to summarize population trends (Ogden, 1978; Erwin and Korschgen, 1979). As indicated above, the most relevant changes to forested wetlands concern the southeastern United States and the lower Mississippi River valley. Ogden (1978) described changing patterns of distribution with herons, egrets and ibises moving northward over the past 30 years, probably recolonizing much of their former range, while the cattle egret has expanded all across the United States. The little blue heron (*Egretta caerulea*) may have declined in many southeastern states as inland wetlands declined; however, it may be increasing along the lower Mississippi River (Ogden, 1978).

In the mid-Atlantic region of the southeast (Maryland and Virginia), extensive loss of forested wetland (Tiner, 1989) in Chesapeake Bay does not seem to have resulted in declines of the great blue heron or other waders

(Erwin and Korschgen, 1979; Erwin and Spendelow, 1991). Even though water quality and associated biotic processes have degraded within the Chesapeake over the past 25 years (Funderburk *et al.*, 1991), there seems to have been little impact on wading bird populations. Apparently an abundance of riparian wetlands and swamps with associated prey remains along many of the major Bay tributaries to sustain wading bird colonies.

Wading bird population changes in the upper midwest and along the Great Lakes shorelines appear to be different from those in the southeast and Chesapeake Bay. Graber *et al.* (1978) and Thompson (1979) suggest that organochlorine pesticides and loss of forested wetlands in Illinois and other midwestern states have resulted in declines of the great blue heron, the great egret, the black-crowned night-heron and possibly other species. They admit, however, that movements between states and regions (north and south along the upper Mississippi River floodplain) confound attempts to conduct regional trend assessment. There remains a large number of colonies and breeding pairs in the upper Mississippi valley where forest loss has been minimal compared with the surrounding uplands that have been mostly converted to agriculture (Thompson and Landin, 1978).

Some population trend information exists for individual states where forested wetland loss has been extreme. In Louisiana, as mentioned above, Martin and Lester (1990) reported at least a doubling of numbers of the great blue heron and the roseate spoonbill from 1976 to 1990, but the reddish egret declined by more than 50% over the same period, and the tricolored heron may also have declined. White ibis populations may have increased dramatically in recent years with the development of the crayfish aquaculture industry (B. Fleury, personal communication). Problems of survey coverage and changes in personnel and methods hamper a critical comparison of populations over the period. In Texas, a current attempt is being made to compare populations from the early 1970s with those in 1990 (Lange, in press). Again, there are logistic problems to overcome in interpreting population comparisons. Nonetheless, Lange (in press) claims that there are reasonably strong data suggesting declines in the ubiquitous great blue heron along the coast, as well as in reddish egret and white-faced ibis, species included on the national list of bird species of concern (OMBM, 1987).

Populations of wading birds in Florida, a state with major losses of both forested and non-forested wetlands, have a long history of interest among biologists (Crowder, 1974; Robertson and Kushlan, 1974; Kushlan and White, 1977; Ogden, 1978; Sprunt *et al.*, 1978; Frohring *et al.*, 1988; Frederick and Collopy, 1988; Kushlan, 1989). Regardless of the exact percentage of declines that may have occurred for many wading birds (Frohring *et al.*, 1988), there have clearly been precipitous reductions in the wood stork, white ibis and other egrets and herons in the state, especially in south Florida (Crowder, 1974; Robertson and Kushlan, 1974; Ogden, 1978; Kushlan and Frohring, 1986; Ogden *et al.*, 1987). However, loss of these species is probably due mostly to changes in the hydrology and productivity of the emergent wetlands in the Everglades region, not to loss of forested wetlands (Ogden, 1978; Kushlan, 1979, 1987, 1989; Frederick and Collopy, 1988). Some evidence is accruing to suggest that coastal mangrove productivity may be declining as a result of reduced freshwater inputs; this, in turn, may result in reduced success of the wood stork and other waders (R. Bjork and G. Powell, National Audubon Society, unpublished data).

Recent interest in waterbirds in Florida has resulted in several atlas efforts (Osborn and Custer, 1978; Nesbitt *et al.*, 1982; Runde *et al.*, 1991). In addition to wading bird declines in south Florida, there seem to be widespread declines of snowy egret, little blue heron, tricolored heron and white ibis in much of the state (Runde *et al.*, 1991). Again, however, one must be cautious in interpreting numbers derived from different inventories.

The endangered wood stork has generated a great deal of interest in Florida and the southeastern United States. Although it has declined in south Florida, it has increased in Georgia and South Carolina in the past decade (Kushlan and White, 1977; Kushlan and Frohring, 1986; Ogden *et al.*, 1987). The same pattern seems to hold for the white ibis, a species that is opportunistic in using both freshwater and brackish-marine wetland habitats (Bildstein *et al.*, 1990). However, in recent years, white ibis numbers have declined in South Carolina (K. Bildstein, personal communication).

In summary, the loss of forested wetlands may have had some negative effects on some waterfowl populations, notably the mallard in the Mississippi River region. However, breeding ground problems in Canada may also contribute. The wood duck, which is the species most dependent on forest wetland, seems to be increasing in North America, probably due at least in part to the popularity of nest-box programs in many states. This increase continues in spite of the heavy hunting pressure on the species. For wading birds, in spite of the heavy loss of forested wetlands in the southeast, there are few indications of population effects (although wood stork declines in south Florida may be an exception). Where there are such indications, the problems seem to be more related to loss or alteration of emergent or other types of wetlands. The effect seems to be greatest for the great blue heron, great egret and other waterbirds in the upper midwest region, although some of the apparent declines may be due to relocation of colony sites. Paradoxically, in many regions of the country, the great blue heron is increasing in spite of wetland loss and increasing human intrusion.

3.5.3 Mexico

(a) *Wetlands*

The wetlands of Mexico are some of the most expansive and diverse in all of Latin America. However, little attempt has been made to conduct national inventories or classifications of wetlands (Olmsted, 1992). Extensive forested wetlands exist in the form of riparian forests, palm thickets, mangroves and inundated low forests, especially on the Yucatan Peninsula (Lot-Helgueras and Novelo, 1990; Olmsted, 1992). The entire coastline of Mexico extends for about 10 000 km and includes extensive lagoons and river deltas with broad expanses of mangrove forest. The largest watershed in the country is in the states of Tabasco and Campeche along the Usumacinta–Grijalva rivers. This is one of the most important regions in North America for waterbirds (Sprunt and Knoder, 1980; Ogden *et al.*, 1988; Correa, 1992; Hartásanchez, 1993). One of the better studied components of this system is the low freshwater coastal forests of the Yucatan (Olmsted and Garcia, in press). These forests have low plant species diversity (sometimes only one tree species), and species distributions are determined by hydroperiod (Olmsted and Garcia, in press).

Correa (1992) states that there are 40 wetland sites in Mexico large and important enough to wildlife to be considered for the List of Wetlands of International Importance under the so-called Ramsar Agreement (Carp, 1972; Ramsar Convention Bureau, 1990); however, at present, only Rio Lagartos in Yucatan has that official distinction. Rio Lagartos has been singled out because it is the only breeding location of the rare greater flamingo (*Phoenecopterus ruber*) in Mexico. Recently, the Sian Ka'an reserve has been designated as a Man and Biosphere Reserve (Lopez-Ornat and Ramo, 1992) because it is the largest wetland complex on the Yucatan Peninsula and harbors large numbers of migrant and resident bird species. Other wetland areas in Mexico have been set aside for wildlife as Faunal Reserves, Refuges, Sanctuaries, National Parks, Biological Reserves and Biosphere Reserves (Scott and Carbonell, 1986).

Scott and Carbonell (1986) list the most important coastal wetlands in Mexico for waterbirds. They only list neotropical wetlands; therefore they exclude much of the higher interior regions of Mexico because they are considered temperate. The most significant forested wetlands are listed in Table 3.7.

The principal threats to Mexican wetlands are: land drainage for agriculture and housing; diversion of water courses; deforestation in the basins of inland lakes; invasion of exotic plants; dam buildings; oil development; pollution from pesticides and fertilizers; and mining (Correa, 1992; Olmsted, 1992). In the Yucatan Peninsula, there is a current threat to the wetlands of the Laguna de Terminos and the Usumacinta–Grijalva delta. Large-scale rice cultivation and shrimp farming are underway, subsidized by the government (Correa, 1992). There are few good statistics on mangroves, but a report from 1974 (Rollet, 1974) estimated that about 660 000 ha of mangroves remained in the country then; much less than that exists today (Correa, 1992).

Table 3.7 The most significant forested wetland complexes for waterbirds and shorebirds in Mexico[a] (from Scott and Carbonell, 1986)

Name	Ref. no.[b]	Approx. size (ha)	Principal species and numbers	Status/threats
Ensenada del Pabellon and Boca de la Barra	8	80 000	>300 000 ducks (winter); 20 000 coots (*Fulica americana*); >100 000 shorebirds	Dam construction threatens populations
Marismas Nacionales	11	200 000	>250 000 waterfowl (winter) (esp. northern pintail); >200 000 shorebirds; ca. 40 000 wading birds	Tourist development (San Blas); squatting
Usumacinta-Grijalva Delta	32	ca. 1 million	250 000 wading birds (50 000 breeding pairs); ca. 250 000 waterfowl (winter) (incl. 50 000 black-bellied whistling ducks); up to 45 *Jabiru mycteria*	Oil field development; irrigation; timbering
Laguna de Terminos	33	300 000	Many species of waterfowl in winter, and migration; important breeding area for waterfowl; important site for rare jabirus	Urban expansion; agriculture

	Number[b]		Description	Status
Marshes north of Campeche	34	120 000	Major migration and wintering area for waterfowl, flamingos, egrets, ibis	Developing petroleum industry around Campeche City
Río de Celestún	35	60 000	Breeding colonies of olivaceous cormorant, black-bellied whistling ducks; feeding grounds for 5000–10000 flamingos; up to 100 000 ducks and 100 000 coots in winter	Faunal Refuge (since 1979); still oil pollution threats, tourist disturbance
Río Lagartos	37	48 000	One of only four sites of breeding Caribbean flamingos in Caribbean Basin; feeding area for numerous wading birds, migrant shorebirds, storks, spoonbills, terns, gulls	Faunal Refuge (1979); tourism disturbance and powerboats; salt industry
Sian Ka'an (Bahía de la Ascensión and Bahía del Espíritu Santo)	40	450 000	Breeding area for pelicans, cormorants, waders; heavy migrant use by shorebirds, waterfowl	Biosphere Reserve; tourist disturbance

[a]Does not include interior highlands of Mexico, which are considered as Nearctic, not Neotropical wetlands.
[b]Number in Scott and Carbonell (1986).

(b)   Wildlife

Because it occupies the juncture of nearctic and neotropical regions, Mexico hosts one of the most diverse collections of fauna anywhere in the New World (Leopold, 1959). Its sport hunting was world renowned in the early part of the century (Leopold, 1959). Mexico has a varied topography and a diversity of wetlands that support large numbers of both resident and migrant waterbirds. Mexico has had a fairly credible record of wildlife inventory and research by biologists from within and outside the country. Since 1937, the USFWS and Mexican counterparts have conducted coopera-tive surveys of wintering waterfowl along both east and west coastlines (Saunders and Saunders, 1981; Brazda, 1988). In addition, the National Audubon Society has supported a number of surveys of wading birds and shorebirds along the east coast since the early 1970s (Sprunt and Knoder, 1980; Ogden *et al.*, 1988). Studies of greater flamingos and other waterbirds have been conducted since the 1970s in the Yucatan region (Scott and Car-bonell, 1986; Correa, 1992) led primarily by Mexican biologists. The orga-nization Ducks Unlimited of Mexico (DUMEX) has supported a number of site-specific studies and surveys of waterfowl (G. Baldassarre, Syracuse Uni-versity, personal communication).

Scott and Carbonell (1986) list 40 wetland sites as being of major sig-nificance to waterbirds in the coastal (neotropical) zone, with another 25 in the interior (nearctic). Seventeen of the 40 sites have extensive forested wetlands. Of these, eight are especially significant to wildlife as largely forested (mangrove) wetland areas (Table 3.7). Of this group of eight, only three have special protective designation (Table 3.7). The Mexican govern-ment and other non-governmental agencies are advocating protective status of at least some areas adjacent to the Usumacinta–Grijalva River deltas and the Laguna de Terminos (I. Hartasanchez, personal communcation; J. Ogden, personal communication). The Usumacinta–Grijalva region, the largest wetland in Mexico, is perhaps the ecological equivalent of Florida's Everglades (J.C. Ogden, personal communication). In terms of both diver-sity and densities of waterbirds, it ranks as one of the most important wetlands in the New World (Ogden *et al.*, 1988; Correa, 1992).

Population changes in waterbirds in Mexico have interested biologists for many years. The status of several waterfowl species that inhabit forested wetlands is given in Table 3.8. In spite of heavy hunting pressure, black-bellied whistling ducks and northern pintails seem to have maintained large numbers at least until the early 1980s (Tables 3.8, 3.9). In western Mexico, however, there may have been a decline in pintails since the mid 1970s as there seems to have been in the United States (USDI, 1986). Mallards are found in very low numbers in Mexico in the winter (Tables 3.8, 3.9, 3.10). The survey data summarized in Tables 3.9 and 3.10 reveal some of the problems associated with long-term surveys. Changes in personnel, methods and area coverage make interpretations difficult. In the Yucatan Peninsula,

Table 3.8 Summary of the status of forested wetland species of waterfowl in Mexico to the 1960s

Species	Period	Status in Mexico	Reference [a]
Black-bellied whistling duck	1937–1964	Formerly numerous in interior of Mexico in winter; underestimation a problem (cryptic)	Saunders and Saunders, 1981
	1950s	'Black-bellied is maintaining its numbers better than the Fulvous Tree Duck' (p. 163)	Leopold, 1959
Muscovy duck	1937–1964	Great decrease from 1950s; heavy hunting pressure; riparian habitat destruction; numbers decreased greatly in many localities	Saunders and Saunders, 1981
	1950	Bottomland loss; heavy hunting during 1930s–1940s; threatened in much of Mexico	Leopold, 1959
Mallard	1937–1964	Rarer now than in late 1800s; only 1% of North American population winters in Mexico; especially uncommon in northeastern Mexico	Saunders and Saunders, 1981
	1950s	Breeds in one region of Mexico (northern Baja California), status unknown	Leopold, 1959
Northern pintail	1937–1964	Most abundant duck in Mexico; many migrate south of Mexico; heavy hunting in interior; most birds winter on coast; no trend information	Saunders and Saunders, 1981

Table 3.9 Summary of wintering waterfowl surveys in eastern Mexico, 1955–1985, for selected species (from Brazda, 1988)

Year[*]	Mallard	Pintail	'Tree ducks'[a]	Total dabblers[b]
1955	28	197,119	12,342	489,526
1956		144,300	26,900	766,360
1958	100	184,000	4,200	557,260
1959		182,070	31,800	390,620
1960	40	43,324	13,180	191,265
1961	383	234,851	2,870	684,809
1962	30	55,633	24,232	199,011
1963	212	32,256	32,644	222,945
1964	57	90,392	29,142	474,178
1965[c]	202	46,732	4,619	248,612
1966[d]	2	21,964	3,389	103,899
1967	105	89,133	12,595	250,390
1970	25	90,236	54,713	537,167
1975	20	57,415	51,035	528,645
1977	340	381,250	24,760	1,096,905
1978	5	150,385	35,725	698,015
1979	90	219,160	81,505	1,139,645
1980	45	207,555	40,355	1,077,095
1981	770	333,390	48,435	1,157,000
1982	10	101,100	38,695	444,110
1985[d]	90	69,500	36,935	306,530

[a]Fulvous whistling duck and black-bellied whistling duck.
[b]Includes species other than those shown here.
[c]Only partial coverage of Tabasco lagoons; Campeche–Yucatan area not covered.
[d]Tabasco lagoons and Campeche–Yucatan lagoons not covered.
[*]There were no surveys in 1957, 1968, 1969, 1971–1974, 1976, 1983, 1984.

Correa (1992) concluded that ducks are recovering from low populations during the 1980s.

With respect to wading bird populations, systematic inventories have not been conducted often enough or during the same seasons (wet–dry) to determine adequate trends (Sprunt and Knoder, 1980; Ogden *et al.*, 1988; Correa, 1992; Hartásanchez, 1993). Correa (1992), referring to the Yucatan, suggests that wading bird species all show large annual variations, but that flamingos and brown pelicans have increased since the 1970s. New information from the Usumacinta–Grijalva delta has recently become available on the nesting and distribution of the rare jabiru, a species once felt to be nearly extinct in Mexico (Knoder *et al.*, 1980; Ogden *et al.*, 1988; Hartásanchez, 1993). No figures are available to assess trends for shorebirds.

Table 3.10 Summary of waterfowl surveys in western Mexico, 1960–1991 (from unpublished reports 'Mexico winter waterfowl surveys', Washington DC and Portland, Oregon)

Year[*]	Tree Ducks[a]	Pintail	Mallard	Dabblers[b]	Total Ducks[b]
1960	–	–	–	–	100,595
1961[d]	32,075	853,730	160	1,433,870	1,609,528
1962[d]	32,455	889,430	45	1,569,705	1,742,155
1963	45,565	445,137	0	1,242,957	1,541,772
1964	46,345	708,280	0	1,438,095	1,670,040
1965	52,353	634,529	0	1,620,487	1,903,411
1966[c,d]	10,255	248,165	0	429,360	522,640
1967[d]	2,551	203,245	0	334,301	374,291
1968[c,d]	1,800	203,850	0	262,650	334,300
1969[c,d]	3,250	469,500	0	716,200	788,250
1970[d]	2,940	410,160	0	701,520	790,250
1971[d]	8,870	159,924	30	343,672	424,126
1972[d]	8,945	614,555	0	999,935	1,088,133
1973[c,d]	6,710	724,700	0	935,420	966,125
1974[c,d]	5,440	205,950	0	335,215	389,866
1975[c,d]	10,655	1,047,980	0	1,612,430	1,871,930
1976	1,457	686,821	2,027	1,310,698	1,487,208
1977	18,005	545,183	172	998,081	1,273,849
1978	32,005	896,650	0	1,759,033	1,917,383
1979	33,135	756,385	35	1,430,585	1,561,290
1980	27,630	576,050	0	1,195,892	1,325,917
1981[d]	35,785	324,365	0	861,460	1,026,990
1982[d]	28,655	421,465	30	1,270,421	1,456,127
1983[c,d]	18,300	303,340	1,040	788,180	897,704
1984[c,d]	11,070	326,980	20	1,004,640	1,079,926
1985[d]	27,100	267,815	210	1,087,190	1,422,062
1988[d]	9,365	205,970	10	635,145	725,405
1991	33,995	324,525	0	1,163,315	1,301,905

[a]Fulvous whistling duck and black-bellied whistling duck.
[b]Includes species other than those shown here.
[c]Incomplete West Coast and Baja Survey.
[d]Incomplete Southwestern coast survey.
[*]There were no surveys in 1986, 1987, 1989, 1990.

Aerial surveys conducted by the National Audubon Society with assistance from Mexico reveal that nesting chronologies for most wading birds are driven by annual wet–dry cycles and that the larger, more stable nesting colonies are found in mangrove forests and along the mangrove–herbaceous marsh ecotone (Ogden *et al.*, 1988).

In spite of the advances that Mexico has made in the conservation arena in recent years, little basic inventory work has been done in some of the major wetlands (see Lopez–Ornat and Ramo, 1992, for example). Further efforts are underway within several federal agencies and non-governmental agencies to include shorebirds with waterfowl surveys in Mexico (E. Cummings, USFWS, personal communication). In addition, Western Hemisphere Shorebird Reserve Network personnel are planning to identify important wetlands for shorebirds in the western hemisphere and to survey areas that have been missed previously (Myers *et al.*, 1987).

3.5.4   Belize

(a) Wetlands

Fortunately for conservation interests, Belize has the lowest human population density of any Central American country (Scott and Carbonell, 1986). Largely for this reason, it supports some of the most pristine wetland habitats and wildlife diversity in the region. It was a British Protectorate until 1981. In that year, legislation was initiated to protect some of the most important areas for parks and wildlife (Zisman, 1989). The most significant act is the National Parks System Act (No. 5, 1981) that defined four types of protected area, using the International Union for Conservation of Nature's (IUCN) categories: Strict Nature Reserves, National Parks, Natural Monuments and Wildlife Preserves (Zisman, 1989). The Belize Audubon Society acted as a catalyst with the Department of Forestry to designate critical wildlife areas. Nineteen important wetland areas had been identified by the time of Scott and Carbonell's book in 1986.

In addition to these wetland areas, other efforts have been made for wildlife area designation and protection. Hartshorn *et al.* (1984) conducted an extensive field study to highlight important natural resource areas as part of an extensive Agency for International Development effort. Also, concern over development on the largest barrier reef in the Western Hemisphere has spurred conservation interest in developing a management plan based on the model provided by Australia's Great Barrier Reef (Perkins, 1983; Perkins and Carr, 1985). Potential loss of the mangrove forests on these barrier islands could be devastating to the reef ecosystem and to associated waterbirds (Poole, 1990). There already is concern over the plight of an endemic species of catbird as development increases (E. McRae, Siwa Ban Foundation, personal communication).

The threats to the wetlands in Belize are minor compared with Mexico and elsewhere in Central America. There is little petroleum development in coastal waters, a limited amount of timber harvesting and, until recently, agricultural development has been limited (Scott and Carbonell, 1986). Concern has been raised over the growing ranks of immigrants from other countries in Central America and their potential impacts on the natural resources.

(b) Wildlife

Per unit area, Belize has one of the most diverse wildlife faunas in the hemisphere. At just one sanctuary, the Crooked Tree Wildlife Preserve, 260 species out of a national total of 392 resident bird species have been recorded, including three nesting pairs of the endangered jabiru (Scott and Carbonell, 1986). The Preserve also contains populations of the endangered jaguar (*Panthera onca*), Baird's tapir (*Tapirus bairdii*) and Morelet's crocodile (*Crocodylus moreleti*) (Zisman, 1989).

The seven largest and most significant forested wetland complexes in Belize are listed in Table 3.11. Several of these have breeding records of rare waterbirds such as the chestnut-sided heron and two tiger heron species. The Crooked Tree Preserve is probably the single most important wetland to waterbirds. It becomes critical to many species during the dry season. Up to 74 species of waterfowl have been recorded (Table 3.11). The most important riparian area for wildlife may be Mussel Creek, which flows through extensive swamp forest. It harbors almost as rich an array of waterfowl as does Crooked Tree to the north and also includes a rich diversity of raptors and kingfishers (Scott and Carbonell, 1986). It has recently been visited by ornithologists to develop a more comprehensive species lists (B. Dowell, NBS, personal communication).

Few extensive and repeated inventories have been done for waterbirds in Belize. Most of the information provided by the Belize Audubon Society was based on a limited number of visits to sites and extensive surveys were not carried out. Poole (1990) conducted a survey of colonial waterbirds along most of the barrier reef to Punta Gorda in March, April and May, 1990. He attempted to compare nesting numbers with census figures from the early 1970s and mid 1980s. He found (as had Weyer, 1982, and others) that nesting was concentrated on cays in the northern half of the reef. Although he concluded that several species of wading bird, frigatebird and booby had declined over the past 8–10 years, he admitted that inconsistent census methods and periods and potential colony shifts to interior wetlands make trend interpretation problematic.

3.5.5 Guatemala

This country has the highest population density of any country in Central America. The wetlands are extensive, with more than 40 river basins, extensive coastal estuaries (ca. 100 km along the Pacific coast) and about 350 lakes and ponds ranging up to the nearly 60 000 ha Lago de Izabal (Scott and Carbonell, 1986). There are five government agencies involved with wetland programs in the country, the most active of which is CECON

Table 3.11 The most significant forested wetland complexes for waterbirds in Belize (from Scott and Carbonell, 1986)

Name	Reference no.[a]	Approx. size (ha)	Principal species and numbers	Status/threats
Marshes along lower New River	2	Unknown (ca. 35 km of river)	Poorly known; incl. boat-billed and tiger herons, whistling ducks	Sugar mill pollution
Crooked Tree Lagoon	8	ca. 10 000	74 spp. of waterfowl recorded; critical during dry season; many waders, incl. wood stork, jabirus, rare Agamia agamia	In 1984 became a 'Wildlife Preserve'; highway construction nearby; hunting
Mussel Creek	10	Unknown (ca. 30 km of creek)	Second only to no. 8 as a rich waterfowl area in Belize; breeding tiger herons; no large ardeid colonies	No major threats
Northern Lagoon	14	3200	Largest white ibis colony in Belize; other waders nest also; fall–winter area for many waterfowl	State-owned; loose protection by Belize Audubon Society (no warden)
Southern (Manatee) Lagoon	15	3200	No colonies but important feeding area for numerous species	Mostly state owned
Punta Ycacos Lagoon	17	9000	Formerly large nesting colonies of egrets on numerous mangrove keys	State owned; major declines due to hunting
Lower Temash River–Temash lagoon	19	> 20 000 ha along 35 km of river	Poorly known but important feeding area; no known large nesting colonies of waders	State owned

[a]From Scott and Carbonell (1986).

(Centro de Estudios Conservacionistas). With regard to the status of forested wetlands, there has been some research on mangroves in the country at several *biotopos*, but no status and trend assessment has been done for the entire country (O. I. Valdez Rodas, CECON, personal communication). Rollet (1974, in Jimenez, 1992) has reported a loss rate of mangroves in Guatemala of 560 ha per year.

The principal threats to wetlands and wildlife in the country are fertilizer pollution, pesticides, human wastes, and mining residues (Scott and Carbonell, 1986). In the interior of the country, sedimentation and erosion from development around lakeshores is the principal problem. There also seems to be an increasing threat from sport hunting in some areas of the country (Scott and Carbonell, 1986).

Twenty-four significant wetland areas were identified in Scott and Carbonell (1986) and the eight most important forested wetlands are listed in Table 3.12. Two of these are on the Pacific coast and three are on the Caribbean, including Lago de Izabal and the associated Golfete and rivers connected with the Bahia de Amatique. The Pacific coast wetlands harbor large numbers of wintering and migrant waterfowl and shorebirds. The black-bellied whistling duck is one of the most abundant species in the Pacific region. The Lago de Izabal region is extensive and diverse and provides habitat to the rare jabiru, wood stork, roseate spoonbill and many shorebird species. Little attempt has been made to conduct large-scale inventories of any waterbirds because of the lack of staff and finances (H. Kihn, CECON, personal communication).

3.5.6 Honduras

This second largest country in Central America has great diversity in both geography and biology (Scott and Carbonell, 1986). Much of the country is mountainous but there are extensive coastal wetlands and mangroves as well. The government is hoping to set aside natural areas in six management categories based largely on Mexico's plan; 24 such areas had been designated as of 1984 (Scott and Carbonell, 1986). However, little attempt has been made to inventory the wetlands or their wildlife on any extensive scale. Only six sites were identified to Scott and Carbonell (1986) for their directory; the five most important forested wetland sites are listed in Table 3.13. The immense coastal lagoon, marsh and mangrove swamp complex known as the Laguna de Caratasca (site 4) merits an inventory of its waterbird populations as it is very remote and virtually unknown.

Major threats to the wildlife and wetlands of Honduras are deforestation and pollution caused by fishermen (Scott and Carbonell, 1986). Recent concern has been expressed over a tourism development plan for the Islas de la Bahia (P. Bacon, University of the West Indies, personal communication).

Table 3.12 The most significant forested wetland complexes for waterbirds in Guatemala (from Scott and Carbonell, 1986)

Name	Reference no.[a]	Approx. size (ha)	Principal species and numbers	Status/threats
Manchon Lagoons	1	13 850	More than 70 species of waterfowl; many migrant nearctic ducks and shorebirds; abundant black-bellied whistling ducks	Mostly state owned; pesticides
Río Acome Estuary	2	3100	Similar to no. 1, but more ardeids, fewer anatids	Pesticides
Lago de Izabal[b]	13	59 000	Many resident and migrant waterfowl; resident jabiru and wood storks, spoonbills; shorebird migrants	Pesticides; nickel mining
Golfete and Río Dulce – Río Chocón	14	ca. 10 000	Several rare waterfowl in Guatamala; storks, egrets, ibis abundant	Intensive agriculture in region
La Graciosa marshes	15	6000	Same as no. 14	State owned
Lago Petén-Itzá and Petenchel[b]	19	10 370	Little known waterfowl information; believed nesting location for many wading birds, rails	None
Rio Escondido and marshes	22	17 900	Similar to no. 19	None
El Tigre lakes[b]	23	16 000	SImilar to no. 19	Oil industry

[a]From Scott and Carbonell (1986).
[b]Most of the area is lake, marsh, etc. but bordered by swamps.

Table 3.13 The most significant forested wetland complexes for waterbirds in Honduras (from Scott and Carbonell, 1986)

Name	Reference no.[a]	Approx. size (ha)	Principal species and numbers	Status/threats
Laguna de Los Micos and Río Ulua Delta	1	55 000	Unknown	Unknown; needs avian surveys
Laguna de Guaymoreto and Río Aguan Delta	2	34 000	Unknown	Unknown; being considered for reserve for manatees
Laguna de Ibans, L. de Brus, Río Platano	3	110 000	Rich for waterfowl, no details	State owned; hunting, forest clearing
Laguna de Caratasca (and nearby lagoons)	4	370 000	Unknown	Unknown; largest wetland complex in country, but poorly known (7 major lagoons)
Golfo de Fonseca	5	71 000	Unknown	State owned; mangrove areas in a Forest Reserve; some illegal forest cutting

[a]From Scott and Carbonell (1986).

3.5.7   El Salvador

The smallest country in Central America, El Salvador is very mountainous and has a 260 km coastline (Scott and Carbonell, 1986). As of 1984, no national parks or reserves had been officially designated; however, five wetland areas were being managed as such (Scott and Carbonell, 1986). No quantitative data were obtained by Scott and Carbonell (1986) for El Salvador in the course of assembling information for their directory, which was based almost solely on personal communication with two biologists.

Of the total of eight wetlands listed by Scott and Carbonell (1986), Table 3.14 lists four that are largely forested. The large coastal lagoon complex at Punta San Juan with its extensive mangrove swamps may represent the richest area for waterbirds but no information is available on its fauna. Shrimp farming may have an impact on the quality of the lagoon (Scott and Carbonell, 1986). In one wetland region (Laguna Jocotal), the natives manage the black-bellied whistling duck population for food.

3.5.8   Nicaragua

Nicaragua is a relatively large country with a low human population of only three million people. The wetlands are large and extensive but have not been specially designated except for an official decree in 1983 to preserve ca. 60 000 ha (Scott and Carbonell, 1986). Loss rates of mangrove forests are high at 385 ha per year (Jimenez, 1992) but not as high as in Guatemala. Lakes Nicaragua and Managua are two of the largest in Central America. In their directory, Scott and Carbonell (1986) list 25

Table 3.14   The most significant forested wetland complexes for waterbirds in El Salvador (from Scott and Carbonell, 1986)

Name	Reference no.[a]	Approx. size (ha)	Principal species and numbers	Status/ threats
Barra de Santiago	3	4800	No information	?
Río Lempa estuary–lagoons	5	11 000	No information	?
Punta San Juan lagoons	6	37 000	No information	?
Golfo de Fonseca	8	11 000	No data, but probably similar to Estero Real (Table 3.15)	?

[a]From Scott and Carbonell (1986).

wetlands of significance, eight of which are associated with Lake Nicaragua. Table 3.15 lists six forested wetlands of major significance to waterbirds.

One of the most significant areas is the Estero Real, part of the Golfo de Fonseca. This represents one of the largest mangrove forests along the Pacific coast of Nicaragua but there is concern about shrimp farm development and mangrove destruction (Scott and Carbonell, 1986). The area is important for breeding pelicans and wading birds, black-bellied whistling ducks and muscovies and is a major migration stop-over area for many shorebirds. This wetland was established as a Nature Reserve in 1983 (Scott and Carbonell, 1986). On the Caribbean side, the Rio Grande de Matagalpa delta is the largest coastal mangrove forest complex and is very important for breeding wading birds, pelicans, etc. as well as migrating shorebirds (Scott and Carbonell, 1986). Farther north, the Miskito coastline has been recently surveyed by aircraft and is reported to have an extremely high density of wading birds – higher than that in the Everglades (P. Frederick, University of Florida, personal communication). Frederick also saw an impressive number of jabiru nests.

3.5.9 Costa Rica

Costa Rica has one of the largest systems of parks and reserves relative to its size of any country in the world; however, its wetlands are less well inventoried and visited than parks and reserves such as Monteverde, La Selva and Guanacaste. Scott and Carbonell (1986) list 12 important wetlands and of these the six most important forested wetlands are listed in Table 3.16. The only large wetland listed is Barra del Colorado in the northeast, but this area is one of the least studied in the country. It is an important wintering area for waterfowl, shorebirds and the declining (in the United States) black tern (*Chlidonias nigra*). The area is a huge complex of mangrove and swamp forests, freshwater lakes and coastal lagoons and marshes, and was proposed as a National Wildlife Refuge 10 years ago (J. Sanchez, Museo Nacional, personal communication). Palo Verde is another important reserve and its birdlife was described qualitatively by Slud (1980). The Rio Tempisque area within the Palo Verde National Park supports a large mixed colony of wood storks, roseate spoonbills and egrets (R. Bjork, personal communication).

Concerns for protecting coastal wetlands have been elevated because of increasing human development and agriculture and its impacts on both the avifauna and the sea turtle populations. The loss rate of mangroves in the coastal regions has been estimated at only 45 ha per year (Jimenez, 1990). Other threats to waterbirds are illegal hunting and pesticide pollution. Hunting may be endangering the muscovy duck (Hartshorn *et al.*, 1982). There is some evidence that pesticides are adversely affecting the wading birds in the lower Tempisque heronries (Hartshorn *et al.*, 1982).

Table 3.15 The most significant forest wetland complexes for waterbirds in Nicaragua (from Scott and Carbonell, 1986)

Name	Reference no.[a]	Approx. size (ha)	Principal species and numbers	Status/threats
Estero Real	1	68 400	'Hundreds or thousands' of breeding species, incl. pelicans, herons, egrets, storks, ibis, whistling ducks; many nearctic migrants, shorebirds	State owned (Nature Reserve); mangrove destruction
Isla del Venado	3	3450	Important breeding and wintering area for waterfowl; breeding by many storks, herons, spoonbills, jabirus, whistling ducks	State owned; mangrove cutting
Laguna de Bismuna	10	48 340	'Very rich area for waterfowl', many breeding waders, migrant waterfowl, shorebirds (similar to no. 1)	State owned; overfishing; deforestation
Laguna de Pahora	11	35 580	Similar to no. 10	State owned; deforestation
Laguna de Wounta	13	35 290	Similar to no. 10	Overfishing, deforestation
Rio Grande de Matagalpa Delta	16	74 240	Similar to no. 10	State owned; overfishing and forest clearing

[a]From Scott and Carbonell (1986).

Table 3.16 The most significant forested wetland complexes for waterbirds in Costa Rica (from Scott and Carbonell, 1986)

Name	Reference no.[a]	Approx. size (ha)	Principal species and numbers	Status/threats
Palo Verde	2	6000	Extremely important area for breeding, migrating, wintering waterfowl; large colonies of wading birds, incl. storks, spoonbills; main area for jabirus in Costa Rica (eight pairs in early 1980s); very important in dry season	National Park and Wildlife Refuge; no known threats
Estero Madrigal	5	300	Important dry season refuge for waterfowl; many shorebird migrants	Pesticides; hunting
Estero Piedras	6	Several hundred	Feeding area; important migrating, wintering area for shorebirds and larids	Road construction; shrimp farming
Laguna Corcovado	8	1200	50 species of breeding, wintering waterfowl, no census data; many shorebirds	Natural Park (1976) (incl. 36 000 ha); no threats known
Tortuguero wetlands	11	15 000	Many migrant waders and shorebirds; manatees and sea turtles	National Park (1970) (Caribbean coast)
Barra del Colorado	12	53 550	No census data but many species at all times of year; manatees also	Road construction (Caribbean coast)

[a]From Scott and Carbonell (1986).

3.5.10   Panama

The wetlands of Panama have been described more thoroughly than those of most Central American countries because of the work of F. Delgado (1983 and others). The best known and studied wetland complex is the Gatun Lake region formed when the Panama Canal was built in 1913. This artificial lake is about 32 000 ha and enjoys the best management and protection in the country (Scott and Carbonell, 1986). Unfortunately, coastal wetlands have received very little attention compared with interior wetlands. As in areas to the north, major threats to the wetlands are oil pollution, mining and pesticides in the western part of the country. Agricultural impacts, shrimp farming and cutting of mangrove forests are threats along the eastern coast.

The waterbirds of Panama reach some of the highest densities of any location in Central or North America. Wading birds in the Rio Grande marshes (site 18), waterfowl in the Golfo de Parita and shorebirds in the Bay of Panama reach extraordinary numbers. Of the 22 sites listed by Scott and Carbonell (1986), eight are important forested wetland areas (Table 3.17). The largest complex is the Rios Sabana, Chucunaque, Tuira and Sambu region (site 22). It has been designated a national park and also a UNESCO Biosphere Reserve (1983) yet little attempt has been made to conduct censuses of birds. No trend information is therefore available for any waterbirds.

3.6   MANAGEMENT RECOMMENDATIONS

3.6.1   North America

The principal goal of management efforts in Canada and the United States should be to preserve the remaining large tracts of bottomland or riparian wetlands especially in coastal British Columbia, the St Lawrence River valley, the Mississippi River corridor, the grazed river corridors of the arid western US and the southeastern coastal plain. In concert with a preservation plan should be restoration of hardwoods in the southern US. Where marginal farmlands are being taken out of production, efforts are now underway to plant oaks and other native hardwood species in the lower Mississippi River valley (NAWMP, 1990).

Other strategies of the Lower Mississippi Valley Joint Venture include expanding the Conservation Reserve Program, public acquisition of priority wetlands, conservation easements on lands held by the Resolution Trust Corporation and the Farm Credit System, and increased technical assistance to private landowners (NAWMP, 1990). Conservation easements represent the most realistic way of setting aside priority forested wetlands with limited federal resources available for fee simple land acquisition.

Table 3.17 The most significant forested wetland complexes for waterbirds in Panama (from Scott and Carbonell, 1986)

Name	Reference no.[a]	Approx. size (ha)	Principal species and numbers	Status/threats
Sansan–Changuinola lagoons	1	6500	No census data but many waterbird species at all time of year; migrant shorebirds	Cattle ranching; banana plantations
Bocas del Toro Archipelago/Laguna de Chiriqui	2	9800	No information	Ranching; agriculture; hunting; pollution
Mangroves-estuaries in David District	6	ca. 50 000	Few data, very important breeding area for pelicans, waders, ducks, wintering waterfowl	Oil pipeline; pesticides; mangrove cutting (rice culture)
Puerto Vidal-Bobi	7	12 000	SImilar to no. 6 above; important for nesting black-bellied whistling duck	Mangrove loss (rice, timber, fuel)
Rio Grande Marshes	18	10 000	Nesting colonies of herons, ibis; only site for spoonbills; no data on ducks	Agricultural drainage; mangrove loss
Gatun Lake	19	32 000	Important year-round area for all waterbirds, esp., herons, spoonbills, pelicans; many wintering larids	Well studied (Smithsonian); Barro Colorado is a Natural Monument; other areas are national parks, reserves
Bay of Panama	20	38 500	Important wintering for shorebirds, some densities are highest recorded anywhere (in millions); also pelicans, herons, larids	Drainage for agriculture, pollution
Ríos Sabana, Chucunaque, Tuira, Sambu	22	433 000	Very rich in waterfowl, some rare species, but little census data	National Park and Biosphere Reserve (1983); no known threats

At state and local (county) levels, several land management models are available that would protect riparian and palustrine forested wetlands. In Maryland, the Chesapeake Bay Critical Areas Law (Maryland Natural Resource Article 8-1801, July 1, 1984) was implemented through county regulations limiting development within 1000 feet of Chesapeake Bay shoreline, with some exemptions made for agriculture. For many tributaries of the Bay, this law has the potential to protect vast expanses of riparian and bottomland forests in a region of rapid urbanization. These areas were recognized as extremely important habitats for many avian species in the Chesapeake watershed (Funderburk *et al.*, 1991). The importance of conserving corridors along streams, rivers and bays has gained wide acceptance, especially among resource managers concerned with migratory birds and large mammals (Anon., undated).

Wildlife resource specialists need to work more closely with landowners of major private holdings such as timber company executives. This will require educating landowners about the value of forested wetlands to wildlife and may require some exchanges of resources. For example, harvesting timber on public lands with low biodiversity could be 'traded' for protecting tracts of old growth timber in a bottomland forest that was privately owned.

Coordination of land acquisition, conservation easements and 'trades' needs to be made within a landscape perspective. For waterbirds, the interspersion of forested wetlands with emergent marshes and even agricultural fields is probably a more effective conservation strategy than is one that attempts to preserve only large forested tracts.

Education, of course, needs to extend to all ages and social strata. Much of the public still perceives forested wetlands as relatively worthless land. Strong public relations efforts such as the Nature Conservancy's 'Last Great Places' campaign could showcase valuable wetlands such as the Okefenokee Swamp and its diverse fauna.

3.6.2 Central America

In Mexico and in Central America, international agencies (e.g. the US Agency for International Development, and the Food and Agriculture Organization) need to assist resource managers with support for basic inventories of their forested wetlands, especially those highlighted in Scott and Carbonell (1986). For many of these important wetlands, there is no current species list, even for birds. Efforts need to be made in working with natural resource biologists in each country to re-evaluate the sitings of the numerous aquaculture facilities and rice production areas. In many Latin countries, rapid mangrove deforestation is taking place for purposes of aquaculture development.

Additional resources need to be put into both designating and prioritizing protection status of the major wetland sites cited by Scott and Carbonell (1986). Once their status is designated, stronger efforts in enforcement are essential. Local protection is best facilitated by educating the local population about the value of the wetland resource. In some cases, sustainable economic development is possible, such as with local market hunting of black-bellied whistling ducks. In some areas, ecotourism has become a major economic force (e.g. Rio Lagartos, Mexico). Such ventures need to be carefully implemented to avoid disturbance to major wildlife concentration areas.

3.7 SUMMARY

Information is synthesized from the published literature, unpublished reports, national databases and experts from a number of countries to provide an assessment of the status of forested wetlands (including palustrine forests, mangroves, estuarine shrubs, riverine forests) and a variety of waterbird species associated with them. The suite of waterbirds considered includes waterfowl, pelicans and allies, and wading birds (herons, storks and allies). A number of species in these groups use forested wetlands for courtship, nest sites, roosts, molting and/or feeding during one or more seasons of the year. In Canada and the United States, the majority of wetland loss since 1900 has occurred in non-forested wetlands, primarily due to agricultural conversion. Since the 1950s, however, losses of palustrine forested wetlands and estuarine shrub–scrub have predominated, especially in the southern United States. Large-scale conversion of bottomland forests along the Mississippi River valley and in the southeastern coastal plain to croplands has been detrimental to a variety of wildlife. Most likely, breeding and wintering wood duck (*Aix sponsa*) has been affected the most, although effects of deforestation and drainage have been masked by the recovery of populations from historic lows in the early 1900s. Some of this recovery is probably due to an expanded nest-box program across the country. Wintering mallards (*Anas platyrhynchos*) in the lower Mississippi region have also been affected by bottomland conversion and many now depend on wet agricultural fields in winter. American black duck (*A. rubripes*) has also declined along the Atlantic Coast, but this decline is probably unrelated to changes in forested wetland habitats; forest area has actually increased in the New England region. In the United States, changes in numbers of wading birds, brown pelicans (*Pelecanus occidentalis*) and double-crested cormorants (*Phalacrocorax auritus*) do not seem to be related to changes occurring in forested wetlands except perhaps in a few regions. In fact, the great blue heron (*Ardea herodias*), an ubiquitous wading bird, appears to be increasing in most areas in Canada and the United States.

In Mexico and Central America, little attempt has been made to quantify wetland change, although there appears to be a growing concern for preserving large wetlands. Losses of mangroves have been widespread in Latin America due largely to expanding dry croplands, rice cultivation and a growing shrimp farming industry. Mangrove destruction throughout Latin America has a much greater potential to reduce waterbird (and many other species) populations than does the loss or alteration of forested wetlands in North America. Mangroves are a critical ecosystem for fish and shellfish as well as wildlife. Population data for most Central American waterbirds are sparse; trend information is nonexistent. A clear need exists to conduct basic inventories for wildlife in many of the extensive wetland complexes left in Central America.

Management needs to focus on: (1) preserving the remaining large tracts of bottomland forests along major river systems across all of North and Central America; (2) restoring areas to previous forested wetlands condition, e.g. by planting hardwoods; (3) ensuring that juxtapositions of open, emergent wetlands and forested wetlands are maintained as complexes; (4) trying to prolong and improve national programs such as the Conservation Reserve Program to improve marginal farmlands for wildlife; (5) working with waterways development agencies to minimize losses and degradation of riparian forests (e.g. dam construction and operations); and (6) improving public education as to the value of forested wetlands to the public good.

3.8   ACKNOWLEDGEMENTS

A large number of people contributed ideas or materials to me during the long gestation period and I thank them: P. Bacon, C. Baxter, M. Bradstreet, A. Brazda, M. Buford, R. Butler, J. Correa, T. Dahl, P. Dugan, V. Glooschenko, I. Hartasanchez, G. Hartshorn, J. Hefner, J. Jeglun, J. Kushlan, E. McRae, J. Ogden, I. Olmsted, K. Reinecke, O. Valdez Rodas and J. Voelzer. W. Manning was very helpful in obtaining library materials, J. Haig helped with editorial assistance and J. Armstrong typed parts of the manuscript. Comments on an early draft were received from J.C. Ogden, J.S. Ramsey, K.J. Reinecke and F.C. Golet.

3.9   REFERENCES

Anon. (1992) Ontario heronry inventory. Long Point Bird Observatory. *Ontario Heronry Inventory Newsletter* 3.

Anon. (undated) *The Forested Wetlands of the Mississippi River: an ecosystem in crisis.* The Nature Conservancy, Baton Rouge, Louisiana.

Aus, P. (1969) What is happening to the wetlands? *Trans. No. Amer. Wildl. and Nat. Resour. Conf.* 34:315–322.

Bacon, P. (1990) Ecology and management of swamp forests in the Guianas and Caribbean region, in *Ecosystems of the world 15: Forested wetlands* (eds A. Lugo, M. Brinson, and S. Brown), pp. 213–250. Elsevier, Amsterdam, The Netherlands.

Bellrose, F. (1980) *Ducks, Geese and Swans of North America.* Stackpole Books, Harrisburg, Pennsylvania.

Bellrose, F.C. (1990) The history of wood duck management, in *Proc. 1988 North Amer. Wood Duck Symposium* (eds L.H. Fredrickson, G.V. Burger, S.P. Havera et al.), St. Louis, Missouri, pp. 13–20.

Bildstein, K., Post, W. Johnston, J. and Frederick, P. (1990) Freshwater wetlands, rainfall, and the breeding ecology of White Ibises in coastal South Carolina. *Wilson Bull.* 102:84–98.

Blus, L., Heath, R., Gish, C. et al. (1971) Eggshell thinning in the Brown Pelican: implication of DDE. *BioScience* 21:1213–1215.

Brazda, A. (1988) Winter waterfowl populations and habitat evaluation – aerial surveys, east coast of Mexico, in *Ecología y conservación del delta de los ríos Usumacinta y Grijalva (Memória)* (pp. 575–594). Inst. Nac. Invest. Recursos Bióticos-Div. Reg. Tabasco, México.

Butler, R.W. (1987) Breeding ecology and population trends of the Great Blue Heron (*Ardea herodias fannini*) in the Strait of Georgia, in *The Ecology and Status of Marine and Shoreline Birds in the Strait of Georgia, British Columbia*, (eds K. Vermeer and R. Butler), pp. 112–117. Canadian Wildlife Service, Sidney, British Columbia, Canada.

Butler, R.W. (1992) Great Blue Heron, in *The Birds of North America*, No. 25, (eds A. Poole, P. Stettenheim, and F. Gill), pp. 1–20. The Academy of Natural Sciences, The American Ornithologists' Union, Philadelphia and Washington, DC.

Carp, E. (ed.) (1972) *Proceedings of the International Conference of Wetlands and Waterfowl, Ramsar, Iran, 30 January–3 February 1971.* International Waterfowl Research Bureau, Slimbridge, England.

Cashin, G., Dorney, J. and Richardson, C. (1992) Wetland alteration trends on the North Carolina coastal plain. *Wetlands* 12:63–71.

Chabreck, R. (1988) *Coastal Marshes: Ecology and Wildlife Management.* University of Minnesota Press, Minneapolis.

Chandler, W.J. (1989) Conserving North American waterfowl: a plan for the future. *Audubon Wildlife Report 1988/89*:219–255.

Clark, J.R. and Benforado, J. (1981) *Bottomland Hardwood Wetlands Workshop – a report.* National Wetlands Technical Council, Washington, DC.

Conservation Foundation (1988) *Protecting America's Wetlands: an action agenda. Final report of the National Wetlands Policy Forum.* The Conservation Foundation, Washington, DC.

Correa, J. (1992) Status of aquatic birds in the coastal wetlands of the Yucatan Peninsula. MSc thesis, Tropical Coastal Management, University of Newcastle-upon-Tyne, United Kingdom.

Cowardin, L., Carter, V., Golet, F. and LaRoe, E. (1979) *Classification of Wetlands and Deepwater Habitats of the United States.* US Fish and Wildlife Service, FWS/OBS-79/31. 131pp.

Crowder, J.P. (1974) *Some perspectives on the status of aquatic wading birds in South Florida.* US Bur. Sport Fish Wildl. South Florida Envir. Proj. Ecol. Rep. DI-SFEP-74-29.

Dahl, T. and Johnson, C. (1991) *Status and Trends of Wetlands in the Conterminous United States, mid-1970s to mid-1980s.* US Fish and Wildlife Service, Washington, DC. 28pp.

Davis, S. and Ogden, J. (eds) (1994) *Everglades: The Ecosystem and its Restoration.* St. Lucie Press, Delray Beach, Florida.

Delgado, F. (1983) Situación actual de las zonas pentanosas de Panamá: Perspectivas de conservación de la avifauna acuática. *Proc. Intern. Waterfowl and Wetlands Research Bureau Symposium*, Edmonton, Alberta, Canada.

DesGranges, J.-L. (1980) A Canadian program for surveillance of Great Blue Heron (*Ardea herodias*) populations. *Proc. 1979 Conf. Colonial Waterbird Group* 3:59–68.

DesGranges, J.-L. and LaPorte, P. (1983) Fourth and fifth tours of inspection of Quebec heronries, 1980–81. *Canadian Wildlife Service Progress Note* 139:1–11.

Dorcey, A.H.J. (ed.) (1991) *Perspectives on Sustainable Development in Water Management: Towards agreement in the Fraser River Basin*, Vol. 1. Westwater Research Centre, University of British Columbia, Vancouver, British Columbia, Canada.

Droege, S. and Sauer, J. (1990) North American Breeding Bird Survey annual summary, 1989. *U.S. Fish Wildl. Serv. Biol. Rep.* 90(8). 22 pp.

Dugan, P. (1991) Future wetland challenges. *Wetlands Programme – Intern. Union Conserv. Nature Newsletter* 3:1.

Dunn, E., Hussell, D. and Siberius, J. (1985) Status of the Great Blue Heron, *Ardea herodias*, in Ontario. *Canadian Field-Naturalist* 99:62–70.

Environment Canada (1986) *Wetlands of the St. Lawrence River region.* Lands Directorate, Environment Canada. Working paper no. 45, Ottawa, Ontario, 29 pp.

Erwin, R.M. and Korschgen, C.E. (1979) *Coastal waterbird colonies: Maine to Virginia, 1977. An atlas showing colony locations and species composition.* US Fish and Wildlife Service FWS/OBS-79-08 647 pp.

Erwin, R.M. and Spendelow, J.A. (1991) Colonial wading birds – herons and egrets, in *Habitat Requirements for Chesapeake Bay Living Resources*, (eds S. Funderburk, S. Jordon, J. Mihursky and D. Riley), pp. 19/1–19/14. Chesapeake Research Consortium, Inc., Solomons, Maryland.

Erwin, R.M., Frederick, P.C. and Trapp, J.L. (1993) Monitoring of colonial waterbirds in the United States: needs and priorities. *Intern. Waterfowl and Wetlands Res. Bur. spec. publ.* 26:18–22.

Finlayson, M. and Moser, M. (eds) (1992) *Wetlands.* International Waterfowl and Wetlands Research Bureau, Facts on File, New York.

Fleming, J. and Clark, D.R. Jr (1983) Organochlorine pesticides and PCBs: a continuing problem for the 1980s. *Trans. No. Amer. Wildl. Nat. Resour. Conf.* 48:186–199.

Frayer, W.E. and Hefner, J.M. (1991) *Florida Wetlands: Status and Trends, 1970s to 1980s.* US Fish and Wildlife Service, Atlanta, Georgia.

Frederick, P. and Collopy, M. (1988) *Reproductive ecology of wading birds in relation to water conditions in the Florida Everglades.* Florida Coop. Fish. Wildl. Res. Unit, Tech. Rep. 30, 259 pp.

Fredrickson, L. and Heitmeyer, M. (1988) Waterfowl use of forested wetlands of the southern United States: an overview, in *Waterfowl in Winter* (ed. M. Weller), pp. 307–324. University of Minnesota Press, Minneapolis.

Fredrickson, L., Burger, G., Havera, S. *et al.* (eds) (1990) *Proc. 1988 North Am. Wood Duck Symposium*, St. Louis, Missouri.

Frohring, P., Voorhees, D. and Kushlan, J. (1988) History of wading bird populations in the Florida Everglades: a lesson in the use of historical information. *Colonial Waterbirds* 11:328–335.

Funderburk, S., Jordan, S., Mihursky, J. and Riley, D. (eds) (1991) *Habitat Requirements for Chesapeake Bay Living Resources*, 2nd edn revised. Chesapeake Research Consortium, Inc., Solomons, Maryland.

Graber, J.W., Graber, R.R. and Kirk, E.L. (1978) Illinois birds: Ciconiiformes. *Ill. Nat. Hist. Surv., Biol. Notes* 109:1–80.

Greene, K., Lynch, J., Sircar, J. and Greenberg, L. (1988) Use of Landsat remote sensing to assess habitat for migratory birds in the Yucatan Peninsula. *Vida Silvestre Neotropical* 1:27–38.

Haramis, G.M. (1990) The breeding ecology of the wood duck: a review, in *Proc. 1988 North Am. Wood Duck Symp.* (eds L. Fredrickson, G. Burger, S. Havera *et al.*), pp. 45–60. St. Louis, Missouri.

Hartásanchez Herrera, I. (1993) Aspectos ecológicos de los humedales alrededor de la Laguna de Terminos, con enfasis en espécies Ciconiiformes en la Delta Usumacinta–Grijalva, Campeche, México. MS thesis, Univ. Nacional, Heredia, Costa Rica.

Hartshorn, G., Hartshorn, L., Atmella, A. *et al.* (1982) *Costa Rica: Country environmental profile*. Tropical Science Center, San Jose, Costa Rica.

Hartshorn, G., Nicolait, L. and Hartshorn, L. (1984) *Belize: Country environmental profile. A field study*. Trejos. Hnos. Sucs. S.A., San Jose, Costa Rica.

Heitmeyer, M. and Fredrickson, L. (1981) Do wetland conditions in the Mississippi Delta hardwoods influence mallard recruitment? *Trans. No. Amer. Wild. Nat. Res. Conf.* 46:44–57.

Jimenez, J. (1992) Mangrove forests of the Pacific coast of Central America, in *Coastal Plant Communities of Latin America*, (ed. U. Seeliger), pp. 259–267. Academic Press, New York.

Johnson, D., Nichols, J., Conroy, M. and Cowardin, L. (1988) Some considerations in modeling the Mallard life cycle, in *Waterfowl in Winter*, (ed. M. Weller), pp. 9–22. University of Minnesota Press, Minneapolis.

Kirby, R. (1990) Wood Duck nonbreeding ecology: fledging to spring migration, in *Proc. 1988 North Amer. Wood Duck Symposium*, (ed. L. Fredrickson, G. Burger, S. Havera *et al.*), pp. 61–76. St. Louis, Missouri.

Knoder, C.E., Plaza, P.D. and Sprunt, A., IV (1980) Status and distribution of the Jabiru Stork and other waterbirds in western Mexico, in *The Birds of Mexico: their ecology and conservation*, (eds P. Schaeffer and S. Ehlers), pp. 58–127. Natl. Audubon Society, Western Educ. Center, Tiburon, California.

Kushlan, J. (1978) Feeding ecology of wading birds, in *Wading Birds*, (eds A. Sprunt, IV, J.C. Ogden and S. Winckler), pp. 249–298. Natl. Audubon Society Res. Report 7.

Kushlan, J. (1979) Design and management of continental wildlife reserves: lessons from the Everglades. *Biol. Cons.* 15:281–290.

Kushlan, J. (1987) External threats and internal management: The hydrologic regulation of the Everglades. *Environ. Manage.* 11:109–119.

Kushlan, J. (1989) Wetlands and wildlife, the Everglades perspective, in *Freshwater Wetlands and Wildlife*, (eds R. Sharitz and J.W. Gibbons), pp. 773–789. US Dept of Energy Symposium Series 61, Oak Ridge, Tennessee.

Kushlan, J. and Frohring, P. (1986) The history of the southern Florida Wood Stork population. *Wilson Bull.* **93**:368–386.

Kushlan, J. and White, D. (1977) Nesting bird populations in southern Florida. *Florida Science* **40**:65–72.

Lange, M. (in press). *Texas Coastal Waterbird Colonies: 1973–1900 census summary, atlas, and trends.* Texas Parks and Wildlife Dept., Texas Colonial Waterbird Society, and US Fish and Wildlife Service, Angleton, Texas.

Le Groupe Dryade (1981) *Habitats propices aux oiseaux migrateurs. Analyse des pertes de végétation riveraine de long du Saint-Laurent, entre Cornwell et Matane (1945–1960, 1960–1976).* Canadian Wildlife Service, Environment Canada. Ste.-Foy, Quebec. Unpublished report.

Leopold, A. (1959) *Wildlife of Mexico.* University of California Press, Berkeley.

Lopez-Ornat, A. and Ramo, C. (1992) Colonial waterbird populations in the Sian Ka'an Biosphere Reserve (Quintana Roo, Mexico). *Wilson Bull.* **104**:501–515.

Lot-Helgueras, A. and Novelo, A. (1990) Forested wetlands of Mexico, in *Forested Wetlands,* (eds A. Lugo, M. Brinson and S. Brown), pp. 287–298. Elsevier Scientific Publ. Co., Amsterdam, The Netherlands.

Lugo, A. (1990) Introduction, in *Ecosystems of the World 15: Forested Wetlands,* (eds A. Lugo, M. Brinson and S. Brown), pp. 1–14. Elsevier, Amsterdam, The Netherlands.

Lugo, A. and Snedaker, S. (1974) The ecology of mangroves. *Annual Review of Ecology and Systematics* **5**:39–64.

Lugo, A., Brinson, M. and Brown, S. (eds) (1990) *Ecosystems of the World 15: Forested Wetlands.* Elsevier, Amsterdam, The Netherlands.

Lynch, J.J., Evans, C.D. and Conover, V.C. (1963) Inventory of waterfowl environments of prairie Canada. *Trans. No. Amer. Wildl. Nat. Resour. Conf.* **28**:93–108.

Lynch-Stewart, P. (1983) *Land use change on wetlands in southern Canada: review and bibliography.* Lands Directorate, Environment Canada. Working paper no. 26, Ottawa, Ontario. 115 pp.

MacDonald, P., Frayer, W. and Clauser, J. (1979) *Documentation, chronology, and future projections of bottomland hardwood habitat loss in the lower Mississippi alluvial plain. Vol. 1. Basic report.* US Fish and Wildlife Service, Div. Ecol. Services, Washington, DC.

Martin, R. and Lester, G. (1990) *Atlas and census of wading bird and seabird nesting colonies in Louisiana, 1990.* Louisiana Dep. Wildl. Fisheries and Louisiana Nat. Heritage Program Spec. Publ. 3. 182 pp.

Mitsch, W. and Gosselink, J. (1986) *Wetlands.* Van Nostrand Reinhold, New York.

Myers, J.P., Morrison, R.I.G., Antas, P.Z. *et al.* (1987) Conservation strategy for migratory species. *Amer. Sci.* **75**:18–26.

National Wetlands Working Groups (1988) *Wetlands of Canada.* Canada Committee on Ecological Land Classification, Ecol. Land Classif. Series, No. 24, Sustainable Development Branch, Environment Canada, Ottawa, Ontario, and Polyscience Publications, Inc., Montreal, Quebec. 452 pp.

Nesbitt, S., Ogden, J., Kale, H. II *et al.* (1982) *Florida Atlas of Breeding Sites for Herons and their Allies: 1976–78.* US Fish and Wildlife Service OBS-81/49. 449 pp.

Nettleship, D.N. and Duffy, D.C. (eds) (1995) *Double-crested Cormorants: A symposium on their ecology and management.* Colonial Waterbirds 18 (Spec. Publ. 1).

Nichols, J. and Johnson, F. (1990) Wood duck population dynamics: a review, in *Proc. 1988 North Amer. Wood Duck Symposium* (eds L. Fredrickson, G. Burger, S. Havera *et al.*), pp. 83–105. St. Louis, Missouri.

NAWMP (North American Waterfowl Management Plan) (1990) *Conserving Waterfowl and Wetlands: the Lower Mississippi Valley Joint Venture.* US Fish and Wildlife Service, Washington, DC.

Odum, W.E. (1971) *Pathways of energy flow in a South Florida estuary.* Univ. of Miami Sea Grant Tech. Bull. 7. 162 pp.

Odum, W.E., McIvor, C. and Smith, J. III (1982) *The ecology of the mangroves of south Florida, a community profile.* US Fish Wildl. Serv. FWS/OBS/81/24.

Office of Migratory Bird Management (OMBM) (1987) *Migratory Nongame Birds of Management Concern in the United States: the 1987 List.* US Fish and Wildlife Service, Washington, DC.

Ogden, J. (1978) Recent population trends of colonial wading birds on the Atlantic and Gulf Coastal plains, in *Wading Birds*, (eds A. Sprunt IV, J. Ogden and S. Winckler), pp. 137–154. Natl. Audubon Soc. Res. Rep. 7.

Ogden, J., McCrimmon, D. Jr., Bancroft, G.T. and Patty, B. (1987) Breeding populations of the Wood Stork in the southeastern United States. *Condor* 89:752–759.

Ogden, J., Knoder, E. and Sprunt, A. (1988) Colonial wading bird populations in the Usumacinta Delta, Mexico, in *Ecológia y conservación del delta de los ríos Usumacinta y Grijalva (Memória)* (pp. 595–606). Inst. Nac. Invest. Recursos Bióticos – Div. Reg. Tabasco.

Olmsted, I. (1992) Wetlands of Mexico, in *Wetlands of the World*, I, (ed. D.F. Whigham), pp. 636–676. Kluwer Academic Publ., Amsterdam, The Netherlands.

Olmsted, I. and Garcia, R.D. (in press) Distribution and ecology of low freshwater coastal forests of the Yucatan Peninsula, in *Freshwater Coastal Forests*, (ed. A. Laderman). Oxford University Press, Oxford, England.

Osborn, R.G. and Custer, T.W. (1978) *Herons and their allies: atlas of Atlantic coast colonies, 1975 and 1976.* US Fish Wildl. Serv. FWS/OBS-77/08. 211 pp.

Palmer, R.S. (ed.) (1962) *Handbook of North American Birds*, Vol. 1. Yale University Press, New Haven, Connecticut.

Parnell, J.F. (1977) Birds, in *Endangered and Threatened Plants and Animals of North Carolina*, (eds J.E. Cooper, S. Robinson and J. Funderburg), pp. 330–384. N.C. State Museum of Nat. History, Raleigh, North Carolina.

Perkins, J. (1983) *The Belize Barrier Reef Ecosystem: an assessment of its resources, conservation status and management.* New York Zoological Society, New York.

Perkins, J. and Carr, A. III (1985) The Belize barrier reef: status and prospects for conservation management. *Biol. Cons.* 31: 291–301.

Poole, A. (1990) *Colonial waterbirds of coastal Belize: ecology and conservation.* Unpublished report, Manomet Bird Observatory, Manomet, Massachusetts.

Portnoy, J.W. (1977) *Nesting colonies of seabirds and wading birds – coastal Louisiana, Mississippi, and Alabama.* US Fish Wildl. Serv. FWS/OBS-77/07. 126 pp.

Ramsar Convention Bureau (1990) *Directory of Wetlands of International Importance. Sites Designated for the List of Wetlands of International Importance.* World Conservation Monitoring Centre, Cambridge, England.

Robertson, W. and Kushlan, J. (1974) The southern Florida avifauna, in *Environments of South Florida: Present and Past*, (ed. P. Gleason), pp. 414–452. Miami Geol. Society Memoir 2.

Rollet, B. (1974) *Ecológia y reforestación de los manglares de México.* FAO Programa de Investigación y Fomento Pesquero, México. FI: SF/Mex 15. Informe Técnico, SRH, CETENAL, Mexico.

Runde, D., Gove, J., Hovis, J. *et al.* (1991) *Florida atlas of breeding sites for herons and their allies: update 1986–89.* Florida Game and Freshwater Fish Comm., Nongame Wildl. Program Tech. Rep. 10. 147 pp.

Rusch, D., Ankney, C.D., Boyd, H. *et al.* (1989) Population ecology and harvest of the American Black Duck: a review. *Wildl. Soc. Bull.* 17:379–406.

Sauer, J. and Droege, S. (1990) Wood duck population trends from the North American Breeding Bird Survey, in *Proc. 1988 North Am. Wood Duck Symp.*, (eds L. Fredrickson, G. Burger, S. Havera *et al.*), pp. 225–231. St. Louis, Missouri.

Saunders, G. and Saunders, D.C. (1981) *Waterfowl and their wintering grounds in Mexico, 1937–64.* US Fish and Wildl. Serv. Resour. Publ. 138.

Savard, J.-P. (1991) Birds of the Fraser Basin in sustainable development, in *Perspectives on Sustainable Development in Water Management: towards agreement in the Fraser River Basin*, Vol. 1, (ed. A.H.J. Dorcey), pp. 189–215. Westwater Research Centre, University of British Columbia, Vancouver, BC, Canada.

Scott, D.A. and Carbonell, M. (compilers) (1986) *A Directory of Neotropical Wetlands.* Intern. Union Conserv. Nature, Cambridge and Intern. Waterfowl and Wetlands Research Bureau. Slimbridge, England.

Seeliger, U. (ed.) (1992) *Coastal Plant Communities of Latin America.* Academic Press, New York.

Scharf, W.C. (1978) *Colonial birds nesting on man-made and natural sites in the US Great Lakes.* US Army Eng. Waterways Exp. Stn., Vicksburg, Miss. Tech. Rep. 0-78-10, and US Fish Wildl. Serv. FWS/OBS-78/15. 136pp. + appendices.

Sharitz, R. and Gibbons, J.W. (eds) (1989) *Freshwater wetlands and wildlife.* US Dept. of Energy Symposium Series No. 61, Oak Ridge, Tennessee.

Slud, P. (1980) *The birds of Hacienda Palo Verde, Guanacaste, Costa Rica.* Smithsonian Contrib. Zool. 292, Smithsonian Inst. Press, Washington, DC.

Smith, G.W. (1995) *A critical review of the aerial and ground surveys of breeding waterfowl in North America.* Natl Biol. Serv. Biol. Sci. Rep. 5. 252 pp.

Snell, E.A. (1987) *Wetland distribution and conversion in southern Ontario.* Land Directorate, Environment Canada. Working paper no. 48, Ottawa, Ontario.

Southworth, C. (1985) Application of remote-sensing data, eastern Yucatan. In *Geology and Hydrogeology of the Yucatan and Quaternary Geology of Northeastern Yucatan Peninsula*, (eds W.G. Ward, A.G. Weidie and W. Back). New Orleans Geol. Soc., Univ. New Orleans, New Orleans, Louisiana.

Sowls, A.L., DeGrange, A.R., Nelson, J.W. and Lester, G.S. (1980) *Catalog of California seabird colonies.* US Fish Wildl. Serv. FWS/OBS-80/37. 371 pp.

Spencer, D. (1984) Beyond the 48, in *Flyways: Pioneering Waterfowl Management in North America*, (eds A. Hawkins, R. Hanson, H. Nelson and H. Reeves), pp. 224–231. US Fish and Wildlife Service, Washington, DC.

Spendelow, J. and Patton, S. (1988) *National atlas of coastal waterbird colonies in the contiguous United States: 1976–82.* US Fish Wildl. Serv. Biol. Rep. 88(5). 326 pp.

Sprunt, A., IV and Knoder, C. (1980) Populations of wading birds and other colonial nesting species on the Gulf and Caribbean coasts of Mexico, in *The Birds of Mexico: their ecology and conservation*, (eds P. Schaeffer and S. Ehlers) pp. 3–16. Natl. Audubon Society, Western Education Center, Tiburon, California.

Sprunt, A. IV, Ogden, J.C. and Wickler, S. (eds) (1978) *Wading Birds*. Natl. Audubon Society Res. Rep. No. 7.

Stewart, R.E., Jr, Krapu, G., Conant, B. *et al.* (1988) Workshop summary: habitat loss and its effect on waterfowl, in *Waterfowl in Winter* (ed. M. Weller), pp. 613–617. Univ. of Minnesota Press, Minneapolis.

Stickel, L. (1973) Pesticide residues in birds and mammals, in *Environmental Pollution by Pesticides*, (ed. C.A. Edwards), pp. 254–308. Plenum Press, New York.

Thompson, D. (1979) Declines in populations of Great Blue Herons and Great Egrets in five midwestern states. *Proc. 1978 Conf. Colonial Waterbird Group* 2:114–127.

Thompson, D. and Landin, M. (1978) *An aerial survey of waterbird colonies along the Upper Mississippi River and their relationship to dredged material deposits*. US Army Corps of Engineers, Waterways Exper. Station, Tech. Rep. D-78-13.

Tiner, R.W., Jr (1984) Wetlands of the United States: current states and recent trends. US Fish and Wildl. Serv. 59 pp.

Tiner, R.W., Jr (1989). Recent changes in the extent of palustrine vegetated wetlands in the Chesapeake Bay region, in *Freshwater Wetlands and Wildlife*, (eds R. Sharitz and J.W. Gibbons), pp. 791–800. US Dept Energy Symposium Series 61, Oak Ridge, Tennessee.

Titus, J., Park, R., Leatherman, S. *et al.* (1991) Greenhouse effect and sea level rise: potential loss of land and the cost of holding back the sea. *Coastal Management* 19:171–204.

Turner, B., Hochbaum, G., Caswell, F.D. and Nieman, D.J. (1987) Agricultural impacts on wetland habitats on the Canadian prairies, 1981–1985. *Trans. No. Amer. Wildl. Nat. Resour. Conf.* 52:206–215.

Turner, R. and Craig, N. (1980) Recent aerial changes in Louisiana's forested wetland habitat. *Proc. Louisiana Acad. Sci.* 43:61–68.

UNESCO (1978) *Tropical forest ecosystems, a state of knowledge report*. Nat. Resour. Res. 14. UNESCO/UNED/FAO Rome, 683 pp.

USFWS (US Fish and Wildlife Service) (1985) Endangered and threatened wildlife and plants; removal of the Brown Pelican in the southeastern United States from the list of Endangered and Threatened wildlife. *Federal Register* 50:4938–4945.

USFWS (US Fish and Wildlife Service) (1986) *1986 Annual Report: Mid-continent Waterfowl Management Project*. US Fish and Wildlife Service, Twin Cities, Minnesota.

USDI (US Fish and Wildlife Service and Environment Canada) (1986) *North American Waterfowl Management Plan: a strategy for cooperation*. US Fish and Wildlife Service, Washington, DC, USA, and Canadian Wildlife Service, Ottawa, Ontario, Canada.

Weller, M.W. (ed.) (1988) *Waterfowl in Winter*. University of Minnesota Press, Minneapolis.

Weyer, D. (1982) Half moon caye: Central America's first marine park. *Parks* 7:5–7.

Zisman, S. (1989) *The directory of protected areas and sites of nature conservation interest in Belize*. Univ. Edinburgh Dep. Geogr. Occas. Publ. 10. 110 pp.

4

Modern forestry and the capercaillie

Kjell Sjöberg

4.1 THE CAPERCAILLIE – A STATIC BIRD IN A CHANGING WORLD?

The large Eurasian woodland grouse, capercaillie *Tetrao urogallus* (Plate B), is a resident species with a breeding range covering a large area within the boreal coniferous zone of the palearctic region. This distribution coincides quite well with the distribution of Scots pine *Pinus sylvestris* (Seiskari, 1962), as well as with the eastern limit of bilberry *Vaccinium myrtillus* (Klaus *et al.*, 1989), two important food species for the capercaillie during winter and summer, respectively. The species' main distribution extends from Scotland in the northwest (where capercaillie have been reintroduced), through Fennoscandia, eastward to northern Siberia to the basin of the lower Kureika River (following approximately the July isotherm of 53°F in the north and 70°F in the south). In Central Europe there are scattered populations in the Pyrenees, the Jura, the Carpathians, the Cantabrian Mountains and parts of the Alps (Voous, 1960; Cramp *et al.*, 1983; Johnsgard, 1983; Klaus *et al.*, 1989) (Figure 4.1).

This chapter reviews published literature about capercaillie (mainly in relation to forestry effects), supplemented with original data about capercaillie chick ecology, particularly in relation to habitat quality and changes in habitat quality as result of forestry – i.e. how they may respond to a changing forest landscape.

During winter when snow covers the ground, the main food is needles from the Scots pine (e.g. Seiskari, 1962; Pulliainen, 1979). Other coniferous species are sometimes used in the southern distribution range, for example in the Cantabrian Mountains the evergreen holly *Ilex aquifolium* is consumed during the winter (Castroviejo, 1975). During spring and summer, bilberry (shoots, leaves, flowers and berries) and a variety of herbs (e.g.

Conservation of Faunal Diversity in Forested Landscapes. Edited by R.M. DeGraaf and R.I. Miller. Published in 1996 by Chapman & Hall. ISBN 0 412 61890 7.

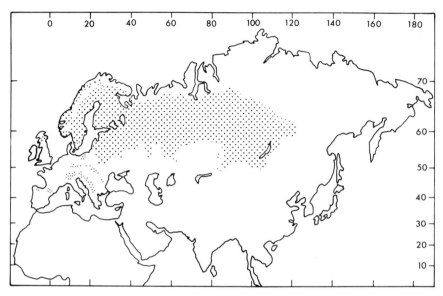

Figure 4.1   Distribution of capercaillie on the Eurasian continent. The capercaillie is not native to the New World. (From Voous, 1960; Crampt *et al.*, 1980; Klaus *et al.*, 1989).

Melampyrum spp.), mosses, *Equisetum*, seeds of *Carex* spp. and other plant material constitute the main food (e.g. Rajala, 1959; Jacob, 1988; Storch *et al.*, 1991; Storch, 1993). The chicks are dependent on invertebrates during the first weeks of life, and the percentage of insects in the stomachs of chicks at the age of 1–2 weeks has been estimated to be up to 80% on a volume basis (Kastdalen and Wegge, 1985; Spidsø and Stuen, 1988; see also Stuen and Spidsø, 1988). Lepidoptera larvae, e.g. geometrids (often utilizing bilberry leaves and stems as food), constitute an important part of the diet (e.g. 76%, Kastdalen and Wegge, 1985) but invertebrates of all sizes are taken. The chicks gradually change from invertebrates to plant material as they grow older, and at the age of about 3–4 weeks, plant material dominates the diet (e.g. Rajala, 1959; Kastdalen and Wegge, 1985; Spidsø and Stuen, 1988).

The capercaillie is regarded as a bird of mature or old-growth forest (e.g. Seiskari, 1962; Eiberle, 1976), and in the northern distribution range it belongs to the group of sedentary taiga birds heavily influenced by modern forestry activities (e.g. Järvinen and Miettinen, 1988, p. 172; Wegge and Rolstad, 1986; Wegge *et al.*, 1990; Helle *et al.*, 1994; but see Rolstad and Wegge, 1987a and b, 1989a; Winqvist and Ringaby, 1989; Picozzi *et al.*, 1992). Within its main distribution range the capercaillie is associated with coniferous forests with a luxuriant cover of ericaceous shrubs, particularly bilberry, and with a tree canopy closure of about 40% (e.g. Gjerde, 1991; Storch, 1993).

In the northwestern part of the breeding range of the capercaillie there has been intensive forestry activities for decades, and since the 1950s clear-cutting has been the prevailing forestry method (e.g. in Fennoscandia). As an effect of clearcutting, the forests have been heavily fragmented, and the area of old-growth forests has decreased (e.g. Esseen *et al.*, 1992). Consequently, capercaillie habitats have also been affected.

The capercaillie is a complex species in many ways. It has the largest size dimorphism of all grouse (the male often more than twice the weight of the female), the sexes to some extent use different habitats during the year, and they shift food from winter to summer (e.g. Seiskari, 1962; Larsen *et al.*, 1982; Rolstad *et al.*, 1988; Rolstad and Wegge, 1989a and b). From an ecological point of view and in many other respects, one could almost regard the two sexes as different species (Seiskari, 1962; Milonoff and Lindén, 1989a). Differences between the sexes are already evident during the chick stage, for example as regards body components like breast muscles and leg size (Milonoff and Lindén, 1989a, b; Lindén and Milonoff, 1991). The chicks switch from an animal diet to plant material as they grow older. Thus, we should not be surprised that the situation may be quite complex as regards the effect of forestry on capercaillie habitat quality.

4.2 POPULATION LEVELS AND TRENDS

In northern Fennoscandia, voles like *Clethrionomys glareolus* and *Microtus agristis* exhibit population peaks with 3–4-year intervals (e.g. Hörnfeldt, 1978; Christiansen, 1983; Hansson and Henttonen, 1989). Predators living on these microtine rodents (such as the red fox, *Vulpes vulpes*) also display regular fluctuations (e.g. Hagen, 1952; Hörnfeldt *et al.*, 1986; Lindström, 1989). There have been several studies on whether the predators on microtines shift to other prey species such as small game like mountain hare (*Lepus timidus* L.) and capercaillie in years when microtine populations crash, and whether the predators thereby also influence the population levels of these species. Thus, the alternative prey should be limited by predation during the decline in the cycle of the main prey. Support for this alternative prey hypothesis was found by Angelstam *et al.* (1984, 1985), Lindström *et al.*, (1987), Marcström *et al.* (1988) and Lindström *et al.* (1994) – although other hypotheses could not be excluded – with the red fox being the crucial factor in limiting the numbers of hare and grouse in autumn, and in conveying the 3–4-year cyclic fluctuation patterns of voles to small game. However, Lindén (1988, 1989) did not find convincing evidence for relations between capercaillie and the vole cycles, and he also concluded that a 6–7-year cycle was normal for capercaillie in Finland.

Regardless of these short-term fluctuations (see also section 4.3.2), in almost all areas where capercaillie populations have been studied there seems to be a long-term population decline. For an overall review, see Johnsgard (1983); for the Nordic countries see Wegge (1979), Marcström (1979), Lindén and Rajala (1981); and for Central Europe see Klaus (1994).

At present, a population size of about 5–10 birds per square kilometer in the autumn seems to be a good population density. The density in southern Norway measured by Wegge and Rolstad (1986) was five birds per km^2. Data from northern Sweden from four different habitats (young managed pine forest on dry soils to mesic mixed pine/spruce forests, i.e. from bad to good habitats) from 1989 to 1994 measured with a line transect method (three-man team; see Rajala, 1974b), varied from 0 to 10 birds per km^2 during August (Sjöberg, unpublished). Judging from inventories in Finland during a period from 1964 to 1985, there has been a gradual decrease in the relative capercaillie population densities from about 10–15 birds per km^2 in central Finland during the 1960s, to about 5 in the 1980s (Järvinen and Miettinen, 1988, based on data from Lindén).

In Finland, the capercaillie population was calculated to be 600 000 individuals as a mean for the period 1964–71 (estimated from censuses in August). In the same period, the mean annual harvest by hunters was 42 000 birds, i.e. 7% (Rajala, 1979). In Sweden, a total population of possibly 100 000 pairs is estimated (Ulfstrand and Högstedt, 1976). The mean number of capercaillies harvested per year in Sweden was about 30 000 birds during the 1980s (Figure 4.2). If the hunting pressure in Sweden is the same as in Finland (i.e. 7%: Rajala, 1979) the autumn population should be about 430 000 birds. However, it is generally considered that the pressure in reality is somewhat lower. If 5%, for example, then the number would be 500 000 birds (see also Lindén, 1991). The breeding range of the capercaillie covers most of Sweden. At least one-third of the 23.5 million ha of forest-land area in Sweden has stands old enough for felling (Svensson *et al.*, 1989) and in these areas there ought to be suitable habitats for capercaillie. A probably somewhat conservative estimate of density would be about 5–6 birds per km^2 in capercaillie habitats in the autumn.

The Norwegian population was estimated to be 300 000–400 000 in the 1960s with an estimated harvest of about 40 000 birds. Since then the population has declined markedly: the harvest during the 1970s was about 10 000 birds (Wegge, 1979).

Changes in climate and disturbance by human activities (e.g. tourism and hunting) have been suggested as reasons for the decline in capercaillies (e.g. Storch, 1993). However, the main reason for the decrease is generally regarded as habitat changes by forestry, and the possible indirect effects on, for example, predation pressure. This is the case at least within Fennoscandia, with a high proportion of forested land, but there the old-growth forests have to a large extent been replaced by young forest stands (e.g. Helle *et al.*, 1994).

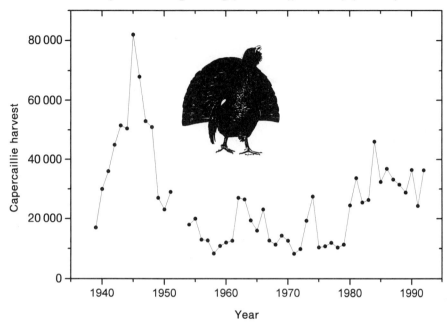

Figure 4.2 Harvest of capercaillie in Sweden during the period 1939–1992 (source: Swedish Hunters Association). Note that the open season has changed slightly during the years, and that the figures during later years also include a period of winter harvest of cocks during January of the following year. (Vignette by Vesa Jussila.)

4.3 POPULATION REGULATING FACTORS AFFECTED BY FORESTRY

4.3.1 Thermoregulation of the chicks and their foraging time – a balance between life and death

Like those of the willow grouse (Aulie and Moen, 1975; Boggs *et al.*, 1977), new-born capercaillie chicks have an insufficient ability to maintain normal body temperature during cold periods. They are sensitive to cold and wet weather during the first part of their life and must depend on the hen for warmth because they have not yet fully developed their thermoregulatory abilities (Höglund, 1955; Marcström, 1956; Hissa *et al.*, 1983). As a consequence, during cold weather the young chicks have to spend quite a lot of time under the hen to get warm, but during a warm day at the same age they can spend more time foraging (Rajala, 1974a). Tests with tame capercaillie chicks in field experiments (Plates C, D and E) have shown that at a temperature of 7°C, chicks about one week old can spend only about 10

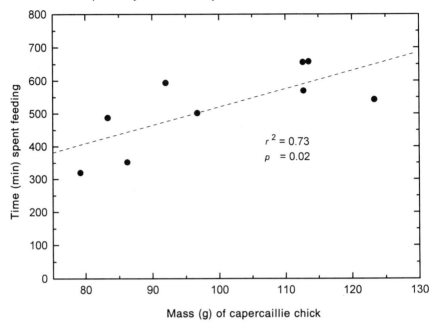

Figure 4.3 Capercaillie chicks are temperature-dependent during the first part of their lives. Total time spent foraging by tame capercaillie chicks of same age but different weights, observed continuously when food-searching from 5 a.m. until 9 p.m. in a moist spruce forest with a developed cover of bilberry. The heavier birds spent more time foraging than the lighter birds.

minutes on food searching before they get cold and need warmth, whereas at 17°C they could spend an entire hour without a break. The actual time seems to depend on the chick's weight (and perhaps condition) to some degree. Among the nine chicks in the test, there was a significant positive correlation between body weight and time spent food searching during a cold day (Figure 4.3). Apparently cold weather prevents the chicks from searching for food, probably causing retarded growth and consequently a higher risk of death (cf. Rajala, 1974a). During cold weather chicks need a lot of additional energy for thermoregulation and so growth may be discontinued (Lindén *et al.*, 1984).

A similar relationship between body size and time spent foraging has been documented for the willow grouse: bigger chicks have longer foraging periods before needing to be warmed (Boggs *et al.*, 1977). The relation between the cold-hardiness of the chick and body size is due to the fact that body temperature to a large extent depends on the size of the breast muscles (Aulie, 1976). The leg and breast muscles are the primary source of heat production even in capercaillie chicks (Hissa *et al.*, 1983). Thus, for the chicks of capercaillie (and other grouse) cold and wet weather causes a rapid drop in body temperature and forces them back to their mother to be warmed (Rajala, 1974a; Pedersen and Steen, 1979; Lindén *et al.*, 1984). The breast muscle of young males is proportionally smaller than in females, and Milonoff and Lindén (1989a,b) suppose that the difference is at least partly responsible for the higher mortality rate of male chicks (cf. Wegge, 1980).

The importance of being able to search for food is illustrated in Figure 4.4, where the imprinted capercaillie chicks used in Figure 4.3 searched for food from 5 a.m. to 9 p.m. for four consecutive days in quite different weather conditions. On 26 June the weather was cold (12–14°C) and rainy. The chicks became cold and frequently needed to be warmed. Two of the chicks became too exhausted to continue the following day (one of them died); the remaining seven birds needed supplementary food during the night to compensate for the weight losses during this day, which is why in Figure 4.4 there is an increase of the mean weight on the next morning (and why the sample size drops from 9 to 7). The next day, in the same habitat

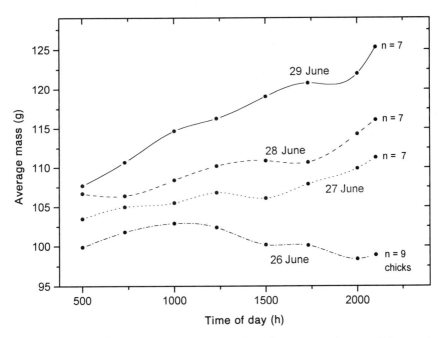

Figure 4.4 Food-searching efficiency varies depending on weather conditions and quality of forest habitats (e.g. amount of insect food available). Results from four days of different weather conditions and different forest habitats: 26 June, rainy and cool (12–14°C); 27 June, same habitat but dry weather, with temperature considerably higher (13–21°C); 28 and 29 June, similar warm weather (12–24 and 13–24°C, respectively), but bad habitat (as regards insects) for 28 June and good insect habitat for 29 June (cubic spline interpolation).

(mature spruce/pine forest rich in bilberry vegetation in the field layer) and with the same chicks, the weather was dry with a temperature of 13–21°C. On the cold and rainy day the chicks had to be warmed for an average of 7 hours and 7 minutes; the following dry and warm day they needed just 1 hour and 11 minutes for warming.

On 28 June, in a habitat with younger pine stands, the weather was sunny and the temperature varied from 12 to 24°C. The next day, the temperature varied from 13 to 24°C, but here the habitat was a mature mixed spruce/pine forest rich in bilberry (i.e. a presumed optimal chick habitat during the time when insects dominate the diet of the chicks). The chicks needed less than an hour for warming (44 and 23 minutes, respectively) early in the morning.

Again, a comparison can be made with published data on willow grouse chicks, where foraging periods were shorter and the brooding longer during poorer weather. Furthermore, foraging periods became longer and brooding periods shorter as the chicks grew older (Boggs *et al.*, 1977; see also Rajala, 1974a; Aulie and Moen, 1975; Erikstad and Spidsø, 1982; Erikstad and Andersen, 1983).

4.3.2   Predation

There is heavy predation pressure on the capercaillie, and it starts at the egg stage. In a study area in southern Norway, Storaas and Wegge (1985) documented an average egg loss of 80% among radio-equipped females during a 5-year period ($n = 35$). Of the hatched chicks, about 50% are killed by predation during the first month (Rolstad *et al.*, 1991). In Finland, Lindén (1981) estimated the nest loss at 34%. Storaas and Wegge (1987) found that nest losses were unrelated to habitat type, forest density, or nest cover, suggesting that predation on nests is similarly high in many habitats.

In a predation study in northern Sweden (Marcström *et al.*, 1988), where red fox (*Vulpes vulpes* L.) and marten (*Martes martes* L.) were removed during a 5-year period, the number of boreal tetraonids increased significantly in the treated area compared with the reference area. The tetraonid brood sizes averaged 5.52 in August and 77% of the hens had chicks, compared with 3.29 and 59%, respectively, in the reference area.

Forestry activities may be expected to influence biological patterns, structures and processes in the forests. As an example of biological processes, predator–prey relationships could be mentioned. The fragmented forest landscape with a high proportion with grass-rich clearcut areas will be able to support a generally higher, but unstable, population of small mammals (Christiansen, 1979; Hansson, 1979; Henttonen, 1989). As a consequence, predators like the weasel and the red fox will increase too, putting a high predation pressure on capercaillie eggs and chicks in years with low small mammal abundance. During such years the local predation losses of capercaillie nests can exceed 90% (Wegge and Storaas, 1990).

4.4 A SMALL-SCALE TO LARGE-SCALE VIEW OF FORESTRY

Forestry influences the conditions for tetraonids in many ways: tree species composition, ground and field layer composition, forest fragmentation, ecological patterns and processes like predator/prey relationships, and nitrogen deposition by air pollution. All influence habitat quality and predation pressure but they may influence the sexes and ages differently.

The capercaillie is a territorial bird during much of the year and it spends most of its time within a limited part of the forest landscape. The adult males are restricted to a part of the lekking site (the cock's communal display ground) during spring, and the lekking sites are often in the same part of the forest from year to year (e.g. Hjort, 1970). Even the hens are dispersed in a territory-like way during spring (Wegge, 1985; Rolstad *et al.*, 1991) but apparently not the hen with brood (Wegge *et al.*, 1992). When undisturbed and in a suitable habitat, the hen and brood are quite stationary on a small area of 10–70 ha (Winqvist, 1988; Sjöberg, unpublished). However, other studies indicate large home ranges of up to 1000 ha after hatching, due to extensive movements during the first 4–6 weeks (Wegge *et al.*, 1982; Wegge *et al.*, 1992) (cf. Sonerud, 1985). During the summer the males move up to 10 km away from the lek to more luxuriant summer habitats, and back again in the autumn. The seasonal movements among the females is more complicated and includes winter habitats which could be 7–8 km away from the breeding and lek area (Rolstad *et al.*, 1991). The young birds seem to move more; for example, cocks aged one and two years move between lekking places before they get established. In good habitats the lekking sites are distributed about 2 km apart (Rolstad *et al.*, 1991; Hjort, 1985, 1994).

Thus, the adult capercaillie or a brood sometimes uses an area no larger than a stand, while at other times of the year they use an area of landscape size. Consequently, a discussion of the ecology of the capercaillie at different scales seems justified: the stand level (hectare size); the landscape level (km^2 size); or even a larger scale – a regional level.

4.5 FORESTRY INFLUENCE AT THE STAND LEVEL

Apparently there is some habitat choice that is specific for the brood, because telemetry studies have shown that females with broods are often present in spruce swamp forests or wet spruce forests, while hens without broods appear more often in younger, dense stands of conifers (Rodem *et al.*, 1984; Rolstad *et al.*, 1988). Forestry activities can cause changes in the quality of the chick habitat regarding both food and susceptibility to predators, and thereby probably also indirectly influence the survival rate.

The abundance of food for chicks can be influenced by forestry both at the early stage (when the chicks feed on an insect diet) and at the later stage (when the chicks feed on a plant diet). Draining swamp forests or wet spruce forests probably decreases the abundance of invertebrates used as food by the chicks. When tame capercaillie chicks searched for food during otherwise similar conditions in wet forests with and without ditches, the growth rate of the chicks (measured as the difference in weight before and after a food searching period) was significantly higher in the habitats without ditches (Sjöberg, unpublished data). Ditching wet spruce forests will increase tree growth and the increased canopy coverage will decrease the light reaching the ground. The result will be a reduced field layer of bilberry and other plants (Plate F). Consequently, the number of invertebrates living in the field layer will decline and the food and growth rate of the capercaillie chicks will be reduced. The same situation (a less favourable field and/or ground layer) is found when comparing young managed plantations with mature forests

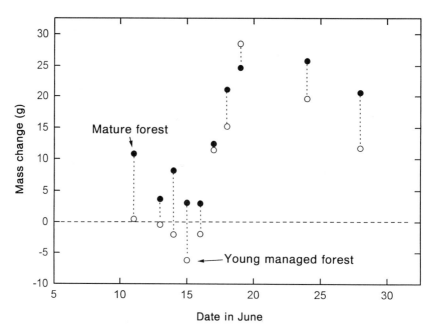

Figure 4.5   The quality of different forest habitats influences the food-searching effi-ciency of capercaillie chicks. Comparisons between habitats with rich and poor bil-berry field layer, respectively: ● = mature forest (mesic mixed spruce/pine), ○ = young managed forest (pine plantations). Difference in foraging efficiency measured as change in chick weights during 2.5 h foraging periods. Repeated paired tests in different forest stands ($P < 0.05$; Wilcoxon signed rank test).

rich in bilberry plants; with tame and imprinted capercaillie chicks the result was the same, i.e. a lower foraging efficiency in the young managed plantations (Figure 4.5). In such habitats with young conifers, the bilberry are to some extent replaced by the grass *Deschampsia flexuosa*.

It is not only forestry which changes the habitat for capercaillie. For example, air pollution may have an effect on the forest as well: airborne nitrogen may alter the competitive balance between the bilberry (which is so valuable to the capercaillie) and the less valuable grass *Deschampsia flexuosa* (Högbom and Högberg, 1991; Rosén *et al.*, 1992; cf. Klaus, 1991; Storch, 1993). Again, the result will be that the substrate for invertebrates living on field-layer plants will decline and reduce the insect-rich foraging areas for capercaillie chicks. Indirectly, there could also be higher predation on eggs and chicks as a result of this less dense field layer.

In forestry management, understory clearing will also influence the chick habitat negatively, and consequently also the capercaillie population. Such clearing will reduce the cover for the broods and thereby increase the predation risk for the broods (Brittas, 1989; cf. Marcström *et al.*, 1982) and probably also for the females. Cover from predators is probably of importance even for the females themselves, because females without broods seem to utilize young forests or plantations in higher proportion than expected (Rolstad *et al.* 1988, 1991). Probably it is a good strategy for the female to be in a habitat with much cover, to avoid predation (Rolstad *et al.*, 1988), but it is also a risk for these heavy birds to maneuver in a dense forest (personal observation of female capercaillie trapped and killed in a forked tree branch, and there is certainly an even higher risk for the heavier males).

4.6   THE INFLUENCE OF FORESTRY AT THE LANDSCAPE AND REGIONAL LEVELS

4.6.1   The natural landscape

In the natural, original boreal forest landscape, fire and wind were dominant large-scale disturbance factors, while for example uprooting of individual trees by wind created small-scale disturbance patterns with gap regeneration in stands or areas with less frequent disturbance by fire (e.g. Zackrisson, 1977; Hytteborn *et al.*, 1987; Kuuluvainen, 1994). The result was a forest with successional stages ranging from open scattered stands with coniferous and broadleaved trees to old-growth forests with uneven, multilayered forest canopies. In this forest landscape, fire refugia with patches of old forests with a long forest continuity and small-scale disturbances were scattered (e.g. Esseen *et al.*, 1992). In such mosaic forests with a mix of habitats with different qualities, the relatively sedentary capercaillie was certainly able to find suitable habitats all times of the year and stages of its life. However, in the present managed coniferous forests, both the start and end of this succession are truncated. In the beginning of the succession, seedlings and bushes of broadleaved trees such as aspen and birch, formerly dominating after a forest fire, are normally removed, and at the end of the succession forests are logged before they reach the old-growth stage. In the stages between, the forest is heavily influenced by changes in tree species composition, converted to one-species, even-aged and single-layer forest stands that are drained, thinned and quite young. It is to such an altered forest landscape that the capercaillie has had to adjust. How well has it succeeded?

In a line transect census of tetraonids covering 21 paired sites with natural and managed forests along about 20-km census lines distributed within each site of *c.* 2000 ha, the relative mean number of observations of capercaillie per 10 km was 4.6 in the natural forests but only 1.6 in the managed forests ($P < 0.05$) (Sjöberg, unpubl. data). This suggests that there are denser capercaillie populations in the natural forests, which is in accordance with the general impression of the capercaillie (especially the cocks) as a bird associated with a landscape mosaic dominated by mature forests (e.g. Marcström, 1977; Rolstad and Wegge, 1989a,b). One reason for the lower densities in the managed forests than in natural forests may be that a proportion of the land area in the managed forests consists of clearcuts and young conifer plantations, which are habitats that the capercaillie do not normally use.

4.6.2   The fragmented landscape

The present forest landscape over large areas of Fennoscandia is heavily fragmented by clearcut areas where conifers are planted after surface sclarification – mainly the native Scots pine, *Pinus sylvestris*, and Norway spruce, *Picea abies*, but also to some extent the introduced North American lodgepole pine, *Pinus contorta*, now covering more than 500 000 ha in Sweden (Skogsstyrelsen, 1992). Plate G shows a forest landscape in Central Sweden with originally mature forests, with Norway spruce as the dominant conifer species, but now succeeded by plantations of *Pinus sylvestris* and *P. contorta*. During a 10-year period the positions of capercaillie in this heavily fragmented area were found by using trained dogs. The distribution of all observations of capercaillie (broods, hens without broods, cocks) is plotted in Figure 4.6. All the observations were made in the remaining mature forests, which gradually decreased in acreage as new parts of the forested landscape were clearcut, or in the remaining forest patches left in the landscape (e.g. wet spruce patches and edges along mires). Hopefully such patches can serve as source areas (cf. Pulliam, 1988) for capercaillie and other birds when the conifers on the previously cut areas have grown older (at present about 17 years).

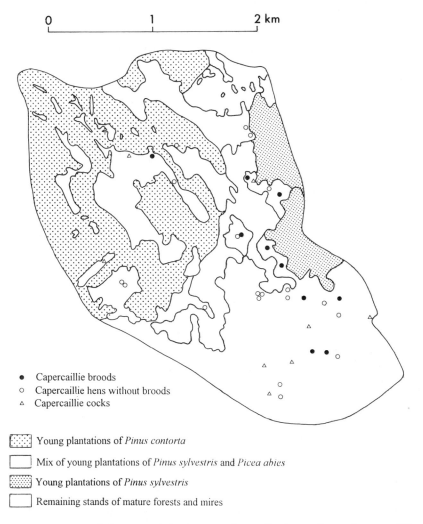

● Capercaillie broods
○ Capercaillie hens without broods
△ Capercaillie cocks

[:::] Young plantations of *Pinus contorta*

[ ] Mix of young plantations of *Pinus sylvestris* and *Picea abies*

[:::] Young plantations of *Pinus sylvestris*

[ ] Remaining stands of mature forests and mires

Figure 4.6 Map of fragmented landscape where all observations of capercaillie during the period 1986–1994 are plotted. The area was surveyed with pointing dogs for two days during early August each year. All capercaillie observations (except one close to the border) were made within old, mature stands or within patches of swamp spruce forests left along mire edges (see Plate H).

Capercaillie broods do not normally use clearcut areas or young planta-tions (Børset and Krafft, 1973) (although the hen often places her nest there). Adult cocks normally begin to use plantations when the trees are 60 years old; females do so a little earlier (Wegge *et al.*, 1992). However, Winqvist

and Ringaby (1989) describe a landscape with a dense capercaillie population, where 30-year-old plantations dominated after heavy storm felling of
the older forest. Patches of the original mature forest were present. Due to a
flat topography, the proportion of unditched wet areas was high. In such a
landscape, despite the presence of plantations, a mosaic of insect-rich wet
habitats suitable for broods was available on lower elevations while on drier
land the young stands provided the females with good protection from predators. Patches with mature, intact forest were suitable for the lekking cocks.
An interpretation of the inventory results from this landscape (deviating from
the results and conditions in normal conifer plantations) is that the
undrained, wet areas are of great importance even in the managed landscape
dominated by plantations of conifers. Such wet areas have to date normally
been lacking or heavily reduced by ditching for commercial forestry. Thus,
even this apparent exception from the rule supports the importance of wet
insect-rich habitats to capercaillie in a forest landscape rather than supporting the idea that the capercaillie can easily adapt to young plantations.

Large areas of Swedish forests were formerly cut by the single tree selection method, which created an open, multilayer forest with an uneven age
distribution. These old habitats seem to have been quite suitable for all the
forest grouse species (e.g. Marcström, 1977). One reason for the extensive
use of the clearcut method since the 1950s is that denser forests with higher
volume of wood could be produced by clearcutting and subsequent
planting. In the study area discussed above, such forest types were
compared as regards abundance of capercaillie with the situation in a
nearby old mixed spruce/pine forest and a large-scale plantation of lodgepole pine and Scots pine. The results are shown in Table 4.1. As expected,
the number of adult cocks was highest in the old-growth mixed spruce/pine
forest, least in the area with large plantations after clearcutting and intermediate in the selectively cut area (cf. Swenson and Angelstam, 1993). The
number of broods was about equal in the old-growth area and open single-
tree selection area, with fewer in the plantation area. Furthermore, those
fewer adults and broods found there were actually seen in areas where parts
of the old forests were left or not yet cut, and along mire edges and swamp
spruce forest patches (Plate H).

Extrapolating to Sweden as a whole, the clearcut forestry method has
created large areas of forested land, which are in an early forest successional stage and largely unsuitable for capercaillie. The question is: will
there be a balance in the future when the capercaillie will again use these
now unsuitable habitats as they mature?

4.6.3 Effects of fragmentation of old forests on capercaillie

It seems to be the cocks who are most strongly associated with old-growth
forests. It is there that they use the traditional lekking places year after year

Table 4.1 Total number of capercaillie broods and adult cocks in three different forest areas measured with help of pointing dogs (each area searched for two days) in August, 1986–1995 (drawings by Martin Holmer)

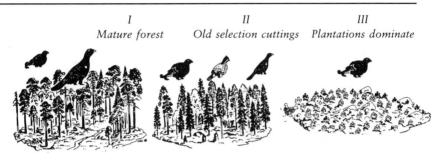

	I Mature forest		*II* Old selection cuttings		*III* Plantations dominate	
	Broods	*Cocks*	*Broods*	*Cocks*	*Broods*	*Cocks*
1986	4	11	1	1	2	1
1987	0	7	6	3	3	2
1988	2	4	3	3	0	2
1989	2	7	2	0	1	0
1990	4	10	1	3	0	0
1991	0	5	0	4	0	1
1992	4	8	3	4	1	0
1993	0	5	1	4	1	0
1994	1	7	2	4	3	1
1995	1	6	2	1	0	2
Total	18	70	21	27	11	9

Area I consists of *Pinus sylvestris* on drier conditions and swamp spruce forests (*Picea abies*) in moist depressions with mixed stands in the areas between, intermingled with aspen (*Populus tremula*) and birch (*Betula* spp.). Almost 50% of the area consists of mires. A forest fire covered the area in the 1860s. In the drier fire-influenced areas the pine forests are 115–130 years old, with single trees surviving the fires 200–350 years old. In the wetter depressions in the landscape, the spruce forest survived the fires. The multilayer stands there are about 200 years old.

Area II consists of mixed, uneven forests with a previous history of single tree selection forestry.

Area III is a heavily fragmented landscape dominated by plantations of *Pinus sylvestris* and *P. contorta*, ca. 17 years old.

(Hjort, 1970; Rolstad and Wegge, 1987a,b). In Finnish studies, the proportion of old forests decreases with distance from the center of the lekking place (Lindén and Pasanen, 1987). When such forests are reduced in size by clearcutting, the effect will be that the birds can adjust to the new situation, but only to a certain degree, and then either move to nearby sections of the forest or enlarge the areas used by individual cocks (Rolstad, 1989; Rolstad and Wegge, 1989b). In the latter case the number of birds on the lek is of course reduced.

Most of the year the capercaillie is related to old or mature forests. The lek is normally found in an older forest (Hjort, 1970; Winqvist, 1983; Rolstad and Wegge, 1987a,b; Helle *et al.*, 1994) and the females with broods often utilize mature wet spruce forests (Rodem *et al.*, 1984) (Plate I). During the rest of the year a wider range of habitats is used (Rolstad *et al.*, 1991) (Plate J). Consequently, it ought to be the lekking cocks and the broods that are most sensitive to forest fragmentation – the cocks by destruction or damage of lekking places, which makes the birds use larger areas, and the broods by increased predation pressure and by decreased habitat quality with subsequent higher risk of death, especially for the male chicks (e.g. Wegge, 1980; Lindén and Milonoff, 1991). In a heavily fragmented area in southern Norway, both broods and subadult males died at a higher rate outside the unfragmented, old forest blocks, due to more favourable conditions for generalist predators in fragmented habitats (Gjerde and Wegge, 1989; Wegge *et al.*, 1990).

Rolstad and Wegge (1987b, 1989b) have developed a model in which they calculate the effects of large- and small-scale fragmentation on capercaillie populations. When the area of each clearcut is small and thus the grain structure of the fragmented landscape is fine, the lekking cocks seem to accept the disturbance on the lekking place. On the other hand, if the clearcut area is large, there will be a coarse-grained fragmentation of the landscape, which makes the cocks move. If the proportion of old forest within the landscape falls below about 30%, the leks are abandoned (Wegge and Rolstad, 1986). At a certain level the population will no longer be self-sustaining; sinks will appear and depend on donor populations in neighbouring areas (Wegge *et al.*, 1992). At longer distances the populations will become isolated, as seems to be the case for many of the capercaillie populations living in Central Europe (Rolstad, 1991; Klaus, 1994).

Predictions can also be made about the effects of forestry on females with broods. They are to a certain extent territorial, but the territory size is smaller than the space needed for a group of cocks (there is, however, contradictory information on the size of the brood raising areas – these need to be estimated better) and they seem to place the nest without relation to the lek area (Hjort, 1994). The brood normally avoids clearcut areas and young plantations (e.g. Figure 4.6) and prefers wet old spruce forests (Plate I), although the nests can be placed in almost every type of forest habitat. The brood is not tied to the place of hatching and can move considerable distances (at least several hundred meters) to good habitats. In younger forests or managed forests with a lower abundance of insect food and/or less cover from predators, the number of fledged birds decreases, especially the number of cocks (Wegge, 1980; Lindén and Milonoff, 1991).

To summarize, females with broods are in theory less sensitive to large-scale landscape fragmentation than cocks. Broods will probably be able to utilize a fragmented landscape, if landscape elements like wet forest patches, forested edges along mires, etc. are available. Thus, both females and females with broods will probably also be able to use the plantations at an earlier stage than the cocks will. It is probably also easier to improve brood habitats than lekking habitats (for example, by not clearing underbrush before clear cutting). However, many of those managed habitats are of lower quality for brood rearing than the natural mosaic landscape in which the capercaillie have evolved.

The age structure of individual forest stands will influence the capercaillie population on a landscape level, as the different sexes and the broods differentially utilize forests stands of different age. For example, because the females start to utilize the young forests at an earlier stage than the males, there will be proportionally larger areas available for females in the future. The landscape arrangement will probably also influence the population structure.

4.7 THE FUTURE FOREST LANDSCAPE

Apparently, changes in the forest area, composition and structure will affect the conditions for capercaillie in the future. For example, what about the proportion of clearcut areas and old-growth forests? In Sweden, according to the National Forest Survey (Svensson *et al.*, 1989; but see Linder and Östlund, 1992, for a longer time perspective), the land area classified as forest land has been almost the same during the survey periods 1923/29 and 1978/82, i.e. 55.5% and 57.1% (23.5 million ha). The area of forests older than 100 years has not changed significantly since the 1920s (Svensson *et al.*, 1989) but the very old age class (> 140 years) has decreased considerably. Compared with the 1950s (the time period when clearcutting began to be the dominant forestry method), the area of stands aged between 3 and 20 years is three times greater now, while the area of the stands between 21 and 60 years of age has decreased gradually. Another trend is the decreasing percentage of mixed forest, whether mixtures of conifers or mixtures of conifers with broadleaved trees. Also, the density of the medium-aged and older forest has increased, leading to a growing stock (volume of standing trees per ha) on forestland at least since the 1920s.

According to Wegge *et al.* (1992), the adult cocks begin to use plantations during spring when the plantations reach an age of about 60 years, and the females somewhat earlier. That means that, in the Norwegian example, the capercaillie can use about 40% of the forests if the percentages of the two oldest age classes are combined. If the actual or ideal age distribution of the forests (from a productivity point of view) will be more even than now in the future Swedish forests, there will be a smaller area suitable for capercaillie. As mentioned above, there is a trend towards increasing area of younger forests at the cost of forests of middle and old age. For example, in northern Sweden the area of forests in the age class 81–120 years has decreased by 30% between 1975 and 1985 (Kempe *et al.*, 1992).

However, from theoretical estimates based on spacing behavior, Wegge *et al.* (1992) calculate that when such a balanced age structure of the forests has appeared, the spring density of males will be reduced to about 40% of its original carrying capacity (compared with the situation when there were 100% old forests around the leks). In fact, Wegge *et al.* (1992) suggest that the reduction could be even greater, as increased predation pressure may prevent the male population from reaching the carrying capacity even when the cover is 40% old secondary forests.

Many of these ongoing changes in the forests may also alter the quality of the capercaillie habitats, which might further depress this calculated value. The effect on bilberry and other quality aspects mentioned above will influence the survival of subadult birds, as well as hens, and not least the chicks. For example, Kardell (1980) has found that the key species, bilberry (cf. Rolstad, 1988; Storch, 1993), does not fully recover in managed forests until after about 80 years. Thus, over a managed stand's total life span of 100–120 years before it is cut again, it will only be reasonably suitable for capercaillie for about 20–40 years.

An increased tree density in the forests will probably also change the balance between bilberry and grasses, with lower-quality habitat for chicks as a result (Figure 4.4). However, it may be of some advantage for females, which seem to prefer dense young stands (probably for shelter from predators).

There are many possibilities for foresters to improve the situation for the capercaillie in the future compared with the last few decades of heavily production-orientated forestry. For example, foresters could avoid or reduce the clearance of underbrush before harvest in order to give the birds shelter from predators; they could reduce or halt the draining of wet forests; they could save patches of swamp forests, to provide the chicks with insect-rich habitats; and they could modify the clearcuts near leks in order to support continuity of social organization and structure of the leks. Many of these considerations are undertaken to some extent under present forestry practices in Sweden, where production and environmental aspects should now be given equal weight in forest management (e.g. Sjöberg and Lennartsson, 1995).

Thus, in the changing world of the capercaillie that is now primarily shaped by modern forestry, the capercaillie is not as static as one might think but it can only adjust to the many changes in the forest landscape to a certain degree. It will certainly be an inhabitant of the boreal forest as long as there is a substantial fraction of mature stands (i.e. as long as sustainable forestry exists). However, because of the direct and indirect effects of forestry, as well as other influences like air pollution, the capercaillie will certainly be found in lower densities in future forests. To what extent the populations will be reduced is much influenced by the foresters themselves.

4.8 SUMMARY

The large woodland grouse, capercaillie (*Tetrao urogallus*), is generally regarded as a species which relies on mature or old-growth forests (especially in the case of the young chicks, and adults except during summer). However, it is a complex species in many ways. There is a large size dimorphism between sexes; the sexes to some extent use different habitats during the year. They also alter their diets from winter to summer. The chicks shift from animal to plant food during their growth period, and during the same period they also change habitat preferences. The quite substantial literature about this important game species clearly indicates that modern forestry influences the capercaillie both directly and indirectly. Direct influence by fragmentation, reduction of mature forest area, etc. influences the lek (with its complex social structure and usually situated on traditional sites). The number of cocks taking part in the lek at each site decreases. Simplification of the habitat structure may directly influence predation pressure, e.g. by reduction of cover in the field layer for the hen with brood.

Indirectly, forestry may influence the capercaillie both by changing the predation pressure and by changing the quality of the habitats and thereby the survival of, for example, the chicks. By reduction of the bilberry field layer, the biomass of the important food items such as lepidoptera larvae may be reduced, which can influence the survival of chicks, especially during periods with wet and cold weather. By forestry's influence on the biomass and distribution of the small mammals (the main food resource for many predatory species in the coniferous forest), the predation pressure may increase by changes in number of predators as well as by possible change in their prey species.

4.9 ACKNOWLEDGEMENTS

This study was financially supported by the Swedish Environmental Protection Agency. I thank Erik Ringaby for invaluable help in the field and for handling the capercaillie chicks used in the foraging experiments. Gunnel Brännström and Marika Hedlund drew the maps, while Vesa Jussila and Martin Holmer made the vignette drawings. Harto Lindén and John P. Ball improved the manuscript by their constructive criticism.

130   *Modern forestry and the capercaillie*

4.10 REFERENCES

Angelstam, P., Lindström, E. and Widén, P. (1984) Role of predation in short-term population fluctuations of some birds and mammals in Fennoscandia. *Oecologia* 62:199–208.

Angelstam, P., Lindström, E. and Widén, P. (1985) Synchronous short-term population fluctuations of some birds and mammals in Fennoscandia – occurrence and distribution. *Holarct. Ecol.* 8:285–298.

Aulie, A. (1976) The pectoral muscles and the development of thermoregulation in chicks of willow ptarmigan (*Lagopus lagopus*). *Comp. Biochem. Physiol.* 53A:343–346.

Aulie, A. and Moen, P. (1975) Metabolic thermoregulatory responses in eggs and chicks of willow ptarmigan (*Lagopus lagopus*). *Comp. Biochem. Physiol.* 51A:605–609.

Boggs, C., Norris, E. and Steen, J.B. (1977) Behavioural and physiological temperature regulation in young chicks of the willow grouse (*Lagopus lagopus*). *Comp. Biochem. Physiol.* 58A:371–372.

Børset, E. and Krafft, A. (1973) Black grouse *Lyrurus tetrix* and capercaillie *Tetrao urogallus* brood habitats in a Norwegian spruce forest. *Oikos* 24:1–7.

Brittas, R., Bohm, S. and Woxlin, H. (1989) Var rädd om undervegetationen (in Swedish). *Skogen* 12/89:30–32.

Castroviejo, J. (1975) El urogallo, *Tetrao urogallus*, en España. *CSIC Monografias de Ciencia Moderna* 84.

Christiansen, E. (1979) Skog- og jordbruk, smågnagere og rev. *Tidsskrift for Skogbruk* 87:115–119 (in Norwegian).

Christiansen, E. (1983) Fluctuations in some small rodent populations in Norway 1971–1979. *Holarctic Ecology* 6:24–31.

Cramp, S., Simmons, K.E.L., Gillmor, R. *et al.* (Eds.) (1983) *Handbook of the Birds of Europe, the Middle East and North Africa. The Birds of the Western Palearctic, Vol. 2.* Oxford University Press, Oxford, pp. 433–443.

Eiberle, K. von (1976) Zur Analyse eines Auerwildbiotops im Schweizerischen Mittelland. *Forstw. Cbl.* 95:108–124.

Erikstad, K.E. and Andersen, R. (1983) The effect of weather on survival, growth rate and feeding time in different sized willow grouse broods. *Ornis Scand.* 14:249–252.

Erikstad, K.E. and Spidsø, T.K. (1982) The influence of weather on food intake, insect prey selection and feeding behaviour in willow grouse chicks in northern Norway. *Ornis Scand.* 13:176–182.

Esseen, P.-A., Ehnström, B., Ericson, L. and Sjöberg, K. (1992) Boreal forests – the focal habitats of Fennoscandia, in *Ecological Principles of Nature Conservation*, (ed. L. Hansson), Conservation Ecology Series. Elsevier Applied Science, London and New York.

Gjerde, I. (1991) Cues in winter habitat selection by capercaillie. I. Habitat characteristics. *Ornis Scand.* 22:197–204.

Gjerde, I. and Wegge, P. (1989) Spacing pattern, habitat use and survival of capercaillie in a fragmented winter habitat. *Ornis Scand.* 20:219–225.

Hagen, Y. (1952) *Rovfuglene og viltpleien.* Gyldendal Norsk Forlag, Oslo.

Hansson, L. (1979) On the importance of landscape heterogeneity in northern regions for the breeding population densities of homeotherms: a general hypothesis. *Oikos* 33:182–189.

Hansson, L. and Henttonen, H. (1989) Rodents, predation and wildlife cycles. *Finnish Game Research* 46:26–33.

Helle, P., Helle, T. and Lindén, H. (1994) Capercaillie (*Tetrao urogallus*) lekking sites in fragmented Finnish forest landscape. *Scand. J. For. Res.* 9:386–396.

Henttonen, H. (1989) Does an increase in the rodent and predator densities, resulting from modern forestry, contribute to the long-term decline in Finnish tetraonids? *Suomen Riista* 35:83–90.

Hissa, R., Saarela, S., Rintamäki, H. *et al.* (1983) Energetics and development of temperature regulation in capercaillie *Tetrao urogallus*. *Physiol. Zool.* 56(2):142–151.

Hjort, I. (1970) Reproductive behaviour in Tetraonidae with special reference to males. *Viltrevy* 7:183–596.

Hjort, I. (1985) The distribution of capercaillie males on leks in relation to the forest structure of the recruiting areas. *Proc. Int. Grouse Symp.* 3:217–235.

Hjort, I. (1994) *Tjädern*. Skogsstyrelsen, 1994 (in Swedish).

Högbom, L. and Högberg, P. (1991) Nitrate nutrition of *Deschampsia flexuosa* (L.) Trin. in relation to nitrogen deposition in Sweden. *Oecologia* 87:488–494.

Höglund, N.H. (1955) Kroppstemperatur, aktivitet och föryngring hos tjädern *Tetrao urogallus* Lin. *Viltrevy* 1:1–87, Stockholm (in Swedish).

Hörnfeldt, B. (1978) Synchronous population fluctuations in voles, small game, owls and tularemia in northern Sweden. *Oecologia (Berlin)* 32:141–152.

Hörnfeldt, B., Löfgren, O. and Carlsson, B.-G. (1986) Cycles in volves and small game in relation to variations in plant production indices in Northern Sweden. *Oecologia* 68:496–502.

Hytteborn, H., Packham, J.R. and Verwijst, T. (1987) Tree population dynamics, stand structure and species composition in the montane virgin forest of Vallibäcken, northern Sweden. *Vegetatio* 72:3–19.

Jacob, L. (1988) Régime alimentaire du Grand Tétras (*Tetrao urogallus*, L.) et de la Gélinotte des bois (*Bonasa bonasia*, L.) dans le Jura. *Acta Oecologica Oecol. Gener.* 9(4):347–370.

Järvinen, O. and Miettinen, K. (1988) *Sista paret ut?* Naturskyddsföreningen, Miljöförlaget, Helsingfors. (In Swedish.)

Johnsgard, P.A. (1983) *The Grouse of the World*. Croom Helm Ltd, Beckenham, Kent, UK.

Kardell, L. (1980) Occurrence and production of bilberry, lingonberry and raspberry in Sweden's forests. *Forest Ecology and Management* 2:285–298.

Kastdalen, L. and Wegge, P. (1985) Animal food in capercaillie and black grouse chicks in south east Norway – a preliminary report. *Proc. Int. Grouse Symp.* 3:499–513.

Kempe, G., Toet, H., Magnusson, P.-H. and Bergstedt, J. (1992) *The Swedish National Forest Inventory 1983–87.* Swedish University of Agricultural Sciences, Department of Forest Survey, Report 51. (In Swedish, with English summary.)

Klaus, S. (1991) Effects of forestry on grouse populations: Case studies from the Thuringian and Bohemian forests, Central Europe. *Ornis Scand.* 22:218–223.

Klaus, S. (1994) To survive or to become extinct: Small populations of tetraonids in Central Europe, in *Minimum Animal Populations*, (ed. H. Remmert). Springer-Verlag, Berlin, Heidelberg.

Klaus, S., Andreev, A.V., Bergmann, H.-H. *et al.* (1989) *Die Auerhühner.* Die Neue Brehm-Bücherei. Ziemsen, Wittenberg Lutherstadt.

Kuuluvainen, T. (1994) Gap distribution, ground microtopography, and the regeneration dynamics of boreal coniferous forests in Finland: a review. *Ann. Zool. Fennici* 31:35–51.

Larsen, B.B., Wegge, P. and Storaas, T. (1982) Spacing behaviour of capercaillie cocks during spring and summer as determined by radio telemetry. *Proc. Int. Grouse Symp.* 2:124–130.

Lindén, H. (1981) Changes in Finnish tetraonid populations and some factors influencing mortality. *Finnish Game Research* 39:3–11.

Lindén, H. (1988) Latitudinal gradients in predator–prey interactions, cyclicity and synchronism in voles and small game populations in Finland. *Oikos* 52:341–349.

Lindén, H. (1989) Characteristics of tetraonid cycles in Finland. *Finnish Game Res.* 46:34–42.

Lindén, H. (1991) Patterns of grouse shooting in Finland. *Ornis Scand.* 22:241–244.

Lindén, H. and Milonoff, M. (1991) The consequences of small breast muscles in juvenile cock capercaillie, in *Global Trends in Wildlife Management. The 18th IUGB Congress, Jagiellonian University, Krakow, Poland; August 1987. Transactions.* Vol. 1. (eds B. Bobek, K. Perzanowski and W.L. Regelin). Swiat Press, Krakow–Warszawa 1991.

Lindén, H. and Pasanen, J. (1987) Capercaillie leks are threatened by forest fragmentation. *Suomen Riista* 34:66–76. (In Finnish, with English summary.)

Lindén, H. and Rajala, P. (1981) Fluctuations and long-term trends in the relative densities of tetraonid populations in Finland, 1964–77. *Finnish Game Research* 39:13–34.

Lindén, H., Milonoff, M. and Wikman, M. (1984) Sexual differences in growth strategies of capercaillie, *Tetrao urogallus. Finnish Game Res.* 42:29–35.

Linder, P. and Östlund, L. (1992) Changes in the Boreal forests of Sweden 1870–1991. *Svensk Botanisk Tidskrift* 86:199–215.

Lindström, E. (1989) The role of medium-sized carnivores in the Nordic boreal forest. *Finnish Game Res.* 46:53–63.

Lindström, E., Angelstam, P., Widén, P. and Andrén, P. (1987) Do predators synchronize vole and grouse fluctuations? – An experiment. *Oikos* 48:121–124.

Lindström, E., Andrén, H., Angelstam, P. *et al.* (1994) Disease reveals the predator: Sarcoptic mange, red fox predation, and prey populations. *Ecology* 75:1042–1049.

Marcström, V. (1956) (Om kroppstemperaturen hos tjäderkycklingar *Tetrao urogallus* Lin. vid kläckningen och omedelbart därefter.) About the body temperature of capercaillie chicks *Tetrao urogallus* Lin. at the time of hatching and immediately thereafter. *Viltrevy* 1:139–149. (In Swedish, with English and German summaries.)

Marcström, V. (1977) Silviculture and higher fauna in Sweden. *XIIIth Congress of Game Biologists,* pp. 401–413.

Marcström, V. (1979) A review of the tetraonid situation in Sweden, *Proc. Int. Grouse Symp.* 1:13–16.

Marcström, V., Brittas, R. and Engren, E. (1982) Habitat use by tetraonids during summer – a pilot study. *Proc. Int. Grouse Symp.* 2:148–153.

Marcström, V., Kenward, R.E. and Engren, E. (1988) The impact of predation on boreal tetraonids during vole cycles: an experimental study. *Journal of Animal Ecology* 57:859–872.

Milonoff, M. and Lindén, H. (1989a) Sexual size dimorphism of body components in capercaillie chicks. *Ornis Scand.* **20**:29–35.

Milonoff, M. and Lindén, H. (1989b) Sexual differences in energy allocation of capercaillie *Tetrao urogallus* chicks. *Ornis Fennica* **66**:62–68.

Picozzi, N., Catt, D.C. and Moss, R. (1992). Evaluation of capercaillie habitat. *Journal of Applied Ecology* **29**:751–762.

Pedersen, H.C. and Steen, J.B. (1979) Behavioural thermoregulation in willow ptarmigan *Lagopus lagopus* chicks. *Ornis Scand.* **10**:17–21.

Pulliainen, E. (1979) Autumn and winter nutrition of the capercaillie (*Tetrao urogallus*) in the northern Finnish Taiga. *Proc. Int. Grouse Symp.* **1**:92–97.

Pulliam, H.R. (1988) Sources, sinks, and population regulation. *Amer. Natur.* **132**:652–661.

Rajala, P. (1959) Metsonpoikasten ravinnosta (The food of capercaillie chicks). *Suomen Riista* **13**:143–155. (In Finnish, with English summary.)

Rajala, P. (1974a) Tjäderns, orrens och dalripans levnadsvanor i ungkullstadiet. *Finsk Viltforskning*, pp. 11–34. (Translation to Swedish from original paper in *Suomen Riista* **15** (1962).)

Rajala, P. (1974b) The structure and reproduction of Finnish populations of capercaillie, *Tetrao urogallus*, and black grouse, *Lyrurus tetrix*, on the basis of late summer census data from 1963–66. *Finnish Game Res.* **35**:1–51.

Rajala, P. (1979) Status of tetraonid populations in Finland. *Proc. Int. Grouse Symp.* **1**:32–34.

Rodem, B., Wegge, P., Spidsø, T. *et al.* (1984) Habitat selection by capercaillie broods. *Viltrapport* **36**:53–59. (In Norwegian.)

Rolstad, J. (1988) Autumn habitat of capercaillie in southeastern Norway. *J. Wildl. Manage.* **52**:747–753.

Rolstad, J. (1989) Effects of logging on capercaillie *Tetrao urogallus* leks. I. Cutting experiments in southcentral Norway. *Scand. J. For. Res.* **4**:99–109.

Rolstad, J. (1991) Consequences of forest fragmentation for the dynamics of bird populations: conceptual issues and the evidence. *Biol. J. Linn. Soc.* **42**:149–163.

Rolstad, J. and Wegge, P. (1987a) Habitat characteristics of capercaillie *Tetrao urogallus* display grounds in southeastern Norway. *Holarctic Ecol.* **10**:219–229.

Rolstad, J. and Wegge, P. (1987b) Distribution and size of capercaillie leks in relation to old forest fragmentation. *Oecologia* **72**:389–394.

Rolstad, J. and Wegge, P. (1989a) Capercaillie habitat: a critical assessment of the role of old forest. *Proc. Int. Grouse Symp.* **4**:no. 33.

Rolstad, J. and Wegge, P. (1989b) Capercaillie *Tetrao urogallus* populations and modern forestry – a case for landscape ecological studies. *Finnish Game Res.* **46**:43–52.

Rolstad, J., Wegge, P. and Larsen, B.B. (1988) Spacing and habitat use of capercaillie during summer. *Can. J. Zool.* **66**:670–679.

Rolstad, J., Wegge, P. and Gjerde, I. (1991) Kumulativ effekt av habitat fragmentering: Hva har 12-års storfuglforskning på Varaldskogen laert oss? *Fauna* **44**:90–104.

Rosén, K., Gundersen, P., Tegnhammar, L. *et al.* (1992) Nitrogen enrichment of Nordic forest ecosystems. The concept of critical loads. *Ambio* **21**:364–368.

Seiskari, P. (1962) On the winter ecology of the capercaillie, *Tetrao urogallus*, and the black grouse, *Lyrurus tetrix*, in Finland. *Papers on Game Research*, **22**.

Sjöberg, K. and Lennartsson, T. (1995) Fauna and flora management in forestry, in *Multiple-use Forestry in the Nordic Countries*, (ed. M. Hytönen), The Finnish Forest Research Institute, Gummerus Printing, Jyväskylä, pp. 191–243.

Skogsstyrelsen (1992) Contortatallen i Sverige – en lägesrapport.

Sonerud, G.A. (1985) Brood movements in grouse and waders as defence against win-stay search in their predators. *Oikos* 44:287–300.

Spidsø, T.K. and Stuen, O.H. (1988) Food selection by capercaillie chicks in southern Norway. *Can. J. Zool.* 66:279–283.

Storaas, T. and Wegge, P. (1985) High nest losses in capercaillie and black grouse in Norway. *Proc. Int. Grouse Symp.* 3:481–492.

Storaas, T. and Wegge, P. (1987) Nesting habitats and nest predation in sympatric populations of capercaillie and black grouse. *J. Wildl. Manage.* 51:167–172.

Storch, I. (1993) Habitat selection by capercaillie in summer and autumn: Is bilberry important? *Oecologia* 95:257–265.

Storch, I., Schwartzmüller, C. and Stemmen, D. von den (1991) The diet of the capercaillie in the Alps: a comparison of hens and cocks. *Trans. Int. Congr. Union Game Biol.* 20:630–635.

Stuen, O.H. and Spidsø, T.K. (1988) Invertebrate abundance in different forest habitats as animal food available to capercaillie *Tetrao urogallus* chicks. *Scand. J. For. Res.* 3:527–532.

Svensson, S.A., Toet, H. and Kempe, G. (1989) *The Swedish National Forest Inventory 1978–82*. Swedish University of Agricultural Sciences, Department of Forestry Survey, Report 47.

Swenson, J. and Angelstam, P. (1993) Habitat separation by sympatric forest grouse in Fennoscandia in relation to boreal forest succession. *Can. J. Zool.* 71:1303–1310.

Ulfstrand, S. and Högstedt, G. (1976) Hur många fåglar häckar i Sverige? *Anser* 15:1–32.

Voous, K.H. (1960) *Atlas of European Birds*. Elsevier, Nelson.

Wegge, P. (1979) Status of capercaillie and black grouse in Norway. *Proc. Int. Grouse Symp.* 1:17–26.

Wegge, P. (1980) Distorted sex ratio among small broods in a declining capercaillie population. *Ornis Scand.* 11:106–109.

Wegge, P. (1985) Spacing pattern and habitat use of capercaillie hens in spring. *Proc. Int. Grouse Symp.* 3:261–277.

Wegge, P. and Rolstad, J. (1986) Size and spacing of capercaillie leks in relation to social behavior and habitat. *Behav. Ecol. Sociobiol.* 19:401–408.

Wegge, P. and Storaas, T. (1990) Nest loss in capercaillie and black grouse in relation to the small rodent cycle in southeast Norway. *Oecologia* 82:527–530.

Wegge, P., Storaas, T., Larsen, B.B. *et al.* (1982) Woodland grouse and modern forestry in Norway. A short presentation of a new telemetry project, and some preliminary results on brood movements and habitat preferences of capercaillie and black grouse. *Proc. Int. Grouse Symp.* 2:117–123.

Wegge, P., Gjerde, I., Kastdalen, L. *et al.* (1990) Does forest fragmentation increase the mortality rate of capercaillie? *Trans. 19th IUGB Congress, Trondheim 1989.*

Wegge, P., Rolstad, J. and Gjerde, I. (1992) Effects of boreal forest fragmentation on capercaillie grouse: empirical evidence and management implications, in *Wildlife 2001*, (ed. D.R. McCullogh and R.H. Barret), pp. 738–749. Elsevier Applied Science, New York.

Winqvist, T. (1983) 100 capercaillie courtship display grounds. *Sveriges Skogsvårds-förbunds Tidskrift* 2/83. (In Swedish, with English summary.)

Winqvist, T. (1988) *Lär känna tjädern*. Svenska jägareförbundet, Stockholm.

Winqvist, T. and Ringaby, E. (1989) Hur utnyttjar skogshönsen skogsbestånd av olika åldrar? (How does the tetraonids utilize forest stands of different age?) *Sveriges Skogsvårdsförbunds Tidskrift* 2-89:39–48. (In Swedish, with English summary.)

Zackrisson, O. (1977) Influence of forest fire on the north Swedish boreal forest. *Oikos* 29:22–32.

—5

Conservation of large forest carnivores

Todd K. Fuller and David B. Kittredge, Jr

5.1 INTRODUCTION

Throughout the world, human occupation and utilization of forested wildlife habitat has caused changes in abundance and distribution of vegetation, wild prey and, not surprisingly, large members of the taxonomic order Carnivora. The level of habitat disturbance has ranged from complete deforestation that allowed for long-term agriculture and permanent human settlement, to sustainable vegetation modification that resulted in negligible human-related development (cf. Chapter 2). In many areas, expanding human populations and their concomitant utilization of wildlife habitat have resulted in decreased animal and plant food availability for carnivores; in combination with direct human-caused mortality, most large-carnivore populations have decreased as a result. However, some large carnivore species have not declined, and others have even prospered, as a result of human-related habitat disturbance. Since increased utilization of forest products is anticipated (Chapter 2), these large carnivore species will continue to be impacted well into the future.

Worldwide, 27 large (i.e. maximum adult weight $\geqslant 20$ kg) carnivore species inhabit forested or semi-forested habitats; these include 9 felids, 7 canids, 7 ursids, 3 hyaenids and 1 mustelid (Table 5.1). These species occur on all continents except Antarctica, some with distributions limited to portions of single continents and others with hemispheric ranges. In addition, species richness varies greatly among geographic areas; only dingoes (see Table 5.1 for all scientific names of carnivores) occur in Australia, whereas Asia accommodates 15 species.

Though we often think of these species as having a common carnivorous

Conservation of Faunal Diversity in Forested Landscapes. Edited by R.M. DeGraaf and R.I. Miller. Published in 1996 by Chapman & Hall. ISBN 0 412 61890 7.

Table 5.1 Large (maximum adult weight ⩾ 20 kg) forest carnivores (Carnivora) of the world

Family	Common name	Scientific name	Adult weight (kg)[a]	Geographic range
Felidae	tiger	*Panthera tigris*	45–320	Ru,As
	lion	*Pantera leo*	120–250	Af,As
	jaguar	*Panthera onca*	36–158	CS
	leopard	*Panthera pardus*	17–130	AF,As,Ru
	puma	*Felis concolor*	25–103	No,CS
	snow leopard	*Panthera uncia*	25–90	As,Ru
	cheetah	*Acinonyx jubatus*	35–72	Af,As
	lynx	*Felis lynx*	8–38	No,Ru,Sc,Eu
	clouded leopard	*Neofelis nebulosa*	16–23	As
Canidae	wolf	*Canis lupus*	10–80	No,Ru,Sc,Eu,As
	red wolf	*Canis rufus*	20–40	No
	African wild dog	*Lycaon pictus*	27–36	Af
	coyote	*Canis latrans*	7–34	No
	maned wolf	*Chrysocyon brachyurus*	20–23	CS
	dingo	*Canis familiaris dingo*	8–22	Pa,As
	dhole	*Cuon alpinus*	10–21	Ru,As
Ursidae	brown bear	*Ursus arctos*	70–780	No,Ru,Sc,Eu,As
	American black bear	*Ursus americanus*	92–270	No
	spectacled bear	*Tremarctos ornatus*	60–175	CS
	Asiatic black bear	*Ursus thibetanus*	47–173	Ru,As
	giant panda	*Ailuropoda melanoleuca*	75–160	As
	sloth bear	*Melursus ursinus*	55–145	As
	Malayan sun bear	*Helarctos malayanus*	27–65	As
Hyaenidae	spotted hyaena	*Crocuta crocuta*	40–86	Af
	striped hyaena	*Hyaena hyaena*	25–55	Af,As
	brown hyaena	*Hyaena brunnea*	37–48	Af
Mustelidae	wolverine	*Gulo gulo*	7–32	No,Ru,Sc

[a]Weights from Ginsberg and Macdonald, 1990; Kitchner, 1991; Nowak, 1991; Seidenstecker and Lumpkin, 1991; Sheldon, 1992.
No = North America; Ru = former USSR; Sc = Scandinavia; Eu = Europe; CS = Central and South America; As = Asia; Af = Africa; Pa = Pacific/Australia.

progenitor (Martin, 1989) and therefore being quite similar ecologically, evolutionary pressures and opportunities have resulted in a diversity of feeding styles and reproductive strategies (Ewer, 1973) and these affect the species' ability to cope with habitat change. All of the large felids maintain a nearly exclusive carnivorous diet but some canids and hyaenas have become partially or predominantly omnivorous. At least one ursid, the sloth bear, is myrmecophagous (feeds on ants and termites) and the giant panda is essentially herbivorous (Ewer, 1973). In addition, reproductive rates, and thus potential rates of increase, vary greatly among species. Litter sizes range from < 2 (giant pandas: Schaller *et al.*, 1985) to > 8 (African wild dogs: Fuller *et al.*, 1992a); first age of reproduction may be delayed until age 6 years (e.g. black bears; Rogers, 1987); and the interval between litters may be 2–3 years (e.g. tigers: Smith *et al.*, 1987). Perhaps more importantly, as a result of large body size and diet, these predators are at the top of the food pyramid and are the major bioaccumulators in ecosystems. As such, the area over which individual members of these species must range is usually quite large (e.g. $\geqslant 1500$ km^2 for individual wolves and African wild dogs: Fuller, 1995a). Thus, overall large carnivore densities usually are low, and susceptibility to population extirpation, from whatever source, is relatively high.

Large carnivores are of interest to humans because of their direct interactions due to sympatric habitation, and the effect of indirect human values (positive and negative). Several carnivore species prey on humans, most compete with humans for prey, many cause significant livestock depredation and several have been valued as a fur resource or trophy. In addition, large carnivores play important roles in some human cultures, they exhibit traits that are visually and behaviorally appealing (thus providing aesthetic value for many people) and they are sometimes valued because of the important role they play as predators and scavengers in the maintenance of diverse and balanced ecological systems (Terborgh, 1990). Recently, increased public concern over the plight of large carnivores – certainly among the most charismatic of megavertebrates (Cox, 1993) – has resulted in heightened legal and financial conservation efforts.

Given these diversities of distribution, natural history traits and human values, it is not difficult to understand why changes in forest habitats due to increasing human pressures can seriously impact large carnivore populations in a variety of ways. This chapter seeks to outline the current status of large forest carnivores worldwide, identify the historic and current factors impacting these species, and identify and recommend solutions for the myriad of conservation problems that have resulted.

5.2 CURRENT STATUS

Much of what is 'known' about large carnivores by the majority of humans is the result of practical experience or stories passed on through word of mouth or the media. Much practical experience has led to accurate knowledge, but the human penchant for the exciting and bizarre has resulted in myths that distort or overwhelm the truth (e.g. Fiennes, 1976; Lumpkin, 1991; Sanders, 1993a). Scientific knowledge is based on unbiased confirmation of practical observations, additional observations, and testing hypotheses derived from these observations. Since technical publications are the product of this scientific research, a crude index of our knowledge of large carnivores and their conservation problems can result from a synoptic review of this research.

As a quick summary of publications concerning large carnivores, we tabulated all publications in which a particular species name occurred in the title of a paper, or was listed as a key word. This was done by searching a current computer data base of wildlife literature (National Information Service Corporation, 1993) that contains over 350 000 citations from five major data bases, and covers the period from 1935 to the present. From this search, it appears that only six of 27 (22%) of large forest carnivore species are well studied (i.e. 32–85 publications/yr during the past 25 years, 1968–1992). A further six of 27 species (22%) have been moderately studied (9–13 publications/yr) and the majority (15 species; 56%) are poorly studied (up to 8 publications/yr).

The most studied species include lynx, wolf, brown bear and wolverine, all of which have holarctic distributions, and puma, coyote and black bear, which occur commonly in North America (though pumas also occur throughout South America). As a result, species in the northern hemisphere can be considered to be moderately to well-studied (0–33% of species/area with < 8 citations/yr), in stark contrast to the carnivores living primarily in the southern hemisphere where a majority of species (50–60%/area) are poorly studied and understood (Table 5.2).

Taxonomically, the distribution, habits, etc. of most felids (except for snow leopard and clouded leopard) are fairly well understood or frequently studied in all areas, but most canids and bears in the southern hemisphere and all of the hyaenas have not been the focus of scientific study (Table 5.3). Relatively recent European colonization (and thus relatively high numbers of carnivore–human interactions) and high per capita income (and thus the finances to study carnivore 'problems' in more modern ways) likely explain the relatively large body of scientific knowledge available from North America.

Scientific knowledge, however, does not necessarily reflect conservation status. Historical distributional changes have occurred for all large carnivore species, regardless of the length of time humans have had impacts or the magnitude of development in the area. Because of the paucity of scientific studies for many species, estimates of current distribution (much less historical distribution) are only approximations and extrapolations. However, dramatic patterns of change are still evident.

Table 5.2 Relative status of knowledge of large forest carnivore species (totalled by taxonomic family) for specific geographic areas, as indicated by mean number of scientific articles published each year during 1968–1992[1] (number of species with < 8 publications/year : total number of species in area)

Geographic area	Felid	Canid	Ursid	Hyaenid	Mustelid	Total	(%)
Primarily Northern Hemisphere							
North America	0:2	1:3	0:2		0:1	1:8	(13)
Former USSR	1:4	1:2	1:2		0:1	3:9	(33)
Scandinavia	0:1	0:1	0:1		0:1	0:4	(0)
Europe	0:1	0:1	0:1			0:3	(0)
Total	1:8	2:7	1:6		0:3	4:24	(17)
(%)	(13)	(29)	(17)	(0)			
Primarily Southern Hemisphere							
C. and S. America	0:2	1:1	1:1			2:4	(50)
Asia	2:6	2:3	4:5	1:1		9:15	(60)
Africa	0:3	1:1		3:3		4:7	(57)
Pacific		1:1				1:1	(100)
Total	2:11	5:6	5:6	4:4		16:27	(59)
(%)	(18)	(83)	(83)	(100)			

[1] Articles in which the species' scientific name appears in the article title or key words list in a > 350 000-citation CD-ROM bibliography (National Information Service Corporation, 1993). Publication rates are for each species throughout its total geographic range. Proportions reflect only this total publication rate and not the subjects addressed in the published studies.

Table 5.3 Status of knowledge (by taxonomic family) of large forest carnivore species, as indicated by number of scientific articles published during 1968–1992 in which the species' scientific name appears in article title or key words list[1]

Total no. of articles	No. of species per category						
	Felid	Canid	Ursid	Hyaenid	Mustelid	Total	(%)
> 1000	1	2	1	0	0	4	(15)
400–1000	1	0	1	0	0	2	(7)
200–400	5	0	0	0	1	6	(22)
< 200	2	5	5	3	0	15	(56)

[1] From articles listed in a > 350 000-citation CD-ROM bibliography (National Information Service Corporation, 1993).

Figure 5.1 Approximate historical (shaded) and current (black) distribution of pumas in North America. (From Hornocker, 1992:55.)

In North America, pumas have disappeared from the entire eastern portion of their range, except for a remnant population in Florida (Figure 5.1) and, concurrently, wolves were extirpated from the entire southern portion of their range (Figure 5.2). Wolves are ecologically dominant over coyotes and aggressively exclude them from their ranges. One consequence of the disappearance of wolves, and the concomitant expansion of human populations, is the dramatic recent increase in the range of the coyote throughout North America (Figure 5.3; Moore and Parker, 1992).

In Europe, wolves were extirpated from many areas hundreds or even

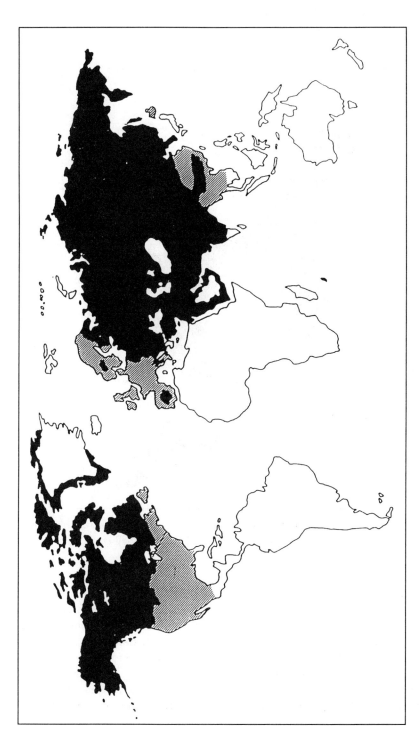

Figure 5.2 Approximated historical (shaded) and current (black) distribution of wolves in the northern hemisphere. (From Nowak, 1983:10; Ginsberg and Macdonald, 1990:38.)

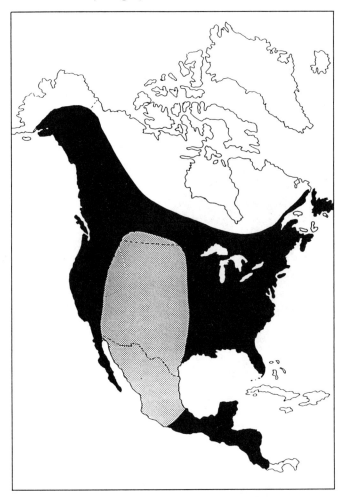

Figure 5.3 Approximate historical (shaded) and current (black) distribution of coyotes in North America. (From Moore and Parker, 1992:25,30.)

thousands of years ago, but viable populations still remain in some countries (Figure 5.2). In Scandinavia, a few individuals remain along the border of Sweden and Norway, but the Finnish population is contiguous with that in Russia and Asia which has changed little in recent times.

In Central and South America, jaguars are still common in remote areas with low human interference (Figure 5.4) but locally are under constant threat from expanding forest clearing for ranching (Swank and Teer, 1989). The same is likely true of pumas (Johnson and Franklin, submitted) in many parts of the continent.

Figure 5.4 Approximate historical (shaded) and current (black) distribution of jaguars in Central and South America. (From Swank and Teer, 1989:16,17.)

The range of African wild dogs has been reduced by more than half (Figure 5.5), a result of human persecution and reduced prey availability (Fanshawe *et al.*, 1991), and in some areas they are found only in huge, relatively intact ecosystems (Mills, 1991). Similar patterns resulting from habitat loss have been identified for cheetahs in Africa (Caro, 1991).

In Asia, tiger distribution has been reduced dramatically (Figure 5.6), and at least three subspecies are now extinct (Tilson and Seal, 1987). Even more dramatic has been the range reduction of the giant panda (Figure 5.7), a species which now resides at dangerously low numbers in mountainous forest fragments in the easternmost part of its former range.

Finally, the range of dingoes, which arrived in Australia only about 4000 years ago and subsequently spread throughout the continent, has recently been reduced somewhat (Figure 5.8) as a result of intense human persecution (Ginsberg and Macdonald, 1990; King, 1994).

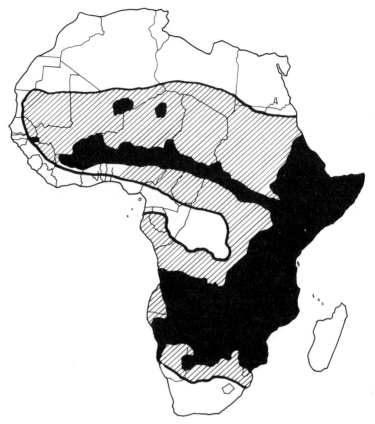

Figure 5.5 Approximate historical (shaded) and current (black) distribution of African wild dogs in Africa. (From Ginsburg and Macdonald, 1990:18.)

Given the difficulty in accurately identifying large forest carnivore distributions, it is not surprising that total population estimates for species are even more difficult to determine. In fact, no estimate apparently has been made for 12 (44%) of the large carnivores species, including all three hyaenas and the wolverine (Table 5.4). All 12 species, except jaguar and wolverine, are poorly studied. The hyaenas, dingo and dhole, like the jaguar and wolverine, are widely, if sparsely, distributed. The Asiatic black bear and spectacled bear also are widely distributed throughout mountainous terrain. Similarly, the clouded leopard and Malayan sun bear from tropical southeast Asian forests, and the maned wolf from South America, have had few, if any, studies conducted on even their basic life history characteristics. An additional six species have total estimated numbers of < 10 000 (Table 5.4), some with critically endangered populations of < 100 (red wolf) to < 1000 (giant panda).

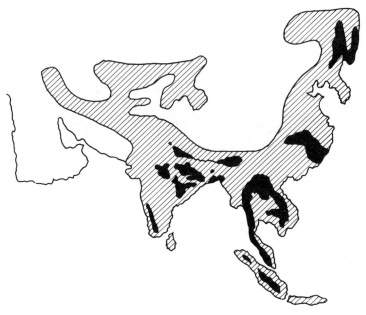

Figure 5.6 Approximate historical (shaded) and current (black) distribution of tigers in Asia and the former Soviet Union. (From Tilson and Seal, 1987:viii; Mills, 1994:53.)

Figure 5.7 Approximate historical (shaded) and current (black) distribution of giant pandas in China. (From Schaller *et al.*, 1985:13.)

Figure 5.8 Approximate historical (shaded) and current (black) distribution of dingoes in Australia. (From Ginsburg and Macdonald, 1990:53; King, 1994:46.)

Table 5.4 Total population status, by taxonomic group, of large forest carnivore species (from Nowak, 1991)

Total population in the wild	No. of species per category						
	Felid	Canid	Ursid	Hyaenid	Mustelid	Total	(%)
> 50 000	3	2	2			7	(26)
10 000–50 000	2					2	(7)
⩽ 10 000	2	2	2			6	(22)
Unknown	2	3	3	3	1	12	(44)

A variety of other felids, canids and ursids have worldwide populations well above 10 000 (Table 5.4) but, even so, a majority of large carnivore species are recognized as being in need of or having some legal protection (Table 5.5) because local or overall populations are threatened or endangered. Regardless of geographic area (except for Africa) or taxonomic family (except for hyaenas), nearly all species are considered to need protection, as indicated by official legal, conservation and/or trade status (Table 5.5; Fuller, 1995b). This broad recognition only emphasizes the worldwide concern for the vulnerability of these species. It also indicates that there are likely to be similar, underlying causes which have resulted in so many precarious populations, and that a review of such causes will probably prove fruitful in guiding future attempts to redress past activities.

Table 5.5 Distribution of large forest carnivore species, by taxonomic group, that are recognized as having some legal protection[1] within all or part of a specific geographic area (no. spp. having some protection: total no. spp.)

Geographic area	Felid	Canid	Ursid	Hyaenid	Mustelid	Total	(%)
Primarily Northern Hemisphere							
North America	1:2	2:3	2:2		1:1	6:8	(75)
Former USSR	3:4	1:2	1:2		0:1	5:9	(55)
Scandinavia	1:1	1:1	1:1		1:1	4:4	(100)
Europe	1:1	1:1	1:1			3:3	(100)
Subtotal	6:8	5:7	5:6		2:3	18:24	(75)
(%)	(75)	(71)	(83)		(66)		
Primarily Southern Hemisphere							
C. and S. America	1:2	1:1	1:1			3:4	(75)
Asia	6:6	1:3	4:5	0:1		11:15	(73)
Africa	1:3	1:1		1:3		3:7	(43)
Pacific		0:1				0:1	(0)
Subtotal	8:11	3:6	5:6	1:4		17:27	(63)
(%)	(73)	(50)	(83)	(25)			
Total	14:19	8:13	10:12	1:4	2:3	35:51	(69)
(%)	(74)	(62)	(83)	(25)	(66)		

[1]Recognized by US Fish and Wildlife Service (1991) as endangered or threatened; by IUCN (1976) as endangered, vulnerable, or rare (Nowak, 1991); and/or by CITES (1973) as listed in Appendix I (no commercial trade) or Appendix II (limited trade by permit) (Nowak, 1991; Sterling, 1993).

5.3 HISTORIC CAUSES OF POPULATION CHANGE

5.3.1 Direct adverse human impacts

Large carnivores and humans have hunted each other ever since they have shared the landscape. Even today, humans are killed by tigers in Asia (McDougal, 1987) and by brown bears (Herrero, 1985) and pumas (Beier, 1991) in North America. The threat of any injuries, let alone mortal ones, inflicted by large carnivores on humans is understandably good reason to fear them, hence the desire for their elimination from areas of human habitation.

The killing of large carnivores also has been carried out to reduce competition for human food resources. Much large ungulate prey killed by large carnivores has been prized as a major food source by humans, as well, and thus less competition meant increased food availability. In recent years, the role of predation in limiting ungulates also harvested by humans is still a major wildlife management concern (Gasaway *et al.*, 1992).

Even in national parks where ungulates are protected, predator control activities have taken place. Earlier in this century, wolves were eliminated from Yellowstone National Park in the United States because they were considered a 'menace' to herds of wild game (Weaver, 1978), just as African wild dogs were officially persecuted in Zimbabwe (Childes, 1988).

Another behavior of large carnivores that leads to their direct killing by humans is the propensity of some individuals to prey on domestic livestock. On every continent, ranchers and farmers fought (and still fight) against depredations by poisoning, shooting and snaring carnivores. Brown bears kill domestic reindeer in Scandinavia (Mysterud and Muus Falck, 1989a), jaguars kill cattle in Belize (Rabinowitz, 1986a) and wolves kill turkeys in Minnesota (Fritts, 1982). As a result of these conflicts, many carnivore species have been extirpated from areas of intensive livestock raising.

Not only do large carnivores kill domestic animals but they also may damage major agricultural crops such as fruit, honey (beehives) and grain. Bears of all species, in particular, cause such damage and extensive efforts have been made to evaluate and manage those problems (e.g. Bromley, 1989).

Large forest carnivores also have been killed for consumption purposes. For example, over 40 000 black bears are harvested annually in North America during sport hunting seasons, providing meat, fur and recreation (Servheen, 1989). Tigers, leopards, cheetahs and jaguars have long been hunted for their pelts, either for garments or decoration, and the resulting endangerment of populations has led to major trade restrictions (Fitzgerald, 1989). Considerable medicinal trade in tiger (Mills, 1994) and bear parts (nearly all species: Mills and Servheen, 1991) has resulted in extirpation of many of their populations throughout Asia and overharvest is a major conservation concern. Sport hunting for snow leopards in Mongolia (O'Gara, 1988) or jaguars in South America (Swank and Teer, 1989) generates sufficient interest such that it has been proposed as a major funding generator for felid conservation activities. Thus legal and illegal harvest of large carnivores still plays an important role in continued maintenance of their populations.

5.3.2   Indirect adverse human impacts

A major indirect factor affecting interactions between humans and large carnivores has been vegetation change. Kittredge (Chapter 2) has related the massive extent of reduction in forest area worldwide and concluded that 'there is no doubt that the amount . . . is decreasing'. The mode of such habitat change, including fuel wood collection, industrial use and clearing for agriculture, often has led to complete elimination of forest cover and thus eradication of wildlife species dependent upon it. As a consequence, carnivore food resources vanish and most carnivores either leave for other areas or try to subsist on human-provided resources (i.e. livestock) which are lethally protected. The reduction of large ungulate populations through habitat change and direct 'overkilling' has been postulated as a major cause of the extinction of five North American large carnivore species at the end of the Pleistocene epoch (Pielou, 1991).

A second consequence of reduced areas of potential large carnivore habitat is population isolation. Because of their large ranges and low densities, large forest carnivore populations are particularly vulnerable to becoming isolated; what we generally think of as 'large' blocks of relatively undisturbed landscapes may only encompass the ranges of several individuals or social groups. Numerically small, isolated populations are susceptible to demographic and genetic effects that result in high extinction rates (Shaffer, 1981). Just by random chance, the age or sex structure of a small population might be such that it could not recover from some serious environmental perturbation. Small populations may become so inbred that further survival is in doubt due to the occurrence of deleterious alleles or susceptibility to lethal disease (e.g. Isle Royale wolves: Wayne *et al.*, 1991; Florida panthers: Roelke *et al.*, 1993).

Additionally, increased reduction in forest habitat often means increased contact with activities of humans that indirectly result in dead carnivores. Diseases transmitted by domestic dogs may be a factor regulating African wild dog numbers, as are vehicle collisions (Fanshawe *et al.*, 1991). Recently, canine distemper has resulted in a major population decline of African lions in the Serengeti area of Tanzania (Morel, 1994). Increased road densities in wolf range in the northcentral United States lead to greater human access and higher intentional and unintentional (e.g. trapping, vehicle collisions, etc.) mortality that may limit wolf numbers (Fuller *et al.*, 1992b). Thus, it is clear that even when humans do not intentionally manipulate large carnivore numbers, adverse impacts do result.

Finally, exaggerated or untruthful depiction, negative and positive, of large carnivore behavior, such as through the 'entertainment' media of books, magazines and film, is common (e.g. wolf attacks on humans in *Little Red Riding Hood* and *White Fang*, or the raising of Kipling's Mowgli by wolves). This affects the public knowledge of carnivores and thus their attitudes (e.g. Kellert, 1985). It also has led indirectly to, and likely continues to result in, calls for the elimination of carnivores in some areas (Lopez, 1978:198) and continued opposition to their conservation or reintroduction (e.g. Weise *et al.*, 1975:26; Bath, 1991). Conversely, in some areas it also may result in 'overprotection' or the legal restriction of any control measures. This can develop into a major problem when carnivore numbers climb so high that conflicts with people become intolerable, attitudes change for the worse and citizens demand action.

5.3.3 Beneficial human impacts

Not all human activities diminish carnivore viability. Many species have great aesthetic appeal to humans, as evidenced by their past and current popularity in zoos and magazines and on television. Wildlife conservation organizations use them as logos (e.g. the giant panda of the World Wide Fund for Nature), as do numerous industrial and commercial businesses (e.g. the Exxon tiger, the MGM lion and the Coca-Cola polar bear). Whether ephemeral or deeply ingrained, this appeal can, via positive media exposure, result in reinforcement of favorable images and increased awareness. As noted above, however, misinformation or exaggeration can lead to misguided public support and thus thwart needed conservation efforts.

Large carnivores also are often highly valued symbols in a cultural context (Lopez, 1978; Lumpkin, 1991; Sanders, 1993a,b). Lions and tigers have symbolized strength and power and were commonly associated with soul transfer and reincarnation. Bears were considered the quintessence of motherly compassion to the Greeks and Romans, and as spiritual guardians by Inuit shamans. Conscious identification with the wolf as a great hunter was a mystical experience for Native Americans of the Great Plains. Even today, professional sports teams identify themselves as large carnivores.

A variety of other values of large carnivores influence people's positive perception of them and may lead to increased protection or support. Little recognized, but still viably argued, is the role large carnivores play in maintaining some sort of ecological balance of nature (Terborgh, 1990). Many problems of overpopulation of large ungulates in suburban and rural areas are the result of the previous elimination of large carnivores and the inadequacy of humans as substitute predators. The restoration of wolves into the Yellowstone ecosystem is largely a public response to a desire to have a 'complete' historical complement of wildlife in the area (US Fish and Wildlife Service, 1993:1–11).

Large carnivores also serve as an important indicator of biodiversity, and are used as 'flagship' species to focus attention on the plight of wildlife and their habitat in particular areas (e.g. giant panda: Schaller, 1993). If areas can sustain viable populations of one or more large carnivores, it is likely that other, less beguiling species and processes also will be conserved.

Although the biodiversity in an area may largely be conserved without the presence of large carnivores, public and governmental attention, funding and support may be much easier to garner if conservation includes large carnivore species. Increased public understanding of this role has led to increased support for large carnivores worldwide.

In more direct terms, wildlife tourism, such as wolf howling excursions in Canada (Strickland, 1983), provides important benefits to carnivores in several ways. It certainly provides a source of revenue, ideally (but usually not) to local people, so that a positive value is associated with maintaining wildlife in areas where they may otherwise only be considered pests or impediments. Tourism may also provide educational experiences that can only be had by direct encounter and observation, and thus are likely to enhance chances of additional monetary support through contributions to governmental and non-governmental conservation organizations. Much wildlife tourism is focused on large forest carnivores, and in places like the Masai Mara in Kenya tourists are disappointed if they do not see lions and cheetahs.

Even habitat change can, in some cases, result in increased carnivore numbers, or sometimes distribution. For instance, early successional stage forests in the central part of North America, the result of intense logging activities a century ago, now provide abundant habitat for white-tailed deer (*Odocoileus virginianus*), a primary prey for wolves. As a result of high ungulate densities, wolf densities there are among the highest anywhere (e.g. Fuller, 1989). North American coyotes, as a consequence of greatly reduced wolf distribution and increased agricultural land, have expanded their distribution to include the forested portion of most of the continent.

Finally, the most obvious beneficial human impacts on large forest carnivores have been legal protection and modern management. When, as a result of enlightened wildlife management policies, bounty laws were finally repealed and carnivores given protected status as game animals or fur-bearers, the decline in numbers and distribution was often halted or reversed (e.g. wolves: Fuller *et al.*, 1992b). Additional protection under endangered species legislation, the development of implementation of recovery plans and adherence to international trade agreements has furthered these trends (e.g. leopards: Bailey, 1993). Revenues generated by sport hunters and trappers have been used in both developed and developing countries (e.g. Lewis *et al.* 1990) for directly funding wildlife management and local community costs. National parks and reserves have been established to protect large carnivore species (e.g. Rabinowitz 1986b). Importantly, increased biological knowledge, often obtained at considerable expense, has been instrumental in providing the factual basis needed to manage and conserve carnivore populations in the face of uninformed or emotional public and political pressures.

5.4 CONSERVATION ACTIONS

As a result of the concern for carnivore control, conservation and management, a myriad of actions recently have been proposed and carried out. Already there are available a number of comprehensive reviews or recovery/conservation plans that focus on species-specific management activities (e.g. Mysterud and Muus Falck, 1989b; Swank and Teer, 1989; US Fish and

Wildlife Service, 1989). In addition, several family-specific reviews (e.g. Servheen, 1989; Ginsburg and Macdonald, 1990) provide comparisons among taxonomically, if not ecologically, similar species. Finally, there are area-specific assessments which provide useful comparisons and more integrative approaches to management on landscape scales (e.g. Stuart *et al.*, 1985; Hummel and Pettigrew, 1991; Mills, 1991).

However, these compendia readily admit that the findings they present are not complete, nor even necessarily sufficient to ensure perpetual conservation of the particular species or group of species on which they focus. Rather, they point out areas of general consensus concerning large carnivore conservation and, perhaps more importantly, areas of concern. They often identify not only what seems to work but also what we need to know to be more sure of the consequences of our conservation actions. In summary, what we need to know and what needs to be done fall into six categories:

1. Increase scientific and management knowledge.
2. Determine conservation goals.
3. Protect areas and manage habitat.
4. Minimize conflicts.
5. Make carnivore conservation worthwhile.
6. Increase law enforcement.

5.4.1 Increase scientific and management knowledge

Not all information pertaining to large carnivores is essential to conservation efforts, but synthesizing what is known and identifying gaps in our knowledge is a fundamental step. First, adequate inventories of species distribution and relative abundance must be conducted to determine where there are viable populations and where species are threatened. Ranges of many well-studied species have been documented through interviews with local people, harvest records, verified livestock depredation reports and a long history of scientific collection and research efforts. In relatively undocumented areas, species presence can also be verified by searching for and identifying species-specific sign such as tracks, digging, scratching, dens, scats, prey remains (e.g. Murie, 1974); collecting hide, skull and hair samples; using food or scent to lure animals to sandy/muddy soil or track boards (e.g. Lindzey *et al.*, 1977; Barrett, 1983); setting up camera stations (e.g. Jones and Raphael, 1993; Kucera and Barrett, 1993); eliciting vocalizations (e.g. Fuller and Sampson, 1988); or live-capturing individual animals (e.g. Schemnitz, 1994). Identifying potential areas in which to expand distribution searches by extrapolating from records of known distribution has recently been augmented by the advent of computerized geographic information systems (GIS) (Koeln *et al.*, 1994). Up-to-date species-wide distribution is often compiled by combining published and unpublished documents with the personal knowledge of professionals (e.g. Stewart, 1993).

Identifying specific types of habitat used by large carnivores is important to the extent that they reflect differences in relative carnivore abundance. For studying most carnivores, identification of both prey and human distribution and abundance are often more pertinent to conservation (e.g. Fuller *et al.*, 1992a) than maps of vegetation types, even though these are often easier to generate or obtain. Specific studies of carnivore habitat (e.g. vegetation) use, while common, are of lesser use, especially if they do not provide tests of hypotheses concerning carnivore abundance or rates of population change.

Also of key importance, and linked with habitat use, are movement patterns as they relate to range size and social organization. The size of an area that an individual or group requires in which to survive and reproduce, and the degree to which individual ranges overlap with one another, are critical factors in estimating minimum areas needed to sustain viable populations (see below). Although such information has been collected through direct observation in a few, unique areas (e.g. the Serengeti Plains: Kruuk, 1972; Schaller, 1972), the most common and elucidative data on this topic are derived from radiotelemetry studies (e.g. White and Garrott, 1990).

In addition, understanding the demography (rates of survival, reproduction and dispersal) of populations and species allows for calculation of the probability of persistence of large carnivores under a variety of scenarios. We know that large carnivore populations are likely impacted most directly through changes in food availability and sources and rates of mortality (e.g. Fuller, 1989; Fuller *et al.*, 1992b). For example, wolf population densities are directly correlated with food abundance; low food availability results in decreased reproductive output (Boertje and Stephenson, 1992) and/or increased juvenile mortality (Fuller, 1989). Also, a limited amount of human-caused mortality can be withstood by some carnivore populations because of relatively high reproductive rates (e.g. wolves: Fuller, 1989) but others are more sensitive to such mortality. Intentional human killing of giant pandas, for example, has limited any population growth (Schaller, 1993). Quantitative estimates of survival and reproduction under various conditions allow for derivation of critical levels of mortality and the potential for population recovery.

It is clear from the earlier review of basic knowledge that much research needs to be focused on the little-known large carnivores. For some species, no good estimate of distribution, much less density, is known. We can guess at age-specific reproductive rates because of information gleaned from captive animals, but natural survival rates are unknown for most species. Additional efforts are warranted to synthesize data for well-known species and to make comparisons among ecologically, taxonomically or behaviorally similar ones (e.g. Gittleman, 1993).

5.4.2   Determine conservation goals

Given the values, both positive and negative, of large carnivores, it is clear that specific goals with respect to carnivore distribution and abundance need to be determined and agreed upon. Depending on conflicting human uses of the landscape, the minimum numbers of individuals needed to sustain a population and the projected changes in human numbers and distribution, such goals can be derived for both local and global situations. Once justified and established, they provide a critical focal point and source of common ground for all interested parties. Effective conservation goals indicate where to focus efforts and establish benchmarks or critical population levels below which viable populations may not successfully sustain themselves.

5.4.3   Protect areas and manage habitat

Land areas of some minimum size are needed to support viable populations of large carnivores. Depending on this population size, the area of concern will be quite large because large-carnivore densities are relatively low. Recognizing or establishing carnivore conservation areas (Hummel and Pettigrew, 1991), whatever the mechanism, is a land-use planning priority for species with limited ranges and small populations, and where conflicts with humans are common. It seems prudent to move quickly to identify and protect 'core' areas that are not immediately threatened, but individual populations on the fringe of a range also may possess unique qualities, traits or genes that are different from those found in the main part of the range. Thus, protection of areas on the fringe should not be disregarded. In all of these areas, efforts should be made to implement natural forest management strategies to mimic natural processes and sustainably reproduce native vegetation, habitat conditions and, of primary importance, prey populations. Throughout the rest of the range(s) of a species or group of species, habitat assessments regarding prey availability and connectivity of adjacent areas are critical; these allow for analysis of subpopulation viability via population growth and dispersal.

Within all identified areas, human-related factors limiting carnivore numbers need to be curtailed or eliminated through strict regulation and enforcement. In most areas, this means increased funding for personnel and administration needed to monitor and control poaching, to monitor prey populations and the impact of human use of those populations, to implement long-term land management plans and to provide for local education regarding conservation efforts.

5.4.4 Minimize conflicts

Differential habitat alteration and human disruption have led to a mosaic of human/carnivore interactions, and carnivore species in many areas are now considered 'endangered pests'. An important first step in minimizing conflicts between large carnivores and humans, and thus enhancing large carnivore viability, is identifying through inventory not only where both groups occur and where conflicts might be common but also identifying through predictive modeling where the groups may occur in the future. Given these data, management decisions can be made on a landscape scale to determine areas where carnivore numbers need to be reduced and where they can or need to be enhanced, thus reducing conflicts. However, population distributions of humans and large carnivores certainly do not need to be allopatric; in many areas they occur sympatrically with little or no conflict. Understanding and clearly outlining what the life requirements and behavior of carnivore species are often enhances their viability. Articulation of the conservation and management goals and processes, hopefully known through legitimate public survey and discussion, will likely go a long way towards improving cooperation amongst private and public landowners and constituents who are concerned with large carnivore populations for a variety of reasons.

Influencing change in human behavior and attitudes, or at least increasing understanding, needs to be addressed by making reliable and correct information available. Successfully imparting such information may be done through schools, through the media (e.g. advertisement, news and magazine articles, television programmes) and through public meetings, seminars and discussions.

5.4.5 Make carnivore conservation worthwhile

A key component of successful carnivore management is clear enumeration of the benefits of having large carnivore populations occurring in specified areas (and not occurring in other areas). Aside from the commercial, aesthetic, recreational and utilitarian values of individual species, their presence can, for example, be easily tied to the sustainability of healthy, balanced ecosystems that need a top carnivore to keep other components 'in line'; balanced ecosystems with balanced predators–prey relationships are 'long term' and sustainable (e.g. Terborgh, 1988). A focus on ecosystem maintenance, via carnivore conservation, is important for long-term survival of many species, including humans, and an increasingly common concern of many people involved with natural resource management.

In the shorter term, however, there are areas where local people literally cannot afford carnivore conservation on their own; livestock depredation may severely impact their livelihoods. Under such circumstances, those who can pay for conservation should do so, via programs such as taxation and subsequent government aid, private tax-deductible donations to non-governmental organizations with proven track records of conservation, and increased import costs on imported natural resources. In local areas, direct payments can help to substitute for alternative land uses and thus sustain local human populations. Host governments often cannot afford conservation on their own but outside funding can be used to create incentives for them to modify or change policies or to implement plans previously without funding. Debt-for-nature swaps have been used in the past, for example, to preserve tropical habitats. Debt forgiveness in return for carnivore conservation might be considered, whereby debts are forgiven as long as carnivore populations are maintained at a certain pre-determined level. Also, funds can be provided from the international community to administer carnivore conservation programs, as long as certain criteria are met. Investing funds beforehand in administration of carnivore programs is more cost-efficient than captive breeding and reintroduction programs (see below).

In addition, there are creative alternatives to habitat destruction resulting in loss of large carnivore habitat/prey. Sustainable management of forests in-place, as opposed to their decimation, exploitation and subsequent conversion to non-forest land-use, can help to ensure perpetuation of large carnivore species and the ecosystems in which they live.

5.4.6   Increase law enforcement

Many countries have already implemented strict wildlife laws and regulations concerning large carnivores, but internal and international illegal trade, including both imports and exports, apparently continues. Thus any progress in stricter enforcement of local, national and international laws will improve the long-term viability of large carnivore populations. In particular, adequate enforcement of CITES regulations to reduce poaching and illegal trade could have significant positive impacts on species like the tiger.

5.4.7   Repopulation efforts

Translocation of large carnivores to new areas, or maintaining them in captivity for their subsequent release into the wild, should be considered last resorts because of the efforts often needed to be successful. Although a variety of large carnivore translocations or reintroduction efforts have been or are currently being made (e.g. Phillips and Parker, 1988; Brocke *et al.*, 1991; Smith and Clark, 1994), such endeavors are not accomplished without extensive considerations and substantial expense (Griffith *et al.*, 1989; Cook, 1993). Particularly for reintroductions, the cost of reclaiming adequate habitats is usually significantly more expensive than their conservation would have been at some time in the recent past.

5.5 CONCLUSIONS

Human populations are inextricably tied to carnivore populations because of competition between them and because of the many differing values of large carnivores to humans. However, our knowledge of a number of large carnivore species is alarmingly sparse and immediate efforts are needed to acquire the minimal information needed to initiate and maintain adequate conservation programs.

It is essential to realize that the 'endangered pest' dilemma is here forever; there will always be places in the world where endangered carnivores co-occur with humans and cause conflict. It is important that such potential for conflict be minimized through active management. In addition, it is essential to identify additional parts of species ranges where conflict is more rare, and to protect those areas from further human encroachment. This will require advanced planning, vision and financial support.

Because large carnivores play a significant role in ecosystem viability, and because such species often can serve as 'flagship' species which attract tremendous financial and emotional support, concern for the well-being of these species should be widespread among those who direct or significantly influence resource management. It is clearly cheaper and easier to maintain functioning ecosystems than it is to attempt to save one or more species at the brink of extinction. We need 'proactive' management for efficiency, especially in a world with dwindling financial resources and expanding needs.

Reduction in worldwide trade of skins and trophies, captive breeding campaigns, expanded attempts to reconstruct complete vertebrate communities in national parks and reserves and increased education efforts have resulted in stabilization or increase in some species numbers and ranges.

Conservation of large carnivore species will continue to be controversial. Actions that are needed to sustain the viability of these species include increased scientific knowledge of the species, increased land protection and management, additional understanding of human/carnivore conflicts, increased education and understanding, assigning and appropriately paying for costs of large carnivore conservation, and perhaps translocation or, as a last resort, captive breeding and reintroduction.

5.6 SUMMARY

Human occupation and utilization of forests has caused changes in the abundance and distribution of the 27 large (i.e. maximum weight $\geqslant 20$ kg) carnivore species (nine felids, seven canids, seven ursids, three hyaenids and one mustelid) that inhabit forests. Forest removal (e.g. for agriculture) or alteration affects distribution and abundance of carnivore prey, and consequently impacts carnivore reproduction and survival. Several carnivore species prey on humans, most compete with humans for ungulate prey, many cause significant livestock depredation, several have been valued as a fur resource or trophy and most are valued for aesthetic purposes. These characteristics often dictate the viability of carnivore populations.

Conservation actions that should be initiated or continued include: gathering and disseminating additional scientific and management knowledge; more clearly identifying conservation goals; increasing protection and management of important habitats; minimizing conflicts between local people and large carnivores; making carnivore conservation worthwhile to local people; increasing local, national and international law enforcement of trade regulations; and assessing the worthiness and viability of restoration programs.

5.7 REFERENCES

Bailey, T.N. (1993) *The African Leopard: Ecology and Behavior of a Solitary Felid.* Columbia University Press, New York. 429 pp.

Barrett, R.H. (1983) Smoked aluminum track plots for determining forbearer distribution and relative abundance. *Calif. Fish and Game* **69**:188–190.

Bath, A.J. (1991) Public attitudes in Wyoming, Montana and Idaho towards wolf restoration in Yellowstone National Park. *N. Am. Wildl. and Nat. Resourc. Conf.* **56**:91–95.

Beier, P. (1991) Cougar attacks on humans in the United States and Canada. *Wildl. Soc. Bull.* **19**:403–412.

Boertje, R.D. and Stephenson, R.O. (1992) Effects of ungulate availability on wolf reproductive potential in Alaska. *Can. J. Zool.* **70**:2441–2443.

Brocke, R.H., Gustafson, K.A. and Fox, L.B. (1991) Restoration of large predators: potentials and problems, in *Challenges in the Conservation of Biological Resources – a practitioner's guide* (eds D.J. Decker, M.E. Krasny, G.R. Goff *et al.*), pp. 303–315. Westview Press, Boulder, Colorado.

Bromley, M. (ed.) (1989) *Proceedings of a symposium on management strategies: bear–people conflicts.* Northwest Terr. Dep. Renew. Resourc., Yellowknife, N.W.T.

Caro, T. (1991) Cheetahs, in *Great Cats: Majestic Creatures of the Wild* (eds J. Seidensticker and S. Lumpkin), pp. 138–147. Rodale Press, Emmaus, Penn. 240 pp.

Childes, S.L. (1988) The past history, present status and distribution of the hunting dog *Lycaon pictus* in Zimbabwe. *Biol. Conserv.* **44**:301–316.

CITES (1973) 27 U.S.T. 1087, T.I.A.S. No. 8249.

Cook, R.S. (1993) Ecological issues on reintroducing wolves into Yellowstone National Park. USDI Natl. Park Serv. Sci. Monogr. NPS/NRYELL/NRSM-93/22, Denver, Colorado. 328 pp.

Cox, G.W. (1993) *Conservation Ecology.* Wm. C. Brown Publishers, Dubuque, Iowa. 352 pp.

Ewer, R.F. (1973) *The Carnivores.* Cornell University Press, Ithaca, NY, 494 pp.

Fanshawe, J.H., Frame, L.H. and Ginsberg, J.R. (1991) The wild dog – Africa's vanishing carnivore. *Oryx* **25**:137–146.

Fiennes, R. (1976) *The Order of Wolves*. Hamish Hamilton Ltd, London. 206 pp.

Fitzgerald, S. (1989) *International Wildlife Trade: Whose business is it?* World Wildlife Fund, Inc., Baltimore, Md. 459 pp.

Fritts, S.H. (1982) Wolf depredation on livestock in Minnesota. *US Fish and Wildl. Serv.* Resourc. Publ. No. 145:1–11.

Fuller, T.K. (1989) Population dynamics of wolves in north-central Minnesota. *Wildl. Mongr.* 105:1–41.

Fuller, T.K. (1995a) Comparative population dynamics of North American wolves and African wild dogs, in *Ecology and Conservation of Wolves in a Changing World* (eds L. Carbyn, S. Fritts and D. Seip). Canadian Circumpolar Institute, Edmonton, Alberta.

Fuller, T.K. (1995b) An international review of large terrestrial carnivore conservation status, in *Integrating People and Wildlife for a Sustainable Future* (eds J.A. Bissonette and P.R. Krausman). Proceedings of the first International Wildlife Management Congress, pp. 410–412. The Wildlife Society, Bethesda, Md.

Fuller, T.K. and Sampson, B.A. (1988) Evaluation of a simulated howling survey for wolves. *J. Wildl. Manage.* 52:60–63.

Fuller, T.K., Kat, P.W., Bulger, J.B. *et al.* (1992a) Population dynamics of African wild dogs, in *Wildlife 2001: Populations*, (eds D.R. McCullough and R.H. Barrett), pp. 1125–1139. Elsevier Applied Science, NY.

Fuller, T.K., Berg, W.E., Radde, G.L. *et al.* (1992b) A history and current estimate of world distribution and numbers in Minnesota. *Wildl. Soc. Bull.* 20:42–55.

Gasaway, W.C., Boertje, R.D., Grangaard, D.V. *et al.* (1992) The role of predation in limiting moose at low densities in Alaska and Yukon and implications for conservation. *Wildl. Monogr.* 120:1–59.

Ginsberg, J.R. and Macdonald, D.W. (1990) *Foxes, Wolves, Jackals, and Dogs: an action plan for the conservation of canids.* IUCN, Gland, Switzerland. 116 pp.

Gittleman, J.L. (1993) Carnivore life histories: a re-analysis in the light of new models, in *Mammals as Predators* (eds N. Dunstone and M.L. Gorman), pp. 65–86. Zoological Society of London Symposia 65, Oxford Science Publications, Oxford, England.

Griffith, B., Scott, J.M., Carpenter, J.W. and Reed, C. (1989) Translocation as a species conservation tool: status and strategy. *Science* 245:477–480.

Herrero, S. (1985) *Bear Attacks: their causes and avoidance.* Nick Lyons Books, Winchester Press, USA.

Hornocker, M.G. (1992) Learning to live with mountain lions. *Natl. Geogr.* 182:52–65.

Hummel, M. and Pettigrew, S. (1991) *Wild Hunters: Predators in Peril.* Key Porter Books Ltd, Toronto, Ontario. 244 pp.

IUCN (1976) *Red Data Book Volume I. Mammalia.* IUCN, Morges, Switzerland.

Johnson, W.E. and Franklin, W.L. (submitted) Ecology of the Patagonian puma in southern Chile. *Conserv. Biol.*

Jones, L.L.C. and Raphael, M.G. (1993) Inexpensive camera systems for detecting martens, fishers, and other animals: guidelines for use and standardization. USDA Forest Serv. Gen. Tech. Rep. PNW-GTR-306, 22 pp.

Kellert, S.R. (1985) The public and the timber wolf in Minnesota. *US Fish and Wildl. Serv. Rep.*, Twin Cities, Minnesota. 175 pp.

King, S. (1994) The down underdog. *BBC Wildlife* 12:52–55.

Kitchner, A. (1991) *The Natural History of Wild Cats.* Comstock Publ. Assoc., Ithaca, NY. 280 pp.

Koeln, G.T., Cowardin, L.M. and Strong, L.L. (1994) Geographic information systems, in *Research and Management Techniques for Wildlife and Habitats,* 5th edn, (ed. T.A. Bookhout), pp. 540–566. The Wildlife Society, Bethesda, Maryland.

Kruuk, H. (1972) *The Spotted Hyaena.* University of Chicago Press, Chicago, Illinois. 335 pp.

Kucera, T.E. and Barrett, R.H. (1993) The Trailmaster camera system for detecting wildlife. *Wildl. Soc. Bull.* 21:505–508.

Lewis, D., Kaweche, G.B. and Mwenya, A. (1990) Wildlife conservation outside protected areas – lessons from an experiment in Zambia. *Conserv. Biol.* 4:171–180.

Lindzey, F.G., Thompson, S.K. and Hodges, J.I. (1977) Scent station index of black bear abundance. *J. Wildl. Manage.* 41:151–153.

Lopez, B.H. (1978) *Of Wolves and Men.* Charles Scribner's Sons, New York. 309 pp.

Lumpkin, S. (1991) Cats and culture, in *Great Cats: Majestic Creatures of the Wild.* (eds J. Seidensticker and S. Lumpkin), pp. 190–203. Rodale Press, Emmaus, Penn. 240 pp.

Martin, L.D. (1989) Fossil history of the terrestrial Carnivora, in *Carnivore Behavior, Ecology, and Evolution,* (ed. J.L. Gittleman), pp. 536–568. Cornell University Press, Ithaca, New York. 620 pp.

McDougal, C. (1987) The man-eating tiger in geographical and historical perspective, in *Tigers of the World: the biology, biopolitics, management, and conservation of an endangered species,* (eds R.L. Tilson and U.S. Seal), pp. 435–448. Noyes Publications, Park Ridge, N.J.

Mills, J.A. and Servheen, C. (1991) *The Asian Trade in Bears and Bear Parts.* World Wildlife Fund, Inc. Baltimore. Md. 113 pp.

Mills, M.G.L. (1991) Conservation management of large carnivores in Africa. *Koedoe* 34:81–90.

Mills, S. (1994) The tiger, the dragon and a plan for the rescue. *BBC Wildlife* 12:52–55.

Moore, G.C. and Parker, G.R. (1992) Colonization by the eastern coyote (*Canis latrans*), in *Ecology and Management of the Eastern Coyote,* (ed. A.H. Boer), pp. 23–37. Wildl. Res. Unit, Univ. New Brunswick, Fredericton, N.B.

Morell, V. (1994) Serengeti's big cats going to the dogs. *Science* 264:1664.

Murie, O.J. (1974) *A Field Guide to Animal Tracks.* Houghton Mifflin Company, Boston. 376 pp.

Mysterud, I. and Muus Falck, M. (1989a) The brown bear in Norway I: subpopulation ranking and conservation status. *Biological Conservation* 48:21–39.

Mysterud, I. and Muus Falck, M. (1989b) The brown bear in Norway II: management and planning. *Biological Conservation* 48:151–162.

National Information Service Corporation (1993) *Wildlife Worldwide* (1935–May 1993). Baltimore, Md.

Nowak, R.M. (1983) A perspective on the taxonomy of wolves in North America, in *Wolves in Canada and Alaska,* (ed. L. Carbyn), pp. 10–19. Can. Wildl. Ser. Rep. Ser. No. 45. Ottawa.

Nowak, R.M. (1991) *Walker's Mammals of the World,* 5th edn. The Johns Hopkins University Press, Baltimore, Maryland. 1629 pp.

O'Gara, B.W. (1988) Snow leopards and sport hunting in the Mongolian People's Republic, in *Proceedings of the Fifth International Snow Leopard Symposium*, (ed. H. Freeman), pp. 215–225. Conway Printers Private Ltd, Bombay, India.

Phillips, M.K. and Parker, W.T. (1988) Red wolf recovery: a progress report. *Conserv. Biol.* 2:139–141.

Pielou, E.C. (1991) *After the Ice Age: the return of life to glaciated North America.* The University of Chicago Press, Chicago. 366 pp.

Rabinowitz, A. (1986a) Jaguar predation on domestic livestock in Belize. *Wildl. Soc. Bull.* 14:170–174.

Rabinowitz, A. (1986b) *Jaguar: Struggle and Triumph in the Jungles of Belize.* Arbor House, New York.

Roelke, M.E., Martenson, J.S. and O'Brien, S.J. (1993) The consequences of demographic reduction and genetic depletion in the endangered Florida panther. *Current Biol.* 3:340–350.

Rogers, L.L. (1987) Effects of food supply and kinship on social behavior, movements, and population growth of black bears in northeastern Minnesota. *Wildl. Monogr.* 97:1–72.

Sanders, B. (1993a) Anthropology, history, and culture, in *Bears: Majestic Creatures of the Wild*, (ed. I. Sterling), pp. 152–163. Rodale Press, Emmaus, Penn.

Sanders, B. (1993b) The bear in literature and art, in *Bears: Majestic Creatures of the Wild*, (ed. I. Sterling), pp. 164–175. Rodale Press, Emmaus, Penn.

Schaller, G.B. (1972) *The Serengeti Lion.* University of Chicago Press, Chicago, Illinois. 480 pp.

Schaller, G.B. (1993) *The Last Panda.* University of Chicago Press, Chicago. 291 pp.

Schaller, G.B., Hu, J., Pan, W. and Shu, J. (1985) *The Giant Pandas of Wolong.* University of Chicago Press, Chicago. 298 pp.

Schemnitz, S.D. (1994) Capturing and handling wild animals, in *Research and management techniques for wildlife and habitats,* 5th edn, (ed. T.A. Bookhout), pp. 106–124. The Wildlife Society, Bethesda, Maryland.

Seidensticker, J. and Lumpkin, S. (eds) (1991) *Great Cats: Majestic Creatures of the Wild.* Rodale Press, Emmaus, Penn. 240 pp.

Servheen, C. (1989) The status and conservation of the bears of the world. 8th Intl. Conf. Bear. Res. Manage. Monogr. Ser. 2, 32 pp.

Shaffer, M.L. (1981) Minimum population size for species conservation. *Bioscience* 31:131–133.

Sheldon, J.W. (1992) *Wild Dogs: the natural history of nondomestic Canidae.* Academic Press, Inc., San Diego, Calif. 248 pp.

Smith, J.D.L., McDougal, C.W. and Sunquist, M.E. (1987) Female land tenure systems in tigers, in *Tigers of the World: the biology, biopolitics, management, and conservation of an endangered species*, (eds R.L. Tilson and U.S. Seal), pp. 97–109. Noyes Publications, Park Ridge, NJ.

Smith, K.G. and Clark, J.D. (1994) Black bears in Arkansas: characteristics of a successful translocation. *J. Mammal.* 75:309–320.

Sterling, I. (ed.) (1993) *Bears: Majestic Creatures of the Wild.* Rodale Press, Emmaus, Penn. 240 pp.

Stewart, P. (1993) Mapping the dhole. *Canid News* 1:18–21.

Strickland, D. (1983) Wolf howling in parks – the Algonquin experience in interpretation, in *Wolves in Canada and Alaska*, (ed. L. Carbyn), pp. 93–95. Can. Wildl. Serv. Rep. Ser. No. 45.

Stuart, C.T., Macdonald, I.A.W. and Mills, M.G.L. (1985) History, current status, and conservation of large mammlian predators in Cape Province, Republic of South Africa. *Biol. Conserv.* **31**:7–19.

Swank, W.G. and Teer, J.G. (1989) Status of the jaguar – 1987. *Oryx* **23**:14–21.

Terborgh, J. (1988) The big things that run the world – a sequel to E.O. Wilson. *Conserv. Biol.* **2**:402–403.

Terborgh, J. (1990) The role of felid predators in neotropical forests. *Vida Silvestre Neotropical* **2**(2):3–5.

Tilson, R.L. and Seal, U.S. (1987) *Tigers of the World: The biology, biopolitics, management, and conservation of an endangered species.* Noyes Publications, Park Ridge, NJ. 510 pp.

US Fish and Wildlife Service (1989) *Red Wolf Recovery Plan.* US Fish and Wildlife Service, Atlanta, Georgia. 110 pp.

US Fish and Wildlife Service (1991) Endangered and threatened wildlife and plants. 50 CFR 17.11 & 17.12. US Fish and Wildlife Service, Washington, DC. 37 pp.

US Fish and Wildlife Service (1993) *The Reintroduction of Gray Wolves to Yellowstone National Park and Central Idaho: draft environmental impact statement.* US Fish and Wildlife Service, Helena, Mont. 374(xxxviii) pp.

Wayne, R.K. Gilbert, D.A., Eisenhawer, A. *et al.* (1991) Conservation genetics of the endangered Isle Royale gray wolf. *Conserv. Biol.* **5**:41–51.

Weaver, J. (1978) The wolves of Yellowstone. US Dep. Inter., Natl. Park. Serv., Nat. Resourc. Rep. No. 14:1–38.

Weise, T.F., Robinson, W.L., Hook, R.A. and Mech, L.D. (1975) An experimental translocation of the eastern timber wolf. *Audubon Conserv. Rep. No. 5*, New York, N.Y. 28 pp.

White, G.C. and Garrott, R.A. (1990) *Analysis of Wildlife Radio-tracking Data.* Academic Press, Inc., San Diego, California. 383 pp.

Part Two

Habitat Change and Wildlife Responses

Changes to many forested landscapes and their wildlife are evident from the preceding chapters. The influence of some of these principal changes to different aspects of the natural histories of disparate forest wildlife species are considered in this next section.

The importance of forest conservation to the status of 36 forest raptor species in North America is considered in Chapter 6. The criteria that determine the status of these species and the current level of threats are reviewed. Forest raptors use different components of a heterogeneous forest landscape for different behaviors on a daily and seasonal basis. It is proposed that forest conservation for a mix of successional stages and a mosaic of non-forest habitat openings will benefit most raptors. Additional sampling of raptor habitat is needed to provide managers with better predictive capability and more information for forest conservation to benefit raptors. Several effective conservation strategies that can be used for raptor species that require special attention are identified.

In Chapter 7 the features of two different scales of temporal dynamics in bird communities of northern boreal forests are appraised. One model invokes year to year variation in boreal bird communities and community stability and the second model involves changes in bird communities in relation to forest succession. Links between these models and practical management and conservation of boreal bird communities are presented.

The many factors that influence wildlife conservation in Japan are presented in Chapter 8. These include problems in agriculture caused by an overabundance of wildlife in plantations. In contrast, the heavy pressure to harvest timber from domestic forests and the impact of this harvest on forest wildlife is discussed. The influence of the rising popularity of wildlife conservation in Japan is also considered. On the other hand, the number of hunters who have taken part in the control of nuisance ungulates has dramatically decreased because they have been wary of being accused of taking part in a cruel sport. The ungulate populations have benefited from this and are increasing in various localities all over the country. Finally, the consequences of policies that lead to both the overdevelopment of forests and the excess protection of ungulates are examined.

The unpredictable impact of natural climatic disturbance phenomena is presented in Chapter 9, which offers a summary of the effects of hurricanes on wildlife populations and the response of wildlife populations to those effects. Management strategies are presented that reduce the vulnerability of endangered terrestrial species to hurricanes.

The trends that have arisen in the dynamic relationship between the moose populations and the forest ecosystems in northern Scandinavia are examined in the final chapter of this section. Moose populations and forestry practices are intimately linked. These moose populations have shown a dynamic pattern characterized by declines, equilibria and expansions during the past 30 years. The authors conclude that the high potential rate of increase in the moose populations is due to low mortality, lack of predators, abundant forage and a regulated hunting system.

6

Forest raptor population trends in North America

Mark R. Fuller

6.1 INTRODUCTION

For decades there was forest wildlife management for game animals and against those species designated as pests. The other wildlife species usually were neglected. However, in the last 20 years, management for nongame animals and biodiversity has become an important, integral part of forest management. Bird conservation has had a strong influence on contemporary forest management strategies and raptors have been especially important, particularly in the United States of America. Raptors, or birds of prey, comprise hawks, falcons, eagles, vultures and owls. Because raptor management influences forest management it is important to understand the population status of these species. Furthermore, there are other reasons why raptors should be included when considering wildlife conservation in forested landscapes.

Raptor conservation in Canada and the United States is dependent on forest conservation because raptors occur throughout North American forests. In this chapter 36 of 63 raptor species are classified as forest raptors (forest inhabitants on the Caribbean and Hawaiian islands are excluded). These species inhabit the complex landscapes of natural forests, including the various successional stages with patches of open land, water and the gradients found through elevation changes. Most raptor species exhibit use of large areas and a diversity of habitat types compared with many other animal groups. Raptors are 'links' among habitats, and they 'connect' ecosystems across the landscape. Therefore, forest raptor conservation requires planning on a landscape scale. In 1979 Thomas (1979:13) noted that large-scale wildlife conservation must be accomplished through timber manage-

Conservation of Faunal Diversity in Forested Landscapes. Edited by R.M. DeGraaf and R.I. Miller. Published in 1996 by Chapman & Hall. ISBN 0 412 61890 7.

ment because timber management affects large areas of wildlife habitat and is well financed; conversely, wildlife management was poorly financed and influenced only a few acres with little effect. Today, wildlife management has a more integral role in timber management but the conservation of wide-ranging raptors remains dependent on landscape-scale forest management.

Birds of prey are important to the forest manager in their own right. A raptor species or population of special concern can dramatically influence forest management. Most notable of recent cases is the effect of spotted owl conservation on forest management in the western United States (Hunter, 1990:75; Yaffee, 1994). Raptors are important, obvious components in forest community food chains and in ecological food webs. They attract a great deal of attention from biologists, and because they are large, attractive predators, the public also is interested in raptors. Consequently, raptors frequently factor in strategies for influencing land use policy (e.g. goshawk: Suckling, 1994) and for natural resource conservation. E.O. Wilson states that conservation of megafauna, including raptors, can act like an umbrella to conserve a diversity of species because megafauna 'require so much land and so complex an environment' (Bourne, 1992). Some raptors are used as indicators of the status of forest systems, such as the northern spotted owl for old-growth Douglas fir forest (Dawson *et al.*, 1987). Some raptors can be indicators of specific perturbations such as those associated with environmental contamination (e.g. Henny, 1977a). Raptors, then, have many characteristics that make them relevant to forest wildlife conservation.

This chapter presents several topics relevant to forest raptor conservation. First, there is an evaluation of the status of each species. Next the great diversity of habitats and large extent of areas used by birds of prey in forest landscapes are emphasized and it is suggested that characteristics of forest raptors, and the lack of information about the 'cause and effects' of forest management on raptor ecology, indicate that raptor conservation is best accomplished by managing for forest biodiversity on a landscape scale. Some examples are given of the effects and importance of forest management for raptors. The chapter concludes with examples of management for spotted owls and northern goshawk on the ecosystem and landscape scale.

6.2 SELECTION OF SPECIES

Those birds of prey for which forest management is likely to affect their population status are included – that is, those species for which changes in the distribution, structure and integrity of trees in the landscape can cause an increase or decrease in raptor population size. Many raptor species use forest habitats for some of their activities and most raptors will use trees if they are available. Those species that depend on forest dominated landscapes for most of their activities have been selected subjectively, along with those species which, in Canada or the United States, exhibit regular breeding-season or wintering-season use of forests by a population segment, even though another segment can exist in non-forested habitat. For example, red-tailed hawks can nest on cliffs, perch on poles or use only the occasional tree in a prairie or agricultural-woodlot landscape. Indeed, these non-forested areas are common red-tailed hawk habitats, but populations also occur in the forests of the Great Lakes and of the Appalachian Mountains so they are included as forest raptors. The prairie falcon (*Falco mexicanus*) and burrowing owl (*Athene cunicularia*), on the other hand, occur predominantly in non-forested habitats; they do not require trees or forests for any activity and thus are not included. Certain species have been included that do not regularly occupy landscapes in which forests are the predominant feature but that do regularly use woodlands for required activities. For example, this category includes the gray hawk, a species that nests in wooded riparian habitat, which occurs in an otherwise non-forested landscape.

Some raptors spend most of their time in the woodland portion of the landscape, others use a variety of vegetation types and structure, and some rely on the forest component for only one or two activities. The species are listed with some of their general forest-associated activities. Also, a few examples are provided of how general behaviors are associated with the forest, and other behaviors with forest openings or adjacent non-forested habitat. The relationships between habitat use, forest management and forest raptor status and trends are discussed for several selected species. This chapter illustrates the diversity of ways in which raptors use the forest landscape.

6.2.1 Basis for determination of status

Status denotes a stage or a relative position, and assignment of current status requires some basis for comparison. The basis for forest raptor status is the population size of each species and specifically whether the size of the total population(s) is stable, increasing or decreasing in relation to some previously determined value. The number of individuals in a population fluctuates on a daily, seasonal, annual and sometimes multi-year basis. Status should reflect stability or change in the total population that, on average, remains within or exceeds these fluctuations, respectively.

The status designations for the birds of prey treated in this account are largely those determined by C.J. Henny, P.B. Wood and the author of this chapter for a report for the US Department of the Interior National Biological Survey (Fuller *et al.*, 1995). We consulted the Federal Register for status determinations by the US Fish and Wildlife Service. The capitalized terms Threatened, Endangered, Category 2 and Category 3 denote a legal status assigned by the US Fish and Wildlife Service under authority of the

Endangered Species Act. For species not classified under this Act, we assigned population status based on our summarization of other people's interpretations and analyses. We did not compile summary statistics or analyze data for any species. We also considered relevant literature and used the 1994 Canadian Species at Risk list (Committee on the Status of Endangered Wildlife in Canada, Birds Subcommittee) and its categories: endangered, threatened and vulnerable. We tried to draw conclusions about the status of each species on a continental scale, but often status has to be qualified because there are local or regional concerns. Local status or concern can reflect habitat modification, persecution or contamination for which information does not exist on a broader scale. Statistical results have been used when available but the status of most raptors must be established by synthesizing information from a variety of sources that used a variety of methods, across a variety of geographic scales. The impressions of local experts and the non-statistical evaluations of numerous types of survey are important components of raptor status assessment.

One convenient source of material about regional and local status is the series of proceedings from regional symposia that were sponsored by the National Wildlife Federation (1988, 1989a,b, 1990, 1991). Also, White (1994) summarized status for western raptors, and Palmer (1988a,b) and Johnsgard (1988, 1990) provide species distribution maps and brief comments about status. Detailed information occasionally occurs in the literature, in reports, and in presentations at meetings and conferences.

6.2.2   Status of forest raptors

The status of eight (22%) species (Table 6.1) is unknown because there are too few counts or estimates of abundance to establish a basis on which to judge status. Six of these species are owls. Stable status has been assigned to 21 (58%) species, but three of these are classified as vulnerable in Canada, indicating concern about recent population decreases or threat of decreases. Eight (22%) raptors are classified Endangered, Threatened or vulnerable, and five species or subspecies have Category 2 status. Raptors with these classifications are assumed to be decreasing or to have undergone a recent decrease, or are likely to be exposed to factors that can cause a threatening population decline. An exception is the Endangered/Threatened bald eagle, which is increasing. The status of increasing was assigned to two other species. Regionally and locally there are many examples of decreases in counts and population estimates, and of concerns expressed about loss of forest habitat as a threat to raptors. It is common to find differences among assessments of status that occur in different locales, and at different temporal and geographic scales.

The sharp-shinned hawk provides an example of differences in assessment of status. During the 1960s and early 1970s local sampling for contaminants

(Snyder *et al.*, 1973) and counts of eastern North American migrants (Nagy, 1977) suggested that this small forest-dwelling hawk was declining in numbers. Later studies of breeding birds in the west (Reynolds, 1989) and midwest and counts of migrants across the continent suggested stable or increasing populations (Bednarz *et al.*, 1990; Titus and Fuller, 1990). Currently, however, there is concern, based on declining counts of migrant sharp-shinned hawk in eastern North America (Kerlinger, 1992). Maturation of eastern forests might be limiting the breeding habitat of this small hawk, which usually occupies earlier forest structure succession than larger raptors. Other factors such as environmental contaminants (e.g. pesticides, acid precipitation) also are being considered (Viverette *et al.*, 1994; P. Kerlinger and C. Viverette, personal communication).

Raptor status often is assessed by considering more than one source of information. Regional and continental information about the status of some raptors is available from the Breeding Bird Survey (BBS), Christmas Bird Counts (CBC), and syntheses of data from counts of migrating birds. The BBS and CBC are conducted across the continent in the contiguous United States and southern Canada, but sampling is better in the east where more BBS routes and CBC circles are covered more consistently. Among raptors the most reliable data are obtained for open country and edge inhabiting hawks. There rarely are useful data for *Accipter* hawks or for owls.

An evaluation of the status of *Buteo* hawks in the northeast United States was based in part on analyses of BBS and CBC data. Titus *et al.* (1989) stratified counts by 'less developed regions' and by 'more developed regions' in which industrialization and suburbanization were likely to have reduced natural habitat (including forests) during the 22-year period of consideration. Red-tailed hawk counts increased in the developed regions, while broad-winged hawks, which are dependent on larger tracts of forest, decreased in more developed strata and were stable in less developed strata. A continental analysis of BBS data revealed that American kestrel counts were stable or increasing in most areas (Fuller *et al.*, 1987), but the analysis of BBS data from the northeast suggested a decrease in the more developed strata (Schueck *et al.*, 1989). The CBC data from the northeast United States were analyzed by regression and suggested an increase in the red-tailed hawk population (Titus *et al.*, 1989) and no change in American kestrels.

Count data from three fall migration and three spring migration concentration sites in eastern North America were pooled and analyzed for trends for periods ranging from 7 to 15 years. Titus and Fuller (1990) concluded that migrant broad-winged hawk counts had declined while red-tailed hawk and American kestrel counts were stable. The recent analyses of counts of migrants from four sites in the western United States indicate stable numbers of red-tailed hawk and American kestrel (S.W. Hoffman, J.C. Bednarz and W.R. De Ragon, personal communication). Data such as these from the BBS, CBC and migrant counts, usually combined with local

Table 6.1   Status and trends of forest raptors in Canada and the United States of America

Species	Status/trend	Comment
Black vulture (*Coragyps atratus*)	stable	Population estimation difficult due to flocking, wide-ranging daily movements and secretive nesting
Turkey vulture (*Cathartes aura*)	stable	
California condor (*Gymnogyps californianus*)	Endangered; extirpated from wild 1987	Captive propagation and release underway
Osprey (*Pandion haliaetus*)	stable	Good information base
American swallow-tailed kite (*Elanoides forficatus*)	stable	Very limited from historic range
Mississippi kite (*Ictinia mississippiensis*)	increasing	Range expansion
Bald eagle (*Halaeetus leucocephalus*)	Threatened, Endangered; increasing	Status reassessment underway
Sharp-shinned hawk (*Accipiter striatus*)	stable	Regional differences
Cooper's hawk (*Accipiter cooperii*)	stable; vulnerable	
Goshawk (*Accipiter gentilis*)	unknown	Category 2; petition to list *A.g. laingi*
Common black hawk (*Buteogallus anthracinus*)	stable	Limited distribution
Gray hawk (*Buteo nitidus* or *Austurina plagiata*)	stable	Category 2, limited distribution
Red-shouldered hawk (*Buteo lineatus*)	stable; vulnerable	Local concern
Broad-winged hawk (*Buteo platypterus*)	stable	Migration count decline in 1980s
Short-tailed hawk (*Buteo brachyurus*)	stable	Northern range limit, < 500 birds
Zone-tailed hawk (*Buteo albonotatus*)	stable	Northern range limit ~100 pairs
Red-tailed hawk (*Buteo jamaicensis*)	increasing	
Golden eagle (*Aquila chrysaetos*)	stable	

Table 6.1   continued

Species	Status/trend	Comment
American kestrel (*Falco sparverius*)	stable	*F.s. paulus*, Category 2
Merlin (*Falco columbarius*)	stable	
Aplomado falcon (*Falco femoralis septentrionalis*)	Endangered	Captive propagation and release underway
Flammulated owl (*Otus flammeolus*)	unknown; vulnerable	Recent surveys reveal more birds, larger range
Eastern screech-owl (*Otus asio*)	stable	
Western screech-owl (*Otus kenniccottii*)	stable	
Whiskered screech-owl (*Otus trichopsis*)	unknown	Northern range limits
Great horned owl (*Bubo virginianus*)	stable	
Northern hawk owl (*Surnia ulula*)	unknown	
Northern pygmy owl (*Glaucidium gnoma*)	unknown	Current survey efforts
Ferruginous pygmy owl (*Glaucidium brasilianum cactorum*)	unknown	*G.b. cactorum*, Category 1. Loss of riparian forest
Elf owl (*Micrathene whitneyi*)	unknown	Current survey efforts
Spotted owl (*S. occidentalis caruina, S.o. lucida*)	Threatened; endangered	*S.o. occidentalis*, Category 2
Barred owl (*Strix varia*)	stable	Western range expansion
Great gray owl (*Strix nebulosa*)	stable; vulnerable	
Long-eared owl (*Asio otus*)	unknown	Local concern
Boreal owl (*Aegolius funereus*)	unknown	Population estimation difficult
Saw-whet owl (*Aegolius acadicus*)	stable	Concern in southeastern Alaska

evaluations (e.g. National Wildlife Federation, 1990), can be used to assess raptor status. Thus, it is concluded that broad-winged hawk status is stable but it is noted that a decline is indicated by some data. The red-tailed hawk status is increasing, and the American kestrel status is stable but with special status (Category 2) of a local population (Table 6.1). Piecing together a status assessment in this way is not conclusive in a statistical sense, and is not satisfying in light of the incomplete geographic coverage of surveys and the incomplete analyses of surveys. Furthermore, commonly there are concerns about status that result from isolated local assessments, of which there are too few to provide conclusive data about population status. The Cooper's hawk provides an example of conflicting evidence.

Rosenfield and Bielefeldt (1993) summarized trends in Cooper's hawk populations as 'common' (and presumably stable) in the west and varying in the east. This forest raptor was on the state of Wisconsin Threatened list, but was removed after general survey and intensive population studies in Wisconsin suggested broad distribution and no decline (Rosenfield and Anderson, 1983; Rosenfield and Bielefeldt, 1993). In 16 other eastern states the species remains on special status lists, 'usually without support of demographic or population research' (Rosenfield and Bielefeldt, 1993). Counts of migrant Cooper's hawk increased in the eastern United States (Titus and Fuller, 1990) and were stable or increasing in the west (S.W. Hoffman, J.C. Bednarz and W.R. De Ragon, personal communication). The conclusion is that, on a continental basis, Cooper's hawk status is stable. Beyond this rough evaluation, carefully designed field surveys are required to determine the real population status of the Cooper's hawk and many other raptor species, especially owls.

Recently, Hayward and Verner (1994) assessed the status of three forest owls: flammulated, boreal and great gray. The flammulated owl was characterized as 'rare' until people began imitating or broadcasting their vocalization and discovered that they occurred over a much larger range and in greater numbers than expected. Despite now being described as locally abundant, their population trends cannot be estimated until a historical basis is established. Fifteen years ago the boreal owl was not recognized as a breeding bird in the contiguous 48 states. Now it is known to breed in several states but its population dynamics are unknown. Similarly there are no regional or local trend data for the great gray owl, which breeds regularly in the southern part of its range, where it is assumed to be 'stable'. However, its breeding fluctuates in the north with prey, and Canada classifies the species as vulnerable. Little effort is expended to monitor or study these species. Hayward and Verner (1994) note that the species' broad distribution, much of it in areas remote from intense human activity, might lead one to assume that the owls' populations are not threatened by current forest management practices. This assumption has no basis in quantifiable population data or in information about the effects of forest management on the owls.

Our ignorance about the distribution and status of forest raptors has been costly. In 1968 there were only 27 records of the northern spotted owl in Oregon, revealing that we knew little about this raptor. Today, because of the most intensive study effort ever undertaken for a forest raptor, there are about 2700 records of sites occupied by this owl in Oregon (E. Forsman, personal communication). Unfortunately, the owl has become classified as Threatened, the forest landscape has deteriorated and the socioeconomics of the region have been disrupted (Dietrich, 1992) during the intervening 25 years. Similarly, in the early 1970s only two or three northern goshawk breeding areas were monitored in the US Forest Service Southwest Region. In 1990 a petition was filed to list the species under the Endangered Species Act. In 1992 about 325 areas were monitored. This increase is not due to an increase in northern goshawks; it is due to more survey effort to find goshawks. In the meantime, forest management has changed dramatically in the Southwest Region in order to conserve northern goshawks (Reynolds *et al.*, 1992). Knowledge of the population status of animals and plants is a key element for wise conservation of our natural resources and for avoiding the costly process of managing Threatened and Endangered species.

6.3 FOREST CHARACTERISTICS ASSOCIATED WITH RAPTORS

Raptor habitat comprises the food, water and shelter (nests, roosts and perches) that the birds require to survive and reproduce, but it is the species and structure (vertical and horizontal) of vegetation in the forest that forest managers work with most often. Habitat data from a variety of spatial scales are useful to forest managers. Hunter (1990) presents material in context of 'The Macro Approach, Managing Forest Landscapes' and 'The Micro Approach, Managing Forest Stands'. Several other authors have discussed wildlife habitat in the context of forest management and presented tables and matrices to relate the needs of selected species to components of the forest (e.g. Thomas, 1979; Verner and Boss, 1980; Hoover and Wills, 1987; DeGraaf *et al.*, 1992). Usually these authors have assigned raptors to forest cover types in an area (e.g. Society of American Foresters, 1964). Across Canada and the United States most raptor species occur in too many cover types to list here. General descriptions of forest raptor habitat associations are given in books such as Palmer's (1988) and Johnsgard's (1988, 1990), and in *The Birds of North America* series edited by A. Poole and F. Gill (e.g. Rosenfield and Bielefeldt, 1993). State/provincial bird books, the National Wildlife Federation Symposia and journal literature provide more detailed descriptions of the flora and structure that are associated with raptor occurrence.

Table 6.2 lists the forest raptors and some of the characteristics of forests with which they are associated. The lack of an entry under a characteristic does not mean that the characteristic is unimportant. Rather it can reflect relatively less importance than other characteristics or a simple lack of information. Much of the information that we do have about raptor habitat use is based on local studies that are not necessarily representative samples from the range of the species. Several generalities about forest raptor habitat can be gleaned from Table 6.2:

- 86% of the species are associated with forests that are in the later succession stages that include large trees;
- 69% of the species use snags, cavities and especially large trees for nesting, roosting or perching;
- 72% of the species are associated with coniferous forests and deciduous forests (rather than only one type) or a mixed forest;
- 94% of the species use forest edges or non-forested habitat in conjunction with forested habitat.

The table reveals that many raptors have some special requirements, such as a nest cavity, but they can find these requirements in a variety of forest landscapes. Also, these forest landscapes often are heterogeneous, comprising more than one successional stage and non-forested habitat.

Individual raptors can use a variety of vegetation cover types and successional stages within their home range. Examples from female and male broad-winged hawks and barred owls illustrate the use of numerous vegetation cover types in central Minnesota (Table 6.3). These data were obtained from radio-marked birds and are based on the predominant habitat in each 0.65 ha grid square in which a location estimate occurred. In the course of 2 months or more, these individuals used from 10 to 13 cover types. Ivlev's Electivity Index (EI) reveals the ranking of use based on proportion of a bird's locations in a type compared with the proportion of the cover type in the total area used. These forest raptors do not use all types with similar intensity but do use a variety of habitats.

There are shifts in habitat use depending on time of day and types of activity. During the daily period when these raptors are more active, they use habitats more extensively than when they are less active (Table 6.4). The signal from each bird's transmitter is an indicator of body movements or gross activity. This activity can be analyzed in the context of habitat use during rest (no movement), activity-in-place, and flight (body movement associated with a change in locations). These analyses re-emphasize that these broad-winged hawks and barred owls make use of a heterogeneous forest landscape (Table 6.5).

As top predators, raptors range over large areas compared with many other groups of terrestrial birds and, generally, bigger species use larger home ranges than smaller species (Newton, 1979:63; Johnsgard, 1990:61).

The broad-winged hawk and barred owl are in the upper third of the range of body masses of North American forest raptors. The area traversed by those individuals for which the habitat use data are given ranged from 150 ha to 565 ha (Table 6.3). As illustrated for the female barred owl, the shape of the range is often irregular (Figure 6.1). Size and shape of home ranges are dependent on the mosaic of cover types available and each bird's habitat requirements. While biologists can generalize about home ranges used by raptors of a certain size, there are no models for predicting habitat types in conjunction with range sizes.

The size of the area used and the habitat composition also change seasonally for many raptors. The extent of seasonal change varies within and among species, and by year. For example, the broad-winged hawk migrates every year from its North American breeding range along forested migration routes and spends the winter in South America. Some sharp-shinned hawks migrate to Latin America while others remain in North America. Northern goshawks leave their northern breeding grounds only periodically, usually in association with reduced prey abundance. In Idaho, boreal owls move to a lower altitude in the winter and use larger areas than during the breeding season (Hayward *et al.*, 1987). The barred owl remains in one area throughout the year.

The types of migration and details about the timing of movements by migrants are presented in literature such as Senner and Fuller (1989) and Kerlinger (1989), respectively. The general forest cover types associated with migrating raptors can be inferred by identifying the habitat at those locations at which migrants are reported (e.g. Journal of the Hawk Migration Association of North America). Forest raptors generally migrate through forested habitat rather than crossing prairies or large bodies of water, and they often become concentrated at topographic and geographic features such as mountain ridges and shorelines (Kerlinger, 1989: Chapter 7). At these concentrations a forested landscape is probably very important, but there are few studies of local habitat use during migration (Holthuijzen *et al.*, 1985) that provide data on which to base forest management. General information is given about migration and, when known, migratory and wintering habitat associations in Palmer (1988) and Johnsgard (1988, 1990). The use of roost and the food habits of wintering bald eagles represent some of the more detailed data about seasonal changes of a migrant raptor (e.g. Stalmaster, 1987; Buehler *et al.*, 1991 a,b).

The information provided above suggests that, generally, raptor–habitat associations involve relatively large areas of heterogeneous forest landscape. Thus, forest management on a landscape scale, with the goal of conserving natural biodiversity, will contribute to forest raptor conservation. Currently, however, forest managers must deal with the consequences of natural disturbances (e.g. fire) and historical practices (e.g. fire suppression and timber harvest) as they plan management to conserve ecosystems in conjunction

Table 6.2 Forest raptor habitat associations in Canada and the United States of America

Species	Forest use	Forest type	Forest success. (sap. pol. lrge)	Canopy cover	Forest edge	Open water	Wet-land	Grass forb	Shrub	Seed-ling	Special feature	Comments
Black vulture (*Coragyps atratus*)	Reg	Mixed	nr								Cavity n	Forage open; nest in dense understory, or in building
Turkey vulture (*Cathartes aura*)	Reg east, Occ west	Mixed	Nr								Cavity, Log N	Forage open; nest in dense understory
California condor (*Gymnogyps californianus*)	Reg	Mixed	Nr								Cliff/ Cavity N	Forage open; contaminants threat
Osprey (*Pandion haliaetus*)	Reg	Mixed	NRP		N	F					Large tree nest	Contaminants threat
American swallow-tailed kite (*Elanoides forficatus*)	Reg	Mixed	Nf		f		F	f			Snag r	Isolated or exposed tree for easy access to nest
Mississippi kite (*Ictinia mississippiensis*)	Reg	Mostly Decid.	Nf		f	f		f			Snag p	Riparian forest important in prairie landscape
Bald eagle (*Haliaeetus leucocephalus*)	Reg	Mixed	NPR		r	F	f				Snag p	Large emergent or accessible tree for stick nest; contaminants threat

Common name (Scientific name)	Occurrence	Forest type						Comments
Sharp-shinned hawk (*Accipiter striatus*)	Reg	Mostly Conif.	N f		f		f	Use forest cover for foraging; biocide threat
Cooper's hawk (*Accipiter cooperii*)	Reg	Mixed	n	Large	nf		f	f	Use forest cover for foraging; biocide threat
Goshawk (*Accipiter gentilis*)	Reg	Mixed	Nr		nf		f	f	Use forest cover for foraging
Common black hawk (*Buteogallus anthracinus*)	Reg	Decid.	Nrp		F		f		Riparian forest in desert landscape
Gray hawk (*Buteo nitidus* or *Austurina plagiata*)	Reg	Decid.	Npf		f				Riparian forest in desert landscape
Red-shouldered hawk (*Buteo lineatus*)	Reg	Mixed	Npf		pf	npf			Larger than average tree for nest; water/wetlands important
Broad-winged hawk (*Buteo platypterus*)	Reg	Mostly Decid.	npf		nf	npf	f		Water/wetlands important
Short-tailed hawk (*Buteo brachyurus*)	Reg	Mixed	NR		F	F	f	f	Woodland nest site proximity to open foraging habitat
Zone-tailed hawk (*Buteo albonotatus*)	Occ	Mixed	N		f		f	f	Large nest tree, near open forage areas, often in mountainous terrain
Red-tailed hawk (*Buteo jamaicensis*)	Occ	Mixed	N		F		f	f	Forest to lone tree for nesting, open for foraging

Table 6.2 continued

Species	Forest use	Forest type	Forest success.			Canopy cover	Forest edge	Open water	Wet-land	Grass forb	Shrub	Seed-ling	Special feature	Comments
			sap.	pol.	lrge									
Golden eagle (*Aquila chrysaetos*)	Occ	Mixed			N		f		f	f	f			Large tree for nest, near open foraging habitat
American kestrel (*Falco sparverius; F.s. paulus*)	Occ	Mixed			n		fp		f	f	f			Hole or cavity nest
Merlin (*Falco columbarius*)	Reg	Mostly Conif.			np		Fp		f	f	f			Proximity to open foraging habitat; riparian forest
Aplomado falcon (*Falco femoralis septentrionalis*)	Occ	Mixed		npr	npr		fp		f	f	f			Riparian forest
Flammulated owl (*Otus flammeolus*)	Reg	Mostly Conif.			NPR	medium	fp							Cavities for nesting
Eastern screech-owl (*Otus asio*)	Reg	Mixed			NR		fp			f	f			Cavities for nesting
Western screech-owl (*Otus kennicotti*)	Reg	Mixed			NR									Cavities for nesting; riparian forest locally
Whiskered screech-owl (*Otus trichopsis*)	Reg	Mostly Decid.			NR		fp							Cavities for nesting
Great horned owl (*Bubo virginianus*)	Occ	Mixed			NR		fp		f	f	f			
Northern hawk owl (*Surnia ulula*)	Reg	Mixed			p		p		f	f	f			Cavities and stubs for nesting

Species										
Northern pygmy owl (*Glaucidium gnoma*)	Reg	Mixed		Nr		FP	f	f		Cavities for nesting, nearby open for foraging
Ferruginous pygmy owl (*Glaucidium brasilianum cactorum*)	Reg	Mostly Decid.		npr		nf	f	f		Cavities for nesting and riparian forest
Elf owl (*Micrathene whitneyi*)	Reg	Decid.		Npr		npf	f	f		Woodpecker cavity for nesting
Spotted owl (*Strix occidentalis*)	Reg	Mixed		NPR	high					
Barred owl (*Strix varia*)	Reg	Mixed		NPR		pf	f			Nest cavities
Great gray owl (*Strix nebulosa*)	Reg	Mixed		npr		pf	f	f	Snag stubs	Deciduous nest trees among conifers, and openings for foraging
Long-eared owl (*Asio otus*)	Reg	Mixed	nR	Npr		pF	f	f		
Boreal owl (*Aegolius funereus*)	Reg	Mixed		NFR			f	f		Nest cavities
Saw-whet owl (*Aegolius acadicus*)	Reg	Mixed	R	Nfr		pR	f	f		Nest cavities from woodpeckers

Key:

Reg = regular; Occ = occasional;

Mixed = mixed coniferous–deciduous; Decid. = deciduous; Conif. = coniferous;

Success. = successional stage; sap. = sapling; pol. = pole size; lrge = large;

N = nest, extensive use; n = nest, less extensive use;

P = perch, extensive use; p = perch, less extensive use;

F = forage, extensive use; f = forage, less extensive use.

Table 6.3 Habitat use by adult barred owls (no. 717b ♀, no. 731 ♂) and adult broad-winged hawks (no. 857 ♀, no. 860 ♂); Ivlev's electivity index (EI = (% observed − % expected)/(% observed + % expected) allows the vegetation types to be ranked (superscripts 1–13) by their use in relation to their occurrence in the birds' range

Habitat type	No. 717b Total study period (25 May–3 August)			No. 731 Total study period (7 March–6 June)			No. 857 Total study period (29 June–24 August)			No. 860 Total study period (10 July–21 August)		
	% total home range	% total locations	EI rank	% total home range	% total locations	EI rank	% total home range	% total locations	EI rank	% total home range	% total locations	EI rank
Agric. field	—	—	—	9.0	0.5	-0.890[10]	5.4	0.9	-0.707[11]	11.9	0.8	-0.882[10]
Undist. field	16.8	1.7	-0.822[9]	13.7	6.2	-0.380[5]	14.8	5.9	-0.430[9]	17.4	6.3	-0.468[7]
Field–forest edge	13.4	6.3	-0.361[7]	15.7	23.5	0.199[2]	11.6	6.4	-0.288[8]	15.8	16.6	0.023[5]
Oak upland	14.2	29.2	0.345[2]	22.4	31.4	0.168[3]	8.9	18.4	0.349[1]	19.0	24.6	0.129[3]
Alder lowland	3.0	2.6	-0.071[5]	12.5	24.9	0.329[1]	15.4	16.4	0.030[5]	13.2	35.2	0.455[1]
White cedar	16.8	9.5	-0.277[6]	0.2	0.0	-1.000[11]	14.2	24.1	0.259[2]	2.2	0.7	-0.521[9]
Tamarack	9.9	9.1	-0.045[4]	2.2	0.3	-0.767[9]	7.6	10.0	0.140[4]	1.7	3.2	0.302[2]
Mixed decid. lowland	10.8	23.7	0.376[1]	10.5	10.5	-0.000[4]	8.6	12.0	0.159[3]	7.1	8.2	0.071[4]
Conif.–decid. upland	10.3	17.3	0.251[3]	0.3	0.1	-0.641[7]	1.3	0.9	-0.156[7]	0.5	T	-0.971[11]
Open marsh	3.9	0.7	-0.704[8]	11.2	2.1	-0.683[8]	9.7	2.9	-0.544[10]	9.5	3.4	-0.475[8]
Oak savannah	—	—	—	—	—	—	0.1	T	-0.879[12]	1.7	1.1	-0.230[6]
Pole white cedar–tamarack	—	—	—	2.2	0.5	-0.627[6]	2.1	2.1	0.019[6]	—	—	—
Open water	0.9	0.0	-1.000[10]	—	—	—	0.4	T	-0.979[13]	—	—	—
Total home range (ha)	150.2			386.9			503.1			565.2		
Total locations	40148			34117			24169			14595		

T = trace.

with mankind's use of forest resources. Additional information about forest raptor ecology and the results of historical practices on raptor occurrence are required to design landscape conservation strategies. There is some information about raptor distribution and habitat use that provides examples and guidance for current management and planning.

6.4 RAPTOR OCCURRENCE AND FOREST CONDITIONS

Our insights about the relationships among forest management practices and raptor population status and conservation largely exist at two levels. At the first level we can document the occurrence of raptor species in association with extensive areas of natural forest cover or regenerated forest. Also we can detect the absence of some species when the amount of natural forest cover is reduced on a regional scale or when the natural mix of several stages is modified. A second level of insight comes from our association of certain characteristics and measurements of forest habitat with the presence of raptors, and from a few cases of relating raptor presence and habitat use to forest management practices.

6.4.1 Distribution and range limits

The distribution and status of raptors can be affected by the distribution and condition of forests on several scales. At the scale of species' range limits, the common black hawk and the gray hawk are examples of species that naturally occur only peripherally in the southwestern United States. Their future distribution here is dependent upon conservation of forested riparian habitat. The Bureau of Land Management (of the US Department of the Interior) is actively managing riparian habitat to sustain biodiversity, including nesting raptors. Several other forest raptors occur in the United States only at the limits of their distribution (Table 6.1), and their continued existence there depends on conservation of forest habitat.

The distribution and status of the swallow-tailed kite provides another example of a relationship between raptor and forest habitat area on the scale of the species distribution. We know that from about 1880 to 1940 the numbers of the swallow-tailed kite were reduced dramatically and it was extirpated from the northern extent of its range along the Mississippi River corridor into central Minnesota (Robertson, 1988). The reduction of forested wetlands, or swamps, in which it nested, and loss of adjacent marshes and grasslands in which it foraged, has led to the swallow-tailed kite's limited distribution in the southeastern United States (Robertson, 1988; Johnsgard, 1990). Its decline is attributed generally to 'loss of habitat', which is a cause of decline frequently cited for many raptor populations.

Within the historic distribution range of the California condor the forested landscape of southern California was the last refuge of this

Table 6.4 Habitat use by adult barred owls (no. 717b ♀, no. 731 ♂) and adult broad-winged hawks (no. 857 ♀, no. 860 ♂) during the daily period of reduced activity (i.e. mostly daytime for owls) and the active time (i.e. mostly night-time for owls)

(a)

Habitat type	No. 717b Reduced activity			No. 717b Active			No. 731 Reduced activity			No. 731 Active		
	% total home range	% total locations	EI rank rank	% total home range	% total locations	EI rank rank	% total home range	% total locations	EI rank rank	% total home range	% total locations	EI rank rank
Agric. field	–	–	–	–	–	–	0.6	0.1	-0.861^8	8.8	1.3	-0.751^{10}
Undist. field	6.2	T	-0.999^7	16.7	6.0	-0.474^8	16.0	6.6	-0.414^5	14.4	6.4	-0.388^6
Field–forest edge	11.1	1.9	-0.703^6	13.2	17.8	0.148^3	19.8	23.6	0.087^3	15.9	25.5	0.233^2
Oak upland	19.8	27.7	0.168^3	14.1	33.0	0.402^1	27.5	33.3	0.096^2	21.5	25.7	0.089^3
Alder lowland	1.2	2.8	0.388^2	3.1	2.1	-0.0184^5	14.1	22.0	0.221^1	12.5	27.4	0.374^1
White cedar	16.1	10.8	-0.194^5	17.2	6.1	-0.478^9	0.3	0.0	-1.000^{10}	0.2	0.0	-1.000^{11}
Tamarack	16.1	10.8	-0.194^5	9.3	4.4	-0.358^7	0.6	T	-0.984^9	2.3	0.7	-0.516^7
Mixed decid. lowland	12.4	28.4	0.394^1	11.0	11.3	0.011^4	10.5	12.5	0.087^4	10.7	8.8	-0.098^4
Conif–decid. upland	17.3	17.4	0.004^4	10.6	16.9	0.230^2	–	–	–	0.4	0.2	-0.307^5
Open marsh	–	–	–	4.0	2.5	-0.233^6	7.0	1.2	-0.702^6	11.4	3.6	-0.519^8
Oak savannah	–	–	–	–	–	–	3.5	0.6	-0.704^7	1.9	0.4	-0.650^9
Pole white cedar–tamarack	–	–	–	–	–	–	–	–	–	–	–	–
Open water	–	–	–	0.88	0.0	-1.000^{10}	–	–	–	–	–	–
Total home range (ha)	52.4			146.9			20.1			367.5		
Total locations	29192			10953			18861			13397		

T = trace.

(b)

No. 857   No. 860

Habitat type	No. 857 Reduced activity			No. 857 Active			No. 860 Reduced activity			No. 860 Active		
	% total home range	% total loca-tions	EI rank	% total home range	% total loca-tions	EI rank	% total home range	% total loca-tions	EI rank	% total home range	% total loca-tions	EI rank
Agric. field	5.6	1.7	-0.501[10]	1.4	0.1	-0.905[10]	12.0	1.3	-0.810[10]	3.0	0.1	-0.968[7]
Undist. field	14.9	6.9	-0.369[8]	7.8	4.9	-0.231[7]	17.5	10.9	-0.230[8]	–	–	–
Field–forest edge	11.6	11.7	0.004[6]	11.0	0.7	-0.885[9]	15.8	13.8	-0.065[6]	6.1	20.3	0.540[1]
Oak upland	8.9	17.3	0.319[1]	10.1	29.6	0.321[1]	19.1	21.4	0.056[4]	25.8	29.1	0.061[3]
Alder lowland	15.5	12.6	-0.104[7]	15.1	20.5	0.141[3]	13.2	26.0	0.325[2]	33.3	47.8	0.178[2]
White cedar	14.2	26.7	0.305[2]	18.8	21.2	0.060[5]	1.8	1.2	-0.216[7]	–	–	–
Tamarack	7.6	7.7	0.007[5]	9.2	12.6	0.158[2]	1.7	4.2	0.417[1]	7.6	1.9	-0.605[4]
Mixed decid. lowland	8.7	10.8	0.110[4]	10.1	13.1	0.129[4]	7.1	13.8	0.318[3]	15.2	0.5	-0.932[6]
Conif.–decid. upland	1.3	1.7	0.142[3]	2.3	0.1	0.921[11]	0.5	T	-0.950[11]	–	–	–
Open marsh	9.6	2.1	-0.634[11]	10.1	3.6	-0.470[8]	9.6	5.9	-0.239[9]	6.1	0.0	-1.00[8]
Oak savannah	0.1	T	-0.781[12]	–	–	–	1.7	1.6	-0.045[5]	3.0	0.4	-0.764[5]
Pole white cedar–tamarack	2.1	0.8	-0.464[9]	3.7	3.6	-0.005[6]	–	–	–	–	–	–
Open water	0.4	T	-0.960[13]	0.5	0.0	-1.00[12]	–	–	–	–	–	–
Total home range	501.2			141.2			562.7			42.7		
Total locations		12565			11598			8424			6169	

T = trace.

Table 6.5 Habitat use by adult barred owls (no. 717b ♀, no. 731 ♂) and adult broad-winged hawks (no. 857 ♀, no. 860 ♂) when at rest (no body movement), when active in place, and in association with flight (a change in location)

(a)

	No. 717b								
	At rest			Active in place			In flight		
Habitat type	% total home range	% total locations	EI rank	% total home range	% total locations	EI rank	% total home range	% total locations	EI rank
Agric. field	–	–	–	–	–	–	–	–	–
Undist. field	16.9	1.6	-0.828[9]	10.5	6.9	-0.207[7]	17.8	6.1	-0.488[8]
Field–forest edge	14.1	6.2	-0.389[7]	14.0	12.0	-0.080[6]	15.2	17.2	0.062[5]
Oak upland	13.6	29.3	0.365[2]	14.0	22.0	0.221[1]	13.6	18.8	0.161[2]
Alder lowland	2.4	2.6	0.055[4]	1.8	2.5	0.178[2]	2.6	1.9	-0.163[7]
White cedar	16.9	9.5	-0.279[6]	19.3	6.3	-0.508[9]	13.6	12.0	-0.063[6]
Tamarack	9.9	9.1	-0.042[5]	8.8	12.0	0.153[4]	8.9	10.6	0.087[4]
Mixed decid. lowland	10.3	23.8	0.395[1]	10.5	14.5	0.158[3]	10.5	14.1	0.148[8]
Conif.–decid. upland	10.8	17.3	0.230[3]	17.5	22.0	0.113[5]	12.0	17.9	0.195[1]
Open marsh	4.2	0.7	-0.726[8]	3.5	1.9	-0.031[8]	4.7	1.4	-0.539[9]
Oak savannah	–	–	–	–	–	–	–	–	–
Pole white cedar–tamarack	–	–	–	–	–	–	–	–	–
Open water	0.9	0.0	-1.000	–	–	–	1.05	0.0	-1.000
Total home range (ha)	137.8			73.8			123.6		
Total locations		39823			159			425	

(b)

No. 731

Habitat type	At rest			Active in place			In flight		
	% total home range	% total locations	EI rank	% total home range	% total locations	EI rank	% total home range	% total locations	EI rank
Agric. field	9.1	0.5	-0.897[10]	3.3	1.1	-0.509[7]	7.6	1.1	-0.751[10]
Undist. field	13.4	6.2	-0.367[5]	12.2	4.4	-0.476[1]	13.5	8.9	-0.203[5]
Field–forest edge	15.8	23.5	0.198[2]	18.9	26.5	0.167[3]	16.2	20.1	0.108[2]
Oak upland	22.5	31.5	0.168[3]	20.6	28.7	0.164[4]	23.2	28.0	0.094[3]
Alder lowland	12.6	24.8	0.327[1]	14.7	25.9	0.273[2]	13.1	23.4	0.283[1]
White cedar	0.2	0.0	-1.000[11]	–	–	–	–	–	–
Tamarack	2.2	0.3	-0.765[9]	0.8	0.0	-1.000[10]	1.1	0.2	-0.660[9]
Mixed decid. lowland	10.6	10.5	-0.003[4]	12.8	9.1	-0.169[6]	10.7	11.4	0.030[4]
Conif.–decid. lowland	0.3	0.1	-0.721[8]	0.6	0.5	-0.012[5]	0.2	0.1	-0.323[6]
Open marsh	11.2	2.1	-0.689[7]	13.1	3.1	-0.614[9]	12.0	5.7	-0.356[7]
Oak savannah	2.2	0.5	-0.637[6]	2.8	0.7	-0.607[8]	2.5	1.1	-0.402[8]
Pole white cedar–tamarack	–	–	–	–	–	–	–	–	–
Open water	–	–	–	–	–	–	–	–	–
Total home range	385.6			232.3			4307.3		
Total locations		33038			735			929	

(c)

No. 857

Habitat type	At rest			Active in place			In flight		
	% total home range	% total locations	EI rank	% total home range	% total locations	EI rank	% total home range	% total locations	EI rank
Agric. field	5.1	0.9	-0.698[11]	3.9	1.7	-0.400[9]	3.4	0.9	-0.566[12]
Undist. field	14.7	5.9	-0.426[9]	15.8	4.2	0.579[11]	16.6	5.4	-0.510[10]
Field–forest edge	11.8	6.2	-0.310[8]	10.0	13.2	0.135[5]	11.8	9.1	-0.127[8]
Oak upland	8.9	18.5	0.349[1]	6.8	16.1	0.404[1]	10.3	14.9	0.183[3]
Alder lowland	15.1	16.5	0.047[5]	14.3	11.0	-0.131[8]	10.6	10.0	-0.029[7]
White cedar	14.4	24.1	0.251[2]	17.7	25.8	0.186[2]	17.5	30.5	0.270[1]
Tamarack	7.7	10.1	0.133[4]	7.7	9.3	0.097[6]	9.1	11.7	0.123[4]
Mixed decid. lowland	8.8	11.9	0.151[3]	9.0	13.0	0.183[3]	8.8	8.7	-0.022[6]
Conif.–decid. upland	1.3	0.9	-0.189[7]	1.9	2.8	0.176[4]	1.9	3.0	0.229[2]
Open marsh	9.7	2.9	-0.541[10]	9.2	1.7	-0.696[12]	6.5	2.0	-0.523[11]
Oak savannah	0.1	T	-0.937[12]	0.2	0.2	-0.077	—	—	—
Pole white cedar–tamarack	2.1	2.2	0.018[6]	2.8	1.1	-0.433[10]	3.0	3.5	0.084[5]
Open water	0.4	T	-0.979[13]	0.6	0.0	-1.00[13]	0.6	0.2	-0.501[9]
Total home range	494.7			303.0			347.1		
Total locations		23544			546			538	

T = trace.

(d)

No. 860

Habitat type	At rest			Active in place			In flight		
	% total home range	% total locations	EI rank	% total home range	% total locations	EI rank	% total home range	% total locations	EI rank
Agric. field	8.1	0.7	-0.852^{10}	6.4	1.6	-0.597^{10}	5.4	1.5	-0.566^{10}
Undist. field	19.9	6.3	-0.517^{8}	14.4	6.1	-0.408^{8}	20.1	8.3	-0.418^{9}
Field–forest edge	13.4	16.9	0.115^{4}	13.9	10.5	-0.138^{7}	13.5	10.8	-0.109^{7}
Oak upland	19.5	24.8	0.120^{3}	22.7	25.1	0.050^{4}	20.8	20.5	-0.009^{5}
Alder lowland	14.7	35.7	0.415^{7}	16.8	21.5	0.120^{3}	15.4	29.8	0.318^{3}
White cedar	1.8	0.7	-0.469^{7}	0.8	0.8	0.005^{5}	0.9	1.2	0.157^{4}
Tamarack	1.5	3.0	0.331^{2}	1.6	6.1	0.582^{1}	1.4	4.4	0.514^{1}
Mixed decid. lowland	7.1	7.8	0.041^{5}	6.7	19.8	0.496^{2}	7.4	15.7	0.360^{2}
Conif.–decid. upland	0.3	0.0	-1.00^{11}	—	—	—	0.4	0.0	-1.00^{11}
Open marsh	11.9	3.2	-0.573^{9}	13.6	5.7	-0.413^{9}	12.6	6.2	-0.343^{8}
Oak savannah	1.8	1.0	-0.277^{6}	2.9	2.8	-0.019^{6}	2.1	1.8	-0.076^{6}
Pole white cedar–tamarack	—	—	—	—	—	—	—	—	—
Open water	—	—	—	—	—	—	—	—	—
Total home range	426.1			242.2			369.7		
Total locations		13952			247			665	

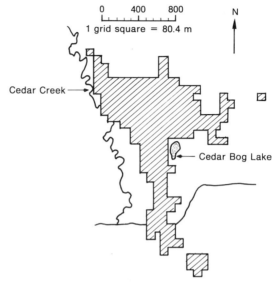

Figure 6.1   Area used by a non-breeding adult femal barred owl (No. 717b) from late May to early August.

Endangered species. These birds bred and roosted in the rugged, forested mountains where there was comparatively little human disturbance. California condors were limited to a landscape of 2.5–4.5 million ha from the 1970s until the remaining birds were removed and placed in a captive propagation and release program in 1987. Persecution, inadvertent poisoning and environmental contamination, and changes in availability of prey (carcasses) in the oak savannah and grassland habitat endangered these condors. As California condors that have hatched in captivity are released, it seems likely that the remote forested landscape of the southwestern United States will be the important refugia for the conservation of California condors.

6.4.2   Regional raptor occurrence

Historically, forests in the Ozark Mountains of Missouri and Arkansas have undergone general conversion from hardwoods to pines and have been modified by fire, grazing, clearcutting and herbicides. In their overview of bird populations in the Ozarks, Smith and Petit (1988) cite a paucity of bird data from before 1900, but for raptors they suggest the following consequences of changes in the forest landscape: red-tailed hawks, great horned owls, barred owls and eastern screech owls remain fairly common; broad-winged hawks breed in small numbers in 'large forest tracts' that presently are threatened by fragmentation; Cooper's hawks and sharp-shinned hawks have declined; red-shouldered hawks, ospreys and wintering bald eagles have declined due to clearing of riparian woodlands.

Douglas fir forests in northern California were cited by Raphael *et al.* (1988) as having the highest total bird abundance of any North American coniferous forest. There the spotted owl, which 'is closely associated with old growth forest', has received a great deal of attention. Before timber harvest, they estimate that about 74% of the forest was mature growth and about 13% occurred in brush/sapling or pole/saw-timber growth. By the 1980s they estimated that about 50% of the mature growth had been cut and was in the earlier growth stages. They expressed concern about the spotted owl, which has since been classified as Threatened. On the other hand, these authors predicted an increased number of sharp-skinned hawks and possibly Cooper's hawks.

The ponderosa pine forest of northern Arizona as described by Brawn and Balda (1988) had, prior to the late 1800s, a rich grassland floor but it has been changed to a landscape of younger, more even-aged stands by grazing, fire suppression and silviculture. These authors conducted spot map surveys for all birds on 15–45 ha plots, but were not able to obtain density estimates for raptors. With the caveat that there were no pre-settlement data about bird abundance, they concluded that 'silviculture activity does not appear necessarily deleterious . . . no large old-growth dependent component of avian communities (excluding raptors) is to be found in ponderosa pine forest'. They noted that it was 'likely that spotted owls and flammulated owls are negatively affected' by general thinning and loss of snags. The Mexican spotted owl (*Strix occidentalis lucida*) has since been classified as Threatened in Arizona. Furthermore, petitions have twice been made for the northern goshawk to be listed under the Endangered Species Act and landscape-scale forest management for the goshawk has been implemented to conserve the species (Reynolds *et al.*, 1992). This was a result of concern about effects of harvest of older trees in southwestern US ponderosa pine forests.

6.4.3 Survey methods and raptor occurrence

It is common that studies of bird abundance in relation to forest types or forestry practices do not provide many data useful for statistically based tests and conclusions about raptors. Because raptors are rare and widely spaced relative to many other bird groups, general bird surveys do not cover large enough areas to provide many raptor data. Askins *et al.* (1990) found only four examples of long-term bird population data in study areas in North America with > 1000 ha of more or less contiguous forest cover. Like Brawn and Balda (1988), a study by Franzreb and Ohmart (1978) provides another example of the limitations of information about raptor–forest relationships. They sampled for bird community composition and density on an unlogged 131 ha and a logged 202 ha plot in northern Arizona. They found flammulated owls only on the unlogged area, American kestrels and great horned owls in both plots, and red-tailed hawks, pygmy owls and saw-whet owls only on logged plots. However, statistically they could not show a difference in the use of logged vs. unlogged; thus we must use the presence/absence data to make inferences about raptor use of these two forested habitats.

Larger areas can be surveyed for raptor presence/absence data that are informative about raptor–forest type associations. For example, McGarigal and Fraser (1984) broadcast owl vocalizations to increase detection of barred owls and great horned owls at samples of old-age (81–212 years) and young (12–72 years) stands of forest in the ridge and valley country of southern Virginia. They sampled 45 old-age stands ($\bar{x}$ 37 ha) and 44 young stands ($\bar{x}$ 64 ha), and had more owl detections at old stands. Forest hawks also can be surveyed for presence/absence over large areas by broadcasting vocalizations from points along roadways (Mosher *et al.*, 1990; Iverson and Fuller, 1991). These techniques have been used to compare hawk occurrence by extent or type of forest cover (Preston *et al.*, 1989) but in Maine the number of detections was low and remained a limiting factor in making comparisons (Duvall, 1990). At the scale of Breeding Bird Survey routes, even when raptor detections are adequate for analyses, the scale of data about the habitat is gross and thus limits interpretation of raptor–forest relationships (e.g. Fuller *et al.*, 1987; Titus *et al.*, 1989). There are raptor survey methods that can be adapted to various objectives and spatial scales (Fuller and Mosher, 1987).

6.4.4   Forest trends and raptor status

Kittredge (Chapter 2) notes that it is difficult to discern global trends in forest cover because of the various methods, definitions and objectives (e.g. economic vs. ecological) associated with forest data. Specifically regarding raptor management, Titus and Fuller (1989) also found that a diversity of habitat classification schemes hampered use of existing data, and that a lack of coordination among bird surveyors and land managers impeded integration of raptor occurrence data with forest cover data. I am aware of no surveys in which the status and trends of forest distribution and raptor populations have been accounted for in a coordinated study design. Therefore, we often are limited to very general results (as described when introducing section 6.4) to infer broad relationships among forests trends and raptor status.

Information from the regional level does reveal changes in distribution and status of raptors in association with changes in forests. In those cases where populations are threatened, we need to be able to modify forest management to conserve raptors. Kittredge (Chapter 2) estimates a three-fold increase in North American forestland that would be protected from harvest, but it represents only about 8% of all forest area and we do not know to what extent this will benefit raptor populations. Furthermore, he predicts a 61% decrease in forest area per capita and a 12% increase in consumption of forest products from the 1990s to 2000 in North America. While this period includes a switch to more sustainable forestry, Kittredge reminds us that the replacement forests can have different implications for wildlife conservation. This certainly can be the case for raptors; an even-aged homogeneous plantation does not provide the same resources as a diverse natural forest. With these changes imminent, we need to know more about raptor habitat requirements and raptors' responses to forest management.

Insight about a raptor's response to forest management largely comes from anecdotal information and isolated cases such as the absence of a nesting pair from a site after tree harvest or road-building. There are few data, and fewer 'experiments', closely linking habitat loss or change to most raptor population dynamics. However, even at a local scale, changes in a number of factors (e.g. vegetation structure, prey, human disturbance) probably cause changes in raptor numbers. We have a poor understanding of how factors interact and of the cumulative effects of changes in factors. Thus, we must piece together information about habitat requirements (e.g. nest site, roost, foraging areas) to manage sites for the local birds, and extrapolate over an extensive enough area to sustain populations. Therefore, it is useful to know the types of data available about habitat use and about some effects of forest management. Below emphasis is given to examples of information about nest habitat and foraging habitat and prey that can be used in devising conservation strategies for forest raptors. Selection of these examples does not imply that other habitat components and environmental factors should be ignored. Nest habitat was chosen because in North America it is the best studied component; foraging habitat, including food, was chosen because of its importance in raptor population ecology (Newton, 1979).

6.4.5   Raptor use of forest habitat

(a)   Vultures

Black vultures and turkey vultures occur in several forest types throughout the eastern portion of their ranges. Hollow trees, stumps and logs are important nest sites. Jackson (1983) assigned nests of these two species to 13 categories (e.g. thicket, cove, hollow log) that he summarizes as 'dark recesses'. These nest sites occur in a variety of forest habitats (e.g. deciduous, wet deciduous, coniferous), but more so in forests in the eastern United States than in the west where cliffs are most important. Additional details about habitat use by vultures are summarized in Palmer (1988a). In the eastern part of the range of black vultures and turkey vultures, Jackson (1983) compared the use of site type through time and found a significant decrease in the use of tree cavity nest by these two vulture species, with an increase in ground nests. He associated this change with a 60% decrease in large hardwoods (≥75 cm iameter at breast height). Ground nests are more vulnerable to predation, flooding and other disturbances, and Jackson notes that the change in forest and change in nest site use were followed by reports of declines in vulture numbers in the southeastern quarter of the United States.

(b)   Ospreys

Ospreys are distributed widely around the world and are associated with forested landscapes in Canada and the United States, where they nest and perch in large trees and forage for fish in salt, brackish and fresh water. They construct stick nests in the tops of tall snags and live trees with flat open tops that allow easy access for these large birds. They often nest within 3–5 km of water (Poole, 1989:88) and they occur in nearly every forest cover type in Canada and the United States.

Osprey nests and potential nest sites were destroyed by harvest of tall trees and removal of snags near (0–5 km) water from which the bird obtained fish (Poole, 1989:165). Furthermore, logging activity and disturbance from recreation can be associated with comparatively low reproduction (Poole, 1989:164). However, the effects of human disturbance to ospreys vary dramatically. In many areas individual birds become habituated to the proximity of people and successfully reproduce (Poole, 1989:159–165). Nevertheless, in remote areas, human activity should be precluded within 500 m from April to September and managers should impose a no-cut zone within 100 m of a nest tree to maintain the site (Henny, 1977b). Nest sites also can be provided by artificial structures (Poole, 1989:184-192, 223–225).

Ospreys capture live fish and conservation of ospreys requires readily available, uncontaminated fish near nests, perches and roosts. In North America osprey declines were detected in the 1960s and the species was classified as Endangered in Canada and placed in the 'undetermined' category in the redbook of Rare and Endangered Fish and Wildlife of the United States (Henny, 1977b). The bioaccumulation of pesticides and other environmental contaminants in the fish prey base was the basic cause of the decline, which ultimately was manifested in greatly reduced reproduction (Henny, 1977a; Poole, 1989:166–178). Forest management with pesticides, herbicides and chemical fertilizer could adversely affect fish in the adjacent drainage. Practices that cause silting, or change aquatic vegetations or spawning habitat, can affect fish abundance, distribution and availability. This, in turn, affects osprey (and bald eagle) foraging and consequently their conservation.

(c) Bald eagle

Bald eagles also require tall trees near water for nesting and they too are susceptible to loss of older forest (especially tall or 'emergent' trees), environmental contaminants and human disturbance (Stalmaster, 1987). Around the Chesapeake Bay, Buehler *et al.* (1991c) found that bald eagles spent little time along shoreline developed with buildings and that they avoided areas with boats and pedestrians. Management of nest sites includes protecting (saving) the nest tree and creating zones of protection around a nest tree (100–200 m) or a nest zone around numerous adjacent nesting pairs (Stalmaster, 1987).

For many forest raptors there are good descriptions of nests and the surrounding habitat, including floristics and measurements that are compared with non-nest plots, thus providing insight about what features the birds select from the available habitat (see Mosher *et al.*, 1987, for methods). While these data are a basis for identifying nest area habitat, they are not representative samples from across a species' range. Therefore, the existing examples must be used cautiously when planning nest area conservation strategies. Planners need to be aware of how large a sample underlies the description of habitat use and they must consider how similar their habitats are to those in which the published descriptions and measurements have been made. Descriptions of foraging habitat are seldom based on observations of hunting and prey capture. Rather, foraging habitat often is assumed to be that area within a home range or within some distance from a nest.

(d) Buteo hawks

Among forest *Buteo* hawks, the nesting habitat of eastern red-shouldered hawks is relatively well described. In Iowa, Bednarz and Dinsmore (1981) found this species restricted to floodplain forests with small marshes adjacent to upland forests. They recommended managing for large contiguous tracts of forest, maintaining a flow of water through the bottomlands, and maintaining small wetland openings in about 15% of the forest. Nearby in Missouri, Kimmel and Fredrickson (1981) also found red-shouldered hawks in bottomland hardwood tracts of > 100 ha that included interspersed wet areas of semi-permanent water. They recommended protecting large trees and maintaining areas free from channelization, deforestation and intensive agriculture. In Arkansas, the species also nests near water, most commonly in oak–gum woodlands (Preston *et al.*, 1989). In New Jersey, red-shouldered hawks also occurred near wetlands; they avoided suburban development and the deciduous trees which commonly are used in other parts of the species' range (Bosakowski *et al.*, 1992). Red-shouldered hawks were absent from areas that had been cleared of most trees for residential development in Ontario, and there Bryant (1986) also found that the species was displaced from forests where selective cutting reduced mean tree density and crown diameters.

Usually, more than one forest hawk species nests in a forest landscape, and because locating nests is time consuming and because management for raptor breeding season habitat must occur over large areas, it is worth gathering and using data from as many species as possible. In central Ontario, Armstrong and Euler (1983) found that broad-winged hawks nested in younger deciduous-dominated stands, while red-shouldered hawks used mature deciduous forest and avoided shorelines where cottages had been built. In Maryland and Wisconsin, Titus and Mosher (1987) concluded that both these *Buteo* hawks placed their nests in large diameter trees, and that broad-winged hawks selected certain tree species in which to nest. The nest site habitat selection is described for Cooper's hawks and red-tailed hawks as well as red-shouldered and broad-winged hawks in the Appalachian Mountain forests of Maryland by Titus and Mosher (1981). These species share many components of the forest but each also exhibits some habitat use that is different from the other cohabitants. Until we know more about raptor ecology and understand how use of different habitat components affects raptor demography, we should assume that the different habitat associations exhibited across a species range are important for their conservation.

Despite the numerous differences revealed by existing samples, there is a basis for useful generalities. For the red-shouldered hawk, I recommend providing large trees for nesting that are near wetlands for foraging and that are not close to human activity. The measurements of nest area habitat at one site in southwest Ontario permitted Morris and Lemon (1983) to predict the selection of mature deciduous forest at a second plot. Mosher *et al.* (1986) used measurements made in two different regions to test predictions (classification) of nest habitat. For example, based on measurements made in Maryland, they could correctly identify red-shouldered hawk nest habitat in Wisconsin 85% of the time and broad-winged hawk habitat 65% of the time. A predictive model and data such as these reveal the potential to plan habitat management for species without having to make extensive surveys to locate and describe raptor nest habitat in every area.

(e) Accipiter hawks

In the western United States there is concern about *Accipiter* hawks, especially the breeding habitat of the northern goshawk. Reynolds *et al.* (1982) found the small sharp-shinned hawk and medium size Cooper's hawk nesting in the same forest landscape in northwestern Oregon; and in eastern Oregon they were joined by the large northern goshawk. Their nest area measurements and those of Moore and Henny (1983) from northeastern Oregon reveal that the northern goshawk nests in older, less structurally variable stands than the two smaller species. Other studies in the western United States generally confirm that there is segregation among these three species based on nesting in younger to older stands. Reynolds and Meslow

(1984) also demonstrated segregation by diet, with sharp-shinned hawks feeding on smaller prey than Cooper's hawks, which eat smaller prey than northern goshawks. Within the size class overlap between Cooper's hawks and northern goshawks, each hawk tended to eat different prey species. Crocker-Bedford and Chaney (1988) and Reynolds (1989) noted that current forest management posed potential challenges for conservation of *Accipiter* hawks where a mix of different successional stages in a landscape was associated with the co-occurrence of the three species.

(f) Owls

Owls, difficult to observe because of their nocturnal activity and secretive roosting behavior by day, pose a challenge for forest wildlife conservation because we know relatively little about them in North America. The three species accounts that follow are derived mostly from the thorough and convenient assessment edited by Hayward and Verner (1994). For flammulated owls, cavities are an 'absolute prerequisite of successful nesting' and woodpecker holes in taller trees probably are selected (McCallum, in Hayward and Verner, 1994). Older, mature trees are characteristic of this species' home range, which often occurs on ridges or southern slopes. McCallum cites dense vegetation for roost sites, and singing sites 'well up in tall trees'. Flammulated owls eat nocturnal arthropods that they capture by 'pouncing' to the ground in more open vegetation, and by hunting in the lower two-thirds of the canopy. The relationship between prey availability and reproductive success has not been studied. On a landscape scale, flammulated owls use exterior and interior forest edges, but they are absent from cut-over areas and clearcut areas, perhaps because of fire suppression and less open, intensively managed regeneration. McCallum concludes that flammulated owls are sensitive to habitat changes but that there is not enough information to develop a conservation strategy for this forest raptor.

Boreal owls (the same species as the well-studied Tengmalm's owl of Fenno-Scandinavia) use a variety of forest types, often from > 1200 m to > 3000 m elevation in the western United States (Verner, in Hayward and Verner, 1994). They nest in woodpecker cavities in large trees or snags in older, mature mixed conifer or aspen stands with a closed canopy and sparse understory. Verner describes roost sites as being numerous throughout the home range, often in conifers such as spruce–fir cover where these owls seek a cool microclimate during the summer. Foraging sites include older spruce–fir forest in which this owl hunts small mammals and birds within about 10 m of a perch. Diets can change by season with changes in prey availability due to migration of prey or protection from snow cover, and there can be yearly variation in diet due to prey population cycles. Verner's assessment includes a valuable section on prey ecology, which provides information about prey habitat. Within the forest landscape of mature spruce, spruce–fir and aspen patches, certain types of openings are important to boreal owls, which require comparatively large home ranges. These owls disappear from areas of large clearcuts but return to some regenerating cuts and appear to remain in areas harvested by ~1.25 ha patch cuts.

Studies of the diet of boreal owl areas are not numerous enough to allow comparisons among areas, or to relate prey bases to the species' population dynamics in North America. There is little information about owl population trends or habitat trends. However, Verner believes that, overall in the western United States quality habitat is declining because of a loss of old, patchy spruce–fir forest and of snags for cavity trees. He notes that harvest of spruce is second only to that of ponderosa pine. Habitat use by boreal owls varies geographically; their movements are complex and probably nomadic on a seasonal basis. Boreal owls are not managed on a regional basis and Verner is concerned about maintenance of metapopulations. Consideration of conservation of this forest owl emphasizes the complexity of raptor ecology and how varied habitats must be within home ranges and across landscapes.

Great gray owls also use a variety of forest types and habitats (Verner, in Hayward and Verner, 1994). They nest in structures such as old stick nests, broken tree tops and mistletoe in forests near wet meadows, muskegs, marshes, lakes and pastures in Canada, and in dry mountain evergreen or deciduous forests in the southern part of their distribution; they roost in dense canopy; and they forage in open, grassy areas for small mammals such as voles and pocket gophers (Duncan and P. Hayward, in Hayward and Verner, 1994). G. Hayward (in Hayward and Verner, 1994) cites an alteration of montane habitat, especially loss of natural meadows and grasslands from fire suppression, grazing and timber harvest, and recommends management for all successional stages for suitable great gray owl habitat.

6.5   MANAGEMENT AND CONSERVATION

Verner (in Hayward and Verner, 1994) learned that most management for great gray owls on forests in the United States amounted only to protection of nest sites. He concluded that this was inadequate and that a broad-based conservation strategy was required. This is the situation for most North American forest raptors. Two current management plans and a habitat management strategy that I believe provide examples of steps toward broad-based conservation strategies are briefly described below.

6.5.1   Northern spotted owl recovery plan

Research and management planning for the northern spotted owl has been greater than that for any other forest raptor. Spotted owls use many habitats and have complex habitat requirements. Northern spotted owls nest, forage and roost in a variety of forest stand conditions but select for mature (> 100 yrs) and old forest (> 200 yrs) stands and select against young and intermediate age stands. Harvest of old-growth timber threatened the existence of the northern spotted owl and this subspecies was listed as Threatened in 1990. The United States Department of the Interior (USDI, 1992) prepared a very detailed recovery plan for the management of northern spotted owls. In addition, an attempt is being made to conserve northern spotted owls in the context of ecosystem management (Forest Ecosystem Management Team, 1993). The ecosystem management approach is required to address owl ecology in a vast, complex ecosystem that is confounded by considerations for local economies that are tied to the forests of the Pacific northwestern United States and British Columbia. It is estimated that the northern spotted owl recovery effort in the United States will require a timber harvest reduction of 2.36 million board feet per year. This is about $8.3 million per year less in timber sales and might cause a loss of about 32 000 local jobs (USDI, 1992).

The primary means of managing the recovery of the northern spotted owl population is by establishing 192 Designated Conservation Areas (DCA) on about 7.6 million acres of federal forest lands. These lands are estimated to include about 46% of the remaining nesting, foraging and roosting habitat for about 1445 known owl pairs on federal lands. The largest DCAs will provide for more than 20 pairs, and the DCAs are arranged spatially to allow for owl dispersal among them. They are combined with younger age forest that should mature into owl habitat that ultimately will support a total population of 2340 pairs. The DCAs also are arranged in conjunction with parks, wilderness areas and other owl habitat or potential habitat, and in consideration of consequences of fire, disease and insect damage to forest habitat.

Northern spotted owl pairs use home ranges of between 1000 and 10 000 acres; consequently management for the owls will include areas used by many other late forest succession species. The authors of the recovery plan expect it to provide benefits for other species while avoiding negative effects on any other species. However, this plan is for northern spotted owls, not all old-growth species. It is applied only to the area thought to be required for recovery of the owl. The northern spotted owl recovery plan is good because of the unprecedented research and planning that has been applied, and because it considers other species, including humans.

The conservation strategy for the northern spotted owl is integrally linked to the planning in *Forest Ecosystem Management: An Ecological, Economic, and Social Assessment* (Forest Ecosystem Management Assessment Team, 1993). With this, the ecosystem conservation strategy includes other species associated with late successional forest habitat, such as the Threatened marbled murrelet and 13 species of fish considered to be at risk because of loss of late succession forest habitat and land use practices. The Forest Ecosystem Management Team (1993) presents management plan options in the first of a multi-phased approach to ecosystem management. The approach involves adaptive management by which planning leads to management action, followed by monitoring of the consequences of management. This is followed by evaluation of monitoring results, which are also evaluated, this forms the basis for the next phase of planning. Conservation by ecosystem management requires good information and it is complex and time-consuming. However, ultimately it should be more effective and efficient than piecemeal, uncoordinated natural resources management.

6.5.2   Northern goshawk habitat management plan

A raptor conservation strategy also has been developed at the forest landscape scale. The northern goshawk (*Accipiter gentilis atricapillus*) breeds in forests in much of North America. Concern about its population status is widespread but there has been special interest in the northern goshawk that breeds in ponderosa pine, mixed species and Engelmann spruce – subalpine fir forests in the southwestern United States. An innovative approach to raptor conservation has been developed and implemented for the southwest region that includes these forest types. Reynolds *et al.* (1992) assessed the information available about goshawk ecology, including special attention to goshawk prey and the ecology of key prey species in the region, as well as the ecology of the forests and local silviculture practices.

The *Management Recommendations for the Northern Goshawk in the Southwestern United States* (Reynolds *et al.*, 1992) prescribes a 30-acre nest area, a 420-acre post-fledgling–family area and an additional 5400-acre foraging area to be managed around existing and historical goshawk nests. The specific management recommendations are designed for returning the present habitats (which have been affected by grazing, fire suppression and timber harvest) to a mix of patches of various successional stages and a relatively open forest dominated by mature trees. The recommendations designate the size and number of areas to manage, the timing of management, where to regenerate and plant, thinning procedures, limits of forage utilization (by game animals, livestock) and treatments of woody debris.

The management recommendations are designed to provide good breeding season habitat for the goshawk and 14 of its prey species. Furthermore: 'recommendations for the northern goshawk in the southwestern United States are recommendations for maintaining biodiversity, with healthy forest, relatively safe from catastrophic fires and pests . . . [and] can be adapted for sustaining productive forests at the landscape level.' The concept of Reynolds *et al.* (1992) could be used as a model, for assessments and strategies in other areas and for other species. The concept is good because it incorporates the best available ecological and management information and considers a variety of species and forest conservation issues.

6.5.3 Management of raptors as a group

Another innovative concept is *Fish and Wildlife 2000 – A Plan for the Future*, used by the US Bureau of Land Management (BLM). This comprises strategy plans for a variety of groups of plants and animals. The Raptor Habitat Management on Public Lands Strategy Plan (Olendorff and Kochert, 1992) provides administrators and land and wildlife managers with an overview of raptors on BLM lands. It is also a guide to implementing management such as inventory and monitoring, land aquisitions and resource management plans for each species on specific areas. It provides guidance for dealing with special status species, and partnerships to achieve objectives, training and identification of research needs. Guidance of this type is especially valuable for those who are not raptor specialists but who must deal with a variety of natural resource issues.

6.6 INFORMATION NEEDS

The authors of reviews and management plans such as those cited identify many topics for which more information is needed. These reviews and plans, and discussions with managers, are good sources of guidance for research. The key to timely acquisition of the data needed for most raptor conservation programs is careful coordination among research and management persons. To quote McCallum (in Hayward and Verner, 1994): 'Piecemeal research would be a terrible mistake.'

The status and distribution of most forest raptors are not well known. Surveys should be based on sampling schemes to provide data that are representative of habitats in which the species exist. Habitat data from surveys should be conveyed to managers at the spatial scales and in the terms with which managers must work. Surveys should be designed to document changes of a magnitude and within a period that managers will acknowledge as important to population management.

We need a better understanding of the basic population dynamics of most raptors to decide when population trends warrant a management response. Research of raptor demographics can be especially valuable if studies are conducted in such a way that comparisons can be made among habitat types or geographic areas, and if the data are from areas where management is planned or is underway.

Diet, hunting behavior and foraging habitat are important components of raptor ecology that need more study in North America. Research of differences in diet, associated with different habitats and different land management, should be related to raptor demographics. These data would be extremely useful for planning conservation strategies.

When raptor ecology research can be conducted in areas and habitats of interest to managers, or in relation to other species of interest (prey, predators, competitors, etc.), most results can be used immediately in conservation. Researchers also should try to take advantage of the methods and manipulations employed by forest managers. Use of management as experiments (e.g. Walters and Holling, 1990; Hayward and Vernon, 1994) places research in the adaptive management 'loop' as described above in relation to *Forest Ecosystem Management: An Ecological Economic, and Social Assessment.*

Synthesis of existing information, such as Reynolds *et al.* (1992) and Hayward and Verner (1994), provides reliable, convenient information in the context of forest conservation. Hayward and Verner include sections about prey and a chapter about the ecology of the forest type(s) that are most relevant to each species. Reynolds and his colleagues develop a conservation strategy. In both volumes the authors identify research needs that will provide results useful for more complete conservation strategies. The challenge of conserving forests (Myers, 1995) and forest raptors in North America is sufficient to occupy the efforts of managers and researchers for some time to come.

6.7 SUMMARY

This chapter classifies 36 raptor species as forest raptors in Canada and the United States. Forest conservation is important for the status of each species. The population status of most species was based on many types of survey data and on the opinions of individuals, rather than on conclusive statistical analyses of counts from a standardized sample design. The status of 22% of the forest raptors is unknown because there was too little information for making a decision. About 58% of the species have a stable population. About 44% have been assigned a special concern status (e.g. Threatened) by the US Fish and Wildlife Service or Canadian Wildlife Service. On a continental scale no surveys reveal any species with declining populations. However, in many cases there are concerns about local population declines. Ignorance about actual population status is costly due to the restrictions on land use associated with special concern status and the expense of intensive management of special concern populations.

Most forest raptors (86%) use late succession stage forests or old, large trees for at least some of their activities. About 70% use snags, holes (cavities) or large trees as nests, perches or roosts. Forest edge habitat and non-forested habitat such as meadows or wetlands are important features in the forest landscape for 94% of the raptors. Thus, forest conservation for a mix of successional stages and a mosaic of non-forest habitat openings will benefit most raptors. Forest raptors use different components of a heterogeneous forest landscape for different behaviors on a daily and seasonal basis.

The loss of forests or major modifications of the flora and structure of forests at a regional scale affect the distribution and abundance of birds, including raptors. Timber harvest, grazing, fire suppression, even-aged stand management and conversion to agriculture are examples of factors that can change raptor occurrence in an area. There are few data about the effects of management treatments on raptor abundance.

Biologists have identified features associated with the breeding season habitat of most species of forest raptor but existing descriptions of habitat use are not representative across a species range. Some measurements of vegetation structure and physiography in nest areas allow predictions of nest area habitat in other parts of the country. Additional sampling of raptor habitat is needed to provide managers with better predictive capability and more information for forest conservation to benefit raptors. Foraging habitat and prey abundance and availability are especially important topics requiring additional study. Until more data are gathered, forest conservation on a landscape scale, with a goal of maintaining natural biodiversity and habitat heterogeneity, will be the most effective way to manage raptors.

When a species or population requires special intensive management, it is suggested that reference is made to three conservation strategies as examples of management at different levels of intensity. For an Endangered species, see the recovery plan for the northern spotted owl (USDI, 1992). For an example of a species for which there is evidence of threat from habitat loss, review the *Management Recommendations for the Northern Goshawk in the Southwestern United States* (Reynolds *et al.*, 1992). For raptors as a group to be managed by an agency, in the general context of wildlife conservation, consider the *Raptor Habitat Management on Public Lands Strategy Plan* (Olendorff and Kochert, 1992).

Finally, it would be very useful to have technical conservation assessments, like those done for three forest owl species (Hayward and Verner, 1994), for all North American raptors. These assessments summarize available information, identify information needs and form the basis for contemporary raptor conservation strategies such as Reynolds *et al.* (1992) or for integration into an ecosystem approach to conservation.

6.8 ACKNOWLEDGEMENTS

C. Rains helped to gather literature for this review from the Raptor Information System at the Raptor Research and Technical Assistance Center and through interlibrary loans. E.D. Forsman, G.D. Hayward, C.J. Henny, R.T. Reynolds and E.E. Starkey provided materials and enlightening discussion.

V.L. Hughes prepared the text and tables. G.J. Niemi and R.M. DeGraaf reviewed the manuscript. My sincere thanks to all.

6.9   REFERENCES

Armstrong, E. and Euler, D. (1983) Habitat usage of two woodland *Buteo* species in central Ontario. *Canadian Field Naturalist* 97:200–207.

Askins, R.A., Lynch, J.F. and Greenberg, R. (1990) Population declines in migration birds in eastern North America, in *Current Ornithology*, Volume 7, (ed. D.M. Power), pp. 1–57. Plenum Press, New York.

Bednarz, J.C. and Dinsmore, J.J. (1981) Status, habitat use, and management of red-shouldered hawks in Iowa. *J. Wildl. Manage.* 45:236–241.

Bednarz, J.C., Klem, D., Jr, Goodrich, L.J. and Senner, S.E. (1990) Migration counts of raptors at Hawk Mountain, Pennsylvania, as indicators of population trends, 1934–1986. *Auk* 107:96–109.

Bosakowski, T., Smith, D.G. and Speiser, R. (1992) Status, nesting density, and macrohabitat selection of red-shouldered hawks in northern New Jersey. *Wilson Bull.* 104:434–446.

Bourne, J. (1992) All creatures, great and small. *Defenders,* September/October: 8–15.

Brawn, J.D. and Balda, R.P. (1988) The influence of silvicultural activity on ponderosa pine forest bird communities in the southwestern United States, in *Bird Conservation 3*, (ed. J.A. Jackson), pp. 3–21. The University of Wisconsin Press, Madison.

Bryant, A.A. (1986) Infuence of selective logging on red-shouldered hawks, *Buteo lineatus,* in Waterloo region, Ontario, 1953–1978. *Can. Field-Nat.* 100:520–525.

Buehler, D.A., Mersmann, T.J., Fraser, J.D. and Seegar, J.K.D. (1991a) Effects of human activity on bald eagle distribution on the northern Chesapeake Bay. *J. Wildl. Manage.* 55:282–290.

Buehler, D.A., Mersmann, T.J., Fraser, J.D. and Seegar, J.K.D. (1991b) Winter microclimate of bald eagle roosts on the northern Chesapeake Bay. *Auk* 108:612–618.

Buehler, D.A., Mersmann, T.J., Fraser, J.D. and Seegar, J.K.D. (1991c) Differences in distribution of breeding, nonbreeding, and migrant bald eagles on the northern Chesapeake Bay. *Condor* 93:399–408.

Crocker-Bedford, D.C. and Chaney, B. (1988) Characteristics of goshawk nesting stands, in *Proc. Southwest Raptor Management Symposium and Workshop,* (eds R.L. Glinski *et al.*), pp. 210–217. National Wildlife Federation Scientific and Technical Series No. 11.

Dawson, W.R., Ligon, J.D., Murphy, J.R. *et al.* (1987) Report of the Scientific Advisory Panel on the spotted owl. *Condor* 89:205–229.

Degraaf, R.M., Yamasaki, M., Leak, W.B. and Lanier, J.W. (1992) *New England Wildlife: management of forested habitats*. Gen. Tech. Rep. NE-144. USDA Forest Service, Radnor, PA. 271 pp.

Dietrich, W. (1992) *The Final Forest*. Simon and Schuster, New York. 303 pp.

Duvall, H. (1990) A theoretical and empirical evaluation of a method of estimating area occupied for breeding woodland hawks in Maine. Master's Thesis. Univ. Maine, Orono.

Forest Ecosystem Management Team (1993) *Forest Ecosystem Management: an ecological, economic, and social assessment.* US Government Printing Office, 1993-793-071, Washington, DC.

Franzreb, K.E. and Ohmart, R.D. (1978) The effects of timber harvesting on breeding birds in a mixed-coniferous forest. *Condor* 80:431–441.

Fuller, M.R. and Mosher, J.A. (1987) Raptor survey techniques, in *Raptor Management Techniques Manual,* (eds B.G. Pendleton, B.A. Millsap, K.W. Cline and D.M. Bird), pp. 37–65. National Wildlife Federation Scientific and Technical Series No. 10.

Fuller, M.R., Bystrak, D., Robbins, C.S. and Patterson, R.M. (1987) Trends in American Kestrel counts from the North American Breeding Bird Survey. *Raptor Res. Rep.* 6:22–27.

Fuller, M.R., Henny, C.J. and Wood, P.B. (1995). Raptors, in *Our Living Resources* (eds E.T. LaRoe, G.S. Farris, C.E. Puckett *et al.*) pp. 65–69. National Biological Service, Department of the Interior, Washington, DC. 530 pp.

Hayward, G.D. and Verner, J. (tech. eds) (1994) *Flammulated, boreal, and great gray owls in the United States: A technical conservation assessment.* Gen. Tech. Rep. RM-253. US Department of Agriculture, Forest Service, Rocky Mountain Forest and Range Experiment Station. Ft. Collins, Co. 214 pp. 3 maps.

Hayward, G.D., Hayward, P.H. and Garton, E.O. (1987) Movements and home range use by boreal owls in central Idaho, in *Biology and Conservation of Northern Forest Owls,* (eds R.W. Nero, R.J. Clark, R.J. Knapton, and R.H. Hamre), pp. 175–184. US Forest Service Gen. Tech. Rep. Rm-142.

Henny, C.J. (1977a) Birds of prey, DDT, and tussock moths in Pacific Northwest. *Trans. N. Am. Wildl., and Nat. Resour. Conf.* 42:397–411.

Henny, C.J. (1977b) Research, management, and status of the osprey in North America, in *World Conference On Birds of Prey, Report of Proceedings,* (ed. R.D. Chancellor) pp. 199–222. International Council for Bird Preservation.

Holthuijzen, A.M.A., Oosterhuis, L. and Fuller, M.R. (1985) Habitat used by migrating sharp-shinned hawks at Cape May Point, New Jersey, USA, in *Conservation Studies of Raptors, International Council for Bird Preservation.* Tech. Publ. No. 5. (eds I. Newton and R.D. Chancellor). Proc. of the 2nd World Conference On Birds of Prey, 1982, pp. 317–327.

Hoover, R.L. and Wills, D.L. (eds) (1987) *Managing Forested Lands for Wildlife.* Colorado Division of Wildlife in Cooperation with USDA Forest Service, Rocky Mountain Region, Denver, Colorado. 459 pp.

Hunter, M.L., Jr (1990) *Wildlife, Forests and Forestry.* Prentice Hall, Englewood Cliff, NJ. 370 pp.

Iverson, G.C. and Fuller, M.R. (1991) Area-occupied survey technique for nesting woodland raptors, in *Proceedings of the Midwest Raptor Management Symposium and Workshop,* (ed. B.G. Pendleton), pp. 118–124. National Wildlife Federation Scientific and Technical Series No. 15.

Jackson, J.A. (1983) Nesting phenology, nest site selection, and reproductive success of black and turkey vultures, in *Vulture Biology and Management,* (eds S.R. Wilbur and J.A. Jackson) pp. 245–270. University of California Press, Berkeley. RMS:07911.

Johnsgard, P.A. (1988) *North American Owls: biology and natural history.* Smithsonian Institution Press, Washington, DC.

Johnsgard, P.A. (1990) *Hawks, Eagles, and Falcons of North America: biology and natural history*. Smithsonian Institution Press, Washington, DC.

Kerlinger, P. (1989) *Flight Strategies of Migrating Hawks*. The University of Chicago Press, Chicago.

Kerlinger, P. (1992) Sharp-shinned hawk populations in a free-fall. *The Peregrine Observer* 15:1–2.

Kimmel, V.L. and Fredrickson, L.H. (1981) Nesting ecology of the red-shouldered hawk in southeastern Missouri. *Trans. Mo. Acad. Sci.* 15:21-27.

McGarigal, K. and Fraser, J.D. (1984) Effect of forest stand age on owl distribution in southwestern Virginia. *J. Wildl. Manage.* 48:1393–1398.

Moore, K.R. and Henny, C.J. (1983) Nest site characteristics of three coexisting accipiter hawks in northeastern Oregon. *Raptor Res.* 17:65–76.

Morris, M.M.J. and Lemon, R.E. (1983) Characteristics of vegetation and topography near red-shouldered hawk nests in southwestern Quebec. *J. Wildl. Manage.* 47:138–145.

Mosher, J.A., Titus, K. and Fuller, M.R. (1986) Developing a practical model to predict nesting habitat of woodland hawks, in *Wildlife 2000: Modeling Habitat Relationships of Terrestrial Vertebrates,* (eds J. Verner, M.L. Morrison and C.J. Ralph), pp. 31–35. University of Wisconsin Press, Madison.

Mosher, J.A., Titus, K. and Fuller, M.R. (1987) Habitat sampling, measurement and evaluation, in *Raptor Management Techniques Manual,* (eds B.G. Pendleton, B.A. Millsap, K.W. Cline and D.M. Bird) National Wildlife Federation Scientific and Technical Series No. 10, pp. 81–97.

Mosher, J.A., Fuller, M.R. and Kopeny, M. (1990) Surveying woodland raptors by broadcast of conspecific vocalizations. *J. Field Ornithol.* 61:453–461.

Myers, N. (1995) The world's forests: need for a policy appraisal. *Science* 268:823–824.

Nagy, A.C. (1977) Population trend indices based on 40 years of autumn counts at Hawk Mountain Sanctuary in north-eastern Pennsylvania, in *Proceedings Report for World Conference on Birds of Prey, 1975*, 2nd edn, (ed. R.D. Chancellor), pp. 243–253. International Council for Bird Preservation, Vienna.

National Wildlife Federation (1988) *Proceedings of the Southwest Raptor Management Symposium and Workshop*. Scientific and Technical Series No. 11, Washington, DC. 395 pp.

National Wildlife Federation (1989a) *Proceedings of the Western Raptor Management Symposium and Workshop*. Scientific and Technical Series No. 12, Washington, DC. 320 pp.

National Wildlife Federation (1989b) *Proceedings of the Northeast Raptor Management Symposium and Workshop*. Scientific and Technical Series No. 13, Washington, DC. 356 pp.

National Wildlife Federation (1990) *Proceedings of the Southeast Raptor Management Symposium and Workshop*. Scientific and Technical Series No. 14, Washington, DC. 248 pp.

National Wildlife Federation (1991) *Proceedings of the Midwest Raptor Management Symposium and Workshop*. Scientific and Technical Series No. 15, Washington, DC. 290 pp.

Newton, I. (1979) *Population Ecology of Raptors*. Buteo Books, Vermillion, South Dakota. 399 pp.

Olendorff, R.R. and Kochert, M.N. (1992) *Raptor Habitat Management on Public Lands: A strategy for the future.* US Bureau of Land Management, Fish and Wildlife 2000 Report, Washington, DC. 46 pp.

Palmer, R.S. (ed.) (1988a) *Handbook of North American Birds, Vol. 4: Diurnal Raptors (Part 1).* Yale University Press, New Haven.

Palmer, R.S. (1988b) *Handbook of North American Birds, Vol. 5: Diurnal Raptors (Part 2).* Yale University Press, New Haven.

Poole, A.F. (1989) *Ospreys. A Natural and Unnatural History,* 1st edn. Cambridge University Press, New York.

Preston, C.R., Harger, C.S. and Harger H.E. (1989) Habitat use and nest site selection by red-shouldered hawks in Arkansas. *The Southwestern Habitat* 34:72–78.

Raphael, M.G., Rosenberg, K.V. and Marcot, B.G. (1988) Large scale changes in bird populations of Douglas-fir forests, northwestern, California, in *Bird Conservation 3,* (ed. J.A. Jackson), pp. 63–82. University of Wisconsin Press, Madison.

Reynolds, R.T. (1989) Accipiters, in *Proc. Western Raptor Management Symposium and Workshop,* (ed. B.G. Pendleton), pp. 92–101. National Wildlife Federation Scientific and Technical Series No. 12.

Reynolds, R.T. and Meslow, E.C. (1984) Partitioning of food and niche characteristics of coexisting accipiter during breeding. *Auk* 101:761–779.

Reynolds, R.T., Meslow, E.C. and Wight, H.M. (1982) Nesting habitat of coexisting accipiter in Oregon. *J. Wildl. Manage.* 46:124–138.

Reynolds, R.T., Graham, R.T., Reiser, M.H. *et al.* (1992) *Management Recommendations for the Northern Goshawk in the Southwestern United States.* Unpublished report. US Dept of Agriculture, Forest Service, Southwestern region. US Government Printing Office 1992-0-673-223/40036. 184 pp.

Robertson, W.B., Jr (1988) American swallow-tailed kite, in *Handbook of North American Birds,* Vol. 4, (ed. R.S. Palmer), pp. 109–131. Yale Univ. Press, New Haven, Conn.

Rosenfield, R.N. and Anderson, R.K. (1983) *Status of the Cooper's hawk in Wisconsin.* Unpublished report, Wisconsin Department of Natural Resources, Madison.

Rosenfield, R.N. and Bielefeldt, J. (1993) Cooper's Hawk *Accipiter cooperii,* in *The Birds of North America,* No. 75, (eds A. Poole and F. Gill). Philadelphia: The Academy of Natural Sciences; Washington, DC. The American Ornithologists' Union. 24 pp.

Schueck, L.S., Fuller, M.R. and Seegar, W.S. (1989) Falcons, in *Proc. Northeast Raptor Management Symposium and Workshop,* (ed. B.G. Pendleton), pp. 71–80. National Wildlife Federation Scientific and Technical Series No. 13.

Senner, S.E. and Fuller, M.R. (1989) Status and conservation of North American raptors migrating to the Neotropics, in *Raptors of the Modern World,* (eds B.-U. Meyburg, and R.D. Chancellor), pp. 33–38. WWGBP, Berlin, London, Paris.

Smith, K.G. and Petit, D.E. (1988) Breeding birds and forestry practices in the Ozarks: past, present, and future relationships, in *Bird Conservation 3,* (ed. I.A. Jackson), pp. 23–50. Univ. of Wisconsin Press, Madison, Wisc.

Snyder, N.F.R., Snyder, H.A., Lincer, J.L. and Reynolds, R.T. (1973) Organo chlorides, heavy metals and the biology of North American accipiters. *Bio. Sci.* 23:300–305.

Society of American Foresters (1964) *Forest Terminology – a glossary of technical terms used in* forestry. 3rd edn. Soc. Amer. Foresters, Washington DC. 97 pp.

Stalmaster, M.V. (1987) *The Bald Eagle.* Universe Books, New York, NY.

Suckling, K. (1994) Can the goshawk save the Tongass. *Wild. Forest Review*, May 1994, pp. 18–21.

Thomas, J.W. (1979) Introduction, in *Wildlife Habitats in Managed Forests: the Blue Mountains of Oregon and Washington,* (ed. J.W. Thomas), pp. 10–21. US Dept of Agriculture, Forest Service, September 1979, Agriculture Handbook No. 553.

Titus, K., and Fuller, M.R. (1989) Western habitats – session summary, in B.G. Pendleton (Ed.) *Proc. Western Raptor Management Symposium and Workshop,* (ed. B.G. Pendleton), pp. 51–57. National Wildlife Federation Scientific and Technical Series No. 12.

Titus, K. and Fuller, M.R. (1990) Recent trends in counts of migrant hawks from northeastern North America. *J. Wildl. Manage.* **54**:463–470.

Titus, K. and Mosher, J.A. (1981) Nest-site habitat selected by woodland hawks in the central Appalachians. *Auk* **98**:270–281.

Titus, K. and Mosher, J.A. (1987) Selection of nest tree species by red-shouldered and broad-winged hawks in two temperate forest regions. *J. Field Ornithol.* **58**:274–283.

Titus, K., Fuller, M.R., Stauffer, D.F. and Sauer, J.R. (1989) Buteos, in *Proc. Northeast Raptor Management Symposium and Workshop,* (ed. B.G. Pendleton), pp. 53–64. National Wildlife Federation Scientific and Technical Series No. 13.

USDI (1992) *Recovery Plan for the Northern Spotted-owl – final draft.* US Department of the Interior, Portland, Oregon. X vols.

Verner, J. and Boss, A.S. (1980) Chapter introduction and scope, in *California Wildlife and their Habitats: Western Sierra Nevada,* (eds J. Verner and A.S. Bass), pp. 1–8. Gen. Tech. Rep. PSW-37. Pac. Southwest Forest and Range Expt. Stat. Forest Service. US Dept. Agric., Berkely, Calif.

Viverette, C., Goodrich, L. and Pokras, M. (1994) Levels of DDE in eastern flyway populations of migrating sharp-shinned hawks and the question of recent declines in number sited [*sic*]. *J. Hawk Migration Assoc. No. Am.* **20**:5–7.

Walters, C.J. and Holling, J.S. (1990) Large-scale management experiments and learning by doing. *Ecol.* **71**:2060–2068.

White, C.M. (1994) Population trends and current status of selected western raptors. *Studies in Avian Biology* **15**:161–172.

Yaffee, S.L. (1994) *The Wisdom of the Spotted Owl: policy lessons for a new century.* Island Press, Santa Barbara, Calif. 350 pp.

—7

Bird community dynamics in boreal forests

Pekka Helle and Gerald J. Niemi

7.1 INTRODUCTION

The northern boreal forest, taiga, is one of the largest biomes of the globe. The biome covers a belt 1000–2000 km wide around the northern hemisphere and is situated between the latitudes of 50 and 70°N. In a few areas the boreal forest extends north of the Arctic Circle (restricted areas in Fennoscandia, Siberia, Alaska and northwestern Canada). Although taiga may be considered relatively uniform, there is marked variation due to climatic differences. This has resulted in various subdivisions of the boreal zone (e.g. Hare, 1954). The yearly mean temperature is about 0°C; winter is long and cold, and the environment is highly seasonal.

The northern boreal forest borders tundra or other arctic vegetation in the north and temperate deciduous forests in the south. Taiga is a relatively young vegetation formation having its origin in the mountains of Siberia about 1.5 million years ago. During the Pleistocene, the present area of northern coniferous forest was largely covered several times with ice. The vegetation advanced and receded in a back and forth movement in association with continental ice. Throughout its short history, the range of taiga has been dynamically changing.

The general physical structure of North American and Eurasian boreal forest is very similar. The canopy is usually one-layered and consists of only a few dominant tree species; the shrub layer is usually sparse. The number of dominant conifer species is highest in Eastern Siberia and lowest in Northern Europe (Fennoscandia), as shown in Table 7.1. The longitudinal sectors in this comparison are not equal in area; hence, the number of tree species listed do not necessarily reflect differences in tree species diversity

Conservation of Faunal Diversity in Forested Landscapes. Edited by
R.M. DeGraaf and R.I. Miller. Published in 1996 by Chapman & Hall. ISBN 0 412 61890 7.

Table 7.1 Main conifers in different parts of the northern boreal zone (Hare and Ritchie, 1972; Larsen, 1980)

Genus	North America	Northern Europe	Western Siberia	Eastern Siberia
Picea (spruce)	*glauca* *mariana*	*excelsa*	*obovata*	*obovata* *jezoensis*
Abies (fir)	*balsamea*		*sibirica*	*nephrolepis* *sachalensis*
Pinus (pine)	*banksiana*	*silvestris*	*sibirica* *silvestris*	*silvestris* *pumila* *cembra*
Larix (larch)	*laricina*		*sibirica* *sukadizewski*	*dahurica*

between areas. There are many plant genera and species common to North American and Eurasian taiga, including deciduous genera such as *Betula* and *Populus*. An interesting feature is that the lower vegetation layer (e.g. low shrubs, mosses and lichens) in forests tend to have more species in common on the two continents. Many species of moss and lichen have a holarctic distribution, whereas none of the conifers are found on both continents (e.g. Kornas, 1972; Larsen, 1980). A compounding and unknown factor, however, is the degree of lumping and splitting that may occur among plant taxonomists with respect to these different plant forms.

Haila and Järvinen (1990) compared the avifaunal composition in different areas of central taiga. The number of species (138 primarily associated with conifer forests) seems to decrease from west to east on both continents:

- Western Canada   63
- Eastern Canada   49
- Finland   63
- Ural Mountains   58
- Central Siberia   49

However, the five areas are not similar in size, which also makes comparisons difficult. Only five species are found in all five areas and less than 10% of the species occur on both continents.

The proportion of migratory species in boreal breeding bird communities varies considerably. Haila and Järvinen (1990) report that about 30% of the species are long-distance migrants (wintering grounds in the tropics). The corresponding figures for short-distance migrants (wintering areas in temperate region) and sedentary species are 25% and 45%, respectively. In general, migrants are more abundant than residents and the proportion of sedentary species is relatively low in the breeding bird community. For instance, Helle and Mönkkönen (1990) report that a typical range is 5–15%, depending on both latitude and habitat type. In data gathered for 80–500 m transects in northern Wisconsin from 1985 to 1992 (Hanowski *et al.*, 1991), the percentage of long-distance migrant species was 56% while short-distance migrants represented 29% of the species and permanent residents only 15% (personal observation).

Fire has played a very important role in the ecology of boreal forests (Wein and MacLean, 1983; Kelsall *et al.*, 1984). Studies from both North America (e.g. Heinselman, 1973), Northern Europe (e.g. Zackrisson, 1977) and Russia (Kuleshova, 1981) show that the average interval between consecutive forest fires on a given site has been about 80–100 years. Because of the central importance of forest fires (in the past) it has been assumed that forest organisms are adapted to habitat alterations and changing habitat landscapes by fire (e.g. Fox, 1983). The natural fire regime is still in place in remote parts of Canada and Siberia but in most areas forest fires are effectively controlled (Niemi and Probst, 1990). In these areas, forestry is largely responsible for forest regeneration.

Forest succession (the change in vegetation from early stages with little or sparse vegetation to mature stages) usually lasts 100–200 years depending primarily on the tree species involved, soil properties and latitude. Forestry plays an important role in the economies of countries in the boreal forest, and a major part of paper, chemical woodpulp, plywood and board produced in the world originates from taiga forests. Despite these changes, the northern forest has maintained its natural state much better than more densely populated areas in temperate regions (e.g. Hämet-Ahti, 1983).

7.2 PROBLEMS OF DYNAMICS

This chapter deals with different scales of temporal dynamics. The main focus is on the general characteristics of bird assemblages. However, communities are made up of individual species, the abundance and importance of which are not equal. Individual bird species and their autecological features play an important role in understanding results derived from general community characteristics. Bird community dynamics is a vast topic (e.g. Wiens, 1989a). Here we concentrate on two major issues.

The first is the amount of year-to-year variation in boreal bird communities compared with more southerly areas. Also, are there differences in community stability in taiga forests in different aged stands or habitat types? Understanding these problems is of crucial importance from a practical perspective. How reliable are results from studies conducted during one breeding season or, alternatively, how many years of study are needed to gain a reliable picture of the structure of a bird community?

The second issue deals with long-term dynamics in which forest succession is the guiding framework. How does the breeding bird community change over time in a given site during forest succession? This view is important for practical forestry management. For instance, can we reliably predict changes that will occur in a given area during succession following clearcutting?

Local bird communities may vary considerably from one year to the next, although the regional population levels may remain stable. In contrast, there may be a marked long-term change in bird populations on a regional scale, but these are not always observed in short-term studies in every locality or in all habitats. Haila and co-workers (Haila, 1983; Haila and Järvinen, 1983; Haila, 1986) have suggested that breeding bird assemblages in the boreal forest are 'samples' from a regional pool. That is, the probability of colonization of a given site by a species can be considered a random (Poisson) process. Both these ideas highlight the importance of stochasticity in northern bird assemblages rather than viewing them as saturated communities close to equilibrium (e.g. Simberloff, 1978; Wiens, 1983).

The problems described above are approached, using selected examples. Most of them originate from studies in Northern Europe, where the tradition in quantitative bird census work is long, but we also present case studies from North America. In North America, ornithological studies have been heavily concentrated in temperate areas with fewer studies in less populated areas of the boreal forest. There are important theoretical problems to be covered in this context but this chapter attempts to find links to practical management when applicable. Information is especially needed in forestry so that management techniques can be developed to benefit wildlife and preserve the biodiversity of northern forests.

7.3 SHORT-TERM DYNAMICS

7.3.1 Concepts

The structure of breeding bird communities varies from year to year. The changes in populations of individual species may occur independently of each other, they may be parallel, or they may be compensatory. Natural variability is an important concept from both a theoretical and a practical perspective. There have been reasonable attempts to measure and understand this phenomenon. The question of community or population variability is conceptually difficult because it can be determined in so many ways (predictability, resilience, persistence, local and regional stability etc.; e.g. Connell and Sousa, 1983; Pimm, 1986). Possible links between stability and equilibrium are also not clear (e.g. Wiens, 1989a, and references therein).

To measure stability, bird censuses are usually carried out at a given site for several years and a comparison is made of the observed changes in numbers or the densities of individual species. Different habitats can be compared in a similar way. At the community level, the most frequently used parameters are total breeding bird density, species richness, species diversity and the densities or proportions of species groups (guilds), based on different ecological classifications (e.g. feeding ecology, nesting ecology, migratory habit, etc.). Methods are also available to measure variability in community composition (e.g. Järvinen, 1979). Variability is usually described by the coefficient of variation (standard deviation divided by mean).

Measuring stability or year-to-year variability is not a simple task. Due to various statistical pitfalls and difficulties, many different techniques have been developed (e.g. Wiens, 1989a; McArdle and Gaston, 1992). A serious concern is that the spatial scale at which variation occurs is not (usually) known; this is especially the case when very different species or even different taxa are compared with each other (McArdle and Gaston, 1992). This is highly relevant to bird studies (Helle and Mönkkönen, 1986; Wiens, 1989b) where spatial scales and population variability are important.

Another problem is stochasticity. As mentioned in section 7.2, stochastic variation may occur in a given site even though the regional population remains relatively stable. It may be a reasonable assumption that population variability in boreal settings is a Poisson variate – that is, the variance of the densities of a species should equal its mean (Svensson *et al.*, 1984; Helle and Mönkkönen, 1986). In this case, the smaller the sample (or the rarer the species) the higher the amount of 'pseudovariation'. This idea is analogous with the 'checkerboard' model introduced by Wiens (1981). Observed population variabilities should be compared with expectations derived from a Poisson distribution. Additional sampling variability is inherent in all bird census techniques because the efficiency of census methods is less than 100%. These problems are less serious in community-level analyses than when variation in individual species populations are compared.

7.3.2 Geographic variation

It is a classic dogma of ecology that animal communities in the north are less stable than those in the south. This is based on the idea that diversity begets stability, e.g. southern communities host more species than more northern ones (MacArthur, 1955; Fischer, 1960). It is true that the population densities of several bird species in the boreal forest fluctuate dramatically. Specialized seed eaters like the redpoll (*Acanthis flammea*), siskin (*Carduelis spinus*), evening grosbeak (*Coccothraustes vespertinus*), crossbills (*Loxia* spp.), great spotted woodpecker (*Dendrocopus major*) and long-tailed tit (*Aegithalos caudatus*) fluctuate greatly in abundance. In some years these species are surprisingly abundant, while the next year they may be absent or found in low abundance. The same applies to many birds of prey depending on the availability of small mammals. Grouse species also

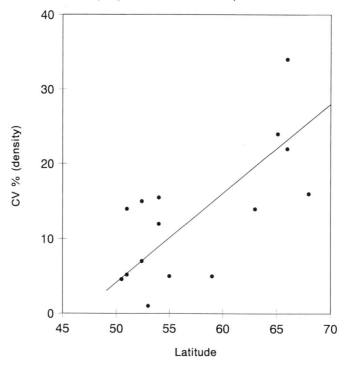

Figure 7.1   Coefficient of variation (CV%) in total breeding bird density in relation to latitude in 15 localities in central and northern Europe. (Drawn from data compiled by Järvinen, 1979.)

show large fluctuations which are often cyclic. Outbreaks of certain insects may have considerable effects on the structure of bird communities such as spruce budworm (*Choristoneura fumiferana*) in North American boreal forests (Kendeigh, 1947) and Epirrita geometrid moth (*Epirrita autumnata*) in Fennoscandia (e.g. Enemar *et al.*, 1984).

Few studies have dealt with this problem using adequate quantitative data. Järvinen (1979) compared different aspects of bird community stability in several Central and Northern European localities (ranging from 50 to 70°N). He showed convincingly that northern communities are less stable than southern ones (Figure 7.1). This result leads to the possibility that northern communities are less predictable than southern communities but the hypothesis has not been rigorously tested. Similarly, other parameters such as species turnover in communities shared the same pattern and likely the same uncertainties in predictive capabilities.

A similar analysis was performed in North America by Noon *et al.* (1985) but they did not find any clear latitudinal patterns. Some signs in community stability parameters suggested the expected result but they were not consistent across different habitat types. Noon *et al.* (1985) concluded that geographic gradients could be masked by scale effects. The areas in which bird censuses are made (usually plots of tens of hectares in size) may be too small compared with the scale at which the processes are occurring. It is also possible that stability of communities is not linearly correlated with latitude. Communities may change rapidly when moving over borders between major vegetational zones. An additional complicating factor when comparing Europe and North America is that data compiled by Järvinen (1979) from Europe covered only 20° in latitude and those from North America covered more than 40° in latitude. Both data sets covered boreal and north temperate areas. It was further discussed by Noon *et al.* (1985) that climatic unpredictability in North America may be less directly dependent on latitude than in Europe. This is at least partly due to the wide longitudinal breadth of North America and the confounding influence of the Rocky Mountains. Järvinen's (1979) data from Europe originate from a relatively narrow longitudinal section.

7.3.3 Habitat effects

Another old idea is that bird communities in simple habitats are less stable than those in more complex habitats (Noon *et al.*, 1985; see also Wiens and Rotenberry, 1981). The same idea is imbedded in the classic 'theory' of succession (e.g. Odum, 1969); early successional habitats are simpler in vegetation structure than mature stages and stability is assumed to increase during succession (because diversity increases with succession).

Helle and Mönkkönen (1986) compared bird community stability in four stages of forest succession in boreal forests of northern Finland (recently clearcut areas, seedling stands, young stands and old conifer stands). The bird community in the earliest stages showed the least variation in basic community parameters over the 6-year study period (for example, the coefficient of variation (CV) in total bird density was about 6%). The other three stages did not significantly deviate from each other (CV% in total density was 29, 25, 29, respectively). It was shown that compensating population fluctuations were important in clear-cut areas, whereas the other stages were characterized by parallel population changes.

Virkkala (1992) compared year-to-year changes in bird communities (7 years' data) between virgin coniferous forests and open fens in northern Finland, about 200 km north of the area studied by Helle and Mönkkönen (1986). In the pooled data, the coefficient of variation in bird density was 7% in fens and 13% in forests. The direction was consistent with the previous data but the difference was not significant. Individual fens (CV% 15–20) and forests (CV% 15–21) showed 'typical' values for northern boreal bird communities (see also Figure 7.1). Interestingly, these data suggested that compensatory fluctuations prevailed in open habitats (fens), while year-to-year changes in forests tended to be independent of each other.

There is a conceptual difficulty when dealing with successional data. If succession is defined as a process whereby early stages exhibit rapid change in community structure and is completed when no major change occurs, stability will by definition increase in the course of succession. To make the test realistic, the variation attributable to directional change toward increasing species diversity or richness must be factored out first and the remaining variance analyzed as a measure of stability.

Głowacinski's (1981) breeding bird data from Poland (Central Europe, 50°N) allow us to make a comparison between boreal (Helle and Mönkkönen, 1986) and temperate forest succession using this approach. Figure 7.2 shows the result concerning relative variation in the total breeding bird density (see Helle and Mönkkönen, 1990, for details). The main results from this analysis are:

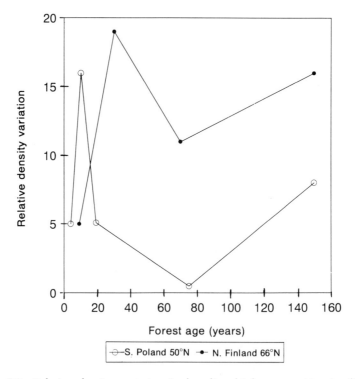

Figure 7.2 Relative density variation in breeding bird communities in different stages of forest succession in Finland (boreal) and Poland (temperate forest). Each dot is based on 4–6 years' data (see text for details).

1. Bird communities in all stages of succession were less stable in Finland than in Poland.
2. There was no clear temporal trend in stability for either area.
3. Climax communities were not the most stable within the succession gradient, in contrast with the classic idea of succession.
4. Stability in the early stages of forest succession was relatively high followed by large annual variation in the next phase, seedling stand.

The first parts of the Finnish and Polish curves were nearly identical, if the considerable differences in growth rates of plants were considered. For example, vegetation height of 5 m is reached in about 7 years in Poland, whereas 20 years are necessary in northern Finland.

Järvinen (1979) did not find major differences in community stability among different habitat types in his study in central and northern Europe. Some differences were observed but the importance of geographical location played a more significant role. Bird communities in grasslands of North America showed more year-to-year variation in some community parameters than forested habitats (Noon *et al.*, 1985; see also Winternitz, 1976). This is consistent with the expectation that simple habitats have more unstable bird communities than complex habitats. As noted above, succession studies from Europe do not support this idea. Interestingly, data from both Europe (Järvinen, 1979, 1981) and North America (Noon *et al.*, 1985) suggest that coniferous forests have larger variations in populations than other forest types.

7.3.4 Factors affecting stability

For a relatively large geographic scale, three hypotheses have been proposed to describe the decreasing gradient in community stability from south to north, as evidenced from central and northern Europe. Stability decreases because:

1. diversity of communities decreases;
2. productivity of habitats decreases; and
3. climatic unpredictability increases.

Järvinen (1979) tested each of these hypotheses and concluded that climatic predictability plays the main role, although the explanations given are not mutually exclusive. Climatic unpredictability is not a reasonable explanation for between-habitat differences in a given locality. In northern boreal conditions, the irregularity of seed crops of trees may play a significant role (Virkkala, 1989). This may partially explain why bird communities in the early stages of succession (or in other open habitats with no mature trees) seem to show less year-to-year variation than those in forests (see also Helle and Mönkkönen, 1990).

Several northern European studies (Svensson, 1977; Solonen, 1981) suggest that populations of tropical migrants are more stable than those of short-distance migrants. Noon *et al.* (1985) suggest that the long-distance migrants in North American breeding bird communities might have a stabilizing effect on community variability. It has been shown further that the proportion of tropical migrants in breeding bird communities increases with increasing latitude (e.g. Europe: Herrera, 1978; North America: Noon *et al.*, 1985). These migrants tend to balance yearly fluctuations in northern bird communities. This could be related to the observation that deciduous forests, with more migrants than coniferous forests, tend to show less year-to-year variation in density and community structure. MacArthur (1959) suggested that the percentage of migrants is highest where the contrast in food resources available for birds between summer and winter (or seasonality) is largest.

7.4   LONG-TERM DYNAMICS: FOREST SUCCESSION

7.4.1   Background

Stand age is among the most important habitat variables affecting bird community structure in boreal settings. This is especially true on a local scale but it also plays a significant role at a landscape level. Many studies have been devoted to describing some aspects of successional patterns in forest bird communities but boreal studies are fewer (e.g. Pospelov, 1957; Martin, 1960; Haapanen, 1965; Ahlén, 1975; Theberge, 1976; Niemi, 1978; Titterington *et al.*, 1979; DesGranges, 1980; Stiles, 1980; Welsh and Fillman, 1980; Fox, 1983; Niemi and Hanowski, 1984; Kelsall *et al.*, 1984; Mönkkönen, 1984; Helle, 1985a,b; Morgan and Freedman, 1986; Probst *et al.*, 1992). For a review of the history of forest bird succession studies, see Helle and Mönkkönen (1990). The specific questions have included various aspects of community structure, interspecific competition, community stability, natural history characteristics of species associated with different stages of succession, turnover rate of succession, degree of specialization of communities and comparisons between forestry and fire effects.

Based on the classic view of succession, Odum (1969) presented a list of trends which should be expected to take place in communities during the course of succession. Some of the results reported here correspond with these 'predictions', while others do not. For specific tests of these ideas, see Głowacinski (1975), Głowacinski and Järvinen (1975), May (1984), Helle (1986), Mönkkönen and Helle (1987) and Helle and Mönkkönen (1990). Mönkkönen (1991) reviewed the existing analyses and data (Table 7.2) and reported a wide variety of patterns in successions of forest birds in western and eastern parts of both Nearctic and Palearctic regions. The general pattern is not yet clear. This is likely due to the many differences in the

Table 7.2 Biogeographical comparison of patterns in breeding bird communities in the course of succession with respect to trends to be expected (Odum, 1969) (after Mönkkönen, 1991)

Bird community characteristic	Trend to be expected	Nearctic		Palaearctic	
		Western	Eastern	Western	Eastern
Total density	?	+	+	+	+
Species diversity	+	+	+	+	?
Average body weight	+	–	=	–	?
Non-passerine/ passerine ratio	+	–	+	–	?
Specialist vs generalist ratio (food)	+	+	+	+–	?
No. eggs per average female	–	?	–	+	?
Density of tropical migrants	?	=	+	+–	+
Proportion of tropical migrants	?	+	+	–	+–

+ Increasing trend along forest succession gradient (not necessarily linear);
– decreasing trend;
+– first increasing then decreasing;
= no clear trend;
? trend not known or studied.

way these studies have been conducted. For instance, few studies have included a wide variety of forest age classes, nor have studies included different latitudes or considered differences in the response of birds to succession following logging compared with natural disturbances.

Although there is no emergence for a unified theory of bird communities in forest succession (see also Horn, 1974), certain regularities and general patterns occur and they may be used to predict changes that will take place after natural or artificial changes (e.g. clearcutting).

7.4.2 Patterns in forest bird succession

In this chapter succession comparisons between North America and northern Europe for successional sequences in forests are presented mainly using published data from different localities. Bird census studies which give quantitative estimates of bird densities from at least four different stages of forest succession have been accepted. The data are mainly from boreal forest but some were gathered from north temperate forests (see Helle and Mönkkönen, 1990, for details).

Forest succession gradients have been divided into four development classes (open after burning or cutting; brush; young stand; mature stand) based on vegetation height (< 1, 1–4, 4–10 and > 10 m, respectively). The division is crude, but it is a relatively simple way to make different studies comparable because vegetation height is usually reported in most published studies. For further elaboration on other methodological problems, see Helle and Mönkkönen (1990). The total data include six successional gradients from northern Europe, three from western North America, four from eastern North America, and data from the mid-North American continent in the Great Lakes region of Wisconsin. The gradients from northern Europe, western North America and eastern North America have been reported previously (Helle and Mönkkönen, 1990), while the data from the mid-North American continent is from an extensive study to examine the potential effects of the US Navy's extreme low frequency (ELF) communication system (Blake *et al.*, 1992, 1994; Hanowski *et al.*, 1993).

Briefly, the mid-North American continent data originate from the northern hardwood/southern boreal forest ecotone in northern Wisconsin. In this study, the line transect method was used in bird censuses. These data were gathered in June and July from 1986 to 1992 using only reference transects that were randomly distributed at least 10 km away from the ELF antenna system (Blake *et al.*, 1991; Hanowski *et al.*, 1991). Along these transects, suitable habitats were identified and classified into one of the four successional groups. Each site sample of the four groups consisted of a 100 m × 500 m (5 ha) portion of the transect and from three to 15 replicates per group.

The main habitats in the study area include forests dominated by aspen (*Populus tremuloides*), paper birch (*Betula papyrifera*), maple (*Acer* spp.), basswood (*Tilia americana*), balsam fir (*Abies balsamea*), pine (*Pinus* spp.) spruce (*Picea* spp.) and tamarack (*Larix laricina*). A typically complex mosaic of habitat types includes primarily second-growth forests following periodic logging activity, with recently cut areas and a variety of non-forested and forested wetlands.

The sampled habitats were divided into four vegetation height categories to match the other data used in the analysis. Basic statistics on the assemblages of breeding birds are given in Table 7.3.

Figure 7.3 shows total breeding bird density and species diversity of communities in different stages of forest succession from the different geographic areas. The densities and the species diversity have been rescaled to make these data more comparable with each other (the highest value within each succession gradient was set to 100 and the other stages calculated as proportions of 100).

Breeding bird density shows a steady increase with forest age in northern Europe, whereas the pattern in North America is different. In North America, density tends to show a two-peak density pattern in the eastern and western portions where the first peak is in the shrub phase and the second in mature stages. In contrast, the mid-continent data show an increase to the young stands and then decreases at the mature stage. Climax boreal forests have breeding bird densities that vary considerably depending on latitude and tree species composition (see also Sanders, 1970; Carbyn, 1971; West and DeWolfe, 1974; Haila and Järvinen, 1990). The two-peak density pattern in forest succession in North America was previously noted by Fox (1983).

The pattern in the mid-continent is more difficult to explain and may be a regional anomaly. However, a possible explanation is that the mature stages included in this study are all second-growth mature forests. In another study conducted in the same region to examine the relationships between old-growth forests and second-growth mature forests (Gokee *et al.*, 1994), the former had significantly more individuals and species per sample unit than the latter. The differences were primarily due to more bird species associated with cavity-nesting, conifers and foliage insectivores found in the old-growth forests as compared with the second-growth mature forests.

There were several reasons for the higher density and diversity in shrub and young forests observed there. Firstly, a high proportion of species associated with early successional, edge and field habitats was observed in the first three phases of the successional sequence but very few individuals were observed in the mature stage. These bird species included northern flicker (*Colaptes auritus*), American robin (*Turdus migratorius*), golden-winged warbler (*Vermivora chrysoptera*), Nashville warbler (*Vermivora ruficapilla*), chestnut-sided warbler (*Dendroica pensylvanica*), and white-throated sparrow (*Zonotrichia albicollis*).

A relatively low number of cavity-dependent species and conifer-associated species were observed in the mature stands relative to the shrub and young forests. For instance, of the cavity-dependent species only the yellow-bellied sapsucker (*Sphyrapicus varius*) was found more commonly in mature stages, whereas the northern flicker (as expected) and the black-capped chickadee (*Parus atricapillus*) were more commonly observed in young forests. Similarly, conifer-associated species such as the yellow-rumped warbler (*Dendroica coronata*) were observed equally as frequently in the young and mature stands, while the chipping sparrow (*Spizella passerina*) was not observed in the mature stand but was found in the other three stages. In this region, as succession proceeds there is generally an increase in larger trees with more cavities and more coniferous trees. In second-growth forests, especially those with relatively short rotations (e.g. 50 years), these elements of the forest do not necessarily develop fully. The combined effects of these two factors may explain the apparent reduction in species diversity and density in second-growth forests of this region. Mature or old-growth forests being relatively rare (because of the relatively short rotation of the second-growth forests in this region), the characteristics of old-growth

Table 7.3 Means of total number of individuals and species as well as most abundant bird species per 100 m transect in four different stages of forest succession in Wisconsin, USA

	Open (< 1 m)	Brush (1–4 m)	Young stand (4–10 m)	Mature stand (> 10 m)
Total individuals	3.83	7.27	7.32	5.54
Total no. species	3.58	5.73	5.84	4.23
Most numerous species:				
Yellow-bellied sapsucker				
Sphyrapicus varius		0.11	0.06	0.19
Northern flicker				
Colaptes auratus	0.03	0.17	0.20	0.02
Least flycatcher				
Empidonax minimus		0.04		0.66
Blue jay				
Cyanocitta cristata		0.11	0.43	0.04
Black-capped chickadee				
Parus atricapillus	0.07	0.22	0.50	0.25
Eastern bluebird				
Sialia sialis	0.17	0.03		
Hermit thrush				
Catharus guttatus	0.17	0.09	0.11	0.22
American robin				
Turdus migratorius	0.17	0.52	0.22	0.08
Red-eyed vireo				
Vireo olivaceus		0.16	0.28	0.74
Golden-winged warbler				
Vermivora chrysoptera		0.11	0.13	
Nashville warbler				
V. ruficapilla	0.14	0.55	1.20	0.18
Chestnut-sided warbler				
Dendroica pensylvanica		0.95	0.68	0.05
Yellow-rumped warbler				
D. coronata		0.10	0.23	0.18
Black-throated green warbler				
D. virens		0.06	0.13	0.45
Ovenbird				
Seiurus aurocapillus		0.27	0.78	1.05
Mourning warbler				
Oporornis philadelphia	0.36	0.41	0.08	0.01
Rose-breasted grosbeak				
Pheucticus ludovicianus	0.14	0.14	0.21	0.15
Indigo bunting				
Passerina cyanea	0.07	0.19	0.16	

Table 7.3 continued

	Open (< 1 m)	Brush (1–4 m)	Young stand (4–10 m)	Mature stand (> 10 m)
Chipping sparrow				
Spizella passerina	0.30	0.13	0.25	0.01
Song sparrow				
Melospiza melodia	0.46	0.37	0.09	
Swamp sparrow				
M. georgiana	0.14	0.09		
White-throated sparrow				
Zonotrichia albicollis	0.56	1.10	0.96	0.03
Dark-eyed junco				
Junco hyemalis	0.20			

forests and vegetation features of these forests may not develop until much later (e.g. >80–100 yr). Bird species diversity may also concomitantly increase during these later stages of forest succession and in a similar fashion to that observed in eastern and western North America.

In general, bird species diversity increases in the course of succession on both continents, as is generally assumed to be the case in successional processes. Middle North America seems to be an exception here; the highest species diversity value is not obtained in the mature stand but in the previous stage. It is unknown whether this result is general or local in nature. A visual inspection of the data by Probst *et al.* (1992) from the same general area suggests the same pattern. The numerical value of species diversity depends on both the number of species and the shape of species-abundance distribution (evenness). The evenness component shows little variation between different stages of succession within the gradients studied. Hence, the number of species plays the most important role here.

In addition to breeding density and species diversity of communities, we also present results concerning the proportion of tropical migrants in breeding bird communities (Figure 7.4). These species and their habitat requirements have received considerable attention, especially in North America (e.g. Keast and Morton, 1980; Hagan and Johnston, 1992). Patterns for Europe and North America are strikingly different. The proportion of tropical migrants generally decreases with increasing vegetation height (forest age) in Europe with the peak in the shrub phase (vegetation height 1–4 m). The opposite pattern is found in eastern and middle North America. The proportion of tropical migrants increases with increasing

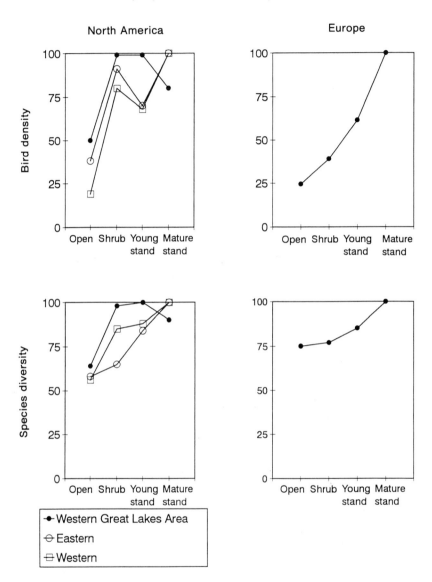

Figure 7.3   Patterns in relative bird density and bird species diversity in four different stages of forest succession in North America and Europe (see also text).

vegetation height; however, data west of the Rocky Mountains show a different trend.

These results are consistent with previous studies (e.g. Whitcomb *et al.*, 1981; Bilcke, 1984; Helle and Fuller, 1988; Mönkkönen and Helle, 1989).

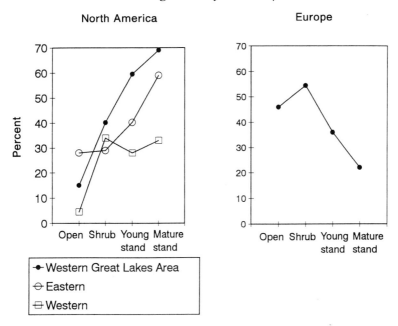

Figure 7.4 Proportion of tropical migrant pairs in breeding bird communities in North American and European (mainly boreal) forest successions (see also Figure 7.3 and text).

The reason for the profound difference between the continents is not quite clear but it is probably associated with habitat availability in the wintering grounds of these species. Tropical Africa is mainly dominated by savanna and other relatively open vegetation, whereas the wintering areas of neotropical migrants in Latin America and the Caribbean area are (or at least have been) tropical forest (Bilcke, 1984; Helle and Fuller, 1986; Mönkkönen and Helle, 1989). Moreover, the continental avifauna of North America and Europe do not have similar evolutionary histories. European avifauna is largely of Palearctic origin with a minor tropical element only, while both Palearctic and tropical elements are considerable in North America (e.g. Mayr, 1946; Snow, 1978). The Pleistocene glaciations had a very different impact on vegetation and avifauna in Europe and North America. For these reasons, the contemporary avifauna of North America has a higher level of specialization than that of Europe. The pattern observed in Figure 7.4 reveals why forestry through its creation of more open habitats has been beneficial for tropical migrants in northern Europe (e.g. Järvinen and Väisänen, 1978) whereas the reverse tends to be true for many North American species (e.g. Whitcomb *et al.*, 1981; see also Mönkkönen and Welsh, 1994).

7.5   TEMPORAL AND SPATIAL VARIATION

Temporal and spatial variation of bird populations and communities are interrelated. Hence, interpretation of results must be on an appropriate scale. Ecological researchers have recently become highly interested in various scales behind population processes (e.g. Wiens, 1981; Dayton and Tegner, 1984). The landscape has become a popular scale, although defining it is not quite clear. The landscape is an intermediate level, somewhere between the size of a territory and the regional distribution of a species (e.g. Dunning *et al.*, 1992). Scale problems of this nature are associated with our present interest from two different perspectives.

The question of bird community dynamics in isolated forest fragments is of central importance. It is a common situation in boreal settings that virgin stands (those that have never been logged) are often islands surrounded by areas of commercial forestry. These fragments do not contain many of the bird species characteristic of these forests over the long term. The reason is that adequate source populations for many of the rare species are not found in the surrounding landscape. These source populations formerly sustained populations of these rarer species through continual colonization, but now many of these fragments have become too isolated from the source areas.

Väisänen *et al.* (1986) give an extreme example of this process. There was a census of breeding birds in a small forest area of northern Finland during the beginning of this century. At that time this was part of a large virgin forest tract. When reassessed in 1981–1983, the local habitat still remained intact but had become surrounded by large areas of managed forest and various open habitats. Census results showed a dramatic change: species of old taiga forest had decreased considerably or were locally extinct, whereas species of managed forests had increased in numbers. A similar example is given by Helle (1985b) from the Oulanka National Park, northern Finland. No species have disappeared (the study period is shorter than in the previous case) but the changes in the protected old forest area parallel those in large surrounding areas dominated by intensive forestry. Area problems of a similar nature have been reported from North America (e.g. Whitcomb *et al.*, 1981; Askins *et al.*, 1987; Blake and Karr, 1987) but data from the boreal forests of North America are lacking.

Another problem is also associated with space and time. We observed that there are general patterns in bird communities in the course of forest succession, which may be used to predict forestry's long-term impacts on birds. A critical assumption here is that bird communities have not changed during the past 50–100 years for other reasons.

Järvinen and Ulfstrand (1980) studied species changes in northern Europe based on published reports from 1850 to 1970. A relatively large change was observed in their study area (Denmark, Sweden, Norway and Finland).

In total 88 colonizations (colonization events per country) and 22 extinctions occurred within the study period. Most of the colonists were species of open habitats; few were true forest species. Note, however, that many species of open or shrub habitats are important components of bird communities in the early stages of forest succession. Many of the species within these early-successional forests and in the grassland/shrub habitat complexes in North America have been decreasing in abundance over the past 25 years (e.g. Niemi and Probst, 1990; Herkert *et al.*, 1993; Johnson and Schwartz, 1993). In the study by Järvinen and Ulfstrand (1980), the most important reasons for the changes observed were habitat changes related to human activities (e.g. forestry, agriculture, eutrophication of lakes, regrowth on grazing land, conifer plantations). This study clearly shows that changes from a variety of causes, especially due to human activities, are occurring over long time scales. On the other hand, we do not know, how representative the northern European example is on a worldwide basis. Quantitative census data from Finland covering several decades also show that population changes (even in larger geographical areas) cannot be explained by regional changes in habitat composition alone (e.g. Järvinen and Väisänen, 1977, 1978).

Helle and Järvinen (1986) compared observed population changes of many common forest bird species in northern Finland from the 1940s to the 1970s. They used predictions based on habitat selection patterns of species and changes that have occurred in forest structure during that time period. Population changes of most species were close to those predicted based on habitat changes. However, many species breeding in spruce forests were an exception, such as siskin (*Carduelis spinus*), robin (*Erithacus rubecula*) and song thrush (*Turdus philomelos*). These species have considerably increased in numbers in northern Finland, although the area of spruce forest has decreased there during the study period. The primary reason appears to be the marked influx of immigrants from southern populations where the area of spruce forests has increased.

Several North American studies have also stressed the importance of small-scale vs. large-scale dynamics, although information from the boreal area is sparse. Whitcomb *et al.* (1981) showed that many forest-dwelling tropical migrants which prefer interior parts of forests have considerably decreased in eastern deciduous forests of North America during the past decades. However, not all studies have shown similar results. For example, Hagan and Johnston (1992) included a collection of articles dealing with population trends of these birds in United States. Moreover, many species that are described as area-sensitive in the eastern United States do not appear to be so in the boreal forest (Welsh, 1987). James *et al.* (1992) reported a very complicated response pattern for neotropical migrants in the southern and southeastern United States from 1966 to 1988. For the eight most common species, two have declined, two have been stable and four have increased. Four species (common yellowthroat, prairie warbler, yellow-breasted chat, Kentucky warbler) showed decreasing trends in uplands, whereas all five forest-dwelling species (including common yellowthroat and Kentucky warbler) have increased in lowland forests.

7.6   CONCLUSIONS

Our present knowledge of various temporal aspects in boreal bird communities is reasonable, with the exception that vast North American and Siberian areas have received little attention. One problem is evident. Although we are able to detect and describe changes that occur in populations, we usually do not know the exact mechanisms involved. Much of our present understanding is based on numbers of birds breeding in given habitats. This usually means the number of singing males because these are most easily and reliably counted. Little is known about the reproductive success of birds in various habitats. Although population density is usually considered a good predictor of habitat quality, this assumption has not been critically tested (van Horne, 1983).

Both short-term and long-term dynamics of bird populations and communities can be detected in local and regional scales. Somewhat independent of these factors, 'mega-scale' issues may also play a role (Haila and Järvinen, 1990). The complexity of this setting makes generalization very difficult. This is extremely problematic when trying to make general recommendations on improvements to forestry practice for the benefit of the overall bird community.

A special problem of nature conservation associated with the boreal forest needs attention. Taiga has been a dynamic ecosystem in the past due to frequent forest fires, glaciation and the general harsh winter conditions. 'Virgin' taiga was not a vast area of old-growth forest but a mosaic of forest patches of various ages and sizes. Due to fire suppression large areas of taiga are no longer naturally dynamic. This is true for most of the protected areas as well. It is understandable that climax stages of forest succession have a high priority in nature conservation. We are rapidly entering into a situation, however, where a high proportion of old forests are protected but future old forests are not regenerating naturally. A majority of younger stages of forest succession originate from silvicultural cutting. In Fennoscandia, for example, it is nearly impossible to find natural forest succession gradients with a sufficient number of different stages for ecological study.

7.7   SUMMARY

Taiga, the northern boreal forest, is a dynamic system. It is relatively young in geological terms and its fauna and flora have been shaped by successive glaciations. Short-term catastrophes such as forest fire have played an important role in the development of its ecology. This chapter considers bird communities of these forests, including their dynamics in various spatial and temporal scales. The overall emphasis, however, is on the general structure of communities using examples from both Eurasia and North America.

The structure of a local bird community exhibits considerable annual variation. Geographic comparisons reveal that this annual variation (instability) increases with latitude in Europe but the pattern is variable in North America. There are also many differences between different habitat types within each geographic area.

Forest succession provides a basic framework for the illustration of long-term dynamics of boreal bird communities. We examine the predictions of classic theory on succession for data gathered in both the western and eastern parts of the Eurasian and North American continents. Our results indicate that total breeding bird density increases with increasing forest age, but some geographic variation is noted. Species diversity also shows a consistent increase with forest age on both continents. However, many detailed characteristics of community structure show gross differences between the two continents. For example, the proportion of tropical migrants in breeding bird communities decreases with increasing forest age in Europe, whereas the pattern observed in North America is the opposite. This is because tropical migrants tend to be found mostly in early successional forests in Europe, while in North America many but not all tropical migrants are found in the later stages of forest succession.

The relationships between temporal and spatial scales are complicated and interrelated. Few data sets have been compiled or gathered to examine the consistency or variability of the bird community on an annual basis or across different spatial scales. Because many of these relationships are largely unknown, they are important issues for future research and are essential in increasing our understanding of conservation biology of these boreal forested ecosystems.

7.8 ACKNOWLEDGEMENTS

We thank Ann Lima for technical assistance in preparing the manuscript and Mikko Mönkkönen and JoAnn Hanowski for constructive comments. Financial assistance was provided by the Minnesota State Legislature from the Environmental Trust Fund as recommended by the Legislative Commission on Minnesota Resources and by the US Department of the Navy through the IIT Research Institute (Contract No. IIT/DO6205-93-C-008). This is contribution number 134 from the Center for Water and the Environment of the Natural Resources Research Institute, University of Minnesota, Duluth.

230 Bird community dynamics in boreal forests

7.9 REFERENCES

Ahlén, I. (1976) Forestry and the bird fauna in Sweden. *Ornis Fennica* 52:39–44.

Askins, R.A., Philbrick, M.J. and Sugeno, D.S. (1987) Relationship between the regional abundance of forest and the composition of forest bird communities. *Biol. Conserv.* 39:129–152.

Bilcke, G. (1984) Residence and non-residence in passerines: Dependence on the vegetation structure. *Ardea* 72:223–227.

Blake, J.G. and Karr, J.R. (1987) Breeding birds in isolated woodlots: Area and habitat relationships. *Ecology* 68:1724–1734.

Blake, J.G., Hanowski, J.M., Niemi, G.J. and Collins, P.T. (1991) Hourly variation in transect counts of birds. *Ornis Fennica* 68:139–147.

Blake, J.G., Niemi, G.J. and Hanowski, J.M. (1992) Drought and annual variation in bird populations: effects of migratory strategy and breeding habitat, in *Ecology and conservation of neotropical migrant landbirds,* (eds J.M. Hagan III and D.W. Johnston), pp. 419–429. Smithsonian Institution Press, Washington DC.

Blake, J.G., Hanowski, J.M., Niemi, G.J. and Collins, P.T. (1994) Annual variation in bird populations of mixed conifer–northern hardwood forests. *Condor* 96:381–399.

Carbyn, L.N. (1971) Densities and biomass relationships of birds nesting in boreal forest habitats. *Arctic* 24:51–61.

Connell, J.H. and Sousa, W.P. (1983) On the evidence needed to judge ecological stability or persistence. *Amer. Nat.* 121:789–824.

Dayton, P.K. and Tegner, M.J. (1984) The importance of scale in community ecology: a kelp forest example with terrestrial analogs, in *A New Ecology: Novel Approaches to Interactive Systems,* (eds P.W. Price, C.N. Slobodchikoff and W.S. Gaud), pp. 457–481. John Wiley & Sons, New York.

DesGranges, J.-L. (1980) *Avian community structure of six forest stands in La Mauricie Natural Park, Quebec.* Occ. Paper No. 41, Canadian Wildlife Service.

Dunning, J.B., Danielson, B.J. and Pulliam, R.H. (1992) Ecological processes that affect populations in complex landscapes. *Oikos* 65:169–175.

Enemar, A., Nilsson, L. and Sjöstrand, B. (1984) The composition and dynamics of the passerine bird community in a subalpine birch forest, Swedish Lapland. A 20-year study. *Ann. Zool. Fennici* 21:321–338.

Fischer, A.G. (1960) Latitudinal variations in organic diversity. *Evolution* 14:64–81.

Fox, J.F. (1983) Post-fire succession of small-mammal and bird communities, in *The Role of Fire in Northern Circumpolar Ecosystems,* (eds R.W. Wein and D.A. MacLean), pp. 155–180. John Wiley & Sons, New York.

Głowacinski, Z. (1975) Succession of the bird communities in the Niepolomice Forest (Southern Poland). *Ekol. Polska* 23:231–263.

Głowacinski, Z. (1981) Stability in bird communities during the secondary succession of a forest ecosystem. *Ekol. Polska* 29:73–95.

Głowacinski, Z. and Järvinen, O. (1975) Rate of secondary succession in forest bird communities. *Ornis Scand.* 6:33–40.

Gokee, A.R., Niemi, G.J., Hanowski, J.M. and Collins, P.T. (1994) Comparison of avian communities in old-growth, secondary growth, and mixed forests in the Lake Superior region. Unpublished manuscript, Natural Resources Research Institute, University of Minnesota, Duluth, MN, USA.

Haapanen, A. (1965) Bird fauna of the Finnish forests in relation to forest succession. I. *Ann. Zool. Fennici* **2**:153–196.

Hagan, J.M. and Johnston, D.W. (eds) (1992) *Ecology and Conservation of Neotropical Migrant Land Birds*. Smithsonian Institution Press, New York.

Haila, Y. (1983) Land birds on northern islands: A sampling metaphor for insular colonization. *Oikos* **41**:334–351.

Haila, Y. (1986) North European land birds in forest fragments: evidence for area effects?, in *Wildlife 2000: modeling habitat relationships of terrestrial vertebrates*, (eds J. Verner, M.L. Morrison and C.J. Ralph), pp. 315–319. Univ. Wisconsin Press, Madison.

Haila, Y. and Järvinen, O. (1983) Land bird communities on a Finnish island: species impoverishment and abundance patterns. *Oikos* **41**:255–273.

Haila, Y. and Järvinen, O. (1990) Northern conifer forests and their bird species assemblages, in *Biogeography and Ecology of Forest Bird Communities*, (ed. A. Keast), pp. 61–85. SPB Academic Publishing bv., The Hague.

Hämet-Ahti, L. (1983) Human impact on closed boreal forest (taiga), in *Man's Impact on Vegetation*, (eds W. Holzner, M.J.A. Werger and I. Ikusima), pp. 201–211. W. Junk Publishers, The Hague.

Hanowski, J.M., Blake, J.G., Niemi, G.J. and Collins, P.T. (1991) *ELF communications system ecological monitoring program: Wisconsin bird studies – Final Report*. Report No. EO6628-2 National Technical Information Service, Springfield, VA.

Hanowski, J.M., Blake, J.G., Niemi, G.J. and Collins, P.T. (1993) Effects of extremely low frequency electromagnetic fields on breeding and migrating birds. *American Midland Naturalist* **129**:96–115.

Hare, F.K. (1954) The boreal conifer zone. *Geogr. Stud.* **1**:4–18.

Hare, F.K. and Ritchie, J.C. (1972) The boreal bioclimates. *Geogr. Rev.* **62**:334–365.

Heinselman, M.L. (1973) Fire in the virgin forests of the Boundary Waters Canoe area, Minnesota. *Quatern. Research* **3**:329–382.

Helle, P. (1985a) Effects of forest regeneration on the structure of bird communities in northern Finland. *Holarctic Ecol.* **8**:120–132.

Helle, P. (1985b) Effects of forest fragmentation on bird densities in northern boreal forests. *Ornis Fennica* **62**:35–41.

Helle, P. (1986) Effects of forest succession and fragmentation on bird communities and invertebrates in boreal forests. *Acta Univ. Ouluensis A* **178**:41 + 93 pp.

Helle, P. and Fuller, R.J. (1988) Migrant passerines birds in European forest successions in relation to vegetation height and geographical position. *J. Anim. Ecol.* **57**:565–579.

Helle, P. and Järvinen, O. (1986) Population trends of North Finnish land birds in relation their habitat selection and changes in forest structure. *Oikos* **46**:107–115.

Helle, P. and Mönkkönen, M. (1986) Annual fluctuations of land bird communities in different successional stages of boreal forest. *Ann. Zool. Fennici* **23**:269–280.

Helle, P. and Mönkkönen, M. (1990) Forest successions and bird communities: theoretical aspects and practical implications, in *Biogeography and Ecology of Forest Bird Communities*, (ed. A. Keast), pp. 299–318. SPB Publishing bv., The Hague.

Herkert, J.R., Szafoni, R.E., Kleen, V.M. and Schwegman, J.E. (1993) *Habitat*

Establishment, Enhancement and Management for Forest and Grassland Birds in Illinois. Natural Heritage Technical Publication 1. Springfield, Illinois. 19 pp.

Herrera, C.M. (1978) On the breeding distribution pattern of European migrant birds: MacArthur's theme reexamined. *Auk* 95:496–509.

Horn, H.S. (1974) The ecology of secondary succession. *Ann. Rev. Ecol. Syst.* 5:25–37.

James, F.C., Wiedenfeld, D.A. and McCullogh, C.E. (1992) Trends in breeding populations of warblers: Declines in the southern highlands and increases in the lowlands, in *Ecology and Conservation of Neotropical Migrant Landbirds*, (eds J.M. Hagan and D.W. Johnston), pp. 43–56. Smithsonian Institution Press, Washington DC.

Järvinen, O. (1979) Geographical gradients of stability in European land bird communities. *Oecologia (Berl.)* 38:51–69.

Järvinen, O. (1981) Dynamics of North European bird communities. *Acta XVII Congr. Int. Ornithol. (Berlin)*: 770–776.

Järvinen, O. and Ulfstrand, S. (1980) Species turnover of a continental bird fauna: Northern Europe, 1850–1970. *Oecologia (Berl.)* 46:186–195.

Järvinen, O. and Väisänen, R.A. (1977) Long-term changes of North European land bird fauna. *Oikos* 29:225–228.

Järvinen, O. and Väisänen, R.A. (1978) Long-term population changes of the most abundant south Finnish forest birds during the 50 years. *J. Ornithol.* 119:441–449.

Johnston, D.H. and Schwartz, M.D. (1993) The conservation reserve program and grassland birds. *Conservation Biology* 7:934–937.

Keast, A. and Morton, E.S. (eds) (1980) *Migrant birds in the Neotropics: Ecology, Behavior, Distribution, and Conservation.* Smithsonian Institute Press, Washington, DC.

Kelsall, J.P., Tefler, E.S. and Wright, T.D. (1984) *The effects of fire on the ecology of the boreal forest, with particular reference to the Canadian north: a review and selected bibliography.* Canadian Wildlife Service, Occ. Paper No. 32.

Kendeigh, S.C. (1947) Bird population studies in the coniferous forest biome during a spruce burworm outbreak. *Ontario Dept. Lands and Forests, Biol. Bull.* 1:1–100.

Kornas, J. (1972) Corresponding taxa and their ecological background in the forests of temperate Eurasia and North America, in *Taxonomy, Phytogeography, and Evolution*, (ed. D.H. Valentine), pp. 37–59. Academic Press, New York.

Kuleshova, L.V. (1981) Ekologickeskie i zoogeographicheskie aspekti vozdejstvijah pozhzarov na lesnyh ptits i mlekopitayushsih. *Zool. Zurn.* 60:1542–1552.

Larsen, J.A. (1980) *The Boreal Ecosystem.* Academic Press, New York, 500 pp.

MacArthur, R.H. (1955) Fluctuating of animal populations and a measure of community stability. *Ecology* 36:533–536.

MacArthur, R.H. (1959) On the breeding distribution pattern of North American migrant birds. *Auk* 76:318–325.

Martin, N.D. (1960) An analysis of bird populations in relation to forest succession in Algonquin Provincial Park, Ontario. *Ecology* 41:126–140.

May, P.G. (1984) Secondary succession and breeding bird community structure: Patterns of resource utilization. *Oecologia (Berl.)* 55:277–281.

Mayr, E. (1946) History of North American bird fauna. *Wilson Bulletin* 58:2–41.

McArdle, B.H. and Gaston, K.J. (1992) Comparing population variabilities. *Oikos* 64:610–612.

Mönkkönen, M. (1984) Metsäkasvillisuuden sukkession vaikutukset Pohjois-Savon metsälinnustoon. *Siivekäs* 5:41–51. (In Finnish.)

Mönkkönen, M. (1991) Geographical patterns and local processes in Holarctic breeding bird communities. *Acta Univ. Ouluensis A* 223:31 + 134 pp.

Mönkkönen, M. and Helle, P. (1987) Avian reproductive output in European forest successions. *Oikos* 50:239–246.

Mönkkönen, M. and Helle, P. (1989) Migratory habits of birds breeding in different stages of forest succession: a comparison between the Palaearctic and the Nearctic. *Ann. Zool. Fennici* 26:323–330.

Mönkkönen, M. and Welsh, D.A. (1994) A biogeographical hypothesis on the effects of human caused landscape changes on the forest bird communities of Europe and North America. *Ann. Zool. Fennici* 31:61–70.

Morgan, K. and Freedman, B. (1986) Breeding bird communities in a hardwood forest succession in Nova Scotia. *Canadian Field-Naturalist* 100:506–519.

Niemi, G.J. (1978) Breeding birds of burned and unburned areas in northern Minnesota. *Loon* 50:73–84.

Niemi, G.J. and Hanowski, J.M. (1984) Relationships of breeding birds to habitat characteristics in logged areas of northern Minnesota. *J. Wildlife Manage.* 48:438–443.

Niemi, G.J. and Probst, J.R. (1990) Wildlife and fire in the upper midwest, in *Management of Dynamic Ecosystems*, (ed. J.M. Sweeney), pp. 31–46. North Cent. Sect., The Wildlife Soc., West Lafayette, Ind.

Noon, B.R., Dawson, D.K. and Kelly, J.P. (1985) A search for stability gradients in North American breeding bird communities. *Auk* 102:64–81.

Odum, E.P. (1969) The strategy of ecosystem development. *Science* 164:262–270.

Pimm, S.L. (1986) The complexity and stability of ecosystems. *Nature* 307:321–326.

Pospelov, S.M. (1957) Birds and mammals in Piceetum Myrtillosum of different ages. *Zool. Zurnal* 36:603–607. (In Russian with English summary.)

Probst, J.R., Rakstad, D.S. and Rugg, D.J. (1992) Breeding bird communities in regenerating and mature broadleaf forests in the USA Lake States. *Forest Ecology and Management* 49:43–60.

Sanders, C.J. (1970) Populations of breeding birds in the spruce–fir forests of northwestern Ontario. *Canadian Field-Naturalist* 84:131–135.

Simberloff, D. (1978) Using island biogeographic distributions to determine if colonization is stochastic. *Am. Nat.* 112:713–726.

Snow, D.W. (1978) Relationship between the European and African avifaunas. *Bird Study* 25:134–148.

Solonen, T. (1981) Dynamics of the breeding bird community around Lammi Biological Station, southern Finland, in 1971–80. *Ornis Fennica* 58:117–128.

Stiles, E.W. (1980) Bird community structure in alder forests in Washington. *Condor* 82:20–30.

Svensson, S. (1977) Population trends of common birds in Sweden. *Pol. Ecol. Studies* 3:207–213.

Svensson, S., Carlson, U.T. and Liljedahl, G. (1984) Structure and dynamics of an alpine bird community, a 20-year study. *Ann. Zool. Fennici* 21:339–350.

Theberge, J.B. (1976) Bird populations in the Kluane Mountains, southwest Yukon, with special reference to vegetation and fire. *Can. J. Zool.* **54**:1346–1356.

Titterington, R.W., Crawford, H.S. and Burgason, B.N. (1979) Songbird responses to commercial clear-cutting in Maine spruce–fir forests. *J. Wildlife Manage.* **43**:602–609.

Väisänen, R.A., Järvinen, O. and Rauhala, P. (1986) How are extensive, human-caused habitat alterations expressed on the scale of local bird populations in boreal forests? *Ornis Scandinavica* **17**:282–294.

van Horne, B. (1983) Density as misleading indicator of habitat quality. *J. Wildlife Manage.* **47**:893–901.

Virkkala, R. (1989) Short-term fluctuations of bird communities and populations in virgin and managed forests in Northern Finland. *Ann. Zool. Fennici* **26**:277–285.

Virkkala, R. (1992) Annual variation of northern Finnish forest and fen bird assemblages in relation to spatial scale. *Ornis Fennica* **68**:193–203.

Wein, R.W. and MacLean, D.A. (eds) (1983) *The Role of Fire in Northern Circumpolar Ecosystems*. John Wiley & Sons, New York.

Welsh, D.A. (1987) The influence of forest harvesting on mixed coniferous–deciduous boreal bird communities in Ontario, Canada. *Acta Oecologica* **8**:247–252.

Welsh, D.A. and Fillman, D.R. (1980) The impact of forest cutting on boreal bird populations. *Amer. Birds* **34**:84–94.

West, G.C. and DeWolfe, B. (1974) Populations and energetics of taiga birds near Fairbanks, Alaska. *Auk* **91**:757-775.

Whitcomb, R.F., Robbins, C.S., Lynch, J.F. *et al.* (1981) Effects of forest fragmentation on avifauna of the eastern deciduous forest, in *Forest Island Dynamics in Man-dominated Landscapes*, (eds R.L. Burgess and D.M. Sharpe), pp. 125–205. Springer-Verlag, New York.

Wiens, J.A. (1981) Scale problems in avian censusing. *Stud. Avian Biol.* **6**:513–521.

Wiens, J.A. (1983) Avian community ecology: an iconoclastic view, in *Perspectives in Ornithology*, (eds A.H. Brush and G.A. Clark), pp. 355–403. Cambridge Univ. Press, Cambridge, UK.

Wiens, J.A. (1989a) *The Ecology of Bird Communities*. Cambridge University Press, Cambridge, UK.

Wiens, J.A. (1989b) Spatial scaling in ecology. *Functional Ecol.* **3**:385–397.

Wiens, J.A. and Rotenberry, J.T. (1981) Habitat associations and community structure of birds in shrubsteppe environments. *Ecol. Monogr.* **51**:21–41.

Winternitz, B.L. (1976) Temporal change and habitat preference of some montane breeding birds. *Condor* **78**:383–393.

Zackrisson, O. (1977) Influence of forest fire on the North Swedish boreal forest. *Oikos* **29**:22–32.

Plate B Capercaillie cock on the lekking ground. (Photograph by Kjell Sjöberg.)

Plate C Newly hatched capercaillie chick. (Photograph by Kjell Sjöberg.)

Plate D Capercaillie chick at age of about two weeks. (Photograph by Kjell Sjöberg.)

Plate E Capercaillie female chick at age of about one month. (Photograph by Kjell Sjöberg.)

Plate F   In ditched swamp spruce forests, as well as in dense plantations of Scots pine, the bilberry field layer vegetation (normally dominant in field layer stands in boreal conifer forests of the region, and a key species for capercaillie chicks) decreases, and thereby also reduces the quality as capercaillie chick habitats. (See also Figure 4.5.)

Plate G   Large areas of old forest are converted to clearcut areas with subsequent planting of seedlings. This is a plantation of introduced lodgepole pine *Pinus contorta*. (Photograph by Kjell Sjöberg.)

Plate H If patches of swamp spruce are left in a managed landscape, both broods and adult capercaillie could still to some extent use a heavily fragmented landscape dominated by pine plantations. (See also Figure 4.6.)

Plate I During the first weeks of life, insects dominate the diet of capercaillie chicks and the broods are often found in this type of insect-rich habitat, i.e. a wet spruce forest, with bilberry (*Vaccinium myrtillus*) dominating the field layer. (Photograph by Kjell Sjöberg.)

Plate J When chicks grow older, they gradually change to a herbivorous diet. At that time the broods are more frequently found in mesic spruce stands or mixed stands with spruce and pine, where the berries of the bilberry plants are ripening in large numbers. (Photograph by Kjell Sjöberg.)

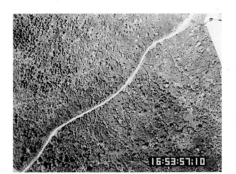

Plate N Examples of scenes to illustrate the primary approach being used with video in the New England Gap Analysis Project: (a) wide-angle scene of a forested area in northern Maine; (b) close-up zoom of center 30 m strip of same scene. Both photographs were derived from original videotapes taken from cameras mounted on wings of aircraft used to videotape the entire New England region.

Terrestrial Ecozones
of Canada

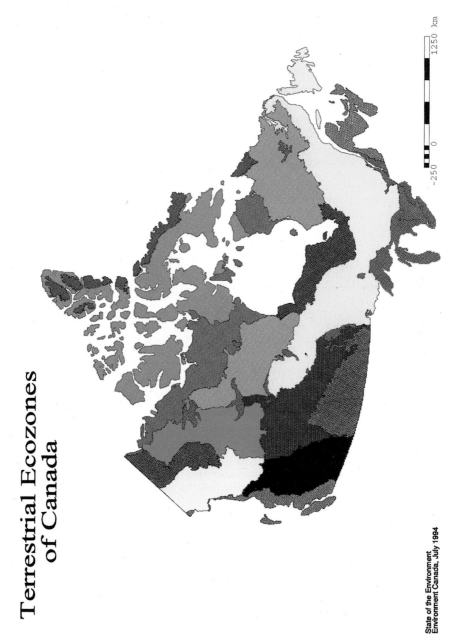

Ecozones
- Arctic Cordillera
- Northern Arctic
- Southern Arctic
- Taiga Plain
- Taiga Shield
- Boreal Shield
- Atlantic Maritime
- Mixedwood Plain
- Boreal Plain
- Prairie
- Taiga Cordillera
- Boreal Cordillera
- Pacific Cordillera
- Montane Cordillera
- Hudson Plain

-250   0   1250   km

State of the Environment
Environment Canada, July 1994

Plate K   The ecozones of Canada. (Courtesy of Ian Marshall, Environment Canada, Ottawa.)

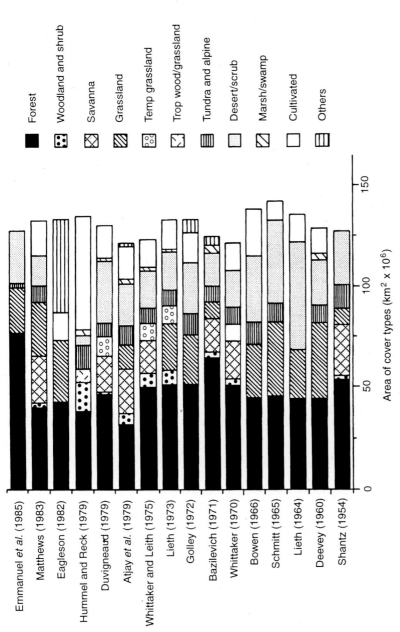

Plate L A demonstration of the variation among estimates of global land-cover classes produced by cartographers between 1954 and 1985. (Reprinted from Estes and Mooneyhan, 1994, by permission from the American Society for Photogrammetry and Remote Sensing; PE&RS, 60:5, 'Of Maps and Myths', pp. 517–524.)

Plate M The Micro Image Processing System (MIPS) as it appears during video interpretation. The five pictured components of the process include: 1) a portion of the Landsat image shown together with the airborne video flight-path as delineated with GPS. The points in this image were generated from differentially corrected Trimble data files. 2) the individual GPS points that can be queried on the screen for the time code; 3) the time code and the position of a selected point. The time code displayed on the video allows the operator to associate individual video frames with specific points on the image. 4) a five pixel data block around a selected point showing the actual raster values; 5) a demonstration of the geographic position of a particular pixel that can be selected on the image. (Chart supplied by D. Slaymaker, GAP Analysis program, University of Massachusetts).

8

The impact of forestry on ungulates in Japan

Naoki Maruyama and Kunihiko Tokida

8.1 INTRODUCTION

The demand for timber production during and after World War II caused large-scale devastation of forests in Japan. After the war, various policies were implemented for recovering the ruined stands and for promoting wood production. The policy on forestry since World War II has been seriously criticized from the viewpoint of nature conservation because it caused conversion of natural forests to artificial plantations, producing large-scale monocultural stands, overharvest of forests by sophisticated techniques, introductions of chemical materials such as fertilizer, pesticide and herbicide, mechanization such as the use of chainsaws, lift-cable systems for timber transportation and motor-powered bushcutters, and construction of large-scale road systems including multipurpose alpine timber roads for tourism.

Artificial plantations, consisting of a few coniferous species such as sugi (*Cryptomeria japonica*), hinoki (*Chamaecyparis obtusa*), pine (*Pinus* spp.) and Japanese larch (*Larix leptolepis*), had increased to 43% of the forest area in 1990 (Forest Agency, 1993). Nevertheless, the forested area still covers a large area – 67% of the Japanese territory – because industrial developments other than forestry have been restricted by the complicated and steep topography (Maruyama, 1989). Some species of wildlife, such as Japanese black bear (*Selenarctos thibetanus*), Ezo brown bear (*Ursus arctos yesoensis*), Japanese monkey (*Macaca fuscata*), Iriomote cat (*Mayailirus iriomotensis*) and Amur cat (*Felis euptilura*), have lost their habitats and decreased in number, while others have met favorably improved habitats and increased. The latter have usually damaged plantations and crops; the

Conservation of Faunal Diversity in Forested Landscapes. Edited by R.M. DeGraaf and R.I. Miller. Published in 1996 by Chapman & Hall. ISBN 0 412 61890 7.

sika deer (*Cervus nippon*) and the Japanese serow (*Capricornis crispus*) are typical such species.

The issues regarding these ungulates and the damage caused by them have illustrated the complicated relations between the socioeconomic aspects of forestry and the growth of nature conservation movements as a new culture in Japan. They are good incentives for finding a reasonable way to counterbalance forestry development with wildlife conservation and for exploring the future of nature conservation so that it may create a new life style never before experienced by Japanese people.

This paper describes the mutual relationship between forestry and ungulates in Japan and proposes a new, interrelated model of forest and game management.

8.2   AFFORESTATION, POPULATION GROWTH OF UNGULATES AND DAMAGE TO PLANTATIONS AND CROPS

8.2.1   Afforestation and damage to the forest caused by ungulates and small herbivores

The rate of afforestation in Japan has rapidly increased since 1950 and attained its first peak of 430 000 ha/year in 1954 (Figure 8.1). It remained around 400 000 ha/year for several years. However, after a second peak of 420 000 ha/year in 1961, it declined. In 1983 it was below 100 000 ha/year, i.e. 23% of that in 1961. The conversion of natural forest to artificial plantations has always been more than 50% of afforestation in national and private forests since World War II. Thus, natural forests were reduced to 57% of the forested area in Japan. In 1961, among the afforested tree species, sugi occupied the top at 39%; hinoki and pine were next at 18% each, with Japanese larch at 16%. In 1983, however, hinoki occupied the top at 42% and all of the others decreased, particularly pine (less than 2%) and larch (5%). These changes were affected by socioeconomic factors such as a decrease in numbers of forest laborers, an increase in the costs of afforestation, and pressure caused by low-priced timber imported from abroad. Therefore, plantations of hinoki trees with high commercial value have been favored by foresters.

The age-class distribution of artificial plantations has been remarkably biased to the younger stages of less than 20 years; the young stands (<20 years) attained 6 730 000 ha, covering about 84% of the whole area of artificial plantation in 1976. This situation could easily provide a favorable habitat for herbivores such as lemmings, voles, hares, sika deer and Japanese serow.

Damage to plantations by lemmings, voles and hares occurred instantly, as their populations increased rapidly in relation to afforestation (Figure 8.1), as it attained its highest peak at 200 000 ha in 1951, after which it soon

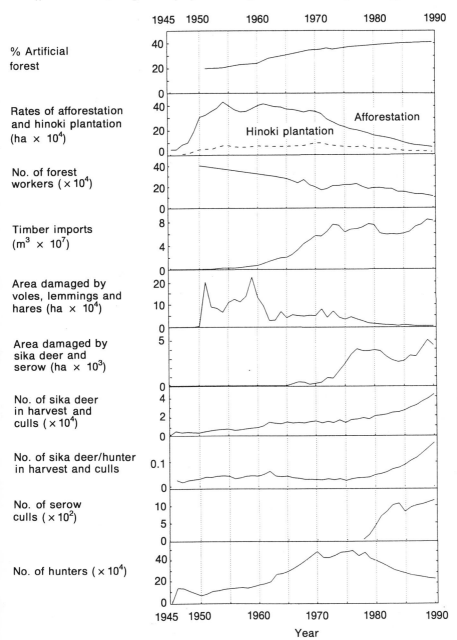

Figure 8.1 Changes in socioeconomic and ecological factors of forestry–ungulate problems in Japan. (Data on forestry from the *Forestry Statistics*, Forestry Agency, 1953–1992; data on game from *Hunting and Game Statistics of Japan*, Department of Agriculture and Forestry, 1922–1971, and Environment Agency, 1972–1990.)

decreased to around 100 000 ha between 1952 and 1958. There was a second peak of 220 000 ha in 1959 but again it rapidly decreased below 100 000 ha. Since then, it has steadily decreased and was less than 10 000 ha in 1990 (Statistic Directories of Forestry in Japan, Rinyakosaikai, 1945–1992).

On the other hand, damage by ungulates (Figure 8.1), sika deer and serow occurred about two decades later than that caused by small herbivores. It has increased since the early 1970s and exceeded 3000 ha in the late 1970s. However, damage by the serow gradually decreased in 1980, while that by the sika showed the contrary: there have been many conflicts between sika deer and forestry in many places. According to a questionnaire survey by the Japan Wildlife Research Center (JWRC) in 1991, damage by sika deer was reported from 506 local public entities (LPEs) in cities, towns and villages. These entities occupied 56% of LPEs inhabited by sika deer (JWRC, 1992).

Differences in period between damage by small herbivores and that by ungulates may be caused by differences in population growth and in reproduction rate, with slower maturity, lower fecundity and greater longevity for ungulates than for small herbivores.

8.2.2   Changes in amount of forage supply caused by forest operations and population growth of ungulates

An increase in forage availability along with increasing afforestation could have been brought about by the increase of young plantations described above. Young plantations with a stand age of less than 15 years are usually rich in forage (Furubayashi, 1980; Maruyama *et al.*, 1985b; Takatsuki, 1990). In particular, miyakozasa (*Sasa nipponica*), a main forage species for sika deer and probably for serow also, usually exceeds 100 g/m^2 dry weight through the year for several years after logging in coniferous plantations (Maruyama *et al.*, 1985b; Takatsuki, 1990). The total forage biomass including forbs, shrubs, graminoides, ferns, trees and liane sometimes attains nearly 600 g/m^2 (dry) in summer in a sugi plantation of about 10 years of stand age (Takatsuki, 1990). After the peak, forage rapidly decreases and does not recover in these evergreen coniferous plantations such as sugi and hinoki until the mature stage (Miyamoto *et al.*, 1980; Kiyono, 1988). Larch plantations, however, recover comparatively rich undergrowth soon after 30 years of stand age (Maruyama *et al.*, 1985b). Fluctuations in the amount of forage are shown in Figure 8.2.

Under the overstocked conditions of young coniferous plantations since the 1970s, sika populations have attained high densities – usually more than 20/km^2 and sometimes over 50/km^2 – in stands with young plantations in various areas all over Japan such as in the Tanzawa mountains (Maruyama and Furubayashi, 1983; Furubayashi, 1985), Nikko (Maruyama, 1981; Maruyama *et al.*, 1985a), Mt Goyo (Ito and Takatsuki, 1985), Hyogo

Prefecture (data of Department of Nature Conservation, Hyogo Prefecture), the Boso Peninsula (Dept. Nat. Cons., Chiba Pref. and JWRC, 1991), Shimane Prefecture (Kanamori *et al.*, 1991), Tsushima Islands (Tokunaga *et al.*, 1986), the Goto Islands (Torisu and Kanematsu, 1983; Doi *et al.*, 1985), Hokkaido (Dept. Nat. Cons., Hokkaido, 1991) (Figure 8.3). The high density of sika deer may be promoted by large-scale logging, which attracts many deer from the surrounding areas and provides an abundance of forage for them. Large-scale clearcut areas of more than 20 ha were frequent before 1970, but have rarely been seen in recent years; since the 1970s, the Forest Agency's guidance is that the size of clearcut areas should be less than 5 ha. In these high density areas corresponding to large clearcut areas, serious forest damage by sika deer has occurred (Maruyama *et al.*, 1971; Tokunaga *et al.*, 1986; Kanamori *et al.*, 1991; JWRC, 1992). These trends in sika population and in damage may be reflected by an increase in the harvest of the animals from less than 10 000 before 1960 to *c*. 30 000 in 1989 (Bobek and Maruyama, 1992; Department of Agriculture and Forestry, 1922–1971); Environment Agency, 1972–1990) (Figure 8.1).

Another aspect of afforestation–deer relationships is the size of logged areas with reference to the sex-segregation of sika deer (Maruyama, 1981), as has been found in various gregarious cervids that are without territories except during the rutting season (Rasmussen, 1941; Dasmann and Taber, 1956; Delap, 1957; Ahlen, 1965; Flook, 1970; McCullough, 1971; Miquella *et al.*, 1992). This phenomenon may be related to the size of the logged areas (Table 8.1). Females and offspring tend to concentrate at a high density in young plantations and logged areas, while males are at a low density in the surrounding forest-covered areas. This distribution might be caused by nutritional and physiological responses to a dispersion of stands; the former group needs highly nutritious forage from open stands for rearing, while males prefer cover with a stable, cool air temperature, avoiding direct sunlight. In Nikko, central Honshu, the female area well overlapped an area of larch plantations with plenty of forage spreading over a comparatively flat mountain slope of several hundred hectares (Maruyama *et al.*, 1985a), while in Tanzawa, central Honshu, where the topography is complicated and where the logging mosaic is small in size, the females' area extended only 20–30 hectares (Maruyama, unpublished data). In the female area, plantations may be seriously damaged and devastation of vegetation may be prominent.

In order to stabilize a sika population at a low density and to regulate the sex segregation in small areas, the stand should be made as small as possible, and logged areas should be dispersed. A small-scale selective logging system is more desirable.

Under such a situation, the serow population has been increasing since 1955, when the serow was designated as a special natural monument by the Law for the Protection of Cultural Properties. The species covered 34 500 km^2 in 1977, when the whole population was estimated to be about

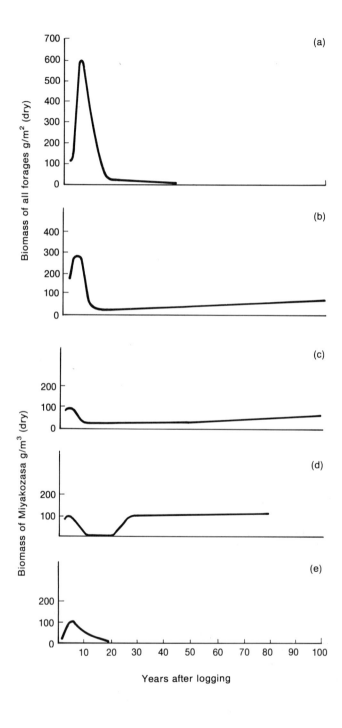

1 Tanzawa Mountains
2 Nikko
3 Mt Goyo
4 Hyogo Prefecture
5 Boso Peninsula
6 Shimane Prefecture
7 Tsushima Islands
8 Goto Islands
9 Gifu Prefecture
10 Nagano Prefecture
11 Aomori Prefecture
12 Odaigahara
13 Kinkasan Island
14 Nakano-shima Island, Toya Lake

Hokkaido

Honshu

Shikoku

Kyushu

Figure 8.3   Location of areas with high densities of sika deer and young coniferous plantations.

Figure 8.2   Changes in sika forage biomass (dry weight) after logging, for different plantations and broadleaved deciduous forests: (a) sugi plantations (summer); (b), (c) broadleaved deciduous forests (summer); (d) larch plantations (winter); (e) sugi and hinoki plantations (winter). Total forage biomass in (a), (b); miyakozasa (*Sasa nipponica*) biomass in (c), (d), (e). (Sources: (a)–(c) Takatsuki, 1990; (d), (e) Maruyama *et al.*, 1985a.)

Table 8.1   Comparison of sex segregation, population density and size of aggregation for sika deer between Nikko and Tanzawa

	Nikko	*Tanzawa*
Size of logging mosaic	Large	Small
Topography	Simple and flat	Complex and steep
Size of female area in sex segregation	Several hundred hectares[a]	20–30 ha[b]
Population density	8–70/km^2 in winter; less than 16/km^2 in summer and fall[a]	32–35/km^{2c}
Size of aggregation	7.6 ± 10.7 (SD) (range 1–80) (n = 82) in winter 1.5 (range 1–4 (n = 94) in August[b]	1.3–2.0 on average (range 1–8 through the year[b]

[a]Maruyama *et al.* (1985a)
[b]Maruyama (1981)
[c]Maruyama and Furubayashi (1983)

70 000–100 000 by the survey of the Environment Agency (Maruyama and Furubayashi, 1979). In the late 1970s, serow were damaging hinoki plantations in Nagano and Gifu Prefectures, central Honshu, and agricultural crops in Aomori Prefecture, northernmost Honshu. Since then, damage has been reported from various places in the eastern part of Honshu. In the second serow census by the Environment Agency in 1983 (unpublished data), the serow population further extended its range and the population size was estimated to be more than 100 000.

The principal agreements on the forestry–serow problems among the three related agencies (Cultural Affairs, Environment and Forest) were:

1. to control the serow populations,
2. to establish 15 serow reserves (900 km^2 on average, 143–2180 km^2), and
3. to give financial support to owners of the damaged forests so that they could replant trees, construct enclosures, etc.

8.2.3   Impact of afforestation on sika winter range

Seasonal migration of sika deer was found along the transitional zone from snowy to snowless areas; in winter the animals are restricted to the area with less than 50 cm of snow depth (Miura, 1974; Maruyama *et al.*, 1976; Maruyama, 1981, 1992; Tokida *et al.*, 1980).

In Nikko, Tochigi Prefecture, central Honshu, the winter ranges are distributed below 1500 m elevation where forest operations have been carried out. Most of the higher half of the winter range between 1000 and 1500 m, except for the steep slopes along valleys, is covered by larch plantations, with miyakozasa (*Sasa nipponica*) communities in the understory (which is a principal winter forage of sika deer). The lower half, between 800 and 1000 m, an emergency area for sika deer in hard winters, is dominated by evergreen coniferous plantations of sugi (*Cryptomeria japonica*) with little undergrowth. Compared with the original conditions dominated by natural broadleaved deciduous forests such as oak (*Quercus monoglica* var. *grosse-serrata*), beech (*Fagus crenata*) and birch (*Betula ermanii*), the carrying capacity may increase in the higher half of the winter range, while it may noticeably decrease in the lower half (Maruyama *et al.*, 1985b) (Figure 8.4).

This means that many deer probably starve when a deer population that has increased on a rich forage in the higher half of the winter range in normal winters without deep snow is forced in a hard winter with deep snow to descend to the lower half of the area that has a poor forage. In practice, mass mortality occurred several times in this area (Ikeda and Iimura, 1969; Maruyama, 1981; Maruyama and Takano, 1985) and it is

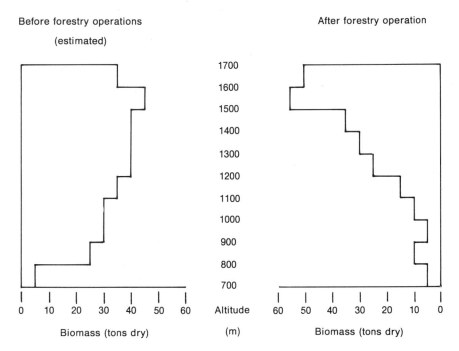

Figure 8.4 Estimated and actual altitudinal distributions of miyakozasa (*Sasa nipponica*) biomass (dry weight total for each altitudinal zone) in the Omote-Nikko winter range for sika deer. (Sources: Maruyama *et al.*, 1985b and in preparation.)

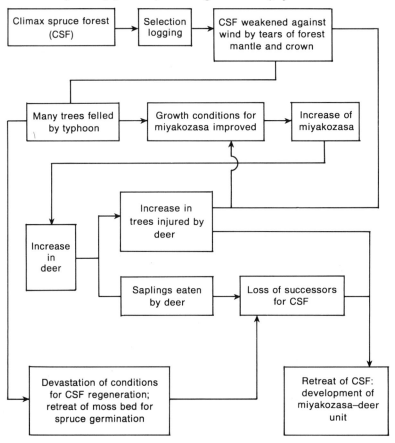

Figure 8.5   A deer problem in a natural stand: retrogression of climax spruce forest to miyakozasa grassland in Odaigahara, central Honshu. (Source: Hoshino *et al.*, 1987.)

concluded that this mortality is a result of the local forest operations that ignore the habitat requirements for sika deer populations and also the silvi-cultural practices that would be appropriate to the climatic and edaphic conditions. If such practices were applied, no permission should be given to forestry operations in the area above 1000 m in altitude, and even in the lower area most of the steep terrain should be given over to wildlife.

8.2.4   Deer problems in natural stands

A deer–forest problem has occurred even in a natural forest at Odaigahara (1500–1600 m in altitude) on the Kii Peninsula, Kinki district, central Honshu (Hoshino *et al.*, 1987) (Figure 8.5). This area was almost totally covered by natural forests including a climax forest of spruce (*Picea yesoensis* var. *hondoensis*). In addition to harvests by selective logging before World War II, a big typhoon through this area in 1959 blew down many trees. Thereafter, the forests were vulnerable to the wind because of gaps in the forest mantles and crowns, and an evergreen bamboograss, miyakozasa (*Sasa nipponica*), a principal winter forage of the animals, invaded the spruce forest and became widespread over the whole area. This contributed to an increase in sika deer by providing enough forage for them and the animals attained a high density of 30/km^2 on average. They peeled the bark off the trees and killed various species of tree such as spruce and fir (*Abies leptolepis*) by eating the formative layer in winter; they also ate the saplings of these species. This increased the number of dead trees and the forests continued to retreat. This retreat resulted in a further increase in the miyakozasa and in the sika deer.

In addition, the process was accompanied by a situation that deterred spruce regeneration. For germination and growth, spruce seeds need wet mossy beds covering fallen tree trunks (Yato, 1963). Under such light and dry environmental conditions, the moss communities soon retreated so that the regeneration of spruce became impossible, and existing saplings also died. Thus, the natural spruce forest will be changed by the miyakozasa communities in the near future and the sika population may increase and further destroy the forests.

Similar cases are easily found in closed habitats such as islands where the dispersal of the animals is prevented. On Kinkasan Island (*c.* 1000 ha), Miyagi Prefecture, northern Honshu, a sika population without predation and without control increased to more than 50/km^2 in the 1970s (Ito, 1985) and has threatened regeneration of a natural beech (*Fagus crenata*) and other forests (Kato and Nishihira, 1972). On the Nakanoshima Islands (524 ha), Toya Lake, Hokkaido, a sika population increased from a male and two females introduced around 1960 to about 300 in 1983. During this period of about two decades, the forest ecosystem drastically changed; the undergrowth community became especially poor and the deer frequently peeled the bark off crown-forming trees so that most of them died (Kaji *et al.*, 1984).

Both the Nakanoshima and Kinkasan populations crashed in the severe winter of 1984 after long-term poor nutrition (Kaji, 1986; Takatsuki and Suzuki, 1985).

8.3 IMPACT OF THE COLLAPSE OF RURAL COMMUNITIES ON UNGULATE–FORESTRY PROBLEMS

The impact of collapses of rural communities on ungulate–forestry problems are important also. Large-scale emigration from rural to urban areas since the 1960s has resulted in a lack of forest laborers for taking care of the large artificial plantations and for promoting afforestation much more in natural stands. The number of forest laborers decreased from 400 000 in 1950 to 110 000 in 1990 (Figure 8.1).

The arable lands and plantations abandoned by the emigrants became favorable habitats for deer, providing nutritious forages; in such areas, deer increased to high densities and they damaged plantations and crops (Maruyama *et al.*, 1971, for Tanzawa mountains; Takatsuki, 1984, Doi *et al.*, 1985 and Doi, 1986 for Nozakishima, Goto Islands, Kyushu). This situation may amplify the remaining local people's unrest directly caused by the collapse of many rural communities (Maruyama *et al.*, 1971). Those whose crops are damaged usually demand control of the sika deer and recompense for the damage. However, no compensation has ever been given by the administration because an article in civil law recognizes a rule of non-ownership of wildlife; in law they belong to neither the government nor the private sector.

8.4 PROGRESS ON UNGULATE–FORESTRY PROBLEMS

The Japanese administration on wildlife management before World War II was concerned mainly with hunting regulation and nuisance control to prevent damage to crops and plantations by wildlife, but there was little concern for wildlife conservation (Forest Agency, 1969). Industrial development was given priority over nature conservation. In the age of the Japanese imperialism from the late nineteenth to the early twentieth century, there was little public demand for wildlife conservation because there was no set of socioeconomic and cultural factors (peace, democracy, industrial development, destruction of environment, and mental and economic richness) in Japan that would create a new culture geared towards and having a desire for 'nature conservation'.

Those five socioeconomic and cultural factors to inspire nature conservation movements came after defeat in World War II and particularly after the 1960s when the Japanese GNP began to increase rapidly. Early nature conservation movements were initially aimed at the protection of vegetation and landscape from various developments such as tourism, construction of hydroelectric plants and cutting of natural forests. The policy of forestry by the Forest Agency had been violently opposed by many non-government organizations (NGOs) and some forestry labor unions. In the late 1960s, NGOs began to show concern also about ungulate–forestry problems.

Among typical examples, one of the earliest was the sika deer–forestry problem in the Tanzawa mountains near Tokyo and Yokohama (Maruyama *et al.*, 1971). In the case of the serow problem since the early 1970s, many national and local NGOs have encouraged serow protection and have criticized local forestry people and national and local forest and wildlife administrations that saw control killing as the only means of resolving the problems of serow damage to young coniferous plantations and crops. Japanese NGOs working for nature conservation greatly increased until in 1983 they numbered 1682 and included about 1 938 353 persons (JSNC, 1989). This had a big political impact on forestry and wildlife conservation administrations. Consequently, the administrations made decisions about not only control killing but also the establishment of 15 serow reserves over the range; further, they provided financial support for forestry people to replant and construct enclosures to protect young plantations from the animals. In these serow reserves, no serow control is permitted even though the animals damage plantations, but industrial activities are allowed there. The establishment of these reserves has been criticized by the NGOs, because reserve locations are biased to the subalpine zone, outside the distribution center of the animals that overlaps the forestry areas. In 1991, the Forest Agency decided to establish more than 20 forest ecosystem protection areas in natural forests in the higher mountain and subalpine zones. Each protection area covers more than 10 km^2 and consists of a core area and a buffer zone. No forest operations are permitted there. This might be a reaction to criticism by the NGOs.

Thus, the political power of the NGOs has surely influenced the administrations not only to control the deer but also to take other actions to resolve the problems. Concurrently, at least initially, these nature conservation movements were apt to lack knowledge on ungulate ecology and to be too emotional, automatically opposing control of the animals. Thus, they caused the wildlife administration to hesitate in its decision on the ungulate–forestry problem because it disliked criticism from the NGOs. Therefore, in many cases the administrations were very slow in taking action and in some cases deer populations increased excessively to a point of self-destruction: malnutrition was a result of habitat deterioration caused by the deer themselves (the Tsushima-sika reserve: Maruyama and Watanabe, 1993).

The number of sport-hunters rapidly decreased after the 1970s because they disliked the criticism that hunting is a cruel sport. At the peak in 1978 there were 505 000 sports-hunters; by 1989 the number had dropped to 250 000 (Bobek and Maruyama, 1992; Department of Agriculture and Forestry, 1922–1971); Environment Agency, 1972–1990). This decrease may help ungulate populations to increase in various areas all over the country. Thus, the wildlife administration encountered a difficulty in controlling the animals because it depended on hunters. A typical case can be seen in Hyogo Prefectures, western Honshu, where the increase in sika deer has exceeded the harvest by usual hunting and culling since 1990.

8.5 PROSPECTS FOR THE FUTURE AND SOME PROPOSALS

Japan has imported lumber and its products mainly from boreal forests in North America and eastern Russia and from tropical rainforests in Southeast Asia. In the near future, however, Japan should depend more upon its domestic forests and will want to develop them more and more, because of the exhaustion of foreign forests and because of strong criticism from domestic and international NGOs. This would provide favorable conditions for ungulate population increases and would lead to increasing conflict between the ungulates and forestry. Thus, there will be a second crisis for forest and wildlife conservation since World War II. How can the crisis be avoided?

1. We should not respond to a demand for timber production by converting more of the remaining natural forests into artificial plantations. The demand should be regulated.
2. Silvicultural practices appropriate to the climatic and edaphic conditions should be strictly adhered to in afforestation. The remaining natural forests in the higher mountain zone and all of the forests at the sub-alpine and alpine zones should be strictly protected for wildlife, including ungulates. These areas should be designated as reserves. Inappropriate plantations should be returned to natural forests.
3. The area of a plantation should be as small as possible. Selective logging systems should be adopted, and the age-class distribution of stands should be uniform in order to regulate large-scale ungulate population fluctuations.
4. Coniferous plantations should be mixed with broadleaved trees and kept at low levels of tree density in order to permit invasion by undergrowth and to maintain the carrying capacity of ungulates in mature stands. This contributes to an increase in the diversity of the wildlife communities (e.g. birds: Yui *et al.*, 1991).
5. A reasonable control system for ungulate populations including sport-hunting should be established by wildlife administrations. In this case, a system of monitoring population fluctuations would be necessary. If hunters further decline and control by hunters becomes impossible, a professional control team should be organized by the administrations, with at least one for each prefecture.
6. The administrations should always promote surveys of the animal populations and should provide the public and the NGOs with scientific information on the relationships between ungulates and forestry so that they can comprehend and make decisions.

In order to conduct the above, national and local forest administrations should introduce ideas of nature conservation to the policy of forest management and they should reflect people's opinions on preserving natural forests for the conservation of wildlife in cooperation with the wildlife administrations. At the same time, they should convince people of the necessity for saving timber resources. The wildlife administrations and the hunting associations, on the other hand, must establish a reasonable harvest system of the ungulates for maintaining the populations in forestry areas, and must endeavor to obtain a public comprehension of it because most Japanese people are not accustomed to sport-hunting and usually feel a spiritual resistance to it. This would also contribute to a budget saving for the ungulate management.

The final target in the future may be to reduce the stress caused by humans on the nature of Japan. This includes the problem of too many people in Japan, who have occupied too much of the wilderness and consumed too much energy and too many resources from all over the world. We also need to make room for the reestablishment of a prey–predator system, as wolves are now extinct in Japan. A forest ecosystem proximate to the primitive one may contribute to a self-control mechanism for stabilizing the ungulate populations. The Polish success in coexistence with wolves and in game management may be a good example for Japan (Bobek *et al.*, 1994).

8.6 SUMMARY

Forests, which cover 67% of Japan, provide wildlife with favorable habitats. However, especially after World War II, artificial plantations have increased to 43% of the forests. These plantations, which consist of only a few coniferous species such as sugi (*Cryptomeria japonica*), hinoki (*Chamaecyparis obtusa*), pine (*Pinus* spp.) and Japanese larch (*Larix leptolepis*), have resulted in many problems for wildlife. For example, problems in young plantations caused by ungulates such as the Japanese serow (*Capricornis crispus*) and sika deer (*Cervus nippon*) cannot be ignored and have been complicated by their relation to nature conservation. Yet the damage caused by ungulates to young plantations have not been as great as those produced by lemmings (*Clethrionomys rufocanus bedfordiae*), voles (*Microtus montebelli*) and hares (*Lepus brachyurus* and *L. timidus*).

The import of large amounts of lumber from tropical and boreal forests has put pressure on domestic forestry to produce low-priced lumber. This has produced a sense of growing crisis among foresters and local people and led to the conversion of natural forests into high-priced hinoki plantations. These new plantations provide the increasing ungulate populations, particularly of the serow, with favorable habitat. The abundant and high quality food made available to the ungulates by these new plantations creates circumstances that result in damage to the environment.

In Japan, citizen movements to promote wildlife conservation have risen in popularity over the last few decades and have opposed the standard forestry policy of control killing. At the same time, the number of hunters who take part in control kills has noticeably decreased because the hunters are wary of being accused of taking part in a cruel sport. Ungulate populations have been helped by this trend and are increasing in various localities all over the country. Policies that lead to the overdevelopment of forests as well as policies that lead to the excess protection of ungulates are both untenable in modern Japan.

8.7 REFERENCES

Ahlen, I. (1965) Studies on the red deer, *Cervus elaphus* L., in Scandinavia (III). *Viltrevy* 3:177–376.

Bobek, B. and Maruyama, N. (1992) The analysis of the harvest of ten game species in Japan for 1948–1984 period, in *Global Trends in Wildlife Management* (eds B. Bobek *et al.*) Trans. 18th IUGB Congress, Krakow 1987, Swiat-Press, Krakow-Warszawa, 389–396.

Bobek, B., Perzanowski, K., Kwiatkowski, Z. *et al.* (1994) Economic aspects of brown bear and wolf predation in southeastern Poland. *Proc. Int. Wildl. Manage. Congr., Sept. 19–25, 1993, San Jose, Costa Rica.* (In press.)

Dasmann, R.F. and Taber, R.D. (1956) Behavior of Columbian black-tailed deer with reference to population ecology. *J. Mamm.* 37:143–164.

Delap, P. (1957) Some notes on the social habits of the British deer. *Proc. Zool. Soc. London* 128:608–612.

Department of Agriculture and Forestry (1922–1971) *Hunting and Game Statistics of Japan.*

Department of Nature Conservation, Chiba Prefecture, and JWRC (1991) *A survey report on management of sika deer in Boso Peninsula, Chiba Prefecture.* 128 pp. (In Japanese.)

Department of Nature Conservation, Hokkaido (1991) *Yaseidobutu bunpu-tou jittai-chosa hokokusho* (A report of status of wildlife in Hokkaido, with special reference to the geographical distribution). 93 pp. (In Japanese.)

Doi, A. (1986) Socio-ecological study of the sika deer in Nozaki Island, the Goto Islands, northwestern Kyushu. *Honyurui Kagaku* (53):39–42. (In Japanese.)

Doi, A., Inakazu, K., Ono, Y. and Kawahara, H. (1985) A preliminary study on the effects of sika deer on natural regeneration of forest. *Bull. Nagasaki Inst. Appl. Sci.* 26:13–18.

Environment Agency (1972–1990) *Hunting and Game Statistics of Japan.*

Flook, D.R. (1970) *Causes and implications of an observed sex differential in the survival of wapiti.* Canadian Wildl. Serv. Rep. Set. 11. 71 pp.

Forest Agency (1969) *Choju Gyosei no Ayumi* (Advance of the Game Administration in Japan). Rinya-kosaikai, 572 pp. (In Japanese.)

Forest Agency (1993) *Ringyo Hakusho* (Annual Report on Japanese Forestry). Jap. Forst. Assoc., 228 pp. (In Japanese.)

Furubayashi, K. (1980) *Shimokita-hantou no nihonkamosika* (Japanese serow in Shimokia Peninsula). Shimokia-hantou Nihonkamosika Kenkyukai, 99–115. (In Japanese.)

Furubayashi, K. (1985) The population dynamics of sika deer in the Tanzawa mountains, Kanagawa Prefecture, in *Dynamics of Larger Mammals (Bear and Deer) and Natural Environment in Japanese Forest.* Nat. Conserv. Bureau, Envir. Agency, Japan, 261–295 (In Japanese.)

Hoshino, Y., Jida, N. and Maruyama, N. (1987) Impacts of sika deer and Japanese black bears on the spruce forests in Ohdaigahara. *Pap. Plant Ecol. & Tax. to*

Mem. Dr. S. Nakanishi, Kobe Geobot. Soc., 367–377. (In Japanese, with English summary.)

Ikeda, S. and Iimura, T. (1969) Ecological analysis of the environment for deer *Cervus nippon centralis* (Kishida) habitat in relation to its hunting and forest utilization. *Bull. Gov. For. Exp. Stat.*, (**220**):60–119. (In Japanese, with English summary.)

Ito, T. (1985) Fluctuations of sika deer in Kinkasan Island. *Monit. Rep. of Protection Plan in Kinkasan Isl.* III, 11–25. (In Japanese.)

Ito, T. and Takatsuki, S. (1985) The population dynamics of sika deer in Mt Goyo, Iwate Prefecture, in *Dynamics of Larger Mammals (Bear and Deer) and Natural Environment in Japanese Forest.* Nat. Conserv. Bureau, Envir. Agency, Japan, 149–213. (In Japanese.)

JSNC (1989) *Waga-kuni no kakushu sizenhogo-katsudo no jittai ni kansuru kenkyu-hokokusho* (A report on status of nature conservation movements in Japan). 176 pp. (In Japanese.)

JWRC (1992) *Choju-gaisei taisaku chosa-hokokusho* (A report on damage by wild birds and mammals). 45 pp. (In Japanese.)

Kaji, K. (1986) Population dynamics and management of sika deer introduced into Nakanoshima Island in Lake Toya. *Honyurui Kagaku* (**53**):25–28. (In Japanese.)

Kaji, K., Ohtaishi, N. and Koizumi, T. (1984) Population growth and its effect upon the forest used by sika deer on Nakanoshima Island in Lake Toya, Hokkaido. *Acta Zool. Fennica* **172**:203–205.

Kanamori, H., Inoue, J. and Suto, Y. (1991) Stem bark damage of coniferous trees by antler rubbing of sika deer, in *Wildlife Conservation* (eds N. Maruyama *et al.*), Proc. Vth INTECOL 1990 Yokohama, 114–115.

Kato, M. and Nishihira, M. (1972) Ecological analysis of the structure of terrestrial ecosystem in Kinkasan Island IX. *Ann. Rep. JIBP/CT-S FY-1971*, 204–219.

Kiyono, Y. (1988) Analysis of factors affecting the dynamics of coverage and number of species in understories in *Chaemacyparis obtusa* plantations. *J. Jap. For. Soc.* **70**:455–460. (In Japanese, with English summary.)

Maruyama, N. (1981) A study of the seasonal movements and aggregation patterns of sika deer. *Bull. Fac. Agric., Tokyo Univ. of Agr. & Tech.* (**23**):85 pp.

Maruyama, N. (1989) The status of Japanese big game and hunting. *Proc. Symp. on Wildl. in Jap.*, CIC & JHA, 73–75.

Maruyama, N. (1992) Ecological distribution of sika deer in Japan, in *Global Trends in Wildlife Management* (eds B. Bobek *et al.*). Trans. 18th IUGB Congress, Krakow 1987, Swiat Press, Krakow–Warszawa, 419–421.

Maruyama, N. and Furubayashi, K. (1979) *Estimation of the range, density and size of the serow population.* Envir. Agen. Jap. 48 pp. (In Japanese.)

Maruyama, N. and Furubayashi, K. (1983) Preliminary examination of block count method for estimating numbers of sika deer in Fudakake. *Jap. J. Mamm.* 9:274–278.

Maruyama, N. and Takano, K. (1985) A mass mortality of sika deer in Nikko caused by the heavy snowfall of 1984 winter, in *Dynamics of Larger Mammals (Bear and Deer) and Natural Environment in Japanese Forest.* Nat. Conserv. Bureau, Envir. Agency, Japan, 248–253. (In Japanese.)

Maruyama, N. and Watanabe, T. (1993) Increase of the Tsushima-sika population in the Tsushima-sika Reserve, Tsushima Islands, in *Deer of China* (eds N. Ohtaishi and S.-I. Sheng). Elsevier Sci. Publ., 258–272.

Maruyama, N., Iimura, T. and Yamagishi, K. (1971) An example of the problems on nature conservation. *Seibutsu-kagaku* 22:135–149. (In Japanese.)

Maruyama, N., Totake, Y. and Okabayashi, R. (1976) Seasonal movements of sika in Omote-Nikko, Tochigi Prefecture. *Jap. J. Mamm.* 6:187–198.

Maruyama, N., Fukushima, Y., Hazumi, Y. and Hazumi, T. (1985a) The geographical distribution of Nikko sika population with reference to the density distribution, in *Dynamics of Larger Mammals (Bear and Deer) and Natural Environment in Japanese Forest*. Nat. Conserv. Bureau, Envir. Agency, Japan, 226–231. (In Japanese.)

Maruyama, N., Hazumi, Y. and Mori, A. (1985b) A winter yard of Nikko sika deer in reference to miyakozasa, in *Dynamics of Larger Mammals (Bear and Deer) and Natural Environment in Japanese Forest*. Nat. Conserv. Bureau, Envir. Agency, Japan, 232–247. (In Japanese.)

McCullough, D.R. (1971) *The Tule Elk*. Univ. Calif. Press, 209 pp.

Miquelle, D.G., Peek, J.M. and Ballenberghe, V.V. (1992) Sexual segregation in Alaskan moose. *Wildl. Mono.* 122:57 pp.

Miura, S. (1974) On the seasonal movements of sika deer population in Mt. Hinokiboramaru. *Jap. J. Mamm.* 6:51–66. (In Japanese, with English summary.)

Miyamoto, M., Tanimoto, T. and Ando, T. (1980) Analysis of the growth of hinoki (*Chaemacyparis obtusa*) artificial forests in Shikoku district. *Bull. For. & For. Prod. Res. Inst. Jap.* 309:89–107. (In Japanese, with English summary.)

Rasmussen, D.I. (1941) Biotic communities of Kaibab Plateau, Arizona. *Ecol. Mono.* 11:231–275.

Takatuski, S. (1984) Ecological studies on effect of sika deer on vegetation. V. *Ecol. Rev.* 20:223–235.

Takatsuki, S. (1990) Changes in forage biomass following in a sika deer habitat near Mt Goyo. *Ecol. Rev.* 22:1–8.

Takatsuki, S. and Suzuki, K. (1985) Mass die-off of sika deer on Kinkasan Island in the spring of 1984. *Monit. Rep. of Protect. Plan in Kinkasan Isl.* III, 27–61. (In Japanese.)

Tokida, K., Maruyama, N. Ito, T. *et al.* (1980) *Geographical distribution of sika deer and its factors. The 2nd Fundamental Survey for Natural Environment of Japan. II Mammals.* JWRC, 38–68. (In Japanese.)

Tokunaga, S., Doi, T. and Ono, Y. (1986) Distribution of Tsushima deer, *Cervus pulchelus Imaizumi*, by the questionnaire method. *Seibutsu-kagaku* 34:175–181. (In Japanese.)

Torisu, C. and Kanematsu, N. (1983) The census of sika deer, *Cervus n. nippon* Temminck, on Nozaki-jima, the Goto Islands. *Bull. Nagasaki Sogo-Kagaku Univ.* 24:249–252. (In Japanese.)

Yato, K. (1963) Ecological studies on the subalpine forest in the Kii Peninsula. II. *Bull. Fac. Agr. Mie Univ.* 28:101–126. (In Japanese.)

Yui, M., Suzuki, Y. and Nakamura, N. (1991) Conservation of forest bird communities, a review and recommendation for Japan's forest, in *Wildlife Conservation* (eds N. Maruyama *et al.*), Proc. Vth INTECOL 1990 Yokohama, JWRC, 107–110.

–9

Effects of hurricanes on wildlife: implications and strategies for management

Joseph M. Wunderle, Jr and James W. Wiley

9.1 INTRODUCTION

Hurricanes, also known as cyclones or typhoons, occur throughout the world and are common in at least six regions encompassing parts of the tropics, subtropics and temperate zone (Nalivkin, 1983). Although the frequency of hurricane occurrence may vary within a region, such storms may occur with sufficient frequency and destructive force (Figures 9.1 and 9.2) to be important factors in determining the structure and species composition of biotic communities (e.g. Wadsworth and Englerth, 1959; Odum, 1970; Doyle, 1981; Weaver, 1986; Crow, 1980). The importance of hurricanes in influencing ecosystems is consistent with the prevailing view that natural systems are largely organized by disturbance (Pickett and White, 1985; Denslow, 1987).

Recent hurricanes which have passed through the Caribbean, Middle America and the southeastern United States have provided an opportunity to study their effects on the local flora and fauna (for recent reviews see Bénito-Espinal and Bénito-Espinal, 1991; Finkl and Pilkey, 1991; Walker, 1991; Pimm et al., 1994; Loope et al., 1994; Smith et al., 1994). Such studies are of particular interest to conservation biologists and managers because of the potential negative impact of catastrophic events on small populations (Ewens et al., 1987). This concern for small isolated populations has increased as a result of habitat loss and fragmentation and is likely to continue increasing if the predicted destructive potential of hurricanes increases with increased global warming (Emanuel, 1987).

Conservation of Faunal Diversity in Forested Landscapes. Edited by
R.M. DeGraaf and R.I. Miller. Published in 1996 by Chapman & Hall. ISBN 0 412 61890 7.

Figure 9.1   Damage one week after Hurricane Hugo, Luquillo Experimental Forest, Puerto Rico, September 1989.

Figure 9.2   The cloud forest at Hardwar Gap, Jamaica, two months after the passage of Hurricane Gilbert in October 1988.

As summarized in our review of the effects of hurricanes on bird populations (Wiley and Wunderle, 1994), it has been demonstrated that hurricanes can have both direct and indirect effects on wildlife. For instance, the high winds and rainfall associated with hurricanes may result in direct mortality of some wildlife. However, after the storm the indirect effects may have even more profound and long-lasting influences on wildlife populations. These include loss of food supplies or foraging substrates; loss of nests and nest or roost sites; increased vulnerability to predation; microclimate changes; and increased conflict with humans. In the short term, before changes in plant succession, wildlife populations may respond to hurricane damage by shifting their diet, foraging sites or habitats, and reproductive patterns. In the long term, wildlife populations may respond to changes in plant succession as vegetation regenerates in storm-damaged old-growth forests.

Knowledge of the effects of hurricanes on wildlife populations and the response of wildlife to such storms is fundamental to the development of management strategies designed to reduce the vulnerability of endangered species. Therefore the purpose of this chapter is to summarize the effects of hurricanes on wildlife populations and the response of wildlife populations to hurricane effects. Here we concentrate on the adverse effects of hurricanes on terrestrial species in an effort to identify the characteristics which make certain populations vulnerable to hurricane damage. We describe the patterns of hurricane effects and responses without giving examples, at least for birds which we have previously reviewed in detail (Wiley and Wunderle, 1994). From these findings we offer some possible management strategies which might reduce the vulnerability of endangered species to hurricanes. Our intent is to stimulate both managers and researchers to devise management strategies that will reduce the vulnerability of endangered wildlife to hurricane damage.

9.2 DIRECT EFFECTS OF HURRICANES

It is often assumed that hurricane winds, rain and wave action directly cause wildlife mortality, although this has been difficult to document. The literature on wildlife mortality from hurricanes is sparse and occasionally contradictory (Cely, 1991). Accounts of avian mortality are confined primarily to waterbirds rather than landbirds, although their larger size could bias mortality surveys. High post-hurricane mortality has been documented for raccoons, rabbits and deer (Ensminger and Nichols, 1958) and for raccoons and rodents (Gunter and Eleuterius, 1971). We are unaware of any published accounts of mass mortality of amphibians and reptiles in the aftermath of hurricanes, although the small size of most species could make detection less likely.

Storms have been cited as responsible for affecting local movements of animals. Birds are particularly vulnerable to displacement by high winds and accounts of storm-displaced seabirds are common.

9.3   INDIRECT EFFECTS OF HURRICANES ON WILDLIFE

Given that hurricane winds strip flowers, fruits and seeds from plants, it is reasonable to expect that populations of nectarivores, frugivores and seed-eaters are more likely to decline in the aftermath of a hurricane than insectivores or raptors. Indeed, several hurricane studies of birds have documented this pattern in sites with varying degrees of vegetation damage in a diversity of habitats. Many insectivore populations might be buffered from hurricane-induced food shortages by relying on a food source characterized by a high level of diversity, resiliency and turnover.

Whereas post-hurricane food loss can explain many population declines, factors such as changes in foliage distribution and vegetation structure can also affect certain species. This is particularly apparent in birds and reptiles (Reagan, 1991) that glean insects off the surfaces of leaves or stems which are vulnerable to hurricane impacts.

Nests, nesting sites, roosting sites and nesting material can be destroyed by high winds, rainfall or flooding. Species requiring large old trees for nesting or roosting are particularly susceptible to hurricane effects, as these trees are often most vulnerable to hurricane damage. Furthermore, wildlife that use storm-damaged old trees for nests or den sites run the risk of losing their nest/den site in future storms.

Animals are likely to be more exposed to predators in defoliated habitats in the aftermath of hurricanes. Unfortunately, we are unaware of any published accounts supporting this probable event, possibly due to the difficulty of observing predation and assessing its impact on prey populations. However, raptors are certainly present and often quite visible after hurricanes and in some cases have not shown population declines after a hurricane.

Many forest-floor or understory species are accustomed to foraging in a dark forest understory and are likely to leave areas with reduced canopy cover. Whether these animals are responding directly to the absence of a canopy or to one of the secondary effects resulting from canopy loss (e.g. change in light level, microclimate, food supply, predator exposure, vegetation structure, etc.) is unknown.

Loss of the forest canopy can have profound effects on the microclimate of the forest understory. Higher temperatures and desiccation may stress organisms adapted to life in the shaded understory of a closed-canopy forest. Whereas microclimatic changes following a hurricane may affect some higher vertebrates, the effect is likely to be greatest upon ectotherms with limited dispersal abilities, such as some of the lower vertebrates (Reagan, 1991; Woolbright, 1991) and invertebrates (Willig and Camilo, 1991).

The negative effects of human activities on wildlife may increase in the aftermath of a major storm. For example, storm-weakened wildlife often provides a readily available and vulnerable source of protein for people who may be in need of food themselves. The tendency for wildlife to wander widely in search of food and cover in the aftermath of storms makes them even more vulnerable, particularly when they move into agricultural or residential areas (Craig *et al.*, 1994). Hurricanes may open primeval forests to human exploitation or increase the likelihood of human-caused fire damage. Opportunistic land settlement by subsistence farmers may encroach on woodlands, where forests have been blown down by hurricanes.

9.4 RESPONSE OF WILDLIFE POPULATIONS TO HURRICANE EFFECTS

One option for some species which have lost their main food source in the aftermath of a hurricane is to shift to another food source. Such post-hurricane diet shifts have been found in several species of bird and in at least one mammal (Spain and Heinsohn, 1973; Heinsohn and Spain, 1974).

If an animal's normal food supply or foraging site has been damaged or destroyed, it can be expected to shift its foraging location shortly after the storm's passage, well before plant successional changes occur. For some species this might involve a shift in foraging site within the same habitat, such as forest canopy-dwelling species foraging near the ground following canopy loss in the wake of a major storm. Other species may respond to hurricane damaged habitats by foraging more within the original home range or by expanding their home range size to forage over a greater area (Willig and Gannon, 1991). Additional foraging site shifts may involve movement into less damaged areas of the same habitat or movement into other nearby habitats, thus accounting for the post-hurricane appearance of some species in atypical habitats. Moreover, altitudinal movements or population shifts are likely in mountainous regions if high altitude sites are more severely damaged than lowland sites or because recovery rates of high altitude vegetation lag behind those of the lowlands.

Some animals breed normally in the aftermath of a hurricane in spite of substantial habitat damage, although the observed reproductive response can be variable. In some instances tropical birds may delay reproduction by several months. In a frog species the clutch size was reduced following a hurricane (Woolbright, 1991), while in rodents an increase in post-hurricane food supplies may have contributed to an increased reproductive output (Rao, 1980).

A major effect of hurricanes is the creation of early successional habitats, by increasing the number and size of forest openings. Second growth and edge species have responded to successional changes in the aftermath of hurricanes by moving into damaged forests where previously they had been rare or absent. Increased forest openings may enable avian brood parasites and some predators to invade extensive forest tracts which they would otherwise avoid. This could be particularly detrimental to the populations of some forest interior birds which have been shown to be highly sensitive to edge-dwelling brood parasites and predators (reviewed in Askins *et al.*, 1990).

In pre-settlement times, hurricanes were undoubtedly beneficial to wildlife diversity by creating early successional habitats in landscapes dominated by extensive forests. These storms create a patchwork or mosaic of communities which influences the abundance of particular species (Levin and Paine, 1974; Whitmore, 1974; Paine, 1979) and community diversity (Loucks, 1970; Connell, 1978; Huston, 1979). No doubt these disturbances provided habitat for edge and second growth species, which otherwise would have been uncommon or rare in those times. However, today most edge or second growth species have an adequate amount of habitat in contrast to species requiring mature forest or forest interior, which face the greatest risk of temporary habitat loss from hurricanes.

9.5 POTENTIAL MANAGEMENT STRATEGIES

Although the study of hurricane effects on wildlife populations is still in its infancy, recent studies suggest some promising management strategies to lessen the impact of hurricanes on wildlife populations in regions where cyclonic storms are expected. These strategies should focus on the stresses facing populations in the aftermath of storms, as a number of studies have indicated that the greatest stress of a hurricane on upland terrestrial wildlife populations occurs after its passage rather than during its impact. Here we list some potential management strategies, although some will require more research for specific sites or species.

1. The greater the geographic range and more widely dispersed a population, the less vulnerable it will be to a single hurricane strike. As suggested by Hooper *et al.* (1990) an effort should be made to ensure that populations of endangered species are uniformly dispersed within their geographic range.
2. Montane forest reserves require lowland forest reserves to which montane frugivores, seedeaters and nectarivores can migrate in the aftermath of a storm. Lowland vegetation, because of its rapid growth rate, can recover faster than montane vegetation and thus can serve as a refugium for montane species in the storm's aftermath (Wunderle *et al.*, 1992). Habitat corridors between montane and lowland reserves will facilitate altitudinal movement (Varty, 1991) and, in the aftermath of a hurricane, likely provide patches of undamaged vegetation.
3. Forest reserves on flat or level terrain should be as large as possible,

while reserves on mountainous or hilly terrain can be smaller. Topographic relief can affect the extent of hurricane damage to vegetation. Forests on level terrain are more likely to receive extensive damage, whereas hilly or mountainous terrain may have protected patches of undisturbed vegetation (e.g. Bellingham, 1991; Bellingham *et al.*, 1992; Wunderle *et al.*, 1992).

4. Encourage food plants with high recovery rates. Because many edge, gap or secondary successional plant species have high recovery rates in the aftermath of hurricanes, patches of these food plants should be available either within or near the reserve. For example, palms such as *Prestoea montana* and *Roystonea* spp. are important foods for West Indian parrots, as well as other frugivores, and often are less likely to be damaged than dicotyledons (Yih *et al.*, 1991; Frangi and Lugo, 1991; Francis and Gillespie, 1993). Palms should be encouraged throughout a range of altitudes and sites within a reserve, as emergency food supply for frugivorous and seedeating birds.

5. Encourage plants which are relatively resistant to hurricane damage. Tree species have been found to differ in their susceptibility to damage (reviewed in Brokaw and Walker, 1991; Francis and Gillespie, 1993) and the more resistant species provide valuable temporary cover and foraging substrate in the immediate aftermath of a storm. Consideration should be given to the possibility of ensuring that patches of resistant vegetation are available. For example, in South Carolina both bald cypress (*Taxodium distichum*) and live oak (*Quercus virginiana*) sustained relatively little hurricane damage in comparison with other species (Putz and Sharitz, 1991; Gresham *et al.*, 1991). On some sites, pines are more likely to suffer hurricane damage than nearby hardwoods (Boucher, 1990; Putz and Sharitz, 1991) but not always (Gresham *et al.*, 1991). We need to know if resistant vegetation can serve as temporary refugia in the immediate aftermath of hurricanes, even for species specialized on different habitat types.

6. Forest reserves and surrounding landscape should have a mixture of different age stands to ensure rapid replacement of hurricane-damaged old growth for wildlife requiring old growth forests. Second growth vegetation is more resistant to structural damage than old growth (reviewed in Brokaw and Walker, 1991) and therefore is likely to recover more quickly and ultimately provide old growth vegetation. Thus forest reserves consisting primarily of old growth forests will likely have a greater time lag for replacement following a hurricane than those forests with a mix of different aged stands.

7. It may be possible to 'hurricane-proof' forests by adjusting the density of trees in a stand, although the appropriate density is likely to vary with a variety of factors (reviewed in Tanner *et al.*, 1991). A direct relationship between pine stand density and hurricane survival suggested to Hooper

et al. (1990) that it might be possible to 'hurricane-proof' pine stands by having more densely stocked stands of pine than is currently practised. Although this may be appropriate for pine stands in South Carolina, for other forests it is unclear if comparatively high stem density makes forests more or less vulnerable to damage (Brokaw and Walker, 1991). The effect of stand density is not always obvious, as found by Reilly (1991) who observed that damage was greatest where neighboring trees were distant in two forest plots, but the reverse in a third. Recently thinned stands were more severely affected by a hurricane than unthinned stands on Puerto Rico, although age/size structure also influenced stand vulnerability (Wadsworth and Englerth, 1959). Finally, dense stands may be more susceptible to damage than open stands as observed after a hurricane in Massachusetts (Foster, 1988). Obviously no single management prescription will exist to 'hurricane-proof' forests by manipulating stand density, but management will depend upon a variety of factors including site, species and age composition.

8. Intensive post-hurricane management intervention may be necessary for some critically endangered species. Efforts concentrated on recovery of damaged or lost nest or roost sites have proven effective. For example, the repair of Puerto Rican parrot (*Amazona vittata*) nest cavities contributed to successful post-hurricane reproduction (Meyers *et al.*, 1993). Hand excavation of artificial woodpecker cavities to replace lost cavity trees is contributing to the post-hurricane recovery of the red-cockaded woodpecker, *Picoides borealis* (Hooper, personal communication). We are unaware of any attempts to provide food to endangered wildlife in the aftermath of hurricanes, and suspect that it might be ineffective for many species given the tendency for food-stressed wildlife to wander widely after hurricanes.

9.6 SUMMARY

Hurricanes, often known as typhoons or cyclones, are common in many parts of the world where their frequent occurrence can have both direct and indirect effects on wildlife. Direct effects include mortality from exposure to hurricane winds, rains and storm surges, and geographic displacement of individuals by storm winds. Indirect effects appear after the storm and include loss of food or foraging substrates; loss of nests, nest or roost sites; increased vulnerability to predation; change in microclimates; and increased conflict with humans. The short-term response of wildlife populations to hurricane damage, prior to changes in plant succession, includes shifts in diet, foraging sites or habitats, and reproductive responses. Wildlife populations may show long-term responses to changes in plant succession as second-growth vegetation increases in storm-damaged old-growth forests.

The greatest stress of a hurricane to most upland wildlife populations occurs after its passage rather than during its impact. Most important is the destruction of vegetation, which secondarily affects wildlife after the storm. Most vulnerable are animals which:

- have a diet of nectar, fruit, or seeds;
- nest, roost or forage on large old trees;
- require a closed forest canopy;
- have special microclimate requirements;
- live in a habitat with a slow recovery rate.

Small populations with these traits are at greatest risk to hurricane-induced extinction, particularly if they exist in small isolated habitat fragments.

Recovery of wildlife populations from hurricane effects is partially dependent on the extent and degree of vegetation damage as well as its rate of recovery. Also, the reproductive rate of the remnant local population and recruitment from undisturbed habitat patches influence the rate at which wildlife populations recover from damage. Potential management strategies which might reduce impacts on endangered wildlife include:

- ensuring that populations are widely dispersed in their geographic range;
- maintenance of lowland reserves as post-hurricane refugia for montane species;
- encouraging food plants which are resistant to hurricane damage and/or have high recovery rates;
- adjusting the density of trees to 'hurricane-proof' a stand;
- post-hurricane intervention, which might include provision of artificial cavities for nesting or supplement feeding.

9.7 REFERENCES

Askins, R.A., Lynch, J.F. and Greenberg, R. (1990) Population declines in migratory birds in eastern North America. *Current Ornithology* 7:1–57.

Bellingham, P.J. (1991) Landforms influence patterns of hurricane damage: evidence from Jamaican montane forests. *Biotropica* 23:427–433.

Bellingham, P.J., Kapos, V., Varty, N. *et al.* (1992) Hurricanes need not cause mortality: the effects of Hurricane Gilbert on forests in Jamaica. *J. of Tropical Ecology* 8:217–223.

Bénito-Espinal, F.P. and Bénito-Espinal, E. (1991) *L'Ouragan Hugo: genèse, incidences géographiques et écologiques sur la Guadeloupe.* Co-editors: Parc National de la Guadeloupe, Délégation Régionale à l'Action Culturelle and l'Agence Guadeloupéene de l'Environnment du Tourisms et des Loisirs. Imprimerie Désormeaux, Fort-de-France, Martinique.

Boucher, D.H. (1990) Growing back after hurricanes. *Bioscience* 40:163–166.

Brokaw, N.V.L. and Walker, L.R. (1991) Summary of the effects of Caribbean hurricanes on vegetation. *Biotropica* 23:442–447.

Cely, J.E. (1991) Wildlife effects of Hurricane Hugo. *J. of Coastal Research* 8:319–326.

Connell, J.H. (1978) Diversity in tropical rain forests and coral reefs. *Science* **199**:1302–1310.

Craig, P., Trail, P. and Morrell, T.E. (1994) The decline of fruit bats in American Samoa due to hurricanes and overhunting. *Biol. Cons.* **69**:261–266.

Crow, T.R. (1980) A rain forest chronicle: a 30-year record of change in structure and composition at El Verde, Puerto Rico. *Biotropica* **12**:42–55.

Denslow, J.S. (1987) Tropical rain forest gaps and tree species diversity. *Annu. Rev. Ecol. Syst.* **18**:431–452.

Doyle, T.W. (1981) The role of disturbance in the gap dynamics of a montane rain forest: an application of a tropical forest successional model, in *Forest Succession: Concepts and Applications* (eds D.C. West, H.H. Shugart and D.B. Botkin), pp. 56–73. Springer-Verlag, New York.

Emanuel, K.A. (1987) The dependence of hurricane intensity on climate. *Nature* **326**:483–485.

Ensminger, A.B. and Nichols, L.G. (1958) Hurricane damage to Rockefeller Refuge. *Proceedings Annual Conference Southeastern Association Game and Fish Commissioners* **11**:52–56.

Ewens, W.J., Brockwell, P.J., Gant, J.M. and Resnick, S.I. (1987) Minimum viable population size in the presence of catastrophes, in *Viable Populations for Conservation* (ed. M.E. Soulé), pp. 59–68. Cambridge Univ. Press, Cambridge.

Finkl, C.W. and Pilkey, O.H. (eds) (1991) Impacts of Hurricane Hugo: September 10–22, 1989. *J. Coastal Res.* Special Issue No. 8.

Foster, D.R. (1988) Species and stand response to catastrophic wind in central New England, USA. *J. Ecol.* **76**:135–151.

Francis, J.K. and Gillespie, A.J.R. (1993) Relating gust speed to tree damage in Hurricane Hugo, 1989. *J. Arboriculture* **19**:368–373.

Frangi, J.L. and Lugo, A.E. (1991) Hurricane damage to a flood plain forest in the Luquillo Mountains of Puerto Rico. *Biotropica* **23**:324–335.

Gresham, C.A., Williams, T.M. and Lipscomb, D.J. (1991) Hurricane Hugo wind damage to southeastern US coastal forest tree species. *Biotropica* **23**:420–426.

Gunter, G. and Eleuterius, L.N. (1971) Some effects of hurricanes on terrestrial biota, with special reference to Camille. *Gulf Research Reports* **3**:283–289.

Heinsohn, G.E. and Spain, A.V. (1974) Effects of a tropical cyclone on littoral and sub-littoral biotic communities and on a population of dugongs (*Dugong dugon* [Muller]). *Biol. Conserv.* **6**:143–152.

Hooper, R.G., Watson, J.C. and Escano, R.E.F. (1990) Hurricane Hugo's initial effects on red-cockaded woodpeckers in the Francis Marion National Forest. *Trans. 55th N. A. Wildl. and Nat. Res. Conf.* **55**:220–224.

Huston, M. (1979) A general hypothesis of species diversity. *Am. Nat.* **113**:81–101.

Levin, S.A. and Paine, R.T. (1974) Disturbance, patch formation, and community structure. *Proc. Nat. Acad. Sci.* **71**:2744–2747.

Loope, L., Duever, M., Herndon, A. *et al.* (1994) Hurricane impact on uplands and freshwater swamp forest. *BioScience* **44**:238–246.

Loucks, O.L. (1970) Evolution of diversity, efficiency, and community stability. *Am. Zool.* **10**:17–25.

Meyers, J.M., Vilella, F.J. and Barrow, W.C., Jr (1993) Positive effects of Hurricane Hugo: record years for Puerto Rican Parrot nesting in the wild. *Endangered Species Techn. Bull.* **18**:1–10.

Nalivkin, D.V. (1983) *Hurricanes, Storms and Tornadoes*. A.A. Balkema, Rotterdam.

Odum, H.T. (1970) Rainforest structure and mineral-cycling homeostasis, in *A Tropical Rainforest* (eds H.T. Odum and R.F. Pigeon), pp. 43–52. National Technical Information Services, Springfield, Virginia.

Paine, R.T. (1979) Diaster, catastrophe, and local persistence of the sea palm *Postelsia palmaeformis*. *Science* 205:685–687.

Pickett, S.T.A. and White, P.S. (1985) *The Ecology of Natural Disturbance and Patch Dynamics*. Academic Press, New York.

Pimm, S.L., Davis, G.E., Loope, L. *et al.* (1994) Hurricane Andrew. *BioScience* 44:224–229.

Putz, F.E. and Sharitz, R.R. (1991) Hurricane damage to old-growth forest in Congree Swamp National Monument, South Carolina, USA. *Can. J. For. Res.* 21:1765–1770.

Rao, A.M.K. Mohana (1980) Impact of cyclone on the rodent population in Andhra Pradesh. *Bombay Nat. Hist. Soc.* 77:502–503.

Reagan, D.P. (1991) The response of *Anolis* lizards to hurricane-induced habitat changes in a Puerto Rican rain forest. *Biotropica* 23:468–474.

Reilly, A.E. (1991) The effects of Hurricane Hugo in three tropical forests in the US Virgin Islands. *Biotropica* 23:414–419.

Smith III, T.J., Robblee, M.B., Wanless, H.R. and Doyle, T.W. (1994) Mangroves, hurricanes, and lightning strikes. *BioScience* 44:256–262.

Spain, A.V. and Heinsohn, G.E. (1973) Cyclone associated feeding changes in the dugong (Mammalia: Sirenia). *Mammalia* 37:678–680.

Tanner, E.V.J., Kapos, V. and Healey, J.R. (1991) Hurricane effects on forest ecosystems in the Caribbean. *Biotropica* 23:513–521.

Varty, N. (1991) The status and conservation of Jamaica's threatened and endemic forest avifauna and their habitats following Hurricane Gilbert. *Bird Cons. Intern.* 1:135–151.

Wadsworth, F.H. and Englerth, G.W. (1959) Effects of the 1956 hurricane on forests in Puerto Rico. *Carib. For.* 20:38–51.

Walker, L.R. (1991) Tree damage and recovery from Hurricane Hugo in Luquillo Experimental Forest, Puerto Rico. *Biotropica* 23:379–385.

Weaver, P.L. (1986) Hurricane damage and recovery in the montane forests of the Luquillo Mountains. *Carib. J. Science* 22:53–70.

Whitmore, T.C. (1974) *Change with time and the role of cyclones in tropical rain forest on Kolombangara, Solomon Islands*. Institute Paper 46, Commonwealth Forestry Institute, Oxford, England.

Wiley, J.W. and Wunderle, J.M., Jr (1994) The effects of hurricanes on birds with special reference to Caribbean islands. *Bird Cons. Intern.* 3:319–349.

Willig, M.R. and Camilo, G.R. (1991) The effect of Hurricane Hugo on six invertebrate species in the Luquillo Experimental Forest of Puerto Rico. *Biotropica* 23:455–461.

Willig, M.R. and Gannon, M.R. (1991) *The effect of Hurricane Hugo on bat ecology with emphasis on an endemic Puerto Rican Bat (Stenoderma rufum) in the Luquillo Experimental Forest*. Report to Southern Forest Experiment Station, USDA Forest Service, New Orleans, LA.

Woolbright, L.L. (1991) The impact of Hurricane Hugo on forest frogs in Puerto Rico. *Biotropica* 23:462–467.

Wunderle, J.M., Jr, Lodge, D.J. and Waide, R.B. (1992) Short-term effects of Hurricane Gilbert on terrestrial bird populations on Jamaica. *Auk* **109**:148–166.

Yih, K., Boucher, D.H., Vandermeer, J.H. and Zamora, N. (1991) Recovery of the rain forest of southeastern Nicaragua after destruction by Hurricane Joan. *Biotropica* **23**:106–113.

–10

Trends in the moose–forest system in Fennoscandia, with special reference to Sweden

Göran Cederlund and Roger Bergström

10.1 INTRODUCTION

Few free-living game species have been so intensively managed on a countrywide scale as the moose (*Alces alces*) in Fennoscandia. The spectacular
increase in moose densities in the 1970s and early 1980s created not only
increased hunting opportunities and an expanded commercial market for
moose meat, but also negative consequences such as damage to commercial
forests (Lavsund, 1987) and a high incidence of moose-related traffic accidents (Lavsund and Sandegren, 1991). During the 1980s approximately 1.5
million moose were killed in Sweden. In 1991, when populations were still
high, approximately 200 000 moose were harvested in Fennoscandia, which
probably exceeds the total legal harvest in North America and former
Soviet Union (Bisset, 1987; Bluzma, 1987; Kelsall, 1987).

Forestry practices in this region, including clearcutting, planting/self-gen-
eration and intensified stand management, have altered the forest landscape
concurrent with this period of increased moose populations. In these countries, the forest industry is an important domestic economic sector and contributes a considerable proportion of annual export.

The objective of this chapter is to show the general trends in moose
population development (and to give some reasonable explanations for
these changes, particularly in the recent decades) and in forest systems, with
special reference to forage production and forest damage due to moose. We
focus on the past 50 years partly because few accurate data exist before that
time and partly because it is unlikely that moose–forest interactions have
ever undergone such dramatic changes in so short a time as during the last

Conservation of Faunal Diversity in Forested Landscapes. Edited by
R.M. DeGraaf and R.I. Miller. Published in 1996 by Chapman & Hall. ISBN 0 412 61890 7.

decades. We concentrate in particular on the situation in Sweden. Finally, we will try to predict the scenario over the next two to three decades.

Moose invaded Fennoscandia soon after the last glaciation, which ended about 10 000 years ago. Ever since, the moose has been a major game species for the human population (see review by Pulliainen, 1987). Although there are no data, historical records indicate low population densities and an uneven distribution (Rülcker and Stålfelt, 1986; Nygrén, 1987), partly due to unregulated and often illegal hunting by the increasing human population. In many areas of Sweden, moose were almost exterminated by the middle of the nineteenth century. Legislation together with substantial predator reduction and regulated harvesting generated a gradual increase in numbers and re-establishment in former ranges. In the early years of the twentieth century moose were found over most of the country. The total population was apparently fairly stable and the annual harvest in Sweden amounted to a few thousand animals.

Assuming harvest levels proportional to the population, it was not until the early 1930s that a gradual increase became evident in Sweden. By that time Finland and Norway had very small populations, but these were beginning to increase slowly (Figure 10.1). For example, after a period of protection the moose population in Finland in 1933 was estimated at

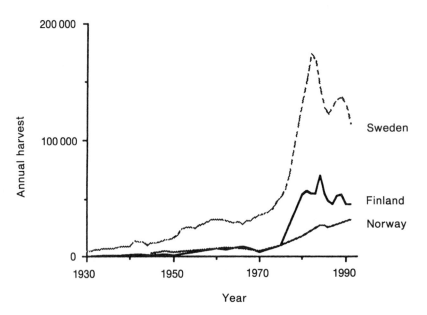

Figure 10.1 General trends in the Fennoscandian moose population, 1930–1991, indexed by total annual harvest.

approximately 3500, which was half the legal harvest in Sweden (Nygrén, 1987). During the following 40 years the moose populations in the three countries developed similarly but at different rates. In the mid 1970s harvest levels increased dramatically and indicated a rapidly growing moose population. During 1970–1980 the total annual harvest in the three countries increased five-fold (from approximately 50 000 to 230 000 moose). The total harvest in the early 1930s was just a fraction of the harvest in 1980. During the 1980s the harvests, and probably also the living populations, were reduced in Finland and Sweden but were still slowly increasing in Norway. Today moose in Europe (excluding the Baltic states and Russia) are found almost throughout Fennoscandia and in northwestern Poland (Figure 10.2).

It must be noted that there are no data based on full-scale surveys of moose populations in Fennoscandia. However, since harvesting is considered to be closely associated with population density, annual kill figures might best indicate population development. It seems plausible to assume that the increase in the moose population during the 1970s and early 1980s developed along a curve that was one or two years ahead of that of the harvest (Cederlund and Markgren, 1987). This was primarily due to a delay in the response between the number of licences issued by the hunting authorities and the actual moose population. During the late 1970s the rate of increase must have gradually dampened. In Sweden the first reduction probably occurred in 1980–81 (Cederlund and Markgren, 1987). Since then harvest quotas have probably exceeded moose production, which has led to a steady decrease in moose numbers (annual statistics from the Swedish Environmental Protection Agency) (Figure 10.1). From 1980 to 1990 the overwintering population of Swedish moose was probably reduced from more than 300 000 animals to 200 000 animals. The conclusion seems to be that the eruptive development of the moose population was primarily due to changes in the management system. However, there are indications that density-dependent changes such as lowered fecundity rates also lowered the potential rate of increase in the moose population during the peak years (Sand *et al.*, in preparation). Changes in reproduction and probably in mortality rates might therefore have had an additive effect on the reduction of the moose population in the 1980s.

Moose populations have developed differently in different parts of Sweden (Figure 10.3). During the 1970s, the most dramatic increase was recorded in the forested counties in the north (e.g. Västerbotten) while the counties dominated by farmland in the south (e.g. Kristianstad) have experienced substantially smaller increases. In the peak years around 1980 the average kill in Västerbotten was almost 2.0 moose/km^2 compared with 0.1 moose/km^2 in the early 1970s. In Västmanland, a county with an intermediate mosaic of agricultural and forested areas, the annual harvest has remained fairly constant (Figure 10.3).

It is evident that Fennoscandian moose populations have a high potential

Figure 10.2   Distribution of moose in northwestern Europe and Poland in relation to vegetation zones. The western limit is indicated by the broken line. (From Bergström and Hjeljord, 1987.)

rate of increase (for definition, see Caughley, 1977), occasionally exceeding 60% annual increase in the southern regions (Stålfelt, 1977). Consistently lower rates of increase, ranging between 15 and 30%, have been reported in North America for moose populations with adequate browse, few or no predators and light hunting pressure (Keith, 1983; Van Ballenberghe, 1983). In a thoroughly investigated population in central Sweden, demographic data indicated a potential rate of increase of 50%, and hence 50% of the winter population could potentially be harvested each autumn if the population was maintained at the same density (Cederlund and Sand, 1991). With a winter density of 1.3 moose/km^2 the density of moose killed was 0.65/km^2, and the annual yield of meat was 75 kg/km^2. Using the current market price in 1990 (US$12.80/kg), each km^2 will provide meat valued at almost US $1000/year.

10.2 FACTORS CONTRIBUTING TO INCREASED MOOSE DENSITIES

A number of factors, separately or in combination, may have contributed to the spectacular development of the Fennoscandian moose population (Figure 10.4).

10.2.1 Habitats

Similar to the situation in other boreal areas, forest fires have, in earlier days, had a considerable effect on the Fennoscandian forest ecosystems. Studies on fire history in northern Sweden have revealed a mean fire frequency of 80 years (Zachrisson, 1977). Other events, such as storms and insect outbreaks, also contributed to the creation of early successional stages. The moose has been a member of the temperate fauna for a long time (Geist, 1987) and it has adapted to the landscape mosaic created by drastic events in Fennoscandia as well as in many other parts of the boreal region of the northern hemisphere (Telfer, 1984). Very small areas are burnt today in Fennoscandia and the creation of moose habitats through forest fires is negligible.

Parallel to this influence, humans utilized forests in different ways. Trees were generally cut selectively (individual trees that had reached proper size were cut and transported out of the forests) instead of harvested *en masse* through clearcutting. This utilization of timber, and later pulp, increased because of human population increases and technical improvements. During the nineteenth century, but more so during the mid twentieth century, new cutting practices appeared (Lavsund, 1987). Selective cutting was gradually replaced by the clearcutting technique. This method of cutting mature forest is still practised. The rotation period varies considerably depending on latitude but is similar to the frequency for wildfires. With a rotation period of 100 years about 1% of the forest is cut each year. The area of annual clearcutting in Sweden was highest during the early 1970s; it decreased thereafter and has been fairly stable since 1980 (Skogsstyrelsen, 1992).

Forest management practices such as clearcutting and subsequent natural regeneration, or planting, have led to a distinct landscape mosaic. The average forest stand size is a few hectares on privately owned land, and

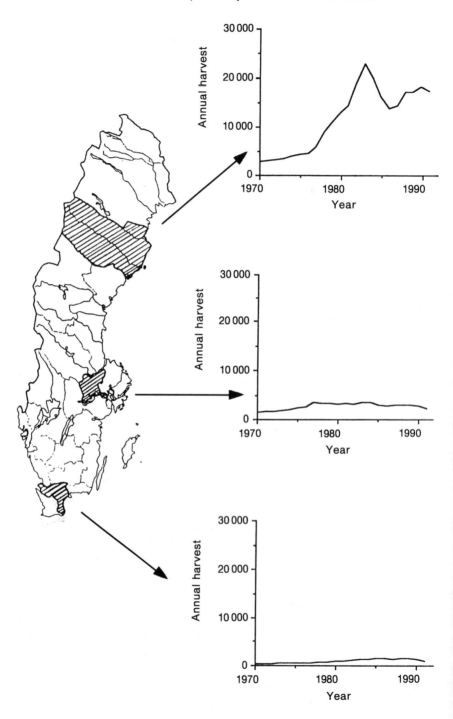

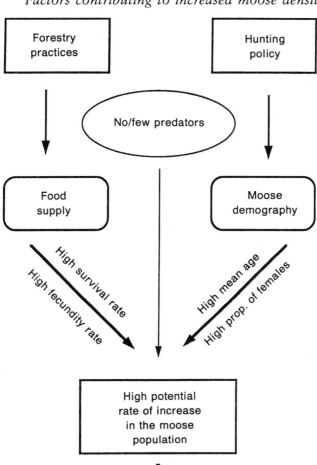

Figure 10.4 Conceptual model of the most important factors contributing to high potential rate of increase in Fennoscandian moose population.

20–40 hectares on land owned by forest companies or the state (Skogsstyrelsen, 1992). The forest stand mosaic created by these clearcutting techniques differs from the mosaic created by fires in, for example, North America, which is more uneven in terms of destruction of plant biomass and geo-

Figure 10.3 Trends in the moose population of three counties in Sweden, 1970–1991, indexed by total annual harvest.

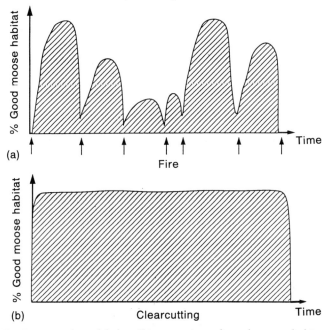

Figure 10.5 Conceptual model describing creation of good moose habitats in Fennoscandia by (a) forest fires and (b) modern forestry. (S. Lavsund, personal communication.)

graphic scale (Figure 10.5; Van Wagner, 1978, 1983). This has led to a consistent high percentage of good moose habitat accompanying clearcutting.

In Sweden, as in the neighboring countries, the environment created by clearcutting techniques is generally regarded as prime summer and winter habitats. After a few years a shrub layer is developed on recently cut areas. The planted or self-generated trees, usually Scots pine (*Pinus sylvestris*) and Norway spruce (*Picea abies*), are parts of this layer together with deciduous species. The pine is a moderately preferred species during winter (Bergström and Hjeljord, 1987), while Norway spruce is seldom touched by moose. The field layer in clearcuts is dominated by grasses (often *Deschampsia flexuosa*), herbs and some dwarf shrubs. The plants in the shrub layer are usually available to moose because of a moderate snow depth. A large part of Sweden has a long-term mean snow cover of less than 0.6 m (Pershagen, 1969). In that respect Sweden differs from many boreal inland areas in Russia and North America.

The mature forests of Sweden and neighbouring countries usually have a poorly developed shrub layer available for moose. Recent large-scale investigations (Bergström *et al.*, unpublished data) have shown that the cover of woody species important as food for moose is between 5 and 10 times higher on clearcuts compared with mature forests. On the other hand, the field layer in mature forests is dominated by dwarf shrubs (mainly *Vaccinium myrtillus, V. vitis-idaea* and *Calluna vulgris*), which are important for moose mostly during spring and autumn (Cederlund *et al.*, 1980). The dwarf shrubs are usually hidden when snow cover exceeds 0.3–0.4 m. The preference for mature forests as winter feeding areas for moose (e.g. Eastman and Ritcey, 1987) has not been verified in Sweden in recent years (e.g. Cederlund and Okarma, 1988).

Clearcuts and young plantations are also used during summer. Moose use clearcuts and plantations more than expected from the availability of these forest stand types (Cederlund and Okarma, 1988). In contrast, older forest were also used but less than expected.

The mean stand size in Sweden is fairly small compared with several other temperate areas, and hence a moose is usually no more than a few hundred meters from cover to avoid predators and unsuitable weather or snow conditions. This might affect the use of of food in clearcuts of different sizes. Andrén and Angelstam (1993) showed that clearcut size did not significantly affect browsing intensity.

10.2.2   Food choice

During normal winters, most of the moose range within Fennoscandia is covered with snow. Bergström and Hjeljord (1987) reviewed the winter diet of moose and presented the following generalized order of preference: mountain ash (*Sorbus aucuparia*) > willows (*Salix* spp.) > aspen (*Populus tremula*) > juniper (*Juniperus communis*) > birch (*Betula* spp.) > Scots pine > alder (*Alnus* spp.) > Norway spruce. However, studies of food choice show that species of medium preference, such as Scots pine and birches, are quantitatively the most important species for the moose (Cederlund *et al.*, 1980; Hagen, 1983; Bergström and Hjeljord, 1987). The average abundance of all these woody species is highest in forest stands with an age of 10–20 years (see review by Bergström and Hjeljord, 1987).

During spring and autumn, dwarf shrubs are very important food for moose. In summer the main foods quantitatively are leaves from deciduous tree species and some herbs, such as fireweed (*Epilobium angustifolium*), growing mostly in young forest stands (Cederlund *et al.*, 1980; Johanson *et al.*, 1984). Thus, young forest stands are the most important feeding habitats during winter and summer, while older stands, with their supply of dwarf shrubs, are important during spring and autumn.

10.2.3   Feeding patterns and forest management

Large herbivores generally do not feed randomly within their habitat but have fine-tuned foraging systems including selection on different levels (Senft *et al.*, 1987). This means that different properties of the environment, of food plants and also of plant parts are important in governing the selection of food by moose. For example, it has been shown that the distance between food plants affects the amount of forage that is taken from each individual tree (Vivås and Saether, 1987). The size of annual shoots of forage plants is also important in the selection process (Danell *et al.*, 1985; Bergström and Danell, 1987). The larger the shoot (i.e. the better the growth of the plants or plant part), the more eaten by moose. The shoot size itself also governs the feeding rate and is correlated to food quality in terms of nutrients, digestibility and secondary compounds. The palatability of forest trees may be influenced by many forest practices, such as the type of seedlings used in regeneration, spacing of seedlings, cleaning and fertilization. It can be concluded that trends in forestry practices during recent decades have not only created optimal habitats but also favoured moose in terms of the accessibility and palatability of individual food plants.

10.2.4 Demography

The rate of increase is primarily a function of the rates of birth and mortality. High productivity among female moose, expressed in terms of high twinning rates and high proportion of fertile yearlings, is common, particularly in the southern parts of the Fennoscandian countries (Stålfelt, 1974; Nygrén, 1983; H. Sand and G. Cederlund, in preparation). It seems that moose with an early age of sexual maturity have been selected for. Rapid increase in body weight may further promote high fecundity rates early in life (Saether and Haagenrud, 1983b), which means that potential productivity is high. Although fecundity is age related, body weight may explain much of its individual variation in female moose (Saether, 1987). Even at the regional level, both age and weight-related fecundity rates may differ (H. Sand, in preparation). In general, there are indications that productivity (e.g. autumn calf recruitment) decreases northward (Karlsson *et al.*, 1988).

Hunting is commonly considered the major cause of mortality among moose. In areas with stable populations, annual harvests are between 30% and 60% of the winter population, which is then considered equal to the potential rate of increase. Consequently, almost no 'natural deaths' occur. However, in northern Sweden early calf mortality reached 20–30% (K. Wallin, personal communication) while the total annual non-hunting mortality rate in a predator-free area with moderate winters in south-central Sweden was less than 7% (Cederlund and Sand, 1991). Locally, car accidents may affect populations: approximately 4100 accidents were reported in Sweden in 1990 (Lavsund and Sandegren, 1991). It should be noted that most of the Fennoscandian moose population is not exposed to predation, although bears kill an unknown number in some northern areas. The total bear population in Sweden is estimated at 600–800 animals (J. Swenson, personal communication).

The rate of increase is strongly dependent on the sex and age composition, in particular the age structure among adult females. It is likely that hunting has a considerable effect in regional variations of the standing age distribution, as there is a significant difference in male–female proportion (Saether, 1987). When compared with most other game species, moose hunting is strongly regulated and very efficient in Fennoscandia. For example, licences are issued and harvest numbers are adjusted to the size of each hunting area and also to adult/calf ratio. The high proportion of calves, often 30–50% of the harvest, has resulted in an increase in the average age of the surviving population and consequently the average productivity of the females has also increased (cf Markgren, 1969; Saether and Haagenrud, 1983a). Heavy hunting pressure on bulls has resulted in a distorted sex ratio among adults, which has positively affected the potential rate of increase in the total population. Moreover, recruitment of bulls has been lowered since the sex ratio among harvested calves has gradually changed towards an even distribution during the last 20 years (K. Wallin, personal communication).

10.3 POPULATION DYNAMICS AND MANAGEMENT OF MOOSE IN NORTH AMERICA AND THE FORMER SOVIET UNION

The historical perspective of moose in North America is rather limited when compared with its European counterpart (Karns, 1987). It is obvious that North American moose populations have undergone fluctuations from time to time. Many reasons have been advanced for these population changes, including forest fires, logging, hunting, predation, weather conditions, plant succession, etc. (Karns, 1987). The fact remains that the major changes in moose populations have been synchronous across the North American continent during the last 80 years.

Although data from the former Soviet Union about moose distribution is limited, there are indications that the population is increasing, at least in Russia and within the borders of the former distribution (Danilov, 1987). The change in moose numbers in the last three decades over large areas is primarily related to alterations in habitat by clearcutting and subsequent succession of forest vegetation.

The presence of one or several large predators, which are more common in both Siberia and North America than in Fennoscandia, not only lowers the hunting yield of moose, but also creates an unstable situation from a management point of view. Theoretically, annual recruitment of moose may vary in a number of ways, as exemplified by a number of case histories (Van Ballenberghe, 1987). Although there has been much progress in recent years in understanding the effects of predation on moose numbers, much work needs to be done. Research is needed on the additive nature of moose mortality caused by predation, hunting and other factors and clarification of changes in such additive relationships as moose density increases (Van Ballenberghe, 1987). The relationship between habitat carrying capacity and moose density is confounded by predation, and managers attempting to provide sustained yields of moose for human use will find predator management a necessity in systems containing naturally regulated predator populations (Ballard and Larsen, 1987).

After questioning a number of management agencies in North America, Karns (1987) concluded that population dynamics based on demographic data was only used occasionally. Hence, few models of population dynamics have been published (Page, 1987). Instead, seasons, quotas and harvest regulations have arrived in a subjective form and most of the estimates lack the precision and accuracy required for precise population management.

Not much is known about moose management in the former Soviet Union. It is obvious that moose have always played an important role in providing people with meat. Therefore, management has a long history in many regions and has included a number of census methods and hunting strategies (Baskin and Lebedeva, 1987). Recently, moose have become interesting targets for sport-hunting although proper management plans are needed as well as rational hunting jurisdictions (Bluzma, 1987).

10.4   FOREST DAMAGE BY MOOSE IN FENNOSCANDIA

During recent decades, dense moose populations, combined with large areas of monocultural stands of young Scots pine, have increased the risk of forest damage in Fennoscandia. Well-developed and fast-growing trees are palatable but relatively resistant to browsing (Danell, 1989). A well-developed young tree has thick annual shoots and branches and the moose cannot remove a very large portion of the tree's needle biomass. On the other hand, a slow-growing young tree can be seriously damaged by browsing because moose can, within a season, remove a large portion of the total needle or leaf biomass. Furthermore, it is easier for moose to break a weak tree, which also is of a vulnerable height during a longer period than a well-developed tree. In general, moose can cause considerable damage to commercially important trees, mostly on less fertile sites.

Damage by moose to forest trees has occurred for many years (Lavsund, 1987). Many types of damage have been documented, such as browsing of lateral twigs and leaders, stem breaking and bark stripping. The latter type of damage may also affect mature trees. Damage may result in growth loss or in technical damage, detectable in the stems of harvested older trees.

Most damage has occurred in Scots pine plantations 1–4 m high. In such plantations browsing, breaking and stripping may take place (Lavsund,

1987). It is a rule, rather than an exception, that moose browse stands that have already been browsed one or several years before. This often seems to be the case even if there are unbrowsed young forest stands of similar height nearby. This suggests that browsed pines are preferred to unbrowsed ones, as has been indicated in recent studies (Löyttyniemi and Piisilä, 1983; Vikberg and Bergström, 1992). It is possible that this system of browsing may partially explain the often geographically patchy browsing and damage pattern.

Most damage has been recorded on traditional moose winter ranges. Such ranges are most often in river valleys where pines are relatively slow-growing because of fairly dry and nutrient-poor soils. These pines are vulnerable to browsing. However, heavily browsed stands usually recover because the mortality of damaged pines is generally low (Vikberg and Bergström, 1992) and trees can recover even after relatively severe damage (Lavsund, 1987, and references therein). Despite recovery, pine plantations may experience diminished growth or technical damage, which can be of economic significance.

The increase in the Swedish moose population during the 1970s and early 1980s is reflected in the amount of forest damage recorded by the Swedish National Forest Survey (Sandewall, 1988). From 1974 to 1984 the proportion of stands with more than 5% of main stems damaged by moose increased approximately eight-fold. Concurrent with the decrease in moose numbers in Sweden in the late 1980s, the proportion of damaged pine plantations decreased (Sandewall, 1988). According to foresters, for economic reasons no more than 5% of the main stems in young forests (1–5 m high) should be allowed to be moderately or severely damaged by moose in Sweden.

10.5 CONCLUSIONS

The moose population in Fennoscandia has been highly productive during the last decades, primarily as a result of abundant food generated by the modern forestry practices and mild winters; also, selective harvest promoted a high average age among adult females, and hence a high proportion of highly fecund females. The potential rate of increase is kept high because of little or no predation in conjunction with high hunting pressure. The risk of a bull of being shot before 5 years of age can be greater than 90% (Cederlund and Sand, 1991), so few moose are likely to die of old age. One can speculate that long-term selective hunting in combination with extremely high hunting pressure (annual turnover rate of approximately 50%) may change one or several qualities which influence, directly or indirectly, the rate of increase, such as onset of fertility, average longevity and rate of increase of body mass. It is obvious that proper management of the large Fennoscandian moose population requires far more detailed knowledge of factors influencing the population dynamics of the moose than is available today. At the population level, we anticipate a management system that will focus on sex and age structure, with bull/cow ratios that represent a compromise between maximum rate of increase with few bulls and maintaining higher genetic variation (Nygrén and Pesonen, 1990).

Clearly, moose and forestry are intimately linked, primarily through forage production and forest damage. Moreover, dense moose populations affect not only Scots pine but also, directly or indirectly, other tree and animal species, which makes the moose a key species in the Swedish boreal forest (Angelstam and Andrén, 1990). The upper limit for moose densities is set by socioeconomic interests and not by biological factors (Haagenrud *et al.*, 1987).

It is difficult to predict future densities in moose populations. Over a period of a few decades many changes may take place in land use. In Sweden, forestry is now aiming at a more diverse and small-scale forest management, probably resulting in more diverse ecosystems. In addition, many farmland areas are being converted to forested land or to land for other uses, such as energy crop production. Some of these changes will benefit moose and some will be negative. On the whole we think that forestry will continue to create habitats with sufficient forage for a large moose population, at least during the next few decades. The annual harvest in Sweden will probably stabilize at a level of approximately 100 000 moose. The future management system will continue to set upper limits for moose densities which may vary from less than 0.1 moose/km^2 in northernmost Fennoscandia to more than 1.2 moose/km^2 in more southerly areas. These goals will be formulated through knowledge of moose population demography, harvest data, surveys of damage on forests and arable crops, and by evaluating browsing pressure.

10.6 SUMMARY

Intensively managed forests have been one of the ultimate factors in development of the moose (*Alces alces*) populations in Fennoscandia during the twentieth century, primarily by providing large amounts of forage. High moose densities have generated high browsing pressure, which in several ways has influenced the forest ecosystem and economically important trees. During periods of high moose densities, intensive browsing has occurred on several tree species. This impact has locally resulted in changes of forest structure. In many areas, especially in moose wintering areas, browsing has led to considerable damage on coniferous species, such as Scots pine (*Pinus sylvestris*).

The Fennoscandian moose population increased dramatically during the early 1970s, after a period of more gradual but steady recovery from a state near eradication early in this century. Although accurate density estimates were lacking, annual harvests indicated a 4–5-fold increase during the 1970s. During the peak years in the early 1980s the total annual harvest was approximately 230 000 moose in Fennoscandia. Since then the population has stabilized or, as in Sweden, been considerably reduced. Here consistent overharvest, perhaps in combination with changes in fecundity, has reduced the estimated winter population size from about 300 000 moose in the peak years to about 200 000 moose in recent years. The potential rate of increase is generally high, between 30 and 60% of the winter population, which is considered by managers to equal the harvest rate. This rate of increase is primarily due to a combination of low mortality, almost no large predators, abundant forage and a regulated hunting system (e.g. protecting highly fecund females and harvesting a large proportion (30–50%) of calves.

10.7 ACKNOWLEDGEMENTS

We appreciate valuable comments on the manuscript from our colleagues Dr Gunnar Markgren and Dr Scott Brainerd, and to Lars Gustafsson for compiling the data.

10.8 REFERENCES

Andrén, H. and Angelstam, P. (1993) Moose browsing on Scots pine in relation to stand size and distance to forest edge. *Journal of Applied Ecology* 30:133–142.

Angelstam, P. and Andrén, H. (1990) The moose as a key species – community effects of high population densities in Swedish taiga forest. *Third Int. Moose Symp., Syktyvkar, 1990.* (In press.)

Ballard, W.B. and Larsen, D.G. (1987) Implications of predator–prey relationships to moose management. *Swedish Wildlife Research*, Supplement 2:581–602.

Baskin, L.M. and Lebedeva, N.L. (1987) Moose management in USSR. *Swedish Wildlife Research*, Supplement 2:619–634.

Bergström, R. and Danell, K. (1987) Moose winter feeding in relation to morphology and chemistry of six tree species. *Alces* 22:91–112.

Bergström, R. and Hjeljord, O. (1987) Moose and vegetation interactions in north-western Europe and Poland. *Swedish Wildlife Research*, Supplement 1:213–228.

Bisset, A.R. (1987) The economic importance of moose (*Alces alces*) in North America. *Swedish Wildlife Research*, Supplement 2:677–698.

Bluzma, P.P. (1987) Socio-economic significance of moose in the USSR. *Swedish Wildlife Research*, Supplement 2:705–724.

Caughley, G. (1977) *Analyses of Vertebrate Populations.* John Wiley and Sons, New York. 243 pp.

Cederlund, G. and Markgren, G. (1987) The development of the Swedish moose population 1970–1983. *Swedish Wildlife Research*, Supplement 1:55–62.

Cederlund, G. and Okarma, H. (1988) Home range and habitat use of adult female moose. *Journal of Wildlife Management* 52:336–343.

Cederlund, G. and Sand, H. (1991) Population dynamics and yield of a moose population without predators. *Alces* 27:31–40.

Cederlund, G., Ljungqvist, H., Markgren, G. and Stålfelt, F. (1980) Foods of moose and roe deer at Grimsö in Central Sweden – results from rumen analyses. *Viltrevy* 11:169–247.

Danell, K. (1989) Vilka tallar väljer älgen att beta? *Sveriges Skogsvårdsförbunds Tidskrift* 2:9–15.

Danell, K., Huss-Danell, K. and Bergstrom, R. (1985) Interactions between browsing moose and two species of birch in Sweden. *Ecology* 66:1867–1878.

Danilov, P.I. (1987) Population dynamics of moose in USSR (literature survey, 1970–1983). *Swedish Wildlife Research,* Supplement 2:503–523.

Eastman, D.S. and Ritcey, R. (1987) Moose habitat relationships and management in British Columbia. *Swedish Wildlife Research,* Supplement 1:101–117.

Geist, V. (1987) On the evolution and adaptations of Alces. *Swedish Wildlife Research,* Supplement 1:11–23.

Haagenrud, H., Morow, K., Nygrén, K. and Stålfelt, F. (1987) Management of moose in Nordic countries. *Swedish Wildlife Research,* Supplement 1:635–642.

Hagen, Y. (1983) *Elgens vinterbeiting i Norge.* Direktoratet for vilt og ferskvannsfisk, Trondheim. Rapport 26, 111 pp.

Johanson, K.J., Bergström, R., Eriksson, O. and Erixon, A. (1994) Activity concentration of ^{137}Cs in moose and their forage plants in Mid-Sweden. *Journal of Environmental Radioactivity* 22:251–267.

Karlsson, F., Thelander, B. and Geibrink, O. (1988) Älgobs har kommit för att stanna. *Svensk Jakt* 126:656–660.

Karns, P.D. (1987) Moose population dynamics in North America. *Swedish Wildlife Research,* Supplement 2:423–430.

Keith, L. (1983) Population dynamics in wolves, in *Wolves: their biology and management in Canada and Alaska,* (ed. L. Carbyn), pp. 66–77.

Kellsall, J.P. (1987) The distribution and status of moose (*Alces alces*) in North America. *Swedish Wildlife Research,* Supplement 1:1–10.

Lavsund, S. (1987) Moose relationships to forestry in Finland, Norway, and Sweden. *Swedish Wildlife Research,* Supplement 1:229–246.

Lavsund, S. and Sandegren, F. (1991) Moose–vehicle relations in Sweden: A review. *Alces* 27:118–126.

Löyttyniemi, K. and Piisilä, N. (1983) Moose (*Alces alces*) damage in young pine plantations in Forestry Board District Uusimaa-Hääme. *Folio Forestalia* 553, 23 pp.

Markgren, G. (1969) Reproduction of moose in Sweden. *Viltrevy* 6:127–299.

Nygrén, T. (1983) The relationship between reproduction rate and age structure, sex ratio and density in the Finnish moose population. *Proc. XVI Congr. Int. Union Game Biol. Vysoke Tatry, Strbske Pleso, CSSR,* 30–42.

Nygrén, T. (1987) The history of moose in Finland. *Swedish Wildlife Research,* Supplement 1:49–54.

Nygrén, T. and Pesonen, M. (1990) The Finnish moose population and its management since 1975. *Third Int. Moose Symp., Syktyvkar, 1990.* (In press.)

Page, R. (1987) Integration of population dynamics of moose management – a review and synthesis of modelling approaches in North America. *Swedish Wildlife Research,* Supplement 2:491–501.

Pershagen, H. (1969) *Snow-cover in Sweden 1931–1960.* Sveriges Meteorologiska och Hydrologiska Institut, Meddelanden. Serie A. No. 5, Stockholm. (In Swedish, with English summary.)

Pulliainen, E. (1987) The moose in the early history of man in Finland – a review. *Swedish Wildlife Research*, Supplement 1:43–48.

Rülcker, J. and Stålfelt, F. (1986) *Das Elchwild*. Verlag Paul Parey, Hamburg. 285 pp.

Saether, B.-E. (1987) Patterns and processes in the population of the Scandinavian moose (*Alces alces*): some suggestions. *Swedish Wildlife Research*, Supplement 2:525–537.

Saether, B.-E. and Haagenrud, H. (1983a) Life history of the moose (*Alces alces*): fecundity rates in relation to age and carcass weight. *J. Mammal.* 64:226–232.

Saether, B.-E. and Haagenrund, H. (1983b) Life history of moose (*Alces alces*): the relationship between growth and reproduction. *Holarctic Ecology* 8:100–106.

Sand, H. (in preparation) Life history patterns in female moose: the relationships between age, body size, fecundity and environmental conditions.

Sand, H. and Cederlund, G. (in press) Individual and geographical variation in age at maturity in female moose (*Alces alces*). *Can. J. Zool.*

Sand, H., Cederlund, G. and Bergström, R. (in preparation) Density-dependent reproductive and body mass variation in female moose.

Sandewall, M. (1988) Osäkert om älg och skador. Sker en ökning? *Skogen* 11:22–23.

Senft, R.L., Coughenour, D.W., Bailey, L.R. *et al.* (1987) Large herbivore foraging and ecological hierarchies. *BioScience* 37:789–799.

Skogsstyrelsen (1992) *Skogsstatistisk Årsbok 1992*. Jönköping, 1992.

Stålfelt, F. (1974) Älgpopulationerna i län med samordnad älgjakt. *Statens naturvårdsverk PM* 485:5–23.

Stålfelt, F. (1977) Älgavskjutningen kan ökas på många håll i landet. *Svensk Jakt* 113:46–47.

Telfer, E.S. (1984) Circumpolar distribution and habitat requirements of moose (*Alces alces*), in *Northern Ecology and Resource Management*, (ed. R. Olson), pp. 145–181. University of Alberta Press, Alberta.

Van Ballenberghe, V. (1983) Rate of increase in moose populations. *Alces* 19:98–117.

Van Ballenberghe, V. (1987) Effects of predation on moose numbers: a review of recent North American studies. *Swedish Wildlife Research*, Supplement 2:431–460.

Van Wagner, C.E. (1978) Age-class distribution and the forest fire cycle. *Canadian Journal of Forest Research* 8:220–227.

Van Wagner, C.E. (1983) Fire behaviour in northern conifer forests and shrublands, in *The Role of Fire in Northern Circumpolar Ecosystems*, (eds R.W. Wein and D.A. MacLean), pp. 65–80.

Vikberg, M. and Bergström, R. (1992) *Skogsskador i Sunnäshägnet. En rapport från älg-vegetationsprojektet*. Swedish Hunters' Association. Mimeo. 37 pp.

Vivås, H.J. and Saether, B.-E. (1987) Interactions between a generalist herbivore, the moose *Alces alces*, and its food resources: an experimental study of winter foraging behaviour in relation to browse availability. *Journal of Animal Ecology* 56:509–520.

Zachrisson, O. (1977) Influence of forest fires on the North Swedish boreal forest. *Oikos* 29:22–32.

Part Three
Effective Conservation Tools and Strategies

This final section presents a number of forest monitoring and management strategies that are used today. These are strategies that are needed both to integrate and to implement forest conservation. Also presented are the most modern technological tools that can be used to implement these strategies.

The initial chapter in this section considers the properties and structures of northern European forested landscapes and habitat types most important to wildlife populations. The forest types and natural disturbance regimes in this area include the boreal forest dominated by fire, temperate deciduous forests dominated by gap-phase dynamics and grazing, and riparian forests dominated by flooding. The forest types, scales of resolution and the studies most germane to the restoration of functioning wildlife communities are discussed. An important conclusion of this chapter is the need for active restoration measures on all geographical scales.

Chapter 12 depicts Australian wildlife management practices in timber production areas. Further modifications of forestry practices and management strategies that will enhance the prospects of long-term conservation goals are recommended.

The effects of forest loss on birds that migrate from breeding grounds in temperate regions to wintering grounds in the tropics are examined in Chapter 13. The populations of a number of migrant bird species in North America that utilize forest resources appear to be declining. It is proposed that alterations to forest habitats play a role in these declines and that this is brought about by removal, degradation and fragmentation of breeding, wintering and/or stopover habitat. Attention is focused upon the identification and protection of these habitats, especially for threatened species.

Chapter 14 comprises a thorough review of a wide variety of wildlife habitat evaluation techniques that are based upon different approaches to modeling these environments. The consideration of these modeling approaches often involves maps, images, photographs and drawings and new technologies for data collection and data management. The use of predictive models in the decision-making process is also considered.

This review of techniques provides a comprehensive spatio-temporal context for addressing issues such as the management of natural resources and for wildlife habitat evaluation. It is proposed that new technologies reduce the level of uncertainty in decision-making by providing a capability for consistent application of rules in the development of decisions. Guidelines are also proposed for a systematic approach to the application of habitat evaluation models; sources upon which such guidelines should be based are cited.

Chapter 15 introduces a spatial strategy for land planning derived from the precepts of landscape ecology. Many varieties of change presented throughout this book are considered in this approach to land planning. The strategy recognizes that many land-use and conservation efforts are focused on too fine a scale. It is proposed that landscape ecology can address key environmental and land-use issues in large land areas. This conservation strategy, based upon principles in landscape ecology, focuses upon the protection of spatial patterns. A classification of spatial patterns is presented which can be used in the consideration of wildlife movement patterns in relation to spatial habitat configurations. Lastly, Chapter 15 presents a range of approaches that can be used for the conservation of vegetated landscapes.

A practical and affordable approach for the identification and assessment of wildlife habitats is still being pursued in the United States. 'Ecosearch' is a methodology for the assessment of wildlife habitats introduced in Chapter 16. This methodology uses an IGIS approach (Davis *et al.*, 1991) for making species/habitat predictions. This methodology predicts the occurrence of suitable habitats derived from aerial photography; it incorporates a habitat template for wildlife species that is not derivable from satellite imagery and represents structural attributes of habitats. This approach is assessed in a forested area of Massachusetts where it is used to associate wildlife species with wildlife habitats using aerial photography.

Chapter 16 highlights perhaps the two primary issues regarding the use of techniques for producing useful maps of forested landscapes and wildlife in today's world. One issue concerns the predictability offered by a map. For example, does the map accurately portray the distribution of a particular forest habitat? The accurate prediction of the distribution of a wildlife species dependent upon this forest habitat is often contingent upon its mapped forest habitat distribution. The methodology presented in Chapter 16 uses aerial photography to produce detailed and accurate representations of forest habitat which permit accurate predictions of the distribution of wildlife species. The second issue addresses the affordability of forest maps produced for a region by a specific project. Maps produced from aerial photographs which cover a large area are currently expensive to produce. Chapter 17 compares the relative costs of some of the principal mapping technologies available today.

The final chapter reviews the current status of methodologies that involve the collection and mapping of more comprehensive information about the forest and wildlife above the local scale. Some promising new approaches and spatial modeling techniques are reviewed that may provide useful tools for more accurately portraying the forest landscape. The issues presented encompass aspects of species, habitats and ecosystems and also applications that can be employed to execute conservation/development policy-making and planning.

Issues are covered that relate to mapping, classification and the delineation of patch boundaries on the forested landscape. The process for classifying forest types and for assessing the accuracy of mapped boundaries of forest patches are considered. Also, some important issues in regard to the resolution of forest data and the scale of forest maps are discussed.

This chapter examines some issues that concern the use of global forest information derived from small-scale maps. Some ideas regarding the functionality of this information are discussed. In regard to small-scale maps, some current map coverages on national, continental and global scales are described. In conclusion, the most promising and practical modern techniques for enhancing regional maps of forests and their wildlife are considered.

Reference

Davis, F.W., Quattrochi, D.A., Ridd, M.K. *et al.* (1991) Environmental analysis using integrated GIS and remotely sensed data: Some research needs and priorities. *Photogrammetric Engineering and Remote Sensing* 57:689–697.

−11

The ghost of forest past – natural disturbance regimes as a basis for reconstruction of biologically diverse forests in Europe

Per Angelstam

11.1 INTRODUCTION

Europe stretches from the Atlantic Ocean to the Ural mountains, and the forests of Europe range from the Mediterranean evergreen forests to the temperate deciduous forests in central Europe, and the boreal coniferous forest in the north (Mayer, 1984). The greatest single factor in the evolution of European civilization has been the removal and modification of the forests that once covered most of the European continent (Brouwer *et al.*, 1991; Patterson and Backman, 1988; Thirgood, 1989). However, the uses of the forests of Europe have a very complex history, and the duration of impacts varies greatly among different regions. Alteration, fragmentation and loss of natural forest habitats and landscapes started in the Middle East (Hobbs and Hopkins, 1990). In the central parts of Europe, forests have been thoroughly altered for more than 5000 years (Amman, 1988; Anderberg, 1991; Rackham, 1988; Berglund, 1991; Jahn, 1991) and most of the present agricultural areas were colonized not later than during the medieval period (Birks *et al.*, 1988; Dodgshon, 1988; Jahn, 1991). At the other extreme, the clearing of original forest habitats both for agricultural purposes and for timber is only now being carried out in the boreal forests of northern Russia.

As a consequence of past changes, many large forest mammals and birds (aurochs, *Bos primigenius*; moose, *Alces alces*, European bison, *Bison bonasus*, capercaillie, *Tetrao urogallus*), top predators (brown bear, *Ursus*

Conservation of Faunal Diversity in Forested Landscapes. Edited by
R.M. DeGraaf and R.I. Miller. Published in 1996 by Chapman & Hall. ISBN 0 412 61890 7.

arctos, wolf, *Canis lupus*; raptors), as well as specialized species such as several species of woodpecker, *Picidae*, and passerines requiring old forests, deciduous trees and dead wood, disappeared from large parts of Europe or declined regionally (Ahlén, 1975a; Aulén, 1988; Angelstam and Mikusinski, 1994; Avery and Leslie, 1990; Boström, 1988; Ekman, 1910, 1922; Haila *et al.*, 1980; Helle and Järvinen, 1986; Heptner *et al.*, 1989; Järvinen *et al.*, 1977; Järvinen and Väisänen, 1977; Ritchie, 1920; Rülcker and Stålfeldt, 1986; Stubbe, 1989; Virkkala *et al.*, 1994). Similarly, many invertebrates, lichens and fungi have declined as a consequence of forest management (Esseen *et al.*, 1992; Berg *et al.*, 1994). The flora has changed both in managed forests and in agricultural landscapes (Hawkesworth, 1974; Malmgren, 1982; Peterken and Game, 1984; Berg *et al.*, 1994).

With a broad perspective, fragmentation is a unifying theme of the history of European forests as well as an explanation for these local and regional extinctions of forest species. Fragmentation occurs when a continuous habitat is transformed into a number of smaller patches of decreasing area, isolated from each other by a matrix of habitats unlike the original. Both these components may cause extinctions; reduction in total habitat area affects population size and thus extinction rates; redistribution of the remaining habitat into more or less isolated patches affects dispersal and thus immigration rates (Schultze and Mooney, 1993).

In general, the fragmentation of forest cover in Europe culminated a century or more ago. In contrast, present dominant patterns are increases in the forest cover and the total timber volume in Europe (Nilsson *et al.*, 1992a,b). From a regional or a local perspective, however, various regions are located at different points along a continuum from areas where the forest fragmentation process has not yet started, to areas where the level of forest fragmentation is being reversed by forest plantations. The increased forest cover in Europe may, if the data on forest cover is interpreted only superficially, be interpreted as an ongoing mitigation of the problems related to wildlife conservation. However, most of the increase in forest cover has resulted from afforestation of former farmland and meadows with conifer plantations (Peterken, 1981; Mayer, 1984; Avery and Leslie, 1990). These forests are often not very well designed for the purpose of hosting species assemblages that are typical for the former natural forest habitats (Hunter, 1990; Avery and Leslie, 1990; Angelstam, 1992; Esseen *et al.*, 1992).

From the viewpoint of wildlife conservation, forest fragmentation is a very crude concept unless the quality of the forest is taken into account. Many species demand specific habitat types produced by various natural disturbance regimes, or certain structures (dead wood, hollow trees, old trees, moist forests, etc.) which are not compatible with intensive timber production. Thus, the transformation of vegetation type and structure that occurs within the forest is very important. In other words the changes take place at several scales of resolution (Figure 11.1):

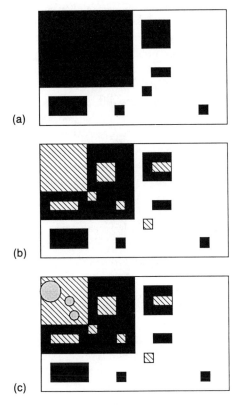

(a)

(b)

(c)

Figure 11.1 Forests are impacted at several scales of resolution. (a) Forest fragmentation in a general sense involves loss of forest habitats altogether. (b) Within the forest there may be several distinct habitat types, each of which may become fragmented from the point of view of a specialist of that habitat type. (c) Finally, within habitat types, certain structures may be reduced beyond some critical level.

- the fragmentation of forest in a broad sense that takes place on the landscape scale;
- the fragmentation of different habitats (successional stages, different specific site types, etc.) that may take place within the forest;
- alteration of habitat quality within stands.

It is likely that forest cover, especially in central parts of Europe, will continue to increase (Nilsson *et al.,* 1992a,b) and that the simplistic structures will persist. New policies that lead to increased proportions of agricultural land being taken out of production will help to speed this process. There is thus a need to obtain and implement knowledge about how to build new near-natural, man-made forests.

Such active management, aimed at safeguarding viable populations of wildlife species, requires studies of the habitat requirements of the species. It is also necessary to find out how much habitat is enough for long-term survival. Knowledge of present habitat use and habitat distribution in the landscape may not be sufficient. Given that landscape changes have already gone very far regarding the loss of quality and quantity of many habitats, as is the case in most of Europe, it may be difficult to ascertain what habitats a given species is adapted to live in. Studies on habitat requirements may not be relevant if the range of possible habitats is restricted due to losses of certain aspects in the managed landscape. In other words, the range of habitat characteristics we can sample in a managed landscape may be too narrow to permit sound conclusions about the habitat requirements of endangered species. It is therefore necessary to gain some understanding about how species were distributed in naturally dynamic landscapes, and what properties and structures are required in different landscape and habitat types.

This chapter first describes the natural disturbance regimes of the three main European forest types: the fire-dominated boreal forest, the temperate deciduous forests where mainly gap-phase dynamics and grazing shaped the dynamics, and finally riparian forests where flooding is the primary disturbance. Secondly, for each type, the chapter reviews how important structural properties affecting wildlife populations have changed over time. Finally, the types of studies and scales of resolution that should be considered in order to restore functioning wildlife communities are discussed.

11.2 NATURAL DISTURBANCE REGIMES IN EUROPEAN FORESTS

The diversity of biotopes in any natural landscape is determined by differences in soil types, topography, climate and access to nutrients and water on the one hand, and by the disturbance regimes on the other. Disturbance regimes vary along continua from the large-scale (fire, hurricanes, insect outbreaks) (Bergeron *et al.*, 1993; Hunter, 1990; Johnson, 1992; Shugart *et al.*, 1992) to the small-scale or localized (gap regeneration, flooding) (Bergeron *et al.*, 1993; Falinski, 1986, 1988; Frelich and Lorimer, 1991; Röhrig, 1991; Brinson, 1990). They also vary along a spectrum of occurrence from frequent to infrequent, as well as from low to high intensity. Based on the pristine tree species composition, the forests of Europe may be grouped into six broad types (Ahti *et al.*, 1968; Jahn, 1991; Mayer, 1984; Larsen, 1980; Polunin and Walters, 1985; Shugart *et al.*, 1992). From north to south these include:

1. boreal coniferous forest;
2. temperate deciduous forest;
3. submediterranean forest;
4. mediterranean forest.

In addition:

5. riverine forests are present in most forest types;
6. mountain forests form a separate type even if they are complex mixtures of the other types.

This chapter is restricted to the forest types found north of the Pyrenees, Alps and the Carpathians since it in these regions that one finds most of the forests used for commercial forestry (Nilsson *et al.*, 1992a,b) (Figure 11.2). These different forest types have different combinations of natural disturbance regimes (Table 11.1).

The first type to be considered (section 11.3) is the circumpolar boreal forest which is the most extensive forest type in the world (Bonan and Shugart, 1989; Kuusela, 1990, Shugart *et al.*, 1992). In these forests, frost may occur in any month of the year and conifers are the main group of tree species in the late successional stages, with spruce dominating in forests with gap-phase dynamics. Scots pine (*Pinus sylvestris*) and Norway spruce (*Picea abies*) form 98–100% of the conifers in the eastern and western part, respectively, of the boreal forest in Europe. East of the Fennoscandian shield, Siberian elements (*Larix sibirica, Pinus sibirica, Abies sibirica*) occur but constitute only a small percentage of the timber volume (Kuusela, 1990; Shugart, 1992). Birches (*Betula pubescens* and *B. pendula*) and aspen (*Populus tremula*) dominate in early and mid successional stages, which are common features of naturally dynamic boreal forests due to large-scale disturbances. This means that in the boreal zone the forest landscape is not always covered by 'forest' in the sense of there always being a closed canopy of large trees.

This chapter also includes the hemiboreal zone immediately south of the boreal zone in order to embrace the distribution of Norway spruce outside the mountainous regions.

Table 11.1 Relative importance of major natural disturbance types in the three main forest types in central and northern Europe

	Boreal	Temperate	Riverine
Fire	X	x	–
Flooding	–	–	X
Wind	x	x	–
Snow	x	–	–
Gap phase	x	X	x
Browsing	x	X	x

X and x indicate, respectively, greater or lesser importance of each type of disturbance in each forest type; – indicates negligible importance.

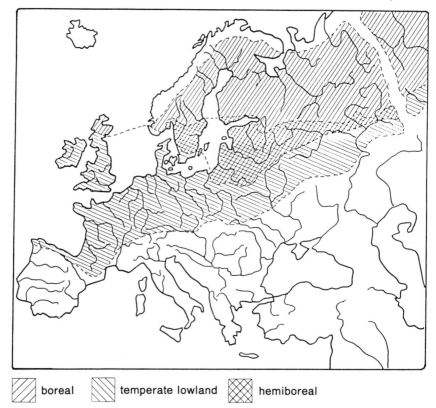

boreal	temperate lowland	hemiboreal

Figure 11.2 Distribution of boreal forest in northern Europe and temperate lowland forest in central Europe. The hemiboreal forest is a transition zone between the two but is included with the boreal forest in this chapter (see also text). Even if only the larger rivers are shown, it is evident that riverine forests are a major component.

The lowland temperate forest zone (section 11.4) is characterized by beech (*Fagus sylvatica*) and oaks (*Quercus petraea* and *Q. robur*). This topographically uniform lowland region extends from the Atlantic Ocean to the Ural mountains. The east–west mountains to the south of the region (Pyrenees, Alps and Carpathians) obstruct the influence of the climate as well as the floral elements of the Mediterranean. The climate is longitudinally uniform in comparison with North America and is usually subdivided into three parts. First there is an Atlantic part in the west which includes the British Isles, France, the Netherlands, Luxemburg and Germany west of the river Elbe. Its eastern border coincides approximately with the January isotherm of 0°C and the eastern distribution border of *Ilex aquifolium* (von Denffer *et al.*, 1976). The strong domination by beech in the west is also typical. Further to the east, in central Europe, winters become increasingly continental and the second part of the region ends at the eastern border of hornbeam (*Carpinus betulus*), beech and *Quercus petraea*. Finally, in the east up to the Black Sea and the River Volga, where the temperate region grades into the steppe region, the Sarmantic province with *Quercus robur* begins, and this extends as far as the Ural mountains. In this subregion Atlantic and subatlantic species are absent. Throughout the three different parts of the temperate forest zone, elm (*Ulmus* spp.), ash (*Fraxinus excelsior*), linden (*Tilia cordata*) and maple (*Acer* spp.) occur in mixed stands, except on dry sites where Scots pine occurs.

Third, the riverine forests (section 11.5) form a forest type that does not constitute a large portion of the area today, but nevertheless is very important for wildlife if present. Here water is the decisive factor; therefore, this forest type does not differ geographically among areas as much as the communities that are climatically determined (Glavac, 1972; Wiegers, 1990). The tree species composition is usually very diverse and structural diversity is high. In dry climates, as in southeastern Europe, riverine forests often constitute continuous corridors of very productive vegetation in a matrix that is without trees. Also, in forested landscapes riverine forests usually form species-rich habitats in a matrix with fewer species.

11.3 BOREAL FOREST

11.3.1 Natural disturbance regimes

The dominating, frequent and large-scale disturbance factor of the natural boreal forest landscape was fire (Kohh, 1975; Zackrisson, 1977a; Engelmark, 1984, 1987). On average, in a large region, 0.5–2% of the landscape was affected by forest fires annually (Bonan and Shugart, 1989). The disturbance intervals in North America are 20 to 200 years, with extremes up to 500 years in humid parts of eastern Canada. In large-scale studies along river valleys in the Swedish taiga, the corresponding values were 40–160 years; depending on vegetation type and topography (Kohh, 1975; Zachrisson, 1977a,b). The variation around these means is, however, very large and the important fire periods creating the major landscape patterns may be several decades apart (Romme and Despain, 1989). Also, few and large burns affect the landscape the most (Hunter, 1990); 97% of all burns larger than 200 ha in Canada from 1980 to 1989 covered only 3% of the forest area (Johnson, 1992). In spite of this, individual areas of 500–2000 km^2 may burn in the main fire years in continental Siberia and Canada (Furyaev and Kireev, 1979; Bonan and Shugart, 1989). In Sweden (Zackrisson and Östlund, 1991) and in eastern Canada (Foster, 1983; Payette *et al.*, 1989), the burned areas were much smaller due to a more complex topography and to more rivers and lakes acting as fire breaks. However, even within large forest fires, many patches with surviving trees remain in large burned areas (Johnson, 1992). Fire refugia also occur in a partly predictable way according to moisture and topography (Zackrisson, 1977b; Engelmark, 1987). Occasionally damage from storms is superimposed on fire patterns (Virkkala *et al.*, 1991; Syrjänen *et al.*, 1994).

It is possible to classify the natural habitats in taiga forests based on the main disturbance regimes in different sites across the landscape (Angelstam, in press; Rülcker *et al.*, 1994). Fire frequencies on firm ground in a naturally dynamic landscape depend on three major factors.

First, the intimate relationship between soil types, soil moisture and the consequent flammability of vegetation affects the probability of fire (Zackrisson, 1977a,b; Engelmark, 1984, 1987; Schimmel, 1993). Wet ground burns very seldom and such sites may be considered as no-fire refugia, unless the climatic fluctuations over centuries are taken into account (Bradshaw and Hannon, 1992; Kuuluvainen, 1994). The fire frequency increases along the gradient moist–mesic–dry, while the intensity declines. Similarly there are parallel gradients from fine-grained soils to coarse soils and from herb-rich vegetation to vegetation without herbs. These relationships hold on local as well as landscape scales (Zackrisson, 1977a,b; Furyaev and Kireev, 1979; Engelmark, 1984, 1987).

Second, fire frequencies vary in the landscape due to differences in topography. Convex parts burn more frequently than concave (Zackrisson, 1977b) and south-facing slopes more than north-facing slopes (Högbom, 1934). Some parts of the landscape, such as islands in lakes and wet mires, are often not reached by fire (Bradshaw and Zackrisson, 1990).

Finally, increasing air humidity reduces fire frequencies to the north (Payette *et al.*, 1989; Johnson, 1992) and with increasing altitude (Zackrisson, 1977b). Similarly, local continentality, i.e. climatic patterns of low rainfall and high summer temperatures, increases fire frequency (Granström, 1993).

Four major types of fire frequencies and associated site types can be recognized (Angelstam, in press) (Figure 11.3a, Table 11.1):

(a) dry sites;
(b) mesic sites;
(c) moist sites;
(d) wet sites.

(a) Dry sites
Fire frequencies on dry sites are high (30–50 years) (Kohh, 1975; Zackrisson, 1977a,b; Engelmark 1984, 1987). In taiga forests west of the Ural mountains, Scots pine is the dominant tree species. Natural forests on dry sites are characterized by complex structures with dead trees, fallen logs, large pine trees that have survived several fires and young pines in gaps, and no (or a poorly developed) deciduous phase. Low-intensity ground fires dominate.

(b) Mesic sites

Fire frequencies on mesic sites are intermediate. Zackrisson (1977a,b) and Bratt *et al.* (1993) found average fire intervals of 91 years and 75 years, respectively. Fires on mesic sites are followed by a well-developed phase with birches and aspen. On some sites the deciduous phase may be kept almost permanent by intermittent fires. Later, Norway spruce dominates.

(c) Moist sites

Fire frequencies on moist sites are low – on average 160 years (Zackrisson, 1977a,b). Such forests are complexes of two disturbance regimes, namely crown fires under extreme droughts, succeeded by a deciduous succession after which follow long periods with internal dynamics. This produces forest stands that are dominated by a continuous and uneven-aged tree cover and usually also a continuous supply of dead wood in different stages of decay. In the Swedish middle and north taiga, reconstructions of the habitat distribution in unmanaged landscape type suggest that forest of this type may have occupied about 30% of the area in the natural state (Zack-risson and Östlund, 1991). Mountain forest bordering the tree-line is another example of this type of forest.

(d) Wet sites

Fire is absent on wet soils, or rather, fire is so rare that within normal eco-logical time scales it may be considered absent (Kuuluvainen, 1994). Internal dynamics prevail as in the previous type. Wet soils are usually situated near water. Such forests therefore often form an important linear landscape component along the edge between water and land. Water level fluctuations and ice produce disturbances in the vegetation (Zackrisson, 1978; Lugo *et al.*, 1990). These effects are generally small-scale and, for example, wood in different stages of decay is continuously present even within a small area. In addition, windfalls are more common on wet than on dry sites (Falinski, 1978), producing a high turnover of dead wood in shore forests. Deciduous forest is often permanent along linear elements that are wetter than average. Such forests thus have a high disturbance frequency and long continuity at the same time. This habitat probably hosts the highest number of species sensitive to traditional clearcutting forestry methods, hence many species in this forest type are now endangered (Esseen *et al.*, 1992; Nilsson and Ericson, 1992; Berg *et al.*, 1994).

11.3.2   History of northern forests from Scotland to the Ural mountains

Reviews of forest vegetation development after the latest glaciation which covered the Fennoscandian shield, but not the areas further east, show that taiga forests with today's species composition has been present only since about 5000 years BP in eastern Finland and 2000 years BP in southern

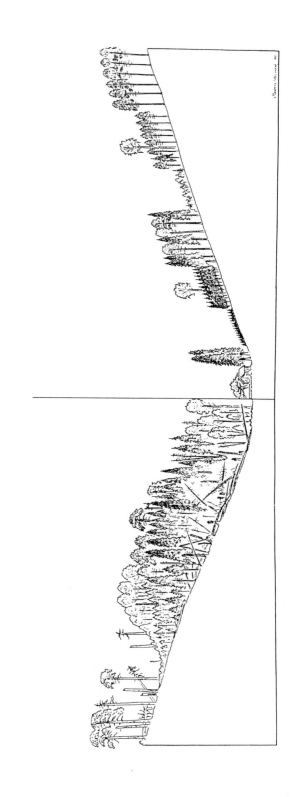

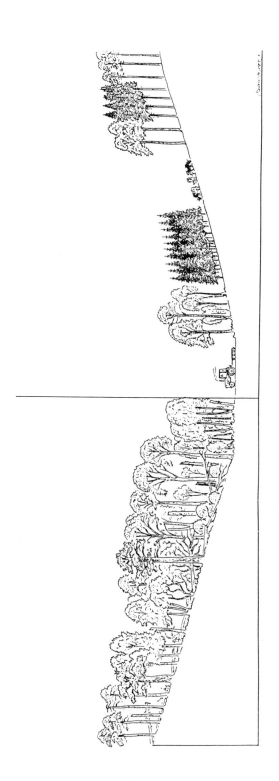

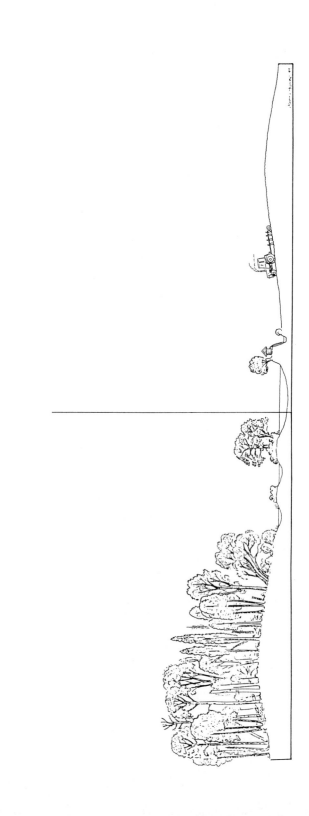

Figure 11.3 Distribution of different forest types (natural on left, managed on right) in: (a) boreal forest; (b) temperate lowland forest; (c) riverine forest. (a) Boreal forest. On dry sites (far left), disturbance by fire is frequent but not intensive, producing uneven-aged Scots pine forests. On fresh and moist sites, fires are increasingly infrequent. Disturbance is followed by a successional development starting with birches and aspen as well as some Scots pine and ending with Norway spruce, or Siberian spruce in the east. In late successional stages and on moist and wet sites there is gap formation. Finally, towards the center of the picture, flooding of the forest occurs. Here Norway spruce and deciduous trees are mixed. The managed forestland (right) was, as a rule, first permanently cleared for settlements in valley bottoms where finer sediments prevail. Other land is managed by clearcutting on all soil types, but with spruce at the moister end of the gradient and Scots pine at the drier end. (b) Temperate forest. On dry sites (far left) there is Scots pine with some oak, and fire may occur. All other forests are dominated by gap phase dynamics with beech, hornbeam, linden, elm, oak, maple and ash, in mixtures depending on the region. In the wet part of the gradient there is ash and alder marsh. The development of agriculture first affected the wetter parts of the gradient. On some intermediate and drier sites, homogeneous Norway spruce and Scots pine plantations were created as the amount of farmland was reduced. Intensive forest management has removed much of the natural structural complexity. (c) Riverine forest. On the left side of the levee there is ash and alder carr; on the right side there are different species of poplar and willow. In the river, flooding rearranges material which also allows primary succession to take place. As indicated in the righthand part of the figure, amelioration, canalization and regulation of the water level in rivers have usually altered this forest type completely.

Sweden, when Norway spruce entered on the scene (Tallis, 1991). The forest histories of Scotland, Sweden and the European part of northern Russia provide good illustrations of the subsequent anthropogenic influence.

(a)   Scotland

Scotland contains the westernmost part of the boreal forest region in the Old World. However, most of the boreal forest disappeared a long time ago. Scots pine associated with birch was the dominant woodland type in the early Caledonian forests (Birks *et al.*, 1988). Six thousand years ago these forests covered almost 15 000 km² (McVean and Ratcliffe, 1962). The occurrence of the endemic bird species (crossbill *Loxia scotica* and red grouse *Lagopus scotica*) still indicates these once extensive forests.

In the Middle Ages the forest resources were diminished and laws were already introduced that new forest should be planted. Ritchie (1920) estimated that 54% of Scotland was forest in prehistoric times compared with 5% at the beginning of the twentieth century. The present remaining native pinewoods have been reduced to less than 1% and lie in remote areas where access and transport remained a problem (Steven and Carlisle, 1959; Bunce and Jeffers, 1977; Rackham, 1990).

Ritchie (1920) reviews the effects of the dramatic reduction in the amount of forest on animal life. Originally the Scottish pinewoods hosted a boreal forest fauna that was gradually reduced. The moose probably disappeared before the Middle Ages. The red squirrel (*Sciurus vulgaris*) first disappeared in the lowlands and later also in the highlands; it was described as 'formerly abundant but now extinct' in 1842 even if it is possible that some individuals survived in remote remnant forests. It was gradually reintroduced in the Scottish lowlands with individuals from England and Scandinavia, and spread rapidly. The capercaillie became extinct in 1770 but was reintroduced in 1837 and 1838. The greater spotted woodpecker (*Dendrocopos major*) disappeared between 1830 and 1840. Ritchie (1920) argues that the reason was lack of suitable nesting trees, since cone-producing small pine trees were already established by replanting at that time. As planted forests grew older the greater spotted woodpecker reappeared voluntarily (Ritchie, 1920), probably after one of the recurrent winter invasions (Salomonsen, 1972).

Sowing and planting of new forest started by 1600 and occurred on a large scale from the eighteenth century. The turning point for British woods, with an all-time low at less than 5% of the country, came in 1919 when the Forestry Commission was set up under the 1919 Forestry Act with a mandate of reafforestation. From 1920 to 1940 the areas of conifers planted in England and Scotland were similar, but since 1960 planting in Scotland has been more than double that in England. In the fiscal year 1985–86 new forest plantations were 6.54 km² England, 7.49 km² in Wales and 210 km² in Scotland. Forest now covers over 15 000 km² and is characterized by poor soils and largely non-native tree species (Anonymous, 1984). The area of planted Sitka spruce (*Picea sitchensis*) has doubled every decade from 1919 to 1980. The forests are also young. In 1980, 77% of the conifers were under 30 years old and only 3% were older than 80 years (Avery and Leslie, 1990). Currently there are attempts to promote natural regeneration of pine by controlling deer and sheep browsing (Bain and Bainbridge, 1988) and to plan management in order to recreate old-growth forests (Peterken *et al.*, 1992).

(b) Sweden

Before the beginning of the industrial era, Sweden was an agrarian society that was mainly founded on a primitive economy. What was needed for household requirements was taken from the forest: firewood, building timber, poles and rods for enclosures and hay-drying racks, etc. In the beginning the ground was more useful than the trees. In the boreal and hemiboreal zone the more fertile parts were cultivated and became arable and pasture land. Even outlying forest was utilized with slash-and-burn methods for cultivation and for pasturage. In regions with vast forests the burning for pasture was often carried out with little control, while in regions that were poor in forests this practice was more limited (Selander, 1957).

Cattle were grazed for as much of the year as possible, which implied the whole year along the western and southern coasts. This, in combination with regular cutting and burning, often resulted in a relatively open forest and the development of heathland. In particular, growth of deciduous shrubs was hampered, since it was attractive to the animals, whereas juniper (*Juniperus communis*) was favoured. The spruce was actively combated. The combined impact contributed to the creation of forests that were entirely different from those of today. Thus, from the Iron Age onwards, the inhabited parts of southern and central Sweden were to a great extent pasture country, and the borderline between farmland and forest was vague. During the nineteenth century, which was a time of strong population growth and food shortage, cultivation and even slash-and-burn agriculture had their widest distribution.

The first use of trees in the forest was for early mining. For a long time (from the seventeenth to the twentieth century) this entailed extensive exploitation of the southern boreal forest (Eriksson, 1955). Firewood was needed when the rock was heated by log-fires, and charcoal was required for the blast-furnaces and the ironwork forges. Wieslander (1936) estimated that in the mining area of Bergslagen, south-central Sweden, mining activities alone consumed 65% of the estimated annual growth of the forests in the 1740s. Wood for households and for construction was estimated to have consumed more than 35% of the annual growth rate. He concludes that the wood resource was overexploited. As a consequence, the shortage of forest soon became severe around the mines in Bergslagen and at the end of the seventeenth century blast-furnaces and ironworks were relocated to regions where there were better supplies of wood. It was easier to transport the ore and pig-iron than the large amount of charcoal that was required. The new foundries and ironworks were established primarily in northern and western Värmland, in Dalecarlia, along the coast of Norrland and in Småland. But even around the new ironworks there was soon a shortage of wood. The consumption of charcoal peaked towards the end of the nineteenth century: in 1885 it was estimated that 20–25% of the cut timber volume was used in making charcoal (Arpi, 1959).

During the eighteenth and nineteenth centuries the burning of wood for potash became an important industry and an important export product for the country. Potash was made chiefly by burning birch wood and then leaching the ash and boiling the lye that was produced. The consumption of wood was very high and this eventually limited production (Borgegård, 1973; Nordstöm *et al.*, 1989). Wood tar, which was made from pine wood by stump grubbing, or by scratching the bark of living trees to increase the tar content of the wood, was one of Sweden's most important export products during the same period. The centre of tar production was moved northwards as shortages of raw materials arose. Stumps on sand and gravel grounds were particularly easy to grub.

During the eighteenth century and the first half of the nineteenth century, the population of Sweden greatly increased. New ground was cultivated and the great partition reforms were carried out. The inland areas of northern Sweden were colonized, and during the first 60 years of the nineteenth century the arable land of Sweden increased from 0.8 to 2.5 million hectares.

Because of the extensive cutting, slash-burn agriculture, hard grazing pressure and the lack of silviculture, forested ground was severely exploited and strongly in retreat in southern and central Sweden during the middle of the nineteenth century, when the timber supply was probably half as great as it is today. Regrowth was left to nature. In the middle of that century, large virgin forests still existed in the interior of Norrland (Juhlin-Dann-feldt, 1959). In many accounts from the seventeenth, eighteenth and nine-teenth centuries there are dramatic reports of the serious scarcity of forest where the population density of people was high (Nordström, 1959). This opinion was to be found early. Sprengtporten (1855) wrote: 'the flat land is short of forests, and the fact that the forest land is wasting its superfluous supplies will easily lead to the conclusion, that is now so often expressed: that the country will come to meet a general forest shortage'.

In a similar way, Larsson (1989) has shown that even in the heavily exploited Småland in south Sweden there were large intact areas and that the talk of forest shortage, in the sense of scarcity of areas covered by forest, was a great exaggeration. According to Ryberg (1982:15), the timber that was floated along the Väster and Österdalälven Rivers during the beginning of the nineteenth century was cut within some 60 km of the rivers. Thus, the perceived forest 'shortage' was largely a consequence of the primitive means and lines of transport and there were plenty of inaccessible reserves of primeval forest (also see Juhlin-Dannfeldt, 1959).

During the middle of the nineteenth century, forestry began to be carried out in a more orderly manner. Forest management plans were established following a model from Germany. Cutting by regulated areas, i.e. clearcutting, was put into practice and also cleaning and thinning in younger forest. Directions were given for the establishment of new forest by means of seed-trees, or by sowing and planting. A considerable importation of tree seeds started and gradually nurseries were founded. The development towards orderly forestry during the latter part of the nineteenth century occurred at the same time as the great sawmill eras; the pulp and paper industry came somewhat later. In some areas, primeval forests that separated settled country along the boundaries of provinces and districts still existed 100 years ago.

In the middle of the nineteenth century a fundamental transformation began in society, as well as in the forest landscape. Industrialization received an impetus and agriculture was intensified. Cultivation became more intense and occupied larger areas of arable land, while slash-burn agriculture and grazing in forests diminished markedly. Instead, the forest acquired a trade value and became a commodity to an extent never previously appreciated. Until about 1830–1840 the export was modest, but then it increased exponentially (Mattson and Stridsberg, 1981, figure 27). The liberal economy led to a great increase in world trade. Between 1850 and 1900, Sweden's export of sawn timber showed ten-fold increase and the harvests reached close to 20 million m^3, resulting in the growth of sawmills along the whole Bothnian coast and exploitation of the primeval forest in the inland areas of Norrland.

The most desirable wood was still heavy sawtimber, which implied that only trees over a certain size were cut and floated down to the coast. However, due to the development of the pulp industry at the turn of the century, smaller timber became usable and the pressure on the forest as a producer of wood increased markedly. Location and lack of people no longer hampered the utilization of the forest.

Thus, from a forest history viewpoint, the latter half of the nineteenth century was an era of heavy exploitation. A good description of forestry of that time is the expression '*occasional industry*' (Mattson and Stridsberg, 1981:144). At the beginning of the twentieth century there was a debate about the prevalent methods of harvesting; clearcutting as well as cutting by size-class was introduced (Wallmo, 1897). This also led to a predominant use of selection systems in the whole country in the 1920s and 1930s. Although clearcutting was limited during this period, it did not disappear entirely. The Swedish climate, coupled with poor selection technique, often resulted in poor regeneration after selective cuts and there was a return to clearcutting in the 1940s (Mattson and Stridsberg, 1981; Lundqvist, 1992).

Today, broadly speaking, all harvesting of old forest is carried out as clearcutting with subsequent planting (Norway spruce and Scots pine), or natural regeneration under seed-trees (Scots pine). The act of rendering transportation more effective through a growing network of forest roads (Mattson and Stridsberg, 1981, figure 72; Karström, 1992) has contributed heavily to the shrinking of the natural forest area in northern Sweden, and the areas with a continuity of unmanaged forest are declining sharply. Today only 0.4% of the productive forest is protected in national parks and nature reserves (Kardell and Ekstrand, 1990). In contrast, more than 32% of the mountain deciduous forests are protected (Nilsson and Götmark, 1992).

(c) Russia

In the European regions of central and northern Russia the human population is traditionally concentrated in the plains and river valleys (Kuusela, 1990). Exceptions exist where industries were developed along the northern rivers (northern Dvina, especially Arkhangelsk, and Pechora) due to the floating of timber, as well as to the development of mineral and gas deposits (Kola peninsula and northeastern Komi). The general trend, however, is gradual settlement and subsequent exploitation of forests at increasing distances from rivers.

The easternmost part of the European taiga is in Komi and the forest history in this area has been reviewed by Galasheva (1961). During the seventeenth century, forest was used only near large cities. Industrial clearcuts along the tributaries of the Dvina started at the beginning of the eighteenth century. From the Komi region, timber was exported to the port at Arkhangelsk, starting in the seventeenth century. Exports consisted only of high quality timber for building ships that were later exported to England and the Netherlands. These resources were quickly depleted and new, more remote areas had to be surveyed. Therefore, investigations of the forests were started very early. Already by 1703, Tsar Peter the Great had ordered that an inventory of the forest should be made for military and ship-building purposes.

During the eighteenth century, lists of forests accompanied by forest maps were produced comparatively often. In 1748 a list was compiled of all the Arkhangelsk forests, on the basis of which people received permission to cut trees. Very soon it was clear that the listed forests could not support the demand. In 1768 the government, being interested in the development and demands of ship-building, sent special expeditions to survey the forests. Interest focused on mast timbers from pine and larch in northwestern Russia all the way to the Urals, and the surveys covered an area of 980 000 km^2. In 1789 the forest department was founded to estimate the State Forest Fund.

Forest land was delineated as building-timber forest, fuel-wood forest and areas with no tree growth. In Komi this survey was made over the following 80–90 years. Timber forests were divided into those suitable for shipbuilding or for saw timber and by age (ripe and over-ripe). The shipbuilding timber was dominated by Scots pine along the Vychegda river, a tributary of the Dvina, and by larch along the Pechora river. These descriptions were very short but nevertheless important; they formed the first reliable forest inventories.

By 1830 the surveyors had reached the Ural mountains; by 1837 all forests were under the control of the Ministry of State Ownership. Before 1850, use of the forest resource was made with the goal of determining the quantity and quality of ship-forests to supply the Arkhangelsk port. Survey methods were very primitive and the areas between the rivers were never investigated (Figure 11.4).

During the second part of the nineteenth century the demand for timber began to increase. This led to an increased interest in the northern forest. Consequently, the old maps and descriptions were no longer sufficient. Scientific research was initiated to establish the importance of different forest regions, and, in addition to the investigation of forest types, studies were conducted on the growth of trees, reasons for lack of forest in the tundra, biology of timber species and the dynamics of the genesis of different forest types.

Logging was done by two methods. First, 80–90% of all the forest within a certain area was partially cut. Second, 10–20% was totally clearcut. By the October Revolution in 1917 only $15\,000$ km^2 (4%) of the $380\,000$ km^2 was organized into state forest enterprises. Altogether $148\,000$ km^2 (38%) were surveyed but not logged. The majority of the land, $223\,000$ km^2 (58%), was untouched.

In 1918 new legislation proclaimed all forests as the property of the people. In the first years after the revolution, the volume of cutting increased rapidly due to lack of coal and oil. All industrial timber logging was stopped in the northern regions so that the labor could be used for cutting fuelwood. In 1925 the first sawmill was established by the Pechora River. In Komi the classification of forests for different uses was completed in 1940.

In connection with the development of the coal and oil industry in the Pechora basin, the volume of logging increased significantly. This demanded new investigations of the forests and new transects were laid out in 1931–32. For the first time ever, not only the river margins were investigated but also the forest interior. In 1947, 37% of the forest land was divided into forest enterprises, 56% was investigated and only 7% remained undescribed. The areas under exploitation, however, were much larger than planned. By 1957, all the forests were organized into forest enterprises. The huge areas of virgin taiga were thus developed like mining.

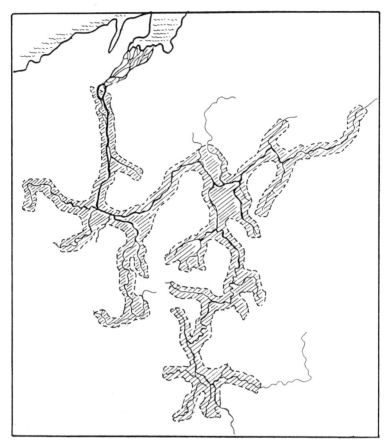

Figure 11.4 Map of northern Komi in the northeastern corner of European Russia, showing areas that were surveyed for timber resources along the Pechora River between 1907 and 1911. (After Galasheva, 1961.)

Today reafforestation is usually done without active management. The exploitative harvesting of timber assortments (in Komi, only timber larger than 15 cm in the top end of the log is taken out of the forest, and about 30% of the wood volume is left in the forest (Pautov, personal communication)) left enormous cut-over areas. Most of the cut areas regenerate naturally (Larin and Pautov, 1991). During the first half of the following succession, pioneer trees such as birches and aspen dominate. From the point of view of industrial raw materials, this is valueless brush and is classified as non-productive forest (Kuusela, 1990:65). On the other hand, from a wildlife conservation point of view, the resulting forest resembles a naturally dynamic landscape (Anufriev and Angelstam, 1992; Angelstam *et al.*, 1995).

Table 11.2 Width of protected zones of rivers, lakes and other water bodies according to the legislation in Komi, Russia (Anonymous, 1990)

Distance from source	Width
< 10 km	15 m
11–50 km	100 m
51–100 km	200 m
101–200 km	300 m
201–500 km	400 m
> 500 km	500 m

Dead trees and large deciduous trees are abundant because of the ways in which forests are managed, and riverine forests are intact because of current legislation (Table 11.2; Anonymous, 1990). Species which are declining in the boreal forests of Fennoscandia, such as several species of woodpecker (Angelstam and Mikusinski, 1994) and some grouse (Swenson and Angelstam, 1993), are doing well under Russian forest management.

To summarize, from 1696 to 1914 the forest cover in European Russia was reduced by 62 million hectares, from 53 to 35% of the total land area (Table 11.3). The major reason was the cultivation of agricultural land which increased by 127 million hectares during that period. After 1914,

Table 11.3 Changes in the forest area (% land area) in some regions of the former USSR (from Tsvetkov, 1957; Isaev, 1991)

Region	1696	1796	1888	1914	1988
Taiga					
North and northwest	70	71	66	64	52
Ural	73	67	56	59	43
Hemiboreal and nemoral					
Central	56	46	36	30	42
Ukraine	–	15	10	9	15
Belorus	–	39	34	27	34
Volga-Vyatsky	80	71	48	39	51
Steppe					
Central Chernozyomny	24	17	12	10	9
Povolzhsky	38	29	18	18	9

there was regional stability but locally there were large changes. Firstly, agricultural collectivization in the 1930s led to a strong movement of peasants from the countryside to the cities. Consequently agricultural land was usually overtaken by forest succession. In the Ukraine, Belorus and Lithuania, the increased forest area was as a result of large-scale afforestation.

Secondly, overexploitation of forest resources led to a decrease of the forest area. This was a result of 'high grading', i.e. taking out the most valuable, productive and easily exploitable stands. Regions with extreme over-harvest in the European part are Arkhangelsk, Komi and Karelia. In the total northern region (Arkhangelsk, Vologda, Murmansk, Komi and Karelia) on average 110% of the mean annual increment was cut in 1988 (Nilsson *et al.*, 1992b). According to Barr (1987), the reserves of commercial forests have a life expectancy of a couple of decades in the south taiga and about 50 years in the middle and north taiga in Arkhangelsk and Komi. Reserves of younger successional stages dominated by deciduous timber are large (Barr, 1983, 1988; Barr and Braden, 1988).

11.3.3   How has management changed forest structures?

The dynamics of the managed forest differ from those of the natural forest in several ways (Angelstam, 1992; Angelstam and Arnold, 1993; Harris, 1984; Hunter, 1990; Esseen *et al.*, 1992), both structurally within a stand and with different disturbance frequencies across the whole landscape (Van Wagner, 1978; Johnson, 1992). Even if large clearcuts sometimes superficially mimic the large-scale character of burns, it is important to note that burned areas have an enormous structural diversity compared with traditional clearcuts. For example, usually less than 10% of the wood is consumed during a fire, leaving large volumes of standing and downed wood (Foster, 1983; Payette *et al.*, 1989), while 95–98% of the wood is removed after clearcutting in Sweden (Hultkrantz and Wibe, 1989). Swedish forest history is well documented and it will be used as a detailed example of how some important qualities in a boreal forest changed with exploitation.

Fire, being a major competitor for wood fibre, has been virtually non-existent in Sweden since 1850, when fire-fighting became efficient (Zackrisson, 1977a). The National Forest Survey started to monitor the growth and growing stock of Sweden's forests in the 1920s. By 1988 the growing stock had increased from 1760 to 2747 million m^3, an increase of 56%. The development varies greatly between the south and the north. In the southernmost, mainly hemiboreal part of Sweden the growing stock was doubled in those 60 years, while in the northern boreal parts of Sweden the growing stock on forest land first dropped by 30–60% in the 1940s, and then returned to almost the initial value (SOU, 1992; Linder and Östlund, 1992).

This difference between southern and northern Sweden is due to very heavy exploitation of the forests of southern Sweden in the past, analogous to that of central Europe. As a consequence of winter grazing by domestic livestock in the south, heather (*Calluna vulgaris*) heathlands developed (Selander, 1957). By the end of the medieval period most of the natural beech forests had disappeared and the heaths reached their maximal distribution around 1850 (Selander, 1957). These and other grazed areas were later planted with spruce, which led to the large increase in timber volume in southern Sweden (Gemmel *et al.*, 1992). The extinction of the great bustard (*Otis tarda*) in Sweden coincided with the disappearance of large heathlands (SOF, 1990). For the same reason the black grouse (*Tetrao tetrix*) is endangered in most of the west European lowland (Cramp and Simmons, 1980).

The age distribution has not changed very much during the period when forests have been surveyed. Forests older than 100 years covered 35% in the 1920s, 34% around 1970 and 34% in 1985 (Mattson and Stridsberg, 1981; Kempe *et al.*, 1992). Nevertheless, rapid changes are expected to take place in the future, leading to lower mean ages (SOU, 1992). According to the official forecasts for Sweden, given the forest policy at the time, Bengtsson *et al.* (1989) estimated that the amount of forests older than 100 years will drop, and then stabilize at well below 20% from 2040 onwards. According to these data it appears that the main changes lie in the future. However, the properties of the forest have already been changed in several ways, both before and during the period when the forests have been surveyed nation-wide. Let us consider how the proportions of large trees, dead standing trees and deciduous trees have changed during the last 100 or more years.

First, selective cutting altered the forests. From 1886 to 1991, in middle taiga forests at the 62nd parallel, trees larger than 34 cm DBH declined from 0.44 to 0.07 trees per km^2 and trees larger than 42 cm DBH declined from 0.14 to 0.01 per km^2 (Linder and Östlund, 1992). These changes have occurred gradually. From 1893–95 to 1952, trees larger than 35 cm DBH declined by 63% (Kohh, 1975). Similar changes are typical for all of central and northern Sweden.

From 1887 to 1930, in central Sweden about 20% of the standing forest biomass consisted of dead trees (Nilsson, 1933). In two areas in the middle taiga, Linder and Östlund (1992) found that the amount of dead standing trees per hectare had decreased as follows: 1890s, 13 m^3/ha; around 1920, 7 m^3/ha; 1953, 0.5 m^3/ha; 1966, 0.1 m^3/ha. Kohh (1975) reports similar changes in central Sweden: 1922, 8.3 m^3/ha; 1938, 2.8 m^3/ha; 1952, 0.8 m^3/ha.

There are no official statistics for fallen dead wood. Linder (1986) found about 70 m^3/ha in mountain spruce forest dominated by internal stand dynamics. Such stands once covered fairly large proportions of the landscape. In northern Sweden, Zackrisson and Östlund (1991) reported proportions of 25–30%. In Norwegian mountain spruce forest, the amount is much higher; in locally continental areas with higher forest fire frequency, the amount is probably much lower.

Comparisons of forest structure in managed Swedish pine forest and in natural Russian pine forest (in Komi on sandy glacifluvial sediments) show that the volume of dead standing trees is 33 times higher, that of dead fallen trees 46 times higher and that of large trees 8 times higher in the latter than in the former (Majewski *et al.*, ms.). Today, the proportion of dead trees in Swedish forests is less than 2% of the stock (Kempe *et al.*, 1992). Present forestry law in Sweden prescribes that not more than 5 m^3 of dead wood per ha may be left after windfalls.

The natural deciduous component consists of three types in boreal forest: the deciduous trees that are found within conifer-dominated stands in naturally dynamic forests; the deciduous forest that forms transient successional stages occurring after a fire; and the deciduous admixture found in more or less permanent corridors along unregulated rivers.

The number of deciduous trees has remained roughly stable during the period when forests have been surveyed. The biomass of large deciduous trees has even increased. Deciduous trees increased from 4 million m^3 in 1938–52, to 8 million m^3 in 1968–72 and 19 million m^3 in 1985–89 (SOU, 1992). The increase is largest in southern Sweden, but in northern Sweden the amount of large trees has increased by 70% since the 1940s. This is due to an increase in deciduous trees on farmland and meadows that were not managed but where forest developed naturally (Ihse, 1993). In the conifer-dominated parts of the landscape there has been a reduction of the deciduous component. However, this decline occurred before the forest survey started in the 1920s. In natural boreal forest landscapes, the deciduous component was high and distributed in a different way from that in modern managed forests (Angelstam and Mikusinski, 1994).

For the second type, transient successional deciduous component, there are no official statistics. According to reconstructions of the pristine landscape based on forest history data, the proportion of transient stands with a high proportion of deciduous trees was 8% in one study in central Lapland (Zackrisson and Östlund, 1991). Today this habitat is almost absent.

Changes in the amount of permanent more or less linear stands are difficult to assess. Drainage of forest land has been intensive (SOU, 1992) and has produced major but not well documented changes.

Woodpeckers and grouse as groups provide good examples of how wildlife is affected by such changes in the forest landscape (Swenson and Angelstam, 1993; Angelstam and Mikusinski, 1994). Species which feed on common plant species and are satisfied by the structures created by managed forest do well, while species that specialize in habitats that are not compatible with intensive forest management (old forest, dead wood, deciduous trees) are declining.

11.3.4 Indirect effects of changing land use

One of the first groups of species that disappeared with human colonization in Europe was the large predators. Several studies show that as herbivore populations are released by lack of predation from large predators, the vegetation structure changes due to increased browsing pressure on pre-ferred food plants, thereby altering future tree species composition (Alverson *et al.*, 1988; Pastor *et al.*, 1988). Because large herbivorous animals are often habitat generalists, this chain of events has consequences that span the whole landscape.

The dominant browser in taiga forest is the moose. In Sweden, moose have increased dramatically during the last 50 years due to low natural mortality and increased areas of young forest providing a good food supply (Strandgaard, 1982). Moose prefer deciduous trees and shrubs to conifers as winter food. According to Ahlén (1975b), moose prefer the main tree species mentioned above in the following order: aspen > birches > pine > spruce. Intensive browsing by moose prevents their preferred deciduous tree species from growing tall and reproducing. This is a serious long-term con-servation problem for hole-nesting birds because aspen is the species in which the majority of nest holes are found in boreal forest. This indirect effect of browsing also adds to the active reduction of the previously common old deciduous trees in taiga forests by forest management. The occurrence of adult deciduous trees is now largely confined to the edge between taiga forest and abandoned farmland (Ihse *et al.*, unpublished). Assuming that today's high browsing pressure is not reduced, species requiring adult rowan and aspen may become threatened regionally.

In boreal forests, beavers (*Castor* spp.) still change the structure and dynamics of streams and thus literally alter the landscape. By cutting wood and building dams, the animals alter stream morphology and hydrology. These activities retain sediment and organic matter, create and maintain wetlands and shallow lakes, and ultimately influence the composition and diversity of plant and animal communities (Naiman *et al.*, 1986).

11.3.5 Conclusions about boreal forest

The development in Scotland, Sweden and northern Russia represent a similar order of events; only the timing and intensity of forest use are different. Since the pristine situation, all three boreal forest landscapes have changed three times, thus forming four distinct periods (e.g. Huse, 1971; Mattsson and Stridsberg, 1981; Angelstam and Arnold, 1993; Bratt *et al.*, 1993).

(a) Virgin boreal forest
Humans were present but did not exploit the forest for purposes other than for fuel and building materials. Forest regeneration was controlled by forest

fires. The area of old forest stands was considerable. Habitat quality was high for species requiring natural features. Structures and processes were intact. In Sweden, this period lasted up to the Middle Ages in the mining districts in south-central Sweden, and to the beginning of the eighteenth century in northern Sweden. In Europe, only parts of northern Russia close to the Ural mountains can be characterized as virgin forest (Taskaev and Timonin, 1993).

(b)   Local exploitation
Forests were burned to improve grazing and agriculture. Production of charcoal, tar and potash occurred and was important in certain areas. Large unlogged areas were still present in the Nordic countries in the middle of the nineteenth century.

(c)   Systematic selective cutting
During the early use of forests, certain important properties were selectively removed (usually large and high quality trees). Control of forest fire became efficient in the Nordic countries at the beginning of the twentieth century.

(d)   Modern forestry
Modern forestry with clearcutting followed by site preparation, planting of nursery-grown trees, pre-commercial thinnings and repeated commercial thinnings was started in the 1950s in the Nordic countries. To maximize conifer wood fiber production, deciduous trees are removed as well as dead and injured trees.

At present the first steps are being taken to integrate new ideas and knowledge in forestry to build more natural features into the managed forest landscapes (Rülcker *et al.*, 1994; Angelstam, in press; Angelstam and Pettersson, in press).

11.4   TEMPERATE LOWLAND FOREST

11.4.1   Natural disturbance regimes

In comparison with the boreal forests, secondary successions in European temperate deciduous forests are not usually triggered by major perturbations such as crown and surface fires. In North American temperate deciduous forests, major disturbances such as fire, hurricanes, ice damage, fungal diseases and insect pests are much more common than in Europe (review in Röhrig, 1991). Instead, the majority of European temperate forests were subjected to local and small-scale disturbances that triggered gap dynamics (Table 11.1), a term describing the processes of regeneration in older stands which are only rarely subjected to major disturbances. Due to death or windthrow, individual dominant canopy trees create gaps which provide space for young trees to grow into the openings. There is, of course, no clear distinction between succession and gap formation.

Because of the long land-use history, very few intact temperate deciduous forests remain in Europe. Not until the twentieth century were nature reserves established with this forest type. They are, however, mainly semi-natural forests (Falinski, 1986; Röhrig, 1991). The following forest types are typical for different sites (Figure 11.3b). On moist and wet ground, windfall followed by increased growth of suppressed saplings is the main successional process (Falinski, 1986; Röhrig, 1991; Szczepanski, 1990). With little disturbance, species that reproduce below a closed canopy, such as linden, ash, maple, beech and spruce, would dominate. In the absence of disturbance, ash and elm would dominate on the most fertile soils. Alder (*Alnus glutinosa*) would be dominant on the wettest soils, linden and maple on intermediate soils and beech and Scots pine on poor soils. The proper place for oak is not clear in this gradient. It seems to need some kind of disturbance that creates large openings for establishment (Jarvis, 1964) but may later persist for several hundred years.

Fire is confined to the eastern part of the region and decreases towards the Atlantic zone in the west; it is never a dominating factor but on dry soils in central and eastern Europe it still affects forests (Kairiukstis, 1968). The extensive use of fire for forest clearance and for maintenance of heathland is probably part of the reason behind the dramatic decline of beech forests in recent centuries (Malmström, 1937). When fire management is no longer practised, beech easily spreads due to dispersal by birds, especially the jay, *Garrulus glandarius* (Nilsson, 1985).

Another important altered process in temperate deciduous forest is grazing and browsing. After the extinction of the big herbivores (mammoth, wood elephant, rhinoceros, European bison, aurochs) at the end of the Weichselian (Nilsson, 1972), the role of humans in keeping the landscape open increased. This applied first to the nemoral forest zone but also to some degree in the boreo-nemoral zone. Elements of the park-like woods of the Pleistocene, such as old trees and open grazed woods, survived in the ancient agriculture where wood-pasturing was practised. In temperate forests, this important habitat is disappearing completely as pure forestry takes over and old-fashioned agricultural methods are abandonded (Andersson and Appelqvist, 1990). As a consequence, species depending on dead wood and old trees disappear. Studies that actually document these long-term changes are rare. Weitnauer and Bruderer (1987) studied the avifauna in a Swiss valley from 1935 to 1985 as several species of birds requiring nest holes and/or dead wood gradually ceased to breed, in parallel with an intensification of agriculture in the area. These included stock dove (*Columba oenas*), hoopoe (*Upupa epops*), little owl (*Athene noctua*), middle spotted woodpecker (*Dendrocopos medius*), lesser spotted woodpecker (*D. minor*) and Wryneck (*Jynx torquilla*). In Sweden there has been a similar development, and several species of lichen and beetle dependent on trees are predicted to become extinct unless active management can produce large, sun-exposed deciduous trees with hollows and dead wood (Nilsson *et al.*, 1994).

11.4.2   From forest to field to forest

(a) Clearing

Forest clearance started about 5000 BP and continued with increasing intensity into the nineteenth century (Behre, 1988; Huntley, 1988; Brouwer *et al.*, 1991). During the Paleolithic period (until approximately 10 000 years BP), European humans lived as nomads. During the Mesolithic, they began to settle but did not alter the landscape directly. Repeated burning and abandonment had already caused extensive heaths on the continental mainland by the Bronze Age. Sheep-grazing and cutting first altered the lowland mixed oak-woods.

Systematic management for coppice occurred before 2000 BC and probably grew out of primitive selective fellings, when coppice sprouted from the stumps and when trees that were too large were left to grow on (Thirgood, 1989). The times of the Romans and the Franconian conquest under the Carolingians were periods of high clearance activities. The impact of the Roman church was especially important because land reclamation became one of the main duties of monastic life (Thirgood, 1989). The eleventh century Benedictines and twelfth century Cistercians initiated a widespread medieval extension of arable and pasture land in England, France, Holland, Belgium, Saxony and Thuringia; hermit monks did the same in Russia.

Woodland regained some of the lost areas during the migration of nations, during 1350–1450 when wars and pestilence reduced the size of the human population. In France, the Hundred Years War (1337–1453) reduced the population by 30–50%. In 1330 the French population reached a peak of about 20 million people but by 1440 the population had fallen to 10 million. After this period, arable land dominated on the better soils. The distribution of field and wood has remained the same since the twelfth century and often coincides with the limit of nutrient-rich soils (Jahn, 1991). During the seventeenth century, the population of France exceeded 20 million for the first time. The Thirty Years War (1618–1648) had a similar impact in Germany. At least a third of the population died; in the state of Württemberg alone, the population fell from 450 000 in 1618 to 166 000 in 1648. As a consequence, land reverted to forest.

There were thus in principle three kinds of land: continuously ploughed or grazed land; land that changed from open to forested; and wooded land. The structure and tree species composition of the remaining woodland were radically altered by woodland pasturing. This type of land use had started to be abandoned by the nineteenth century, but is still continued in some regions (northern Spain, the Carpathians). The tree species composition was largely determined by how different species were exploited. Oak and beech were promoted due to the important production of mast for pigs. Hornbeam, ash and elm were lopped, i.e. the twigs were cut and dried for winter fodder (Rackham, 1988; Jahn, 1991). Pasturing gradually advanced deeper into the interior of the woodland as the population increased.

As salt-works, iron foundries, glass-works and lime-kilns were developed, the demand for construction and timber for fuel increased. For charcoal-burning, complete woodlands were clearcut and coppice woodlands developed extensively. It has been estimated that in 40 days one furnace could level the forest for a radius of almost 1 km (Jahn, 1991). Forests were then managed for further charcoal production by coppicing without and with standards. Coppice management implies that the forest was clearcut every 10–30 years. In Rheinland as well as in western Britain, oak coppice forests were sometimes developed for the production of tanning bark for the leather industry (Jahn, 1991). 'Coppice with standards' means that some saplings were left for timber production. This management type still persists in parts of France and England. Coppicing produced a marked alteration of the species composition by promoting species that sprout – maple, hornbeam, ash, linden, oak and hazel (*Corylus avellana*) – and hence reducing the beech, spruce and pine.

As a consequence of these complex and extensive changes, the unmanaged forest area decreased greatly. The mountain areas in Germany had the largest remaining forests. In northern Germany, large areas had been transformed to heaths. The remaining forests were in poor condition from a forestry point of view, with low timber volumes per unit area and low quality trees. Finally, the ground had been altered through soil changes, draining of wet forests and alteration in the water regime of rivers.

(b) Reafforestation

Already in the Middle Ages it was noted that removal of forest was a problem. The rate of deforestation culminated in the eighteenth century. Later, the pressure on forests decreased due to the introduction of the potato, which made panning and lopping less profitable. Meadows produced higher yields than woodland pasture, and sheep herds became unprofitable as wool could be imported from elsewhere at a lower cost. Fertilizers in agriculture, the use of hard coal and the use of steel and concrete further relieved the pressure on forests.

The beginnings of modern forestry in central Europe came in the middle of the eighteenth century and were developed during the nineteenth century. Wood became more valuable than game in the forests, and interest in developing the economy of the wood resource increased. Wood also became a strategic material – the petroleum of its age. For example, the building of a single ship required the felling of about 24 ha of oak woodland (Thirgood, 1989). This development was followed by reafforestation of large areas where forests had disappeared and restoration of existing forests. On waste-land, spruce, pine and/or larch was planted with the intention of replacing them with broadleaved trees later. Pine was planted in the plains and spruce in the mountains. In the remaining forests natural regeneration and additional planting in gaps occurred. The closing of forest thus promoted beech, which grows well in shadow, at the expense of oak, which requires more light to thrive. The dominant trend was, however, an increase in the amount of conifers. As an example, in southwest Moravia in the Czech republic, the proportion of *Picea* increased from 5–30% between 1600 and 1800 to 64–85% in the mid twentieth century (Mayer, 1984:129). Figure 11.5 describes, in a generalized form, the changes from 1300 to 1900 in Germany (Mayer, 1984:128). In the Netherlands almost all forests

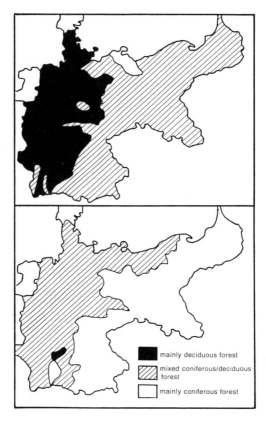

Figure 11.5 Broad changes in the tree species composition in Germany, from 1300 to 1900. (Redrawn after Mayer, 1984:128.)

originate from reafforestation and more than 80% are coniferous. The most striking example of bare land afforestation is the coastal Landes and Girond area in southwest France where almost 2000 km^2 of maritime pine (*Pinus pinaster*) has been planted since 1803, producing resin and timber on what was marsh and shifting sand.

Thus, very little of the ancient woodlands remain in central Europe. Wildwoods disappeared remarkably early and completely in England and Ireland. The last period where the land could still have been covered by forest to 50% was in the Iron Age. The *Domesday Book* of 1086 shows that forests in England then covered only 15% of the land. Much of that woodland was to disappear in the period up to the great plague of 1349. In Ireland, the disappearance of forests was even more rapid; in the 1660s only 2.6% of Ireland was woodland (Rackham, 1986). In England conversion of woodland to farmland or to plantations continued after World War II. The ancient woodland now remaining is 1.5% in England and 0.1% in Ireland (Rackham, 1988).

In England, Scotland and Wales the amount of woodland increased from a total of 11 000 km^2 in 1895 to 20 000 km^2 in 1980 and the amount of coppice woods were reduced from 2000 km^2 in 1905 to 300 km^2 by the late 1960s (Peterken, 1981; James, 1990). According to Kirby (1988), development of forests has produced four types. The dominant type is the recent plantations that cover about 13 000 km^2. Semi-natural stands cover 1000 km^2; plantations on ancient woodland sites about 3000 km^2; and about 3000 km^2 are ancient semi-natural stands. The remaining ancient wood patches which have had forest cover since at least 1600 are very small, with a mean of less than 0.1 km^2 (Kirby, 1988).

In areas which were originally dominated by deciduous forest, the present proportion of coniferous trees is 81% in Ireland, 65% in Great Britain, 77% in the Netherlands, 47% in Belgium and 35% in France (Mayer, 1984). This is in contrast with, for example, New England in the northeastern part of the United States, where naturally regenerated deciduous forests have replaced meadows and farmland by natural succession over very large areas during the last 100 years (Williams, 1989).

The long-term changes in forest cover along a north–south gradient in the European part of the former Soviet Union show a distinct pattern in the forest cover (Table 11.3). In the northern taiga region there has been a continuous decline in forest cover, mainly due to over-harvesting. In the southern region, with natural meadows and grasslands as the original vegetation type, the same pattern of continuous decline is evident due to the increased proportion of agricultural land. Finally, in the intermediate region, the decline in forest cover halted at the beginning of the twentieth century and forest cover increased thereafter. For example, in Poland the forest cover increased from 20.1% at the end of World War II to 27.9% at present (Andrzejewski and Baranowski, 1993). In Lithuania

Table 11.4 Exploitable forest area, estimated decline effect due to pollution and expected expansion of forest area on former agricultural land in Europe (data from Nilsson *et al.*, 1992a: 24, 126, 128)

Country	Exploitable forest area (ha × 10³)	Decline effect due to pollution (% of potential harvest)	Annum expansion of forest area (ha × 10³)
Finland	19 335	8	0
Norway	5 184	4	0
Sweden	23 365	8	14
Belgium	584	18	3
Denmark	434	12	6
France	13 231	6	44
FRG	7 477	24	28
Ireland	271	11	20
Italy	4 787	15	15
Luxembourg	34	26	0.5
Netherlands	221	22	1.5
UK	1 924	28	33
Austria	2 831	20	7
Switzerland	1 092	32	12
Greece	1 947	4	2
Portugal	1 475	20	25
Spain	5 605	0	60
Turkey	15 877	10	0
Yugoslavia	8 028	12	16
Bulgaria	3 196	26	2
CSFR	4 159	40	0
GDR	2 461	34	0
Hungary	1 503	35	7
Poland	7 938	36	28
Romania	6 207	9	0
Europe	139 424	19	324

the corresponding figures are 19.7 and 27.9% (Balciauskas and Angelstam, 1993).

In the future, more land will be taken out of agricultural production. The regional differences are, however, very unevenly distributed across Europe (Table 11.4). Another important future change is the effect of airborne pollution on forests (Nilsson *et al.*, 1992a; Table 11.4).

11.4.3 Conclusions about temperate lowland forest

Human colonization of new land was the main cause of the transformation of forest to field. The following pattern in two steps is characteristic at the border between good/wet and poor/dry soils (see also Emanuelsson, 1988). From ancient times to the point when the amount of farmland peaked, there was a more or less steady increase in the amount of farmland. During this time many species of non-forest vascular plants colonized the agricultural landscape, became abundant and gradually formed partly new assemblages of species (Hawkesworth, 1974; Malmgren, 1982). In Sweden this period ended in the 1880s when the rationalization of the iron industry and the intensification of agriculture started. In Russia it started the 1860s, i.e shortly after the cessation of serfdom in 1861.

Subsequently, farming intensified. As a consequence, biodiversity was reduced not only to the level prior to the start of forest fragmentation but also below that level. Later, farmland on less fertile soils started to be abandoned; forest trees first started to invade the edge between forest and farmland and were later sometimes planted. Rationalization of farming has been made by combining several units, whereby land with a lower average productivity has been taken out of production and usually afforested. This pattern can be seen on the scale of a farm, in regions and in countries.

Currently, all over Europe, the edge between arable land and forest continues to change with the systems of subsidies to farmers, and programs to repatriate and privatize previously state-owned land.

11.5 RIVERINE FOREST

11.5.1 Natural disturbance regimes

Riverine forests are those that owe their dynamics, structure and composition to processes of inundation, transport of sediments or the abrasive and erosive forces of water and ice movement (Table 11.1). The power and frequency of inundation are inversely proportional, and inundations range from those that determine geomorphic patterns persisting for centuries to those that occur annually. In this context, only with the latter will be considered. In contrast with boreal and temperate forests, where secondary successions prevail, albeit on different scales, naturally dynamic riparian forests are dominated by primary succession.

Generally, riparian communities have a vertical zonation from no vegetation or herbaceous vegetation near water to shrub communities and then to forest that is affected only at the high-water level (Mayer, 1984; Malanson, 1993). The tree species composition in different areas varies considerably. White willow (*Salix alba*) is one of the most common tree species in central European riverine forests and is found in lowlands except for northern and northwestern Europe. In Scandinavia and the Alps, grey alder (*Alnus incana*) is the most widespread species, while in southern and eastern Europe white poplar (*Populus alba*) often dominates. In the flood-plain communities, three types of forest may occur: elm–oak forest on the temporarily flooded banks of large rivers; ash–alder woodlands in the plains and depressions on boggy soils; and wet alder-woods in nutrient-rich swamps.

In the natural landscape, habitats subject to periodic flooding along the large river systems formed enormous areas of riverine woodlands (Imboden, 1987; Kuhn, 1987; Wiegers, 1990; Yon and Tendron, 1981). Plains dissected by numerous rivers and their tributaries at an altitude of less than 100 m above sea level cover about 30% of central Europe.

11.5.2   The decline of riverine forests

The process of confining rivers within artificial boundaries had already started in the Middle Ages (Wiegers, 1990). Today, the riverine woodlands are limited to a small portion of their former extent. Riparian forests were usually the first forest habitats to be influenced since humans used waterways for transport, and the soils near water are usually the most productive ones. Most of the rivers in Europe have been regulated and now the banks are usually artificial and reinforced to stop flooding. Extensive areas in the plains were drained and wet woodlands with alder and ash were converted to meadows. Wet alder-woods are, therefore, rare both in the Atlantic and the central European parts of the temperate deciduous forest. The remaining forests are strongly altered. This applies in particular to the oak–elm communities which are on sites better suited for agricultural use because they are not flooded as often as the *Salix–Populus* communities on the lower banks.

Floodplain forests represent the most productive and also one of the the most endangered forest types in central and western Europe. The Rhine is a good example. Until the beginning of the last century, the floodplain forests of the the upper Rhine valley were a complex network of tributaries and oxbow lakes (Spitznagel, 1990). Extreme floods in lowland areas changed the course of the river, creating new open areas. The water regime was characterized by spring floods after snowmelt in the Black Forest and the Vosges, and the usually more pronounced summer flood following snowmelt in the Alps. The regular flooding fertilized the alluvial soils, and disturbed sites were recolonized by vegetation. Forest use mainly consisted of coppicing and grazing. The upper Rhine was made navigable between 1825 and 1879. The great meanders were cut and dams were built. As a consequence, groundwater levels dropped, the flora and vegetation changed and forestry became feasible. First, coppicing under the canopy of the remaining timber was practised. In the early years of the twentieth century, free development of these forests resulted in a complex, three-layered structure. After World War II, large-scale clearings were replanted mainly by fast-growing and non-native tree species (*Populus* × *euroamerina, Acer pseudoplatanus, Quercus rubra, Robinia pseudoacacia, Catalpa bignonioides* and *Juglans nigra*).

In the taiga, the water level is usually lowest in late winter, rapidly increasing to its annual peak at snowmelt and then decreasing, although with much fluctuation during the rest of the year. The strong effect of flooding on species richness in natural boreal riparian habitats is evident from comparisons of the flora in regulated and natural rivers. Nilsson *et al.* (1991) showed that along the regulated Ume river local species richness was only 72% of the unregulated Vindel river. Vegetation cover in the regulated riparian zone was only 16% of that in the natural riparian zone. For many animals the riparian forests are preferred to other forests, both for foraging (Nilsson, 1992) and as travel or dispersal corridors (Harris, 1984). Riparian forest corridors are usually the most species-rich forest habitat in a landscape (Décamps *et al.*, 1987; Wesolowski, 1987; Tomialojc, 1993).

If riparian habitats are left to develop freely, they reach a natural appearance at 60–80 years, i.e. much quicker than the 200 years or more it takes for forest on drier soils to develop similar characters (Falinski, 1986; Szczepanski, 1990).

Among the major habitats in Europe, riverine forests have the greatest diversity of bird species in general and the highest density of birds, as well as rare and endangered species in particular (Reichholf, 1987; Wesolowski, 1987; Tomialojc, 1993). The reason is probably that this forest type is structurally very complex, including several vertical strata as well as a very high degree of patchiness on a horizontal plane due to a complex mosaic of habitat conditions.

Though riverine and floodplain forests were once widespread, especially in central Europe, only small patches remain. Many changes took place so long ago that no detailed data exist to illustrate this process. What happened in Poland may serve as an example. Throughout that country, most of the willow–poplar forests were destroyed. They are almost extinct today and the ash–elm and ash–alder forests are also very scarce (Wesolowski, 1987). The reason for these changes was direct destruction due to the regulation of rivers, but also indirect changes in soil moisture due to lowered watertables or separation from the river by embankments which produced a gradual transition towards oak–hornbeam associations (Tomialojc, 1993). Also, forest managers do not favour alder, willows and birches. Finally, this is the most endangered of the forest types dealt with in this chapter and it often requires active re-creation (Tomialojc, 1993).

The changes in animal life have had effects on the structure and dynamics of riverine forests. As a consequence of the alteration of the dynamics of rivers, as well as of hunting, the beaver became extinct in the Netherlands (in 1825) and also in Germany, except for a single colony (Wiegers, 1990). As beavers are now coming back after reintroduction, for example in the Baltic republics, their dynamic effects on vegetation are reappearing.

11.6 DISCUSSION

11.6.1 Roles of environmental concern and the market

Even in prehistoric times, the gradual development of a cultural landscape probably created altered conditions for the fauna and flora (Selander, 1957; Ahlén, 1966; Birks *et al.*, 1988; Mönkkönen and Welsh, 1994). It is likely that certain species expanded after accidental or deliberate introductions, while new ones immigrated from adjacent regions at the expense of others which decreased in range and number, or even disappeared. Some probably survived under entirely changed conditions. As regards the total species richness, it can be presumed that north European landscapes, thanks to the rich variation, had the greatest number of species just before the commencement of industrialization. There were primeval forests, untouched swamps, heaths, pasture woods, grazed wet meadows, oak groves, hazel brush, hay-fields and wet meadows (Selander, 1957; Bruzewitz and Emmelin, 1985; Kirby, 1988). After this point the diversity was reduced to below the natural level. In western Europe the landscape had been thoroughly altered long before.

If we want to restore important structural properties and balance important processes, it is necessary to distinguish between the pre-settlement or unmanaged forest (i.e. the wildwood), the present forest and the potential near-natural forest (Angelstam and Arnold, 1993; Ellenberg, 1988; Ellenberg Jr, 1988; Hansson and Angelstam, 1991; Jahn, 1991; Peterken, 1981). These distinctions make it easier to decide how forests may be restored on different sites, since certain forest types may have disappeared altogether (Figure 11.3). Of course attention must also be paid to alteration of the soil due to deterioration, erosion, peat-cutting, draining, deep-ploughing, fertilization and pollution. Europe has been a cultivated landscape for a very long time.

In the future, the structure and amount of forest in Europe will depend on at least three factors (e.g. Nilsson *et al.*, 1992a):

1. The level of air pollution. At present this factor reduces the potential timber harvest by up to 40% (Table 11.4), and the areas suffering from direct and indirect effects will increase (Angelstam, 1992; Ellenberg, 1988).
2. The anticipated expansion of the forest landbase. It is estimated that before 2030 about 30% of the land used presently for growing cereal will be taken out of agricultural production. The increase is estimated at 29 million m^3 per year over 100 years, assuming a constant landbase and no forest decline due to pollution. About two-thirds of this increase will take

place in the EEC-9 region. As a consequence, the Nordic region will be challenged as the leading wood supplier in Europe (Nilsson *et al.*, 1992a).

3. Changing policies for environmental concern. Recent traditional European forest management practices focus on timber production and usually involve mono-specific, even-aged stands, which are mainly coniferous. Partly due to heavy criticism (Gamlin, 1988; Laurell, 1990; Anonymous, 1993), there is presently an increasing emphasis in Europe and North America towards multiple-use with less emphasis on timber production and more emphasis on non-timber values (Swanson and Franklin, 1992; Liljelund *et al.*, 1992; Angelstam, 1993; Angelstam and Pettersson, in press; Rülcker *et al.*, 1994). Achieving this balance will be difficult but it is possible, given that appropriate knowledge exists about how to mimic natural disturbance regimes. Management and set-aside for preservation of biological diversity will naturally reduce the harvestable volume of timber but maybe not the total value.

11.6.2 Need for management at all scales

The reduction of forest cover, as well as the structural simplification of the managed forests in Europe, has reduced the distribution, ranges and abundance of many species of forest-living birds, mammals, insects, lichens and fungi. The effects on different groups of species are scale dependent; what is considered as fragmentation from the perspective of an organism with small area requirements may be a reduction of the quality in a corner of the home range of a species with large area requirements. Also, knowledge about the proportions of forest habitats in general, and/or particular habitats or structures in the forest in particular, that are necessary for survival of viable populations is important because there are thresholds in the response of populations to habitat loss. As the proportion of a given habitat declines, the chances of successful dispersal decreases stepwise. The effects of fragmentation is thus not linear (Figure 11.6). Percolation theory models (Gardner *et al.*, 1987; Gardner and O'Neil, 1991) show that continuous habitats break down to fragments when about 60% of the original habitat remains in the landscape. Franklin and Forman (1987) modeled landscape dynamics and found that the first patches became isolated when 70% of the original habitat was left. As fragmentation proceeds, more patches become isolated; when 30% of the original habitat is left, all of the original habitat is in fragments. Continued habitat loss leads to an exponential increase in the distance between patches (Gustafson and Parker, 1992; Andrén, 1994).

As shown in a review of the roles of random sampling vs. true effects of habitat isolation for species richness at a community level by Andrén (1994), the random sample hypothesis (Connor and McCoy, 1979) was a good predictor for landscapes with a high proportion (more than 40%) of

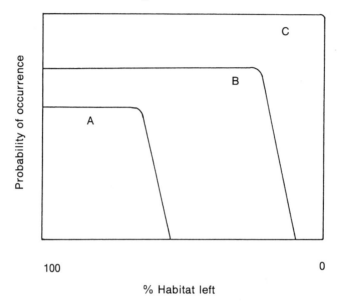

Figure 11.6   Model of how different types of species differ in their response to habitat losses. The three curves represent species with different requirements: A = species with narrow habitat and large area requirements and which cannot disperse through the matrix; B = species with moderate area requirements and a limited dispersal ability through the matrix; C = generalist species with very small area requirements and can disperse across the matrix.

suitable habitat. The negative effects of reduced patch size and isolation tended not to occur until less than 20–30% of the suitable habitat remained (see also Wilcove *et al.*, 1986). The consequences for individual species such as spotted owl (*Strix occidentalis*) (Thomas *et al.*, 1990) and capercaillie (*Tetrao urogallus*) (Rolstad and Wegge, 1987, 1989) support these models, i.e. habitat tracking is replaced by faster losses than expected by the decline in the amount of habitat when 30% or less of the original habitat is left.

When applying this landscape-scale view on population dynamics, it is important to separate the individual-scale effects of, for example, absence of sufficiently large forest patches for a pair from the population-scale of fragmentation and altered mosaic structure (Haila *et al.*, 1993). Andrén (1994) found that surprisingly many studies referred to the theory of island biogeography or metapopulation dynamics, although the effect of habitat fragmentation occurred at the individual scale. Out of 36 studies only 11% had studied patches larger than 1000 ha, an area which generally can hardly be said to host a bird or a mammal population. Moreover, migratory birds were involved in 47% of the studies. Only two of the 36 studies had studied a range of patch sizes which were relevant to the species selected.

When trying to evaluate the effects of fragmentation at different geo-graphical scales (Figure 11.1), the problem with thresholds should be eval-uated at each scale. As an illustration I propose to identify parts of Europe where the effects of forest fragmentation at the landscape scale will be espe-cially important. Figure 11.7 indicates areas where, according to the ideas about thresholds in fragmentation, forest fragmentation should not be a general problem (>40% forest), where it may be a problem (>10% to <40% forest left), and where it is likely to be a general problem (<10% forest left). Based on data presented in Mayer (1984) and ESA 1992, the conclusion is that the proportion of forest in natural taiga landscapes is above this level, most of central Europe is below 40%, while the Atlantic

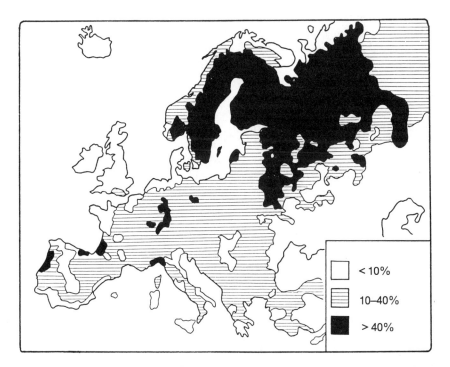

Figure 11.7 Forest cover in different parts of Europe, showing areas where forest fragmentation in the broad sense is unlikely to occur (>40% forest), where it may occur (10–40% forest) and where it is likely that it occurs (<10% forest cover). The proportion of forest in pristine Europe was 80–90%. East of the Ural moun-tains the low forest cover is due to a large proportion of mires and wetlands. Note, however, that habitat fragmentation may occur within all forest types and that dilu-tion of important properties may occur universally. (After Mayer, 1984; ESA, 1992.)

coast and UK have a very low proportion of forest. In Europe, over the whole period of forest alterations, the forest cover has decreased from 80–90% in the natural state to less than 40% over large regions (Mayer, 1984). Species that would react to such changes are forest generalists such as moose and black woodpecker (*Dryocopos martius*), both of which respond positively to increases in forest cover (Cuisin, 1985).

However, if the forest cover is above 40%, there is no guarantee that fragmentation is not a problem. Many properties within a forest (such as certain habitats and structures) may also become diluted and are subject to similar thresholds. It is therefore also important to find out under what circumstances fragmentation of distinct habitats and structures within a forest is a problem for wildlife species (Figure 11.1; Enoksson *et al.*, 1995).

11.7 CONCLUSIONS

Forest fragmentation in general is presently not the major problem for wildlife populations in Europe. Instead fragmentation of certain forest habitats and reduction of structural properties within forests (dead wood, old trees, forest structure, certain tree species) are the main problem.

To solve the consequences of past forest fragmentation as well as present habitat fragmentation and reduction of structural properties, active restoration measures at all geographical scales are necessary. This involves creating structurally diverse stands with native species, arranging the juxtaposition of stands to preserve both the individual species and species' interactions, and restoring connectivity among different habitat types and properties in forest landscapes. This requires the selection of suitable indicator species for research on critical thresholds regarding all different important habitats and structures with regard to:

- selection of the natural forests that should be protected (in most regions with a long forest history this means all natural or near-natural forests);
- conservation of the semi-natural forests;
- improvement of artificial forests;
- re-creation where biotopes and structures are lacking.

When farmland is taken out of production, there is a unique opportunity to design stand size, structure and context so as to mimic the natural landscape that was once present in the region. In large parts of central Europe it is therefore vital to take into account also land that no longer has forest.

It is important to carry out research in reference areas and work with historical ecology in order to understand how natural disturbance regimes and species of different taxa were distributed on different sites in different regions.

Research is needed on whether or not reversion of open land back to forest is possible and, if so, under what circumstances. A first step is to study the colonization rates of different taxa in open fields and meadows after abandonment. By using historical maps and documents, it is possible to date the age of today's forests as well as quantify many changes that occurred (Tack *et al.*, 1993; Östlund, 1993). With a time scale for forests of different ages we can measure the time it takes for species from different taxa to colonize and thus study how rapidly different forest communities reassemble.

These steps require international cooperation between managers and researchers both in areas where forest landscapes need to be restored and in areas which can provide knowledge and inspiration as reference areas where the fragmentation process has not advanced very far or even where it has not started (Angelstam *et al.*, 1995; Angelstam, in press).

11.8 SUMMARY

This chapter considers both the changes in forest cover and the fragmentation that characterizes European forested habitats. The most important habitat types for wildlife populations and the disturbance regimes impacting them are described and discussed. These types and their dominant natural disturbance regimes include the boreal forest dominated by fire, temperate deciduous forests dominated by gap-phase dynamics and grazing, and riparian forests dominated by flooding. Changes in important structural properties affecting wildlife populations are considered for each forest type. The forest types, scales of resolution, and the studies most germane to the restoration of functioning wildlife communities are presented. An important conclusion of this chapter is the need for active restoration measures at all geographical scales.

11.9 ACKNOWLEDGEMENTS

I thank Dick DeGraaf for stimulating me to learn more about forest dynamics and write this chapter, and Sanoat Katanova for translating and reading Russian works for me. Martin Holmer drew the vegetation profiles based on sketches developed together with Vladimir Anufriev, Ian Bainbridge, Olivier Biber, Rob Fuller, Ludwik Tomialojc and Tomasz Wesolowski during a nice stay in Bialowieza in Poland. Rob Fuller, Ludwik Tomialojc, Henrik Andrén, Kris Jasinski and two anonymous reviewers provided valuable comments on the text.

11.10 REFERENCES

Ahlén, I. (1966) Landskapets utnyttjande och faunan. *Yearbook of Sveriges Natur* 1966:73–99.
Ahlén, I. (1975a) Forestry and the bird fauna in Sweden. *Ornis Fennica* 52:39–44.

Ahlén, I. (1975b) Winter habitats of moose and deer in relation to land use in Scandinavia. *Swedish Wildlife Research* 9:45–192.

Ahti, T., Hämet-Ahti, L. and Jalas, J. (1968) Vegetation zones and their sections in northwestern Europe. *Annales Botanica Fennici* 5:169–211.

Alverson, W.S., Waller, D.M. and Solheim, S.I. (1988) Forests too deer: edge effects in Northern Wisconsin. *Conservation Biology* 2:348–358.

Ammann, B. (1988) Palynological evidence of prehistoric anthropogenic forest changes on the Swiss plateau, in *The Cultural Landscape – Past, Present and Fugure*, (eds H.H. Birks, H.J.B. Birks, P.E. Kaland and D. Moe), pp. 289–299. Cambridge University Press.

Anderberg, S. (1991) Historical land use changes: Sweden, in *Land Use Changes in Europe*, (eds I.F.M. Brouwer, A.J. Thomas and M.J. Chadwick), Kluwer, Dordrecht.

Andersson, L. and Appelqvist, T. (1990) The influence of the pleistocene megafauna on the nemoral and the boreonemoral ecosystems. *Svensk Bot. Tidskr.* 84:355–368.

Andrén, H. (1994) Effects of habitat fragmentation on birds and mammals in landscapes with different proportions of suitable habitat: a review. *Oikos.* 71:355–366.

Andrzejewski, R. and Baranowski, M. (eds) (1993) *The State of the Environment in Poland*. Panstwowa Inspeckcja Ochrony Srodowiska. Warszawa. (In Polish.)

Angelstam, P. (1992) Conservation of communities – the importance of edges, surroundings and landscape mosaic structure, in *Nature Conservation by Ecological Principles – a Boreal Perspective*, (ed. L. Hansson), pp. 9–69. Elsevier, London.

Angelstam, P. (1993) Nature conservation and forest management in boreal forest: ideas for establishing a new compromise, in *Skoglig geologi*, (ed. H. Minell), pp. 11–18. Skogsstyrelsen.

Angelstam, P. (in press) Fire-dynamics of taiga landscapes as a model for combining forestry and biodiversity, in *Fire in Ecosystems of Northern Eurasia*, (eds J. Goldammer and V. Furyaev).

Angelstam, P. and Arnold, G. (1993) Contrasting roles of remnants in old and newly cleared landscapes – lessons from Scandinavia and Australia for restoration ecologists, in *Reconstruction of Fragmented Ecosystems: global and regional perspectives*, pp. 109–125. Surrey Beatty and Sons, Chipping Norton, New South Wales, Australia.

Angelstam, P. and Mikusinski, G. (1994) Woodpecker assemblages in natural and managed boreal and hemiboreal forest – a review. *Annales Zoologici Fennici* 31:157–172.

Angelstam, P. and Pettersson, B. (in press) Ecological principles in forestry planning in Sweden – a review. *Proceedings of FAO/ECE/ILO International Forest Seminar, Prince George*.

Angelstam, P., Majewski, P. and Bondrup-Nielsen, S. (1995) West–east cooperation in Europe for sustainable boreal forests. *Water, Air and Soil Pollution* 82:3–11.

Anonymous (1984) *Census of Woodlands and Trees 1979–82*. Forestry Commission Bulletin 62.

Anonymous (1990) *Clause about zones of rivers, lakes and water bodies of Komi under protection*. Confirmed Decision of the Government of Komi ASSR 15 October 1990. No. 212. (In Russian.)

Anonymous (1993) Plünderer in Norden. *Der Spiegel* 46:244–247.

Anufriev, V. and Angelstam, P. (1992) The effects of anthropogenic transformation of middle and north taiga forests on bird communities, in *Ecological problems of the European North*, (eds V.J. Maslov, V.D. Kozlova, E.A. Kozlova *et al.*), pp. 127–129. Institute of the Ecological Problems of North, Archangelsk. 218 pp. (In Russian.)

Arpi, G. (ed.) (1959) *Sveriges skogar under 100 år*. Ivar Häggströms boktryckeri. Stockholm. 680 pp.

Aulén, G. (1988) Ecology and distribution history of the white-backed woodpecker *Dendrocopos leucotos* in Sweden. Dissertation. Department of Wildlife Ecology. Report 14. Uppsala.

Avery, M. and Leslie, R. (1990) *Birds and forestry*. T. & A.D. Poyser.

Bain, C. and Bainbridge, I. (1988) A better future for our native pinewoods. *RSPB Conservation Review* 2:50–53.

Balciauskas, L. and Angelstam, P. (1993) Ecological diversity: to manage it or restore it. *Acta Ornitologica Lithuanica* (7–8):3–15.

Barr, B.M. (1983) Regional dilemmas and international prospects in the Soviet timber industry, in *Soviet natural resources in the world economy*, (eds R.G. Jensen, T. Shabad and A.W. Wright). University of Chicago Press, Chicago.

Barr, B.M. (1987) Alternatives in Soviet timber management, in *Environmental Problems in the Soviet Union and Eastern Europe*, (ed. F. Singleton), Lynne Rienner Publishers, Boulder, USA.

Barr, B.M. (1988) Perspectives on deforestation in the USSR, in *World deforestation in the 20th century*, (eds J.F. Richards and R.P. Tucker). Duke University Press, Durham, NC.

Barr, B.M. and Braden, K.E. (1988) *The Disappearing Russian Forest: a Dilemma in Soviet Resource Management*. Rowman and Littlefield, Totowa, N.J.

Behre, K.E. (1988) The role of man in European vegetation history, in *Vegetation History*, (eds B. Huntley and T. Webb III) pp. 633–672, Kluwer Academic Publishers, 803 pp.

Bengtsson, G., Holmlund, J., Lundström, A. and Sandevall, M. (1989) *Avverkningsberäkning 1985, AVB 85*. (Long-term forecast of timber yield in Sweden, AVB 85). 329 pp. Swedish University of Agricultural Sciences, Department of Forest Survey. Report 44. (In Swedish.)

Berg, Å., Ehnström, B., Gustafsson, L. *et al.* (1994) Threatened plant, animal and fungus species in Swedish forests – distribution and habitat associations. *Conservation Biology* 8(3):718–731.

Bergeron, Y., Bradshaw, R. and Engelmark, O. (1993) *Disturbance Dynamics in Boreal Forest*. Opulus Press.

Berglund, B. (1991) The cultural landscape during 6000 years in southern Sweden – the Ystad project. *Ecological Bulletins* 41. Munksgaard International booksellers and publishers.

Birks, H.H., Birks, H.J.B., Kaland, P.E. and Moe, D. (ed) (1988) *The Cultural Landscape – Past, Present and Future*. Cambridge University Press. 521 pp.

Bonan, G.B. and Shugart, H.H. (1989) Environmental factors and ecological processes in boreal forests. *Ann. Rev. Ecol. Syst.* 20:1–28.

Borgegård, L.-E. (1973) *Tjärhanteringen i Västerbottens län under 1800-talets senare hälft*. Kungliga Skytteanska Samfundet, Umeå. 279 pp.

330   Natural disturbance regimes and reconstructed diversity

Boström, U. (1988) Fågelfaunan i olika åldersstadier av naturskog och kulturskog i norra Sverige. *Vår Fågelvärd* **47**:68–76.

Bradshaw, R. and Hannon, G. (1992) The disturbance dynamics of Swedish boreal forest, in *Responses of Forest Ecosystems to Environmental Changes*, (eds A. Teller, P. Mathy and J.N.R. Jeffers), pp. 528–535. Elsevier.

Bradshaw, R.H.W. and Zackrisson, O. (1990) A two thousand year history of a northern Swedish boreal forest stand. *Journal of Vegetation Science* **1**:519–528.

Bratt, L., Cederberg, B., Hermansson, J. *et al.* (1993) Särnaprojektet. Inventeringsrapport från en landskapsekologisk planering. (The Särna projekt. Report of inventories made for a landscape ecological forestry planning). *Dalanatur* **10**:5. 216 pp. (In Swedish).

Brinson, M.M. (1990) Riverine forests, in *Ecosystems of the World 15 – Forested Wetlands*, (eds A.E. Lugo, M. Brinson and S. Brown), pp. 87–141. Elsevier, Amsterdam. 527 pp.

Brouwer, F.M., Thomas, A.J. and Chadwick, M.J. (1991) *Land use changes in Europe*. Kluwer Academic Publishers, Dordrecht. 528 pp.

Brusewitz, G. and Emmelin, L. (1985) *Det föränderliga landskapet*. Esselte Herzogs, Uppsala.

Bunce, R.G.H. and Jeffers, J.N.R. (eds) (1977) *Native Pine Woods of Scotland*. ITE, Cambridge.

Connor, E.F. and McCoy, E.D. (1979) The statistics and biology of the species–area relationship. *American Naturalist* **113**:791–833.

Cramp, S. and Simmons, K.E.L. (1980) *The Birds of the Western Palearctic. Volume II. Hawks to Bustards*. Oxford University Press. Oxford. 695 pp.

Cuisin, M. (1985) Range expansion of black woodpecker *Dryocopus martius* in Western Europe. *British Birds* **78**(4):184–187.

Décamps, H., Joachim, J. and Lauga, J. (1987) The importance for birds of the riparian woodlands within the alluvial corridor of the river Garonne, SW France. *Regulated rivers: research and management* **1**:301–316.

von Denffer, D., Schumacher, W., Mägdefrau, K. and Ehrendorfer, F. (1976) *Strasburger's Textbook of Botany*. Gustav Fischer Verlag, Stuttgart.

Dodgshon, R.A. (1988) The ecological basis of highland peasant farming, 1500–1800 AD, in *The Cultural Landscape – Past, Present and Future*, (eds H.H. Birks, H.J.B. Birks, P.E. Kaland and D. Moe), pp. 139–151. Cambridge University Press.

Ekman, S. (1910) *Norrlands jakt och fiske*. Norrländskt handbibliotek 4, Uppsala.

Ellenberg, H., Jr (1988) Floristic changes due to nitrogen deposition in central Europe, in *Critical loads for sulphur and nitrogen*, (eds J. Nilsson and P. Grennfeldt), pp. 375–383, Miljörapport 1988:15. Nordic Council of Ministers.

Ellenberg, H. (1988) *Vegetation Ecology of Central Europe*. Cambridge University Press. 741 pp.

Emanuelsson, U. (1988) A model for describing the development of the cultural landscape, in *The Cultural Landscape – Past, Present and Future*, (eds H.H. Birks, H.J.B. Birks, P.E. Kaland and D. Moe), pp. 111–121. Cambridge University Press.

Engelmark, O. (1984) Forest fires in the Muddus national park (northern Sweden) during the past 600 years. *Canadian Journal of Botany* **62**:893–898.

Engelmark, O. (1987) Fire history correlations to forest type and topography in northern Sweden. *Annales Botanici Fennici* **24**:317–324.

Enoksson, B., Angelstam, P. and Larsson, K. (1995) Deciduous trees and resident birds – the problem of fragmentation within a coniferous forest landscape. *Landscape Ecology* 10(5):267–275.

Eriksson, G.A. (1955) The decline of small blast-furnaces and forges in Bergslagen after 1850 with special reference to enterprises in the valley of Kolbäck river. *Jernkontorets Berghistoriska Skriftserie* 15. (In Swedish, with English summary.)

ESA (European Space Agency) (1992) *Remote Sensing Forest Map of Europe.* ESA/ ESTEC, Nordwijk. 18pp. and map.

Esseen, P.-A., Ehnström, B., Ericsson, L. and Sjöberg, K. (1992) Boreal forest – the focal habitats of Fennoscandia, in *Ecological Principles of Nature Conservation*, (ed. L. Hansson), pp. 252–325. Elsevier, London.

Falinski, J.B. (1978) Uprooted trees, their distribution and influence in the primeval forest biotope. *Vegetatio* 38:175–183.

Falinski, J.B. (1986) *Vegetation dynamics in temperate lowland primeval forests.* Dr. W. Junk publishers, Dordrecht. 537 pp.

Falinski, J.B. (1988) Succession, regeneration and fluctuation in the Bialowieza forest (NE Poland). *Vegetatio* 77:115–128.

Foster, D.R. (1983) The history and pattern of fire in the boreal forest of south-eastern Labrador. *Can. J. Bot.* 61:2459–2471.

Franklin, J. and Forman, R.T.T. (1987) Creating landscape patterns by forest cutting: ecological consequences and principles. *Landscape Ecology* 1:1–18.

Frelich, L.E. and Lorimer, C.G. (1991) A simulation of landscape-level stand dynamics in the northern hardwood region. *J. Ecol.* 79:223–233.

Furyaev, V.V. and Kireev, D.M. (1979) *A landscape approach in the study of post-fire forest dynamics.* Nauka, Novosibirsk. 159 pp. (In Russian.)

Galasheva, V.A. (1961) *Forest and Forest Industry in Komi ASSR.* Akademia Nauk SSSR Komi Filial. Goslecbomizdat, Moskva. 395 pp. (In Russian.)

Gamlin, L. (1988) Sweden's factory forest. *New Scientist* 117:41–47.

Gardner, R.H. and O'Neill, R.V. (1991) Pattern, process and predictability: the use of neutral models for landscape analysis, in *Quantitative Methods in Landscape Ecology*, (eds M.G. Turner and R.H. Gardner), pp. 289–307. Springer, New York.

Gardner, R.H., Milne, B.T., Turner, M.G. and O'Neill, R.V. (1987) Neutral models for analysis of broad-scale landscape patterns. *Landscape Ecology* 1:19–28.

Gemmel, P., Peterson, G. and Remröd, J. (1992) *Framtida skogsbruk och angelägen skogsforskning i södra Sverige.* Sveriges Lantbruksuniversitet Info/Skog, Rapport 9.

Glavac, V. (1972) *Über Höhenwuchsleistung und Wachstumsoptimum der Schwarzerle auf vergleichbaren Standorten in Nord-, Mittel- und Südeuropa.* Schriftenr. Forstl. Fak. Univ. Göttingen 45.

Granström, A. (1993) Spatial and temporal variation in lightning ignitions in Sweden. *Journal of Vegetation Science* 4:737–744.

Gustafson, E.J. and Parker, G.R. (1992) Relationships between landcover proportions and indices of landscape spatial pattern. *Landscape Ecology* 7(2):101–110.

Haila, Y., Järvinen, O. and Väisämen, R.A. (1980) Effects of changing forest structure on long-term trends in bird populations in SW Finland. *Ornis Scandinavica* 11:12–22.

Haila, Y., Hanski, I.K. and Raivio, S. (1993) Turnover of breeding birds in small

332 Natural disturbance regimes and reconstructed diversity

forest fragments: the 'sampling' colonization hypothesis corroborated. *Ecology* 74:714–725.

Hansson, L. and Angelstam, P. (1991) Landscape ecology as a theoretical basis for nature conservation. *Landscape Ecology* 5(4):191–201.

Harris, L.D. (1984) *The Fragmented Forest. Island Biogeography Theory and the Preservation of Biotic Diversity*. University of Chicago Press, Chicago. 211 pp.

Hawkesworth, D.L. (ed.) (1974) *The changing Flora and Fauna of Britain*. Academic Press, London.

Helle, P. and Järvinen, O. (1986) Population trends of North Finnish land birds in relation to their habitat selection and changes in forest structure. *Oikos* 46:107–115.

Heptner, V.G., Nasimovich, A.A. and Bannikov, A.G. (1989) *Mammals of the Soviet Union*. E.J. Brill. 1147 pp.

Hobbs, R.J. and Hopkins, A.J.M. (1990) From frontier to fragments: European impact on Australia's vegetation, in *Australian Ecosystems – 200 years of utilization degradation and reconstruction*, (eds D.A. Saunders, A.J.M. Hopkins and R.A. How), Proc. Ecol. Soc. Austr. vol. 16, pp. 93–114. Surrey Beatty & Sons: Chippimng Norton, NSW.

Högbom, A.G. (1934) *Om skogseldar förr och nu: och deras roll i skogarnas utvecklingshistoria*. Almquist och Wiksells förlag. Stockholm.

Hultkrantz, L. and Wibe, S. (1989) *Skogsnäringen: Miljöfrågor, avreglering, framtidsutsikter*. Bilaga 8 till Långtidsutredningen 1990. SOU. Finansdepartementet, Stockholm.

Hunter, M.L. (1990) *Wildlife, Forests and Forestry. Principles of managing forests for biological diversity*. Prentice-Hall, New Jersey. 370 pp.

Huntley, B. (1988) Europe, in *Vegetation History*, (eds B. Huntley and T. Webb, III), pp. 341–383. Kluwer Academic Publishers. 803 pp.

Huse, S. (1971) *Forstlig historiogram for Norge*. Skogsbruk, jakt og fiske 9–16. Norsk skogbruksmuseum. Elverum. (In Norwegian.)

Ihse, M. (1993) Naturvårdsanpassad planering i skogsbruket med hjälp av flygbilder i IR-film. (Planning for nature conservation in forestry using infra red aerial photography). *Skogsstyrelsen Rapport* 93(1):19–32. (In Swedish.)

Imboden, E. (ed.) (1987) *Riverine Forests in Europe – Status and Conservation*. International Council of Bird Preservation, Cambridge, 64 pp.

Isaev, A.S. (ed.) (1991) Forestry on the border of the 21st century. *Ecologiya* I, II. Moscow. (In Russian.)

Jahn, G. (1991) Temperate deciduous forests of Europe, in *Temperate Deciduous Forests*, (eds E. Röhrig and B. Ulrich), pp. 377–402. Elsevier.

James, N.D.G. (1990) *A History of English Forestry*. Basil Blackwell.

Järvinen, O. and Väisänen, R. (1977) Long-term changes of the North European land bird fauna. *Oikos* 29:225–228.

Järvinen, O., Kuusela, K. and Väisänen, R. (1977) Effects of modern forestry on the number of breeding birds in Finland 1945–1975. *Silva Fennica* 11:284–294.

Jarvis, P.G. (1964) The adaptability of *Quercus petraea* (Matt.) Liebl. *J. Ecol.* 52:545–571.

Johnson, E.A. (1992) *Fire and Vegetation Dynamics. Studies from the North American Boreal Forest*. Cambridge University Press, Cambridge. 129 pp.

Juhlin-Dannfeldt, M. (1959) Skogarna och deras vård i södra Sverige under 100 år, in *Sveriges skogar under 100 år*, (ed. G. Arpi), pp. 263–313. Ivar Häggströms boktryckeri. Stockholm. 680 pp.

Kairiukstis, L. (1968) *Lietuvos TSR misku ukis.* (Forestry in Lithuania). Leidykla "mintis", Vilnius. 217 pp. (In Lithuanian.)

Kardell, L. and Ekstrand, A. (1990) *Skyddad skog i Sverige. 1. Areal och virkesförråd inom nationalparker, naturreservat och domänreservat.* (Protected forest in Sweden. 1. Area and amount of timber in national parks, nature reserves and state forest reserves). Report 48. Department of Forest Landscape Management, Uppsala. (In Swedish.)

Karström, M. (1992) Steget före i det glömda landet. *Svensk Botanisk Tidskrift* **86**:115–146.

Kempe, G., Toet, H., Magnusson, P.-H. and Bergstedt, J. (1992) *Riksskogtaxeringen 1983–87.* Institutionen för skogstaxering, rapport 51.

Kirby, K.J. (1988) Conservation in British woodland – adapting traditional management to modern needs, in *The Cultural Landscape – Past, Present and Future*, (eds H.H. Birks, H.J.B. Birks, P.E. Kaland and D. Moe), pp. 79–89. Cambridge University Press.

Kohh, E. (1975) Studier över skogsbränder och skenhälla i älvdalsskogarna. (Studies of forest fires and hardpan in the Älvdalen forests). *Svenska skogsvårdsförbundets tidskrift* **71**(3):299–360. (In Swedish.)

Kuhn, N. (1987) Distribution, general ecology and characteristics of European riparian forests, in *Riverine Forests in Europe – Status and Conservation*, (ed. E. Imboden), pp. 7–15. International Council of Bird Preservation, Cambridge. 64 pp.

Kuuluvainen, T. (1994) Gap disturbance, ground microtopography, and the regeneration dynamics of boreal coniferous forests in Finland: a review. *Annales Zoological Fennici* **31**:35–51.

Kuusela, K. (1990) *The Dynamics of Boreal Coniferous Forest.* Finnish National Fund for Research and Development. Helsinki. 172 pp.

Larin, V.B. and Pautov, Yo.A. (1991) *Formation of Young Coniferous Forest on Clearcuts.* Nauka, Leningrad. 144 pp.

Larsen, J.A. (1980) *The Boreal Ecosystem.* Academic Press. 500 pp.

Laurell, M. (1990) *Framtidsskogen.* Naturskyddsföreningen. Katarinatryck AB.

Liljelund, L.-E., Pettersson, B. and Zackrisson, O. (1992) Skogsbruk och biologisk mångfald. *Svensk botanisk tidskrift* **86**(3):227–232.

Linder, P. (1986) *Kirjesålandet – en skogsbiologisk inventering av ett fjällnära urskogsområde i Västerbottens län.* Swedish University of Agricultural Sciences, Umeå. 108 pp.

Linder, P. and Östlund, H. (1992) Förändringar i norra Sveriges skogar 1897–1991. *Svensk botanisk tidskrift* **86**(3):199–217.

Lugo, A.E., Brinson, M. and Brown, S. (1990) *Forested Wetlands.* Elsevier, Amsterdam. 527 pp.

Lundqvist, L. (1991) Some notes on the regeneration of Norway spruce on six permanent plots managed with single-tree selection. *Forest Ecology and Management* **46**:49–57.

Malanson, G. (1993) *Riparian Landscapes.* Cambridge University Press, Cambridge. 296 pp.

Malmgren, U. (1982) *Västmanlands Flora.* Borgströms, Motala. 669 pp.

Malmström, C. (1937) Tönnersjönatens försökspark i Halland. Ett birag till känne-

334   *Natural disturbance regimes and reconstructed diversity*

dom om sydvästra Sveriges skogar, ljunghedar och torvmarker. *Meddelanden från Statens skogsförsöksanstalt* **30**(3):323–528.

Mattson, L. and Stridsberg, E. (1981) *Skogens roll i svensk markanvändning.* Swedish University of Agricultural Sciences, Department of Forest Economics. Report 32abc. Umeå.

Mayer, H. (1984) Wälder Europas. Gustav Fischer Verlag. 691 pp.

McVean, D.N. and Ratcliffe, D.A. (1962) *Plant Communities of the Scottish Highlands.* HMSO (Nature Conservation Monograph 1). London.

Mönkkönen, M. and Welsh, D.A. (1994) A biogeographical hypothesis on the effects of human-caused landscape changes on the forest bird communities of Europe and North America. *Annales Zoologici Fennici* **31**:61–70.

Naiman, R.J., Melillo, J.M. and Hobbie, J.E. (1986) Ecosystem alteration of boreal forest streams by beaver (*Castor canadensis*). *Ecology* **67**:1254–69.

Nilsson, C. (1992) Conservation management of riparian communities, in *Nature Conservation by Ecological Principles – a Boreal Perspective*, (ed. L. Hansson), pp. 352–372. Elsevier, London.

Nilsson, C. and Götmark, F. (1992) Protected areas in Sweden: is natural variety adequately represented? *Conservation Biology* **6**(2):232–242.

Nilsson, C., Ekblad, A., Gardfjell, M. and Carlberg, B. (1991) Long-term effects of river regulations on river margin vegetation. *Journal of Applied Ecology* **28**:963–987.

Nilsson, J.E. (1933) Kort historik över Hamra kronopark för åren 1884–1930. (A short history of the Hamra state forest 1884–1930). *Norrlands Skogsvårdsförbunds Tidskrift* **1933**:321–348. (In Swedish.)

Nilsson, S., Sallnäs, O. and Duinker, P. (1992a) *Future Forest Resources of Western and Eastern Europe.* Parthenon Publishing Group, Carnforth, UK. 496 pp.

Nilsson, S., Sallnäs, O., Hugosson, M. and Shvidenko, A. (1992b) *The Forest Resources of the Former European USSR.* Parthenon Publishing Group, Carnforth, UK. 407 pp.

Nilsson, S.G. (1985) Ecological and evolutionary interactions between reproduction of the beech *Fagus silvatica* and seed-eating animals. *Oikos* **44**:157–164.

Nilsson, S.G. and Ericsson, L. (1992) Conservation of plant and animal populations in theory and practice, in *Ecological Principles of Nature Conservation*, (ed. L. Hansson), pp. 71–112. Elsevier, London.

Nilsson, S.G., Arup, U., Baranowski, R. and Ekman, S. (1994) Tree-dependent lichens and beetles in old-fashioned agricultural landscapes. *Svensk Botanisk Tidskrift* **88**:1–12. (In Swedish, with English summary.)

Nilsson, T. (1972) *Pleistocen.* Lund. 508 pp. (In Swedish.)

Nordström, L. (1959) Skogsskötselteorier och skogslagstiftning, in *Sveriges skogar under 100 år*, (ed. G. Arpi), pp. 241–260. Ivar Häggströms boktryckeri, Stockholm. 680 pp.

Nordström, O., Larsson, L., Käll, J. and Larsson, L.-O. (1989) *Skogen och smålänningen.* Smålands museum, Växjö.

Östlund, L. (1993) *Exploitation and structural changes in the north Swedish boreal forest 1800–1992.* Dissertations in Forest Vegetation Ecology 4. Swedish University of Agricultural Sciences, Umeå.

Pastor, J., Naiman, R.J., Dewey, B. and McInnes, P. (1988) Moose, microbes, and the boreal forest. *BioScience* **38**:770–777.

Pattersson III, W.A. and Backman, A.E. (1988) Fire and disease history of forests, in *Vegetation History*, (eds B. Huntley and T. Webb III), pp. 603–632. Kluwer Academic Publishers. 803 pp.

Payette, S., Morneau, C., Sirois, L. and Desponts, M. (1989) Recent fire history of the northern Quebec biomes. *Ecology* 70:656–673.

Peterken, G.F. (1981) *Woodland Conservation and Management*. Chapman & Hall, London. 328 pp.

Peterken, G.F. and Game, M. (1984) Historic factors affecting the number and distribution of vascular plants in the woodlands of central Lincolnshire. *Journal of Ecology* 72:155–182.

Peterken, G.F., Ausherman, D., Buchenau, M. and Forman, R.T.T. (1992) Old-growth conservation within British upland conifer plantations. *Forestry* 65:127–144.

Polunin, O. and Walters, M. (1985) *A Guide to the Vegetation of Britain and Europe*. Oxford University Press.

Rackham, O. (1988) Trees and woodland in a crowded landscape – the cultural landscape of the British Isles, in *The Cultural Landscape – Past, Present and Future*, (eds H.H. Birks, H.J.B. Birks, P.E. Kaland and D. Moe), pp. 53–77. Cambridge University Press.

Rackham, O. (1990) *Trees and Woodland in the British Landscape*. J.M. Dent, London. 234 pp.

Reichholf, J.H. (1987) Composition of bird fauna in riverine forests, in *Riverine Forests in Europe – Status and Conservation*, (ed. E. Imboden), pp. 16–21. International Council of Bird Preservation, Cambridge. 64 pp.

Ritchie, J. (1920) *The Influence of Man on Animal Life in Scotland*. The University Press, Cambridge. 550 pp.

Röhrig, E. (1991) Vegetation structure and forest succession, in *Temperate Deciduous Forests*, (eds E. Röhrig and B. Ulrich), pp. 35–49. Elsevier.

Rolstad, J. and Wegge, P. (1987) Distribution and size of capercaillie leks in relation to old forest fragmentation. *Oecologia (Berl.)* 72:389–94.

Rolstad, J. and Wegge, P. (1989) Capercaillie *Tetrao urogallus* populations and modern forestry – a case for landscape ecological studies. *Finnish Game Research* 46:43–52.

Romme, W. and Despain, D. (1989) Historical perspective on the Yellowstone fires of 1988. *BioScience* 39(10):695–699.

Rülcker, C., Angelstam, P. and Rosenberg, P. (1994) *Ecological forestry planning: a proposed planning model based on the natural landscape*. Forestry Research Institute of Sweden. Report 8. 47 pp. (In Swedish, with English summary.)

Rülcker, J. and Stålfeldt, F. (1986) *Das Elchwild*. Verlag Paul Parey, Hamburg. 285 pp.

Ryberg, S. (1982) *Stora Kopparbergs skogar genom tiderna*. Stora Kopparbergs Bergslags AB, Falun. 141 pp.

Salomonsen, F. (1972) *Fugletraecket og dets gåder*. Munksgaard, København.

Schimmel, J. (1993) *Fire behavior, fuel succession and vegetation response to fire in Swedish boreal forest*. Dissertations in forest vegetation ecology, 5. Swedish University of Agricultural Sciences, Umeå.

Schultze, E.D. and Mooney, H.A. (1993) *Biodiversity and Ecosystem Function*. Ecological studies 99, Springer-Verlag. 525 pp.

Selander, S. (1957) *Det levande landskapet i Sverige.* (The living landscape in Sweden). Bokskogen, Göteborg. 492 pp. (In Swedish.)

336 Natural disturbance regimes and reconstructed diversity

Shugart, H.H., Leemans, R. and Bonan, G.B. (1992) *A Systems Analysis of the Global Boreal Forest.* Cambridge University Press, Cambridge. 565 pp.

SOF (1990) *Sveriges fåglar,* 2nd edn. Stockholm. 295 pp.

SOU (1992) Skogspolitiken inför 2000-talet. *Statens offentliga utredningar* 1992:76.

Spitznagel, A. (1990) The influence of forest management on woodpecker density and habitat use in floodplain forests of the upper Rhine valley, in *Conservation and Management of Woodpecker Populations,* (eds. A. Carlson and G. Aulén), pp. 117–145. Swedish University of Agricultural Sciences, Department of Wildlife Ecology. Report 17.

Sprengtporten, J.W. (1855) *Om skogsvård och skogslagstiftning.* Stockholm.

Steven, H.M. and Carlisle, A. (1959) *The Native Pinewoods of Scotland.* Oliver and Boyd, Edinburgh.

Strandgaard, S. (1982) *Factors affecting the moose population in Sweden during the 20th century with special attention to silviculture.* Swedish University of Agricultural Sciences, Department of Wildlife Ecology, Report 8.

Stubbe, H. (1989) *Buch der Hege. Band 1 Haarwild.* VEB Deutscher Lantwirtshaftsverlag, Berlin. 706 pp.

Swanson, F.J. and Franklin, J.F. (1992) New forestry principles from ecosystem analysis of pacific Northwest forests. *Ecological Applications* 2(3):262–274.

Swenson, J.E. and Angelstam, P. (1993) Habitat separation by sympatric forest grouse in Fennoscandia in relation to forest succession. *Can. J. Zool.* 71:1303–1310.

Syrjänen, K., Kalliola, R., Poulasmaa, A. and Mattson, J. (1994) Landscape structure and forest dynamics in continental Russian European taiga. *Annales Zoologici Fennici* 31:19–34.

Szczepanski, A.J. (1990) Forested wetlands of Poland, in *Ecosystems of the World 15 – Forested Wetlands,* (eds A.E. Lugo, M. Brinson and S. Brown), pp. 437–446. Elsevier, Amsterdam. 527 pp.

Tack, G., van den Bremt, P. and Hermy, M. (1993) *Bossen van Vlaanderen – Een historische ecologie.* Davidsfonds/Leuven. 320 pp. (In Flemish.)

Tallis, J.H. (1991) *Plant Community History.* Chapman & Hall. 398 pp.

Taskaev, A.I. and Timonin, N.I. (1993) *List of protected nature in the Komi republic.* Russian Academy of Sciences, Syktyvkar. 190 pp. (In Russian.)

Thirgood, J.V. (1989) Man's impact on the forests of Europe. *Journal of World Forest Resource Management* 4:127–167.

Thomas, J.W., Forsman, E.D., Lint, J.B. *et al.* (1990) *A Conservation Strategy for the Northern Spotted Owl.* USDA Forest Service. 427 pp.

Tomialojc, L. (1993) *Nature and Environment Conservation in the Lowland River Valleys of Poland.* Inst. Ochr. Przyrody PAN, Krakow. 233 pp. (In Polish.)

Tsvetkov, M.A. (1957) *Changes in forest cover in European Russia from the end of the XVIIth century to 1914.* Nauka, Moscow. 312 pp. (In Russian.)

Van Wagner, C.E. (1978) Age-class distribution and the forest fire cycle. *Canadian Journal of Forest Research* 8:220–227.

Virkkala, R., Heinonen, M. and Routasuo, P. (1991) The response of northern taiga birds to storm disturbance in the Koilliskaira National Park, Finnish Lapland. *Ornis Fennica* 68:123–126.

Virkkala, R., Rajasärkkä, A., Väisänen, R.A. *et al.* (1994) Conservation value of nature reserves: do hole-nesting birds prefer protected forests in southern Finland? *Ann. Zool. Fennici* 31:173–186.

Wallmo, U. (1897) *Rationell skogsafverkning*. Örebro. 288 pp. (In Swedish.)

Weitnauer, E. and Bruderer, B. (1987) Veränderungen der Brutvogel-Fauna der Gemeinde Oltingen in den Jahren 1935–85. *Der Ornitologische Beobachter* 84(1):1–9.

Wesolowski, T. (1987) Riverine forests in Poland and the German Democratic Republic – their status and avifauna, in *Riverine Forests in Europe – Status and Conservation*, (ed E. Imboden), pp. 48–54. International Council of Bird Preservation, Cambridge. 64 pp.

Wiegers, J. (1990) Forested wetlands in western Europe, in *Ecosystems of the World 15 – Forested Wetlands*, (eds A.E. Lugo, M. Brinson and S. Brown), pp. 407–436. Elsevier, Amsterdam. 527 pp.

Wieslander, G. (1936) The shortage of forest in Sweden during the 17th and 18th centuries. *Sveriges Skogsvårdsförbunds Tidskrift* 34:593–633. (In Swedish, with English summary.)

Wilcove, D.S., McLellan, C.H. and Dobson, A.P. (1986) Habitat fragmentation in the temperate zone, in *Conservation Biology: The Science of Scarcity and Diversity*, (ed. M.E. Soulé), pp. 237–256. Sinauer Associates, Sunderland, Massachusetts. 584 pp.

Williams, M. (1989) *Americans and their Forests – a Historical Geography*. Cambridge University Press, Cambridge.

Yon, D. and Tendron, G. (1981) *Alluvial Forests in Europe*. Council of Europe, Nature and Environment series No. 22, Strasbourg.

Zackrisson, O. (1977a) Influence of forest fires on the north Swedish boreal forest. *Oikos* 29:22–32.

Zackrisson, O. (1977b) *Forest fire and vegetation pattern in the Vindelälven valley, N. Sweden during the past 600 years. The river valley as a focus of interdisciplinary research*. Proceedings of an international conference to commemorate Maupertuis' expedition to the river Tornio, Northern Finland, 1736–37.

Zackrisson, O. (1978) Vegetational successions on a shore at the lake Storvindeln, N. Sweden, during the past 200 years. *Svensk Bot. Tidskr.* 72:205–226. (In Swedish, with English summary.)

Zackrisson, O. and Östlund, L. (1991) Branden formade skogslandskapets mosaik. (Fire shaped the mosaic structure of the boreal forest landscape). *Skog och Forskning* 1991(4):13–21. (In Swedish.)

–12

Conservation and management of eucalypt forest vertebrates

Harry F. Recher

12.1 INTRODUCTION

Less than 5% (42 million ha) of Australia is forested and the nation relies heavily on imported timber and wood products (Resource Assessment Commission (RAC), 1992). Regardless, the domestic timber industry is a significant component of the national economy with an estimated annual value of A$8.5 billion (1% of GDP) and employing more than 8000 people (Australian Bureau of Statistics (ABS), 1990; Dargavel *et al.*, 1985; Forest Use Working Group (FWG), 1991; RAC, 1992). Despite the industry's economic and social importance, there is significant pressure from environmental groups to end logging in native forests (e.g. Kirkpatrick *et al.*, 1990; Sutton, 1991). Already more than 70% of the commercially important tropical, subtropical and warm temperate rainforests in New South Wales and Queensland have been reserved from logging for nature conservation (Lunney, 1991a; RAC, 1992; Shields, 1992). Of the much more extensive eucalypt forests 19% is in conservation reserves and additional large areas are proposed for wilderness and national park dedication (Lunney, 1991a; RAC, 1992). Of the total forest estate, 33.5 million ha (~75%) are considered capable of timber production (RAC, 1992). Much of this is unavailable for logging and the RAC estimated a 'net production forest' area of only 13.4 million ha. Thus, there is little scope to accommodate national demands for forest products, meet the resource requirements of the timber industry and satisfy the demands of environmentalists for forest conservation.

There are many reasons why Australian environmentalists would like to end logging in native forests. The preservation of wilderness, aesthetics, an

Conservation of Faunal Diversity in Forested Landscapes. Edited by
R.M. DeGraaf and R.I. Miller. Published in 1996 by Chapman & Hall. ISBN 0 412 61890 7.

almost religious identification with old growth forests and the conservation of forest wildlife figure importantly in environmental efforts to restrict logging. Wilderness and a personal identification with trees and undisturbed forests are fundamentally incompatible with logging. Accommodation and compromise between the timber industry and those opposed to logging can only be reached through land-use allocations in which logging is excluded from certain areas, while continuing elsewhere. Unfortunately, while the reservation of forest as wilderness and national park is important for wildlife conservation, it is not sufficient to guarantee the long-term survival of Australia's forest biota (Lunney and Recher, 1986; Lunney, 1991a; Recher, 1976a, 1990; Recher and Lim, 1990; Kirkpatrick *et al.*, 1990). On the assumption that the commercial exploitation of Australia's forests will continue for the foreseeable future, the long-term survival of Australia's forest biota can only be assured by fully integrating the management and conservation of wildlife with logging and other forest management practices (Shaw, 1983; Loyn, 1985; Recher, 1985a, 1986; Davey, 1989; FWG, 1991).

The integration of wildlife management with the management of forests for wood production in Australia is a recent initiative. The first efforts were made in the 1970s and were intuitive and based on faunal surveys and patterns of animal distribution, rather than founded on research and experimentation. As knowledge of the ecology of forest wildlife and the effects of management on fauna improved through the 1980s, management prescriptions were refined and extended. Many are now based on detailed quantitative studies but their long-term effectiveness remains unproven.

This chapter reviews the development of wildlife management prescriptions for Australian forests and discusses the directions that forest management should take in the 1990s, if the conservation of the continent's forest biota is to succeed. The emphasis is on eucalypt forests (i.e. forests dominated by the genus *Eucalyptus*) and on vertebrates. There are reasons for this emphasis. Firstly, most of the work on the effects of forest management on wildlife has been done in eucalypt forests and concerns vertebrates. Secondly, eucalypt forests are the most extensive of Australia's forests. They are where logging is most intensive and where industry efforts to expand pulpwood logging, if successful, will greatly reduce the area of mature and mixed-age forests over the next two decades. Thirdly, eucalypt forests are rich in birds, mammals, frogs and reptiles, many of which depend on eucalypts for food and shelter. These dependent species are adversely affected by intensive logging. Fourthly, the last decade has seen significant reductions in the amount of rainforest logging in Australia with much of the northern tropical and subtropical rainforests reserved. The need to integrate wildlife management with forest management is therefore greatest in eucalypt forests.

12.2   POLITICS AND FOREST MANAGEMENT

This is an appropriate time to review management practices for forest fauna in Australia. Since 1989, there have been two major reviews of Australia's forests and forestry practices. These culminated in the adoption in 1992 of a National Forest Policy signed by the Commonwealth and all states except Tasmania (Commonwealth of Australia (CWA), 1992). The genesis of the reviews and the National Forest Policy can be traced to the 1989 election statement of the then prime minister, Bob Hawke, on the environment, *Our Country, Our Future*, in which he committed the nation to 'ecologically sustainable development (ESD)' (Hawke, 1989). A review of forests and forestry, as part of Hawke's (1989) commitments, was inevitable given the high level of public conflict over forest management and conservation in the preceding decade.

The reviews took two forms. The first was conducted by the Resource Assessment Commission which was directed to hold 'an inquiry into options for the use of Australia's forest and timber resources' (RAC, 1992: iii). The second was conducted by the Ecologically Sustainable Development Working Group into Forest Use as one of a series of reports commissioned by the prime minister in 1990 to 'provide advice on future policy directions and focus on measures which will encourage the integration of environmental considerations into decision-making processes' (FWG, 1991: iii). Both groups reported to the government in 1991, the RAC with a draft report (the final report being released in 1992) and the ESD Groups in November 1991. Both reports provide an overview of Australia's forests and forest resources, with the RAC report collating for the first time a complete set of data on Australia's forests, forest resources and the conservation status of those forests. A draft National Forest Policy Statement, *A New Focus for Australia's Forests*, was released for public comment in July 1992 and adopted in December 1992.

Parallel documents exist for each of the states. Tasmania, for example, in declining to sign the National Policy Statement committed itself to the management of its forests as set out in the *Tasmanian Forests and Forest Industry Strategy* (CWA, 1992). Victoria adopted a *Timber Industry Strategy* in 1986 and a *Code of Forest Practices* in 1989 (Wilson, 1991). These documents provide the framework on which to integrate wildlife management with timber production and emphasize the sustainable use of forest resources. The emphasis in state forests in New South Wales is also on 'ecologically sustainable forest use' (Drielsma, 1992). Smith (1991), however, is critical of Victoria's application of the principles outlined in the state's forest policy documents, pointing to the gap between stated intentions and application. He argues that this 'credibility gap' leads to conflict when public expectations are not realized. A similar situation exists in New South Wales where the state's forest management authority is continually embroiled in public conflict over the management of state forests and the conservation of forest fauna.

Although the reports of the RAC and the FWG are important sources of information on Australia's forest resources and collectively provide an insight into where forest management and conservation might proceed in the next century, neither critically assesses the status of Australia's forests, their fauna, nor the impact of forestry practices on forest ecosystems. In a critical review, Lunney (1993) pointed out that the Resource Assessment Commission brought together the available information on the impact of logging and forestry practices on forest fauna, but failed to analyze the data collected or to use it to recommend ways in which forest wildlife conservation and management could be improved. This chapter goes a small way towards correcting this omission.

12.3   AUSTRALIAN FORESTS

A forest was defined by the RAC (1992) as a plant community dominated by trees greater than 15 m in height and with 30% or greater canopy cover. Somewhat confusingly, the RAC's definition and its analysis of the area of forest in Australia encompasses substantial areas of woodland, which is generally considered to be a tree-dominated community with 10–30% canopy cover (Specht *et al.*, 1974). Tables 12.1 to 12.3 present information extracted from the RAC (1992) report on the area of forest by forest type, in different land tenures, indicating the proportions of logged or unlogged forest. The RAC (1992) accepted 'unlogged forest' as 'old-growth forest'.

Forest types in Australia are described in a variety of ways. Tables 12.1 and 12.2 use descriptive names, such as rainforest or mangrove forest, and geographical location. Table 12.3 uses a structural definition of vegetation formations based on projected canopy cover and average canopy height (Specht *et al.*, 1974). In this classification, rainforest is described as 'closed forest' (70–100% canopy cover). Eucalypt forests are 'open-forest' with 'tall open-forest' equivalent to 'wet sclerophyll' (eucalypt forest with a rainforest understory) and 'open-forest' equivalent to 'dry sclerophyll' (eucalypt forest with a sclerophyllous shrub layer). In Tasmania, eucalypt forest with a rainforest understory is referred to as mixed forest, while eucalypt forest with soft broadleaved shrubs is wet sclerophyll (Taylor, personal communication).

On the definition of a forest as used by the RAC (1992), about 10% of Australia (70–80 million ha) was forested at the time of European settlement in 1788. Since 1788, about half of this (32–40 million ha) has been cleared for agriculture, urban development and forest plantations (usually of exotic pine species) (Wells *et al.*, 1984). Most (> 90%) of the remainder is fragmented and modified to a greater or lesser extent by logging, changed fire regimes, grazing by domestic stock, and introduced plants and animals.

Table 12.1 Area (ha × 10³; % of total in parentheses) of forest in Australia by forest type, location and conservation status (adapted from RAC, 1992, Tables 3.5 and 3.6, pp. 146–148)

Forest type	Area		Area in conservation reserves		Area of unlogged forest (old growth)		% unlogged forest in conservation reserves	
	area	(%)	area	(%)	area	(%)	(%)	area
Rainforest	2433	(5.8)	1395	(57)	913	(37.5)	38.8	(543)
northern[a]	1432		1011	(71)	354	(24.7)	24	(243)
southern[b]	1001		384	(38)	559	(55.8)	78	(300)
Mangrove	172	(<1)	56	(31)	170	(99)	98	(55)
Swamp[c]	1060	(2.5)	918	(87)	931	(87.8)	86	(789)
Eucalypt forest	26415	(62.5)	4936	(18.7)	4995	(18.9)	48.9	(2415)
southwest[d]	2405		455	(19)	433	(18)	36	(164)
southeast[e]	14735		3019	(20)	3280	(22.2)	52.5	(1586)
northeast[f]	9192		1379	(15)	1231	(13.4)	43.8	(604)
River red gum[g]	83	(<1)		?	51	(61.4)		?
Woodland[h]	12168	(28.7)	1469	(12	9240	(76)	87	(1278)
Total	42247		691	(21)	16249	(38.5)	58	(5028)

[a]Tropical, subtropical and monsoonal rainforests (closed forest) of northern Australia.
[b]Temperate and cool temperate rainforests of eastern and southeastern Australia.
[c]Swamp eucalypt and paperbark (*Melaleuca*) forests; mainly in northern and eastern Australia.
[d]Southwestern Western Australia.
[e]Southeastern New South Wales, Victoria, South Australia and Tasmania.
[f]Northeastern New South Wales and eastern Queensland.
[g]Widespread forest type along inland river floodplains; dominated by *Eucalyptus camaldulensis*.
[h]Includes cypress pine (*Callitris*), mulga (*Acacia aneura*), brigalow (*A. harpophylla*), as well as malee and tropical woodlands dominated by eucalypts (*Eucalyptus*).

In describing more than 90% of Australia's forest as fragmented, forest fragmentation is considered to be the result not only of clearing but also of the isolation and separation of forest habitats by roads, including internal forestry roading, as these can restrict the movements of forest animals. More than 60% (26.4 million ha) of the remaining forests are dominated by eucalypts with an additional 29% (12.2 million ha) in woodlands dominated by either eucalypts or wattles (*Acacia* spp.) (Table 12.1) (RAC, 1992). Only 6% of forest (2.4 million ha) is rainforest.

Table 12.2 Tenure (ownership) of forest outside conservation reserves (area ha × 10^3, with % unlogged in parentheses) (adapted from RAC, 1992, Table 3.4, p. 160)

Forest type[a]	State forest		Other public land		Private forest land	
Rainforest	651	(28)	225	(56)	162	(45)
northern	286	(13)	43	(26)	92	(65)
southern	365	(40)	182	(63)	70	(19)
Mangrove	0		46	(100)	69	(100)
Swamp	?		48	(100)	95	(100)
Eucalypt forest	9723	(22)	5020	(8)	6736	(? <1)
southwest	1443	(19)	?		507	(?)
southeast	5243	(29)	3001	(6)	3472	(<1)
northeast	3037	(19)	2019	(11)	2757	(?)
River Red Gum	?		?		?	
Eucalypt woodland	1134	(17)	5284	(80)	4280	(83)
Total	11508	(22)	10623	(45)	11342	(34)

[a]See Table 12.1 for explanation of forest types.

Table 12.3 Area of eucalypt forest according to land tenure outside the Northern Territory (ha × 10^3, with % total in parentheses) (adapted from RAC, 1992, Appendix F.3.li)

Forest type	State forest		Other public land		Conservation reserve		Private land		Total
Total area									
Tall open-forest (wet sclerophyll)	4510	(58)	1149	(15)	770	(10)	1335	(17)	7764
Open-forest (dry sclerophyll)	3399	(28)	2227	(18)	3030	(25)	3524	(29)	12180
Total	7909	(40)	3376	(17)	3800	(19)	4859	(24)	19944
Area of unlogged (old growth)[a]									
Tall open-forest (wet sclerophyll)	1263	(64)	356	(18)	331	(17)	13	(< 1)	1963
Open-forest (dry sclerophyll)	442	(20)	379	(17)	1424	(63)	0		2245
Total	1705	(40)	735	(17)	1823	(43)	13	(< 1)	4208

[a](% of total unlogged forest in tenure category).

Extensive forest clearing accompanied the expansion of agriculture across Australia and between 1890 and 1930 the area under agriculture increased from 2.2 million to 9.9 million ha (RAC, 1992). Between 1869 and 1988 Victoria's forest cover was reduced from 88% to 35% (Woodgate and Black, 1988). The greatest impact was on forests on the most productive soils (Lunney and Leary, 1988; RAC, 1992; Recher, 1982; Wells *et al.*, 1984). Near Sydney and in the Bega Valley in southeastern New South Wales the richest eucalypt forests were cleared by 1900 (Lunney and Leary, 1988; Recher *et al.*, 1993). The extensive subtropical rainforests located on the nutrient-rich volcanic soils of the Illawarra district of New South Wales just south of Sydney and along the Clarence, Richmond and Tweed River valleys in northeastern New South Wales were cleared for dairy farming in the late nineteenth century and early twentieth century; only remnants survive (Frith, 1976, 1982; Date *et al.*, 1991; RAC, 1992). There was also extensive clearing of subtropical rainforest on the Atherton Plateau and of lowland tropical rainforest in northeastern Queensland (Frith, 1982). Less than 50% (2.4 million ha) of Australia's original rainforest survives.

Deforestation for agriculture continues and it is estimated that 0.22–0.27% (150–190 thousand ha) of forest and woodland is cleared annually throughout Australia (RAC, 1992). For example, in Tasmania, an average of 15 000 ha of forest was cleared or inundated annually between 1972 and 1980 (Kirkpatrick, 1991). This was reduced to 6000 ha annually between 1980 and 1988. Almost all clearing in Australia occurs on private land, including forest purchased by state forest authorities for plantation establishment, and affects mainly eucalypt forest types. In a simplistic projection, the RAC calculated that at current clearing rates all of Australia's forest and woodland will be cleared in 175 to 250 years (Lunhey, 1991a).

Of the remaining eucalypt forest, 5 million ha (19%) is unlogged (RAC, 1992). Twenty-one per cent of forest is protected in conservation reserves (RAC, 1992). This includes 57% of the surviving rainforest and 19% of eucalypt forest (Table 12.2), but does not include forest excluded from logging in production forests. According to the RAC (1992), 40% of the reserved rainforest and 50% of the reserved eucalypt forest is unlogged (Table 12.1). Exclusive of conservation reserves, 66% (22.1 million ha) of forest is publicly owned, while 11.3 million ha (34%) is private (freehold) forest (Table 12.2) (RAC, 1992).

In terms of fauna conservation the relatively large area of forest in conservation reserves (8.7 million ha) is misleading. Most forest conservation reserves in Australia are at high elevation in mountainous or rugged terrain and/or are on nutrient-poor soils (e.g. sandstone, coastal sands). This includes extensive reserves of unlogged forest on nutrient-poor and topographically rugged sandstones in New South Wales. Forest at middle to low elevations with gentle topography and on well-watered nutrient-rich soils are poorly represented in the reserve system (Table 12.3) (Hall, 1988; Lunney and Recher, 1986; Recher, 1976a; Whitehouse, 1990). These are the forests with the richest and most diverse fauna (Braithwaite, 1983; Braithwaite *et al.*, 1984; Majer *et al.*, 1992; Recher *et al.*, 1980, 1987, 1991; Recher, 1985a) and include some of the continent's most diverse old-growth communities.

12.4 OLD-GROWTH FOREST

Old-growth forest is at the centre of most forest conservation debates in Australia. Indeed, if logging old growth ceased, then many of the objections to logging native forests would probably also cease. While it can be argued that it is the components of old-growth forest that are critical for the conservation and management of forest wildlife, the management of old growth is central to wildlife management and conservation issues (Kirkpatrick *et al.*, 1990; Loyn, 1985; RAC, 1992; Recher, 1985b). For the purposes of this chapter, an old-growth forest is defined as one that has not been substantially disturbed or changed by Europeans. In particular, it is an unlogged forest (RAC, 1992). It may be a young forest regenerating after wildfire or storm, but most often is multi-aged and characterized by a high proportion of mature and old trees (Kirkpatrick *et al.*, 1990; Recher, 1992). Associated with a high proportion of mature and old trees are large volumes of standing dead wood (e.g. snags, dead branches), logs, large woody debris and litter (Scotts, 1991).

Whether old-growth eucalypt forests sustain different ecological processes (e.g. competitive or predatory interactions, pollination, commensalism) or are functionally different from logged forests (that is, have different hydrological or nutrient cycles) is not known (Recher, 1992). There is evidence, however, that some forest animals are dependent on components of old-growth ecosystems. Among vertebrates, the sooty owl (*Tyto tenebricosa*) and powerful owl (*Ninox strenua*), a number of parrots and lorikeets (Psittacidae, Loriidae), the crested shrike-tit (*Falcunculus frontatus*), greater glider (*Petauroides volans*), yellow-bellied glider (*Petaurus australis*), Leadbeater's possum (*Gymnobelideus leadbeateri*), tiger quoll (*Dasyurus maculatus*), potoroos (Potoroidae), most species of small insectivorous bats and a number of lizards and frogs either reach their greatest abundance in old growth or in forests where substantial components (e.g. large trees, senescent trees, large logs, an abundance of large woody debris, a complex vegetation structure) of old-growth ecosystems remain (Kirkpatrick *et al.*, 1990; Lindenmayer *et al.*, 1990; Lunney, personnel communication; Saunders *et al.*, 1982; Scotts, 1991: Smith and Lindenmayer, 1988; Smith *et al.*, 1992). The needs of these animals are often complex and may require a particular combination of young and old trees. The endangered Leadbeater's possum, for example, requires a mosaic of old growth and regeneration including large standing snags as den sites. Such conditions are the result of intense wildfires (Smith and Lindenmayer, 1988). The endangered Tasmanian sub-species of the wedge-tailed eagle (*Aquila audax fleayi*) (Garnett, 1992) requires a minimum area of 10 ha of old-growth forest around its nest tree and is intolerant of disturbance when nesting (Mooney and Holdsworth, 1991). Kavanagh and Bamkin (1995) have shown that species such as the greater glider, yellow-bellied glider, powerful owl and sooty owl, which are widely considered to be dependent on old-growth components, can survive in a mosaic of logged and unlogged forest.

While the evidence is equivocal and a matter of dispute among Australian forest ecologists, it is my view that significant components of the forest fauna will be shown to be restricted to old-growth forest. Most of these will be forest invertebrates, particularly species that inhabit the forest litter and the bark of eucalypts, but some forest vertebrates may also prove to be dependent on large expanses of unlogged ecosystems. Unfortunately, the final impact of current logging practices and the loss of old-growth forest will not be evident for some years. Until then, caution seems advisable.

Although the information available on the relationships between old-growth forest ecosystems and forest wildlife is species specific, the conservation of old-growth dependent communities almost certainly requires the protection of remnant old growth including the establishment of buffer zones to protect remnants from environmental changes in adjacent disturbed ecosystems (*sensu* Harris, 1984). The need for caution when planning the logging of old-growth forest is highlighted when the distribution and conservation status of the remaining old-growth eucalypt forest is analyzed.

Table 12.3 excludes forests from the Northern Territory where little logging occurs: almost all (98%) the Territory's 9 million ha of forest, which includes 7.8 million ha of tropical woodland, is unlogged (RAC, 1992). The vertebrate fauna of the Territory is also different from that of eucalypt forests in eastern and southwestern Australia with little overlap in species. About 20 million ha of eucalypt forest occur in eastern and southern Australia (Table 12.3). Of this, 21% is unlogged (RAC, 1992). Of the unlogged forest, 43% is included in conservation reserves, while the remainder is in state forests (40%) and on other public lands (17%) and available for logging. Logging is excluded from conservation reserves in all states. Less than 1% of private forest is unlogged.

Although nearly equal areas of old growth are in conservation reserves and production forests (Table 12.3), 78% of the unlogged forest in reserves is dry open-forest and occurs mainly on nutrient-poor soils at relatively high elevations. Such forests are unproductive and have low vertebrate species abundances (Braithwaite, 1983; Braithwaite *et al.*, 1983, 1984; Majer *et al.*, 1992, in press; Recher, 1985a; Recher *et al.*, 1980, 1991, 1993). In contrast, 74% of the unlogged forest in state forests are the more productive and species-rich tall open-forests at lower and middle elevations. Twenty-five per cent of tall open-forest is unlogged (Table 12.3), of which 64% occurs in state forests and only 17% is in conservation reserves. The situation is reversed for the less productive open-forest (Table 12.3). On better soils at middle to low elevations the proportion of old growth in areas such as northeastern New South Wales may be less than 3% of the original forest (Recher, 1992). Little of this occurs in patches greater than 500 ha and most of these are at risk of logging, changed fire regimes and the effects of roading.

The conclusions and recommendations of the Resource Assessment Commission (1992) and the Forest Use Working Group (1991) with respect to old-growth forests were incorporated in the National Forest Policy. The Policy identifies old-growth forest as having important aesthetic, cultural and nature conservation values (CWA, 1992). Under this policy the states and the Commonwealth agreed to assess the conservation values of old growth and to avoid activities, such as logging, that may compromise these values until the assessment is completed. The Policy provides for the establishment of a representative reserve system to protect old-growth forest on public lands by the end of 1995 and on private land by 1998 (CWA, 1992). In 1994, the assessment of old growth had commenced but logging continued in old-growth eucalypt forests in all states. This includes logging in forests that have already been identified as having high and largely irreplaceable nature conservation values, such as those in the Coolangubra State Forest in southeastern New South Wales (Jenkins and Recher, 1990; Joint Scientific Committee, 1990; Mosley and Costin, 1992; Pyke and O'Connor, 1991).

12.5 FOREST VERTEBRATE FAUNA

Australia has a rich terrestrial vertebrate fauna: 260 species of mammals, 460 species of birds, 700 species of snakes and lizards and 180 of frogs (data from Cogger, 1992; Strahan, 1983; Pizzey, 1980). A significant proportion of these are forest and woodland species: 170 species of mammals, 315 birds, 320 reptiles and 85 frogs. Of the mammals, 130 species inhabit eucalypt forest. They include more than 60 species of bat, 39 species of which are restricted to eucalypt forest (Richards, 1991). Another 34 species of mammals, including 12 species of bat (Richards, 1991), occur exclusively or predominantly in rainforests. A similar number of mammal species occur exclusively or predominantly in eucalypt woodlands. One hundred and seventy species of reptiles and 30 of frogs occur in eucalypt habitats. Of birds, 235 species are most abundant in eucalypt forests and woodlands and 80 species are most abundant in rainforests. Almost 50% of the continent's terrestrial vertebrate diversity can be found in the 5% of the continent that remains forested. Only reptiles are more abundant in the arid inland.

12.5.1 Status of forest vertebrates in Australia

With the exception of the Tasmanian tiger (*Thylacinus cynocephalus*) and possibly the white-footed rabbit-rat (*Conilurus albipes*) there is no evidence that any forest vertebrate has become extinct as a result of European settlement (Bureau of Rural Resources, 1990; Strahan, 1983). However, as a result of forest clearing and other European activities, all forest and woodland vertebrates are now less abundant and more patchily distributed than originally. The intensification of forestry operations since the 1960s, with an emphasis on clearfelling, plantation establishment and broad area prescription burning (also known as hazard reduction or controlled burning) is a particular cause of concern for the conservation of forest vertebrates (Bureau of Rural Resources, 1990; RAC, 1992). Other threats to forest wildlife include the continued clearing of forests for agriculture and plantations, exotic species, grazing by domestic and feral stock, and recreational use of forests (RAC, 1992).

Although few forest vertebrates have become extinct, there have been significant regional extinctions and a high proportion of the fauna is threatened (Garnett, 1992; Kennedy, 1990; Norton, 1988; Recher and Lim, 1990; Recher, 1990, 1994). The greatest concern has been for mammals, with Australia having the highest extinction rate for mammals (22 species), as a result of European activities, of any continent (Kennedy, 1990; Recher and Lim, 1990). Of 130 mammal species recorded from New South Wales, 77 (59%) are endangered or extinct in the state (Lunney *et al.*, ms. a). While the concern for mammals is warranted, recent studies show that the impact of European settlement is at least as great on birds, reptiles and frogs, albeit with fewer extinctions (e.g. Lunney *et al.*, ms. a,b; Recher 1994). Of the 231 species of terrestrial birds in Victoria, 48% have declined in abundance and distribution with many regional extinctions (Robinson, 1991, personal communication). Forty (17%) of the 231 species are endangered. This pattern of decline in the avifauna is repeated throughout Australia (Recher, in preparation). Among the birds most affected are ground-dwelling, ground-nesting and ground-foraging species and birds that are obligate tree-hole nesters, e.g. treecreepers (Climacteridae) and parrots (Garnett, 1992; Lunney *et al.*, ms. b, Recher, ms; Robinson, 1991, personnel communication). This includes many forest species that require old trees for feeding or shelter (Loyn, 1985). There are many reasons for the decline in abundance of native vertebrates, but habitat loss and fragmentation, changed fire regimes, predation by the European fox *Vulpes vulpes*, and loss of habitat trees for nesting are important (Christensen, 1980a,b; Garnett, 1992; Hamilton, 1892; Loyn, 1985; RAC, 1992; Recher *et al.*, 1987; Recher 1992).

A significant and rapid decline in most of the smaller and medium sized (50–4000 g) mammals in the forests and woodlands of southwestern Western Australia since the early 1970s is associated with the spread of the

European fox (Christensen, 1980b; Wardell-Johnson and Nichols, 1991). Evidence for the impact of fox predation in Western Australia comes with the recovery of populations when foxes are controlled by poisoning and when fire regimes are modified to provide suitable cover. Ground-dwelling forest mammals in the same size range in eastern Australia declined earlier, but reports since 1991 from field observers suggest that the decline is continuing with mammals, such as bandicoots (Peramelidae) and brush-tailed possums (*Trichosurus vulpecula*), appearing to be much less common than in the 1970s.

Habitat loss and modification are major factors in the continuing decline of forest vertebrates in eastern Australia, but fox predation is a significant factor (Hamilton, 1892; Lunney and Leary, 1988; Recher, 1990, 1994; Recher and Lim, 1990). Conservationists opposed to the extension of logging into remnant old-growth forests have argued that internal roading facilitates the movement of foxes into forests and increases the impact of predation on ground-dwelling forest vertebrates. Post-logging burns and prescription or hazard reduction burning reduce the cover available to ground fauna, as does grazing by domestic stock, increasing the risk of predation and leading to competition from edge and open-forest species (Christensen, 1980b; Lunney and O'Connell, 1988; Newsome *et al.*, 1975; Smith *et al.*, 1992). Smith *et al.* (1992) considered that the impact of post-logging grazing and burning regimes on forest vertebrates was potentially more severe, with longer-term consequences than the logging itself. This was particularly the case in low site quality forest where logging intensities tended to be low (> 50% canopy retention).

12.5.2   Forest resources for vertebrates

Many eucalypt forest vertebrates depend on tree hollows, cavities and crevices for shelter and nesting (Cowley, 1971; Tyndale-Biscoe and Calaby, 1975; Ambrose, 1982; Norton, 1988). Thirty per cent of eucalypt forest mammals, including 24 species of bats, require tree hollows, cavities or crevices for shelter (Strahan, 1983). Of 13 arboreal marsupials, nine are obligate tree-hollow users (Bureau of Rural Resources, 1990; Strahan, 1983). Tree hollows and logs are also important for many species of small, ground-dwelling mammals (Dickman, 1991). Thirty-five per cent (45 species) of forest birds use hollows or cavities for nesting and 80% of these (32 species) are obligate hole nesters (Ambrose, 1982; Disney and Stokes, 1976; Recher *et al.*, 1980; Wardell-Johnson and Nichols, 1991). Tree hollows are also used by reptiles and frogs (Cogger, 1992). On a regional scale, the dependence of the fauna on tree hollows can be considerable. For example, in the depauperate forests of southwestern Western Australia, 18 species of birds, 9 bats and 10 arboreal marsupials use tree hollows for shelter or nesting (Dickman, 1991; Wardell-Johnson, personnel communication).

Tree hollows are therefore an important resource for Australian forest vertebrates, but unlike other continents there are no primary excavators of tree hollows. Hole-using species can only enlarge or prepare a hollow that has developed as a result of fire or termite and fungal damage. Generally, hollows only appear in eucalypts that are 100 or more years of age and may only become suitable for use by vertebrates in trees greater than 150 years old (Ambrose, 1982; Lindenmayer *et al.*, 1990; Mackowski, 1984). In Western Australia hollows suitable for arboreal marsupials in jarrah–marri (*E. marginata–E. calophylla*) forests may not develop until trees are more than 400 years old (Inions *et al.*, 1989). The requirements for tree hollows are complex and there are pronounced differences in the size, type and position of hollows used by different species of vertebrates (Haseler and Taylor. 1993; Lindenmayer *et al.*, 1990; Saunders *et al.*, 1982; Smith and Lindenmayer, 1988). Individuals and social groups may use two or more hollows and the type or position of the chosen hollow may change seasonally. Loyn (1985) stressed the importance of the large old trees associated with hollows as a foraging resource. Eighty-six per cent of above-ground foraging observations made by Dickman (1991) of 12 species of small ground-dwelling mammals were in large mature trees with complex branching structures. Taylor and Haseler (1993) found that the nest trees used by four species of hollow-nesting birds in Tasmania were larger and had a greater number of potential hollows and a greater proportion of large hollows than a representative sample of trees with hollows. Manna or ribbon gum (*E. viminalis*) was disproportionately selected as a nest tree.

Although tree hollows are important for forest vertebrates, other resources are required for nests, shelter and foraging and must be considered when managing eucalypt forest faunas (Dickman, 1991; Loyn, 1985; Recher, 1991; Scotts, 1991; Smith, 1984, 1985). Birds such as fantails (*Rhipidura*), robins (*Petroica*), cuckoo-shrikes (*Corinna*), sittellas (*Daphoenositta*), thornbills (*Acanthiza*), warblers (*Gerygone*) and honeyeaters, including *Melithreptus* and *Meliphaga*, use spider web as a primary nesting material (Recher, 1991, unpublished data). Lichen is frequently used on the exterior of the nest for camouflage. Lichen, mistletoe and other epiphytes are prolific on large mature or overmature trees and are important foraging substrates for vertebrates. Particularly in woodlands, mistletoe is an important source of nectar and clumps of mistletoe support greater numbers of invertebrates than comparable amounts of eucalypt foliage (Recher, unpublished). Spider web is often most abundant on dead wood, among debris and in tree cavities, crevices and hollows (Recher, 1991).

Eucalypt forest and woodland birds differ in the sites selected for nest location (Recher, unpublished) but about 10% (14 species) place their nests on dead wood, either a vertical branch (e.g. orange-winged sittella, *D. chrysoptera*) or a horizontal one (e.g. satin flycatcher, *Myiagra cyanoleuca*; black-faced cuckoo shrike, *C. novaehollandiae*). Some hole- and cavity nesters may place their nests behind loose bark (e.g. grey shrike thrush, *Colluricincla harmonica;* buff-rumped thornbill, *Acanthiza reguloides*). Bark is also an important foraging substrate (Kavanagh, 1987; Goldingay and Kavanagh, 1991; Noske, 1985; Recher, 1991; Recher *et al.*, 1985).

All eucalypts shed bark from terminal branches and species in the sub-genus *Symphyomyrtus* shed bark annually along main branches and the trunk. This loose or decorticating bark harbors an abundant, and distinctive invertebrate fauna (Recher *et al.*, 1983; Monaghan and Recher, unpublished data) including many insects (e.g. Homoptera, Hemiptera) that produce exudates rich in sugars and complex carbohydrates. These exudates, as well as the insects themselves, are exploited by birds and mammals. Energy-rich carbohydrates on foliage and/or bark in the form of manna (secretions by the plant where insect damage has occurred), lerp (secretions of psyllid insects that form a protective covering over the insects), honeydew (exudates from sap-sucking insects) and sap are a feature of eucalypt forests and an important resource for arboreal mammals (e.g. sugar glider, *Petaurus breviceps;* yellow-bellied glider; feathertail glider, *Acrobates pygmaeus*) (Kavanagh, 1987; Goldingay and Kavanagh, 1991; Recher, personal observation) and carbohydrate-dependent birds (e.g. honeyeaters, Meliphagidae; Australian warblers, Acanthizidae) as well as other insects (Ford and Recher, 1991; Paton, 1980).

Depending on fire frequency, decorticating bark accumulates on the base of the tree as a 'stocking' or on the ground near the tree. This debris may harbour an invertebrate fauna which then re-colonizes the trunk and branches of the tree as new bark is shed. It may therefore be an important source of invertebrates for forest birds regardless of whether they forage on bark or among the foliage. Decorticated bark, other woody debris and logs also provide shelter and foraging sites for ground-dwelling or ground-foraging mammals, birds, reptiles and frogs (Scotts, 1991). Of 12 species of ground-dwelling mammals, logs were the main shelter for echnida (*Tachyglossus aculeatus*), little long-tailed dunnart (*Sminthopsis dolichura*) and dusky antechinus (*Antechinus swainsonii*) (Dickman, 1991). Logs may be a limited resource, the availability of which is affected by logging and fire, including post-logging and hazard reduction burns (Recher, personal observation). Logs, woody debris and litter are important resources for 60 species (30%) of forest reptiles and 12 species (15%) of forest frogs (Cogger, 1992).

12.5.3   Effects of forest management

Forest management includes logging, firewood cutting, wildflower picking, silviculture, plantation establishment, prescribed burning, roading, mining, regulation of streams and rivers, provision of recreation facilities and grazing by domestic stock including commercial honeybees (*Apis mellifera*). Exploitation of forest resources is not restricted to production forests and, with the exception of logging and plantation establishment, also occurs in conservation reserves. Mining, for example, is permitted in some conservation reserves in Western Australia, while prescription burning is standard practice in conservation reserves throughout Australia.

There are clear structural differences between old growth and forests disturbed by logging or repeated burning (Davey, 1989; Lindenmayer *et al.*, 1990; Recher, 1985a, 1992). Logging reduces the proportion and changes the spatial distribution of old and large trees; the abundance of standing dead wood, logs and woody debris is reduced (Lindemnayer *et al.*, 1990; Recher, 1985a). Logging also affects the species composition of the canopy and understory vegetation (Floyd, 1962; Henry and Florence, 1966; Kellas *et al.*, 1988; Loyn *et al.*, 1983; Recher, 1985a). Intensive selection of preferred timber or firewood trees may rapidly increase the abundance of non-preferred species (Recher, personal observation; Traill, 1991). Non-preferred species may (e.g. woollybutt, *E. longifolia*) or may not (e.g. red stringybark, *E. macrorhyncha*) be species that form hollows used by forest vertebrates. Litter accumulates rapidly after fires but repeated burning, whether by accident or design, reduces the amount of large woody debris and logs and affects the size and species composition of the shrub understory (e.g. Christensen and Abbott, 1989). There is a tendency for the litter that accumulates in repeatedly burnt areas to be looser and drier than in similar habitats with longer fire frequencies. Logs that survive fires may be heavily charred, drier and less useful as habitat for small ground-dwelling mammals, such as brown antechinus (*A. stuartii*) and bush rat (*Rattus fuscipes*) (Recher, unpublished data). Although the debris left after logging includes tree heads and butts and may increase the number of logs and amount of large woody debris, ultimately the size of logs recruited into the ecosystem declines, while post-logging burns reduce their value as wildlife habitat.

Changes in plant species composition is exacerbated by the introduction of exotic tree species in plantations or in reafforestation after mining. Often different species of eucalypts from those of the original forest are used. For example, in Western Australia eucalypts from eastern Australia are used to establish plantations and revegetate mine sites, and for reafforestation after logging (Curry and Nichols, 1986; Tacey, 1979).

The forest itself is fragmented by mining and forestry operations with similar aged patches isolated by older and younger regeneration or by cleared lands, roads and plantations. As fragmentation and roading increase, the extent of edge habitat increases. Edges facilitate the intrusion of edge as well as non-forest or open-country species of plants and animals, including introduced predators (Recher *et al.*, 1987). Grazing contributes to changes in the structure and composition of understory and ground vegetation. These changes are accentuated by repeated burning to promote grass growth for cattle (Smith *et al.*, 1992). Movements of domestic stock facilitate the introduction and spread of weeds. Commercial apiaries remove nectar required by native fauna and feral honeybees may compete with native fauna for tree hollows, although limited data suggest that bees use hollows that would not be suitable for birds or manimals (Recher, unpublished data).

Whether changes in forest structure result in changes in ecological processes or functions may not be known, but logging, changed fire regimes, grazing and other forms of forest management result in changes in the distribution and abundance of forest vertebrates (Kirkpatrick *et al.*, 1990; Loyn, 1985; Lunney, 1991b; McIlroy, 1978; Recher *et al.*, 1980; Smith *et al.*, 1992). Hollow-dependent species, those that use logs, dead wood and bark as foraging substrates or for shelter, ground-dwelling and ground-foraging animals and animals with special resource requirements, such as spider web or lichens, decline in abundance according to the type and extent of management activities (Christensen and Abbott, 1989; Loyn, 1985; Recher *et al.*, 1980, 1987; Robinson, 1991; Tyndale-Biscoe and Calaby, 1975).

For most of the 1970s and 1980s, it was considered that the impact of logging was greatest on old-growth dependent fauna and species that required large mature trees for shelter and feeding (e.g. Recher *et al.*, 1980; Tyndale-Biscoe and Calaby, 1975). Thus there was special concern for arboreal marsupials (e.g. greater glider, Leadbeater's possum), forest owls (e.g. powerful owl) and cockatoos (e.g. glossy black cockatoos, *Calyptorhynchus lathami*). It was recognized that other vertebrates were affected but there was a widely held view that these effects were transitory and that, as the forest regenerated after logging, recolonization would occur (e.g. Loyn, 1985; Recher, 1976b; Recher *et al.*, 1980). Studies of birds showed initial declines in abundance of many passerines after logging or fire followed by a gradual recovery in numbers (e.g. Kavanagh *et al.*, 1985; Smith, 1984, 1985). Similar results were obtained from post-logging and post-fire studies of small ground-dwelling mammals and reptiles (e.g. Lunney and Ashby, 1987; Lunney *et al.*, 1987, 1991). However, Abbott and Van Heurck (1985) did not find any significant changes in bird abundance or species composition following thinning in jarrah forest that removed about half the canopy and all large bull banksia (*Banksia grandis*) from the understory. More recently it has been recognized that the changes to the vegetative structure of forest ecosystems from logging, fire and grazing have wider and longer lasting effects on forest vertebrates than realized and that the interaction of these effects threatens the survival of populations (e.g. Kirkpatrick *et al.*, 1990; Lunney *et al.*, 1991; Recher, 1991; Smith *et al.*, 1992). Stochastic events, such as drought, are critical compounding factors.

12.5.4   Wildlife management prescriptions

Presently all state forest authorities incorporate wildlife conservation and management prescriptions in their logging management plans. However, with the exception of Tasmania, there are no requirements to apply wildlife management prescriptions to logging operations on private land and few, if any, private logging operations in Australia make any special effort to conserve and manage forest wildlife. In Tasmania, under the Forest Practices Act (1985) logging operations on private, as well as public, lands are required to be conducted in an 'environmentally acceptable manner' (Forestry Commission, Tasmania, 1993). Appropriate guidelines, including those for wildlife conservation, are provided in the *Forest Practices Code*. However, the only mandatory prescriptions that apply in Tasmania are those relating to rare or threatened species (Taylor, personal communication). The *Forest Practices Code* was worded so that the retention of wildlife habitat strips and habitat trees would only apply 'on [private] native forest proposed for fauna maintenance by the owner' (Forestry Commission, Tasmania, 1993: 54).

In New South Wales, the Threatened Species (Interim Protection) Act (1991) requires all logging operations to be licensed by the National Parks and Wildlife Service and, if necessary, a Fauna Impact Statement (FIS) to be prepared before the license is granted. An FIS is required if it can be shown that there are endangered fauna on the site to be logged. The Act only pertains to terrestrial vertebrates but has served to raise public awareness and the need to manage logging operations to avoid adverse effects on endangered species. This may result in the application in private forest logging operations of wildlife management prescriptions used in state forests. Prescriptions for the management and conservation of eucalypt forest wildlife fall into three categories. The first is the protection or reservation of critical habitat or resources ranging from the dedication of conservation reserves to the protection of individual trees. The second is the adoption of procedures to create or modify habitats and control predators or competitors. This includes manipulation of fire regimes. The third involves the reintroduction or translocation of fauna and may involve captive breeding. These procedures may be used together.

12.6 HABITAT AND RESOURCE CONSERVATION

The most widely applied prescriptions for the conservation and management of forest fauna in Australia are based on habitat and resource protection. Forest conservation reserves in the form of national parks, flora preserves and nature reserves are an integral part of forest wildlife conservation throughout Australia. Although there is an intent to create a system of reserves throughout Australia that comprehensively samples continental biodiversity and includes areas large enough to be self-sustaining (Australian Academy of Science, 1968; CWA, 1992; RAC, 1992), reserve establishment is seldom based on ecological surveys and most reserves are established for reasons other than nature conservation (Frawley, 1988). Achieving a representative system with adequate replication to protect against catastrophic events and allow for long-term climate change is no longer possible in Australia without significant, and politically unacceptable, social disruption and economic cost. If the goal of conservation is the long-term survival of fauna and the ecosystems they form, it will be necessary to conserve and manage habitats and the resources required by forest fauna on all forested lands irrespective of ownership and competing land uses (Recher, 1985a; Davey, 1989). Wildlife management is especially important where the primary objective is timber production. Not only does logging have a lasting impact on forest ecosystems but also production forests are extensive (Table 12.2), providing important linkages between conservation reserves.

In most forestry operations in Australia 5–25% of the forest cannot be logged (e.g. Horne *et al.*, 1991; Loyn, 1985; Mosley and Costin, 1992; Wardell-Johnson and Nichols, 1991). This is comprised of filtration or buffer strips along water courses and in gullies to protect water quality, scenic corridors along roads and on skylines, steep slopes, areas that are too rocky or that are prone to erosion, and where the timber is not commercially viable or access is difficult (Loyn, 1985; Smith *et al.*, 1992). Unlogged patches or strips of forest of this type contribute to the conservation of forest wildlife but are often small and/or narrow, with high edge to area ratios, and commonly have steep environmental gradients so that each habitat type is of limited extent. Many patches that cannot be logged because of steep or rocky conditions are of low productivity. Forest retained along roads has relatively poor conservation value, with high predation rates on birds and ground-dwelling vertebrates by automobiles (Recher, unpublished data). Despite such defects, these patches and strips are identified by forestry authorities as having high wildlife conservation and management values (e.g. Dobbyns and Ryan, 1983; Loyn, 1985; JSC, 1990; Kavanagh and Bamkin, 1995). Other initiatives taken by forestry authorities for wildlife management are the establishment of management priority areas including wildlife corridors and the retention of habitat trees.

12.6.1 Management priority areas

Commencing with the establishment of Fauna Priority Areas in Western Australia during the early 1970s (Christensen, 1973), all states identify areas of high wildlife value for special management (Drielsma, 1992; Horne *et al.*, 1991; Shields, 1992; Taylor, 1991a,b,c; Smith *et al.*, 1992; Wilson, 1991).

The establishment of Fauna Priority Areas in Western Australia had two major objectives (Christensen, 1973:8):

1. The conservation and management of the total forest environment with particular reference to the fauna.
2. To use the area as a centre of research aimed at establishing the basic principles for sound fauna management in forest areas.

Fauna Priority Areas were established in Western Australia because most forest national parks were small and had limited value for nature conservation. In contrast, state forests were extensive and rich in birds, reptiles and frogs (Christensen, 1973). In the establishment of Fauna Priority Areas, areas of state forest were selected that retained populations of endangered fauna and were exceptionally rich in species. Within these zones, management was adapted to suit wildlife requirements without necessarily excluding logging or other commercial activities. Since then, the most important Fauna Priority Areas have been given nature reserve status.

An important component of wildlife management in Fauna Priority Areas and nature reserves in Western Australia is the diversification of fire regimes to encourage a mosaic of burnt and unburnt vegetation to provide cover and feeding areas for forest vertebrates (Christensen, 1973; Christensen and Abbott, 1989). In a program initiated in the 1950s, 70% of forests in Western Australia are burnt on a regular cycle using cool burns in spring (Wardell-Johnson and Nichols, 1991). Jarrah forests are burnt on a 5–6-year cycle, while the moister karri (*E. diversicolor*) forests are burnt every 7–9 years (Christensen and Abbott, 1989). When using fire for fauna management, fire frequency is varied to provide a broad range of seral stages in the post-fire regeneration. Occasional hot fires are used to encourage the development of shrub thickets by stimulating the germination of seeds (e.g. *Gastrolobium, Acacia*). Following extensive wildfires in January 1994, the New South Wales government has proposed a program of hazard reduction burning similar to that in use in Western Australia.

As in Western Australia, wildlife priority areas in the eastern states are selected because they are unusually rich in species, contain endangered species or species sensitive to the effects of logging, or have unusually high population densities of particular species (Drielsma, 1992; Smith *et al.*, 1992; Taylor, 1991c). The Preferred Management Priority classification (PMP) in New South Wales is analogous to a zoning system such as used by town planning authorities (Drielsma, 1992; Horne *et al.*, 1991). Areas given a PMP classification for flora and fauna protection are judged to require wildlife management prescriptions exceeding those routinely applied during logging operation (Horne *et al.*, 1991). These might include establishing wider filtration strips as wildlife corridors, reserving corridors to link catchments, greater canopy retention in logging operations, and modified fire regimes.

Ideally, zoning for flora and fauna protection is based on pre-logging wildlife surveys designed to sample total forest ecosystems and their associated flora and fauna. In practice the emphasis is on vertebrates, primarily birds and mammals, and specific species. Most surveys are brief (a few weeks or months at best) and are designed to locate rare and endangered fauna and to sample those groups of animals (e.g. arboreal marsupials) known to be sensitive to the effects of logging. Commonly the sampling design precludes quantitative or statistical analysis and long-term monitoring.

In Tasmania, Wildlife Priority Areas (WPAs) totalling 9987 ha have been created for just six species: giant and blind velvet worms (*Tasmanipatus barretti* and *T. anophthalmus*); a marsupial, the Tasmanian bettong (*Bettongia gaimardi*); a bird, the forty-spotted pardalote (*Pardalotus quadragintus*); and two species of fish, the swan and Clarence galaxias (*Galaxias fontanus* and *G. johnstoni*) (Taylor, 1991a,c). A WPA is proposed for the grey goshawk (*Accipiter novaehollandiae*). As in Western Australia and New South Wales, depending on the sensitivity of species and their habitat, logging may be permissible in WPAs. For example, logging is permissible in the velvet worm WPAs but not in those protecting the pardalote. No more than 30% of older age classes can be logged in the bettong WPAs to ensure adequate fruiting of the underground fungus *Mesophellia*, a major food of the bettong. Bettongs and their fungal food resources appear to respond to limited logging and post-logging regeneration burns but are disadvantaged by clearfelling (Taylor, 1991a,c). Only cool regeneration burns are allowed in the velvet worm WPAs.

Boundaries and locations of management priority areas can shift as vegetation matures or the environment changes. For example, foraging habitat for Tasmanian grey goshawks is characterized by a closed canopy and an open subcanopy; regrowth forests are not used (Mooney and Holdsworth, 1991). Blackwood (*Acacia melanoxylon*) swamps are favoured habitats, but these develop as the result of disturbance and are ultimately replaced by tea tree (*Leptospermum*) or rainforest. WPAs for grey goshawk will therefore need to be re-located as existing habitats decline in value and new areas develop. Protection for wedge-tailed eagles in Tasmania centres on their nest trees (Mooney and Holdsworth, 1991). Each pair of eagles may have several nests that they alternate between. The trees selected for nesting are tall and therefore invariably located in old-growth forest. To protect the nest tree from exposure and to prevent disturbing nesting birds, the recommended practice in logging operations is to leave a minimum of 10 ha around each nest tree and preferably more (Mooney and Holdsworth, 1991; Taylor, 1991c). Unfortunately, existing recommendations only allow for the establishment of new nest trees (by reserving appropriate trees and sized patches of forest from logging) within a territory if all existing nest trees have been destroyed (Mooney and Holdsworth, 1991). They are also not binding on private forest owners but rely on cooperation.

12.6.2 Buffer strips and wildlife corridors

Although the long-term value of corridors for the conservation of forest fauna has yet to be demonstrated (Lindenmayer *et al.*, 1993), the delineation of a network of wildlife corridors has become a standard wildlife management procedure in state forests since the 1970s (Purdie, 1990). The retention of unlogged forest along water courses and in gullies is standard erosion mitigation practice in Australian forestry operations (Clinnick, 1985; Cornish, 1989; Purdie, 1990). These buffer or filtration strips are usually < 10–20 m to either side of the water course with selective logging allowed in an outer zone of the wider strips. Although narrow, as an integral and accepted part of forestry operations, this net of minimally logged forest provides a base on which to develop wildlife management prescriptions and has been especially useful in clearfelling operations and plantations (Recher, *et al.*, 1980; Taylor, 1991b). On a different scale, the National Forest Policy recommends that corridors link reserves across a range of altitudes and geographic variation to allow for long-term climate change (CWA, 1992).

In Western Australia about 20% of karri forest blocks are left unlogged and the remainder clearfelled (Anon., 1988; Wardell-Johnson and Nichols, 1991). Forest blocks are about 5000 ha in area. The unlogged forest is left as a network of corridors along streams, rivers and roads linking national parks, nature reserves and forest regeneration of different ages. These management procedures were adopted in the 1970s with the establishment of a karri woodchip industry and the advent of broad area clearfelling (Anon., 1988). River and stream zones were established because of their high biological diversity and recreation value and to protect water quality. Road zones were designed to preserve visual amenities for travellers (Anon., 1988). Corridors of unlogged forest also had value as fire breaks. Initially relatively wide corridors were retained: 100–200 m to either side of streams or rivers and 400 m to either side of roads. Logging was excluded but not all streams or roads had corridors.

The establishment of the Shannon National Park in 1987 reduced the amount of timber available to the timber industry and led to a review of the corridor system (Anon., 1988; Borg *et al.* 1988). Based on research and experience since the 1970s (e.g. Borg *et al.*, 1988; Clinnick, 1985; Recher *et al.*, 1980, 1987), the review concluded that corridor width could be reduced to 100 m to either side of rivers, 50 m on streams and 200 m along roads, but that corridors should be established along streams and some roads that lacked corridors. The objective was to maintain the same total area of forest in stream and amenity corridors but to compensate for the loss of timber from the dedication of the Shannon National Park by allowing selective logging in road reserves and the outer 50 m of corridors along rivers (Anon., 1988). These areas tended to have high volumes of timber. As part of the new procedures, the intensity of logging may be graded from clearfelled areas to the zones from which logging is excluded. In a dissenting view, Wardell-Johnson and Nichols (1991) argued that no evidence had been provided supporting the contention that the original corridors were wider than required for wildlife conservation.

Wardell-Johnson *et al.* (1991) made a number of recommendations for the optimization of wildlife values in the retention of unlogged forest along roads, rivers and streams in Western Australia. These included variation in the width of zones along drainage lines according to their importance as streams so as to extend the reserve system to all drainage lines, the use of 'ecological boundaries' to define stream zones, protection of seepage sites and valley head-waters, linking of stream zones between catchments, reallocation of reserved areas from roadside to streams, no harvesting of timber within designated stream zones and restriction on the movement of machinery across streams, the protection of unique habitats, such as granite outcrops, and the retention of groups of trees and logs within logged areas.

The original river and stream corridors in Western Australia provided a model for the management of wildlife in the Eden Management Area of southeastern New South Wales. State forests in the Eden Management Area have been managed for the integrated production of sawlogs and pulpwood (integrated logging) since 1968 (Anon., 1982; Bridges, 1983). Canopy retention varies but prior to 1991 it was commonly 10% or less. Since 1991, canopy retention rates have varied between 15 and 35% (Kavanagh, personal communication). Logging is by alternate coupes or compartments. The size of logged areas has been variable and subject to change for environmental and political reasons (Anon., 1982; Humphreys, 1977; Recher *et al.*, 1987). The average size of a coupe in 1989 was 42 ha (Mosley and Costin, 1992), but the original logging plans were based on clearfelling 4000 ha blocks separated by unlogged (but prescription burnt) forest of up to 1.6 km in width. The first logging cycle will be completed by 2020, when the forest will be a mosaic of variable sized coupes aged between 0 and 50 years post-logging. Dispersed among these will be corridors and patches of unlogged or substantially unlogged forest. It is estimated that, depending on harvesting practices and the length of cutting cycles, after several cycles 5–30% of the forest in production areas will be in excess of 200 years old (Mosley and Costin, 1992). This is in addition to the 18% (64 000 ha) of the region's forests in conservation reserves (Mosley and Costin, 1992).

The southeast region of New South Wales has a rich fauna with a substantial number of species dependent on mature or old-growth forest or on the components of old growth (JSC, 1990; Jenkins and Recher, 1990; Kavanagh and Bamkin, in press; Pyke and O'Connor, 1990; Recher *et al.*, 1980). Recher *et al.* (1980) recommended the establishment of a system of wildlife corridors based on the filtration strips retained along water courses as a way to retain mature forest throughout the logged area for sensitive or dependent fauna (Figure 12.1). The corridors were intended to link catchments, national parks and nature reserves. Thus, they had several functions: the provision of essential resources (e.g. tree hollows, standing dead timber) for fauna using regenerating forests; facilitating the dispersal of fauna between blocks of unlogged forest and regeneration of different ages; acting as a refuge for wildlife that depended on mature forest; and retaining popu-

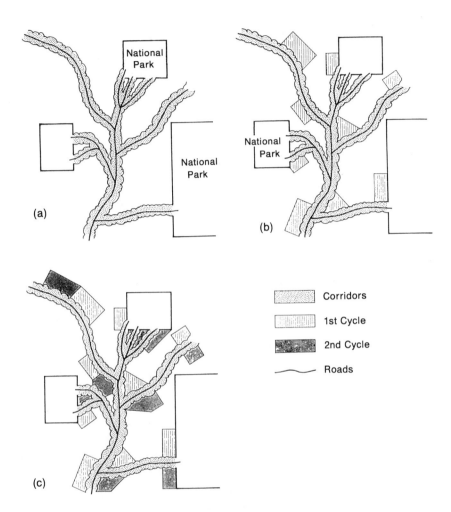

Figure 12.1 (a) The existing system of creek reserves and wildlife corridors at Eden was designed to protect riparian habitats particularly rich in species and to link large areas of mature forest in conservation reserves. (b) To improve the values of the corridors system for wildlife which may not be able to adapt to a linear environment or which require resources found on slopes or along ridges, it was proposed that logging coupes, selected for their wildlife value or because they contained critical resources for wildlife, should be logged on much longer cutting cycles than the remainder of the forest. (c) As logged coupes matured, and developed the resources (e.g. tree holes) required by wildlife, the coupes originally reserved from logging might be logged and replaced by a second set contributing the same management values. (Adapted from Recher *et al.*, 1987.)

lations of plants and animals, including invertebrates, throughout the region to recolonize logged forest as it regenerated.

The retention of strips of forest for fauna conservation in Tasmanian state forests commenced in 1989/90. These are called Wildlife Habitat Strips, to emphasize their importance as habitat rather than implying that they facilitate movement and dispersal (Taylor, 1991b). As in southeastern New South Wales, Wildlife Habitat Strips have several functions: to protect invertebrates that may be currently unknown; to retain all types of old-growth forest throughout production areas; to act as refuges for fauna that will recolonize logged forest as it regenerates; to link larger reserves; and to provide shelter and nesting sites for animals that can otherwise feed in the regenerating forest (Taylor, 1991b).

Corridor width was not specified by Recher *et al.* (1980) but the minimum widths of filter strips in the Eden Management Area at the time of the recommendations was 5 m to either side of the water course where the catchment was < 40 ha, and 20 m where the catchment exceeded 40 ha (Anon, 1982; Purdie, 1990). Instead, Recher *et al.* (1980) recommended that corridor boundaries be based on ecological boundaries. The implementation of these recommendations (Dobbyns and Ryan, 1983) meant that some corridors along streams at low elevations (< 300 m asl) were greater than 100 m in total width but rarely exceeded 40 m at higher elevations. Wider corridors were retained along rivers to protect scenic quality, provide recreation and conserve wildlife (Dobbyns and Ryan, 1983). Buffer zones of unlogged or selectively logged forest were also retained around sensitive vegetation (e.g. rainforest).

The Tasmanian Forest Practice Code (Forestry Commission, Tasmania, 1993) provides for Wildlife Habitat Strips of 100 m width and spaced every 3–5 km. Strips should connect any large areas of forest that will not be logged. As in New South Wales and Western Australia, the Wildlife Habitat Strips are based on the streamside reserve system used to protect water quality. Streamside reserves in Tasmania vary from 20 to 40 m to either side of any river or lake important as a source of drinking water or for recreation. On water courses that only carry water intermittently (class 4 streams) machinery is restricted from approaching closer than 10 m to the stream course, but it is only necessary to retain the undergrowth intact (Forestry Commission, Tasmania, 1993).

The establishment of wildlife corridors along water courses in the Eden Management Area of southeastern New South Wales was emphasized as a means to protect forest habitats with the greatest species diversity and population densities (Recher *et al.*, 1980, 1987, 1991). There is a high correlation in eucalypt forests between foliage nutrients and the richness and abundance of fauna (Braithwaite, 1983; Braithwaite *et al.*, 1983, 1984; Kavanagh and Lambert, 1990; Majer *et al.*, 1992; Recher, 1985a). Foliage nutrients are correlated with soil fertility which differs according to the underlying parent material and which increases from ridge to gully (McColl, 1969; McColl and Humphreys, 1967; Turner and Lambert, 1988). Wildlife corridors retained along water courses protect the greatest number of species but do not fully sample the fauna of the region (Loyn, 1985; Recher *et al.*, 1987). In particular, species restricted to ridges or dry forest are excluded (Claridge *et al.*, 1991). Because they are long and relatively narrow, corridors may disadvantage some species which require the range of resources provided over the entire ecological gradient from ridge to gully, those that have large home ranges or territories, and those with changing seasonal requirements (Lindenmayer *et al.*, 1993; Lindenmayer and Nix, 1993; Recher *et al.*, 1987).

Based on studies of the fauna that survived in corridors of eucalypt forest that had been retained in areas cleared for the establishment of pine plantations, Recher *et al.* (1987) concluded that corridors less than 100 m in total width would not sustain the full complement of forest fauna. Narrower corridors lost forest-dependent birds and mammals and were dominated by edge and open-country species. The effect was greater when corridors were burnt. The results fit the predictions of the theory of central place foraging (Recher *et al.*, 1987) (Figure 12.2). They therefore recommended that the wildlife corridor system in the southeastern forests be modified in the following ways. The minimum total width of corridors should be 100 m and unlogged patches of forest should be retained at intervals along the corridors, creating a system analogous to stringing pearls on a necklace (Figure 12.1). The unlogged patches should be greater than 100 ha to provide broad areas of habitat for species with large home ranges or territories. They would also provide a range of resources from the ridgeline to the gully and allow for some seasonal change in resource availability and location.

Lindenmayer *et al.* (1993) surveyed the abundance of arboreal marsupials in wildlife corridors retained in mountain ash (*E. regnans*) and alpine ash (*E. delegatensis*) forests in the Central Highlands of Victoria. A minimum width of 40 m to either side of a stream is presently retained in production ash forests (Macfarlane and Seebeck, 1991) but the total width of corridors surveyed ranged from 30 to 264 m. Lindenmayer *et al.* (1993) found that the number of animals in corridors was lower than predicted by models of habitat suitability and that species such as Leadbeater's possum and sugar glider, which are social and depend on dispersed food sources (e.g. wattle gum, nectar), were particularly disadvantaged. Corridors with numerous hollow trees and which included a range of habitats from ridge to gully were most likely to support arboreal marsupials. The results of this work support the conclusions and recommendations of Recher *et al.* (1987): to be effective, wildlife corridors cannot be based solely on linear strips but must include wider sections linking habitats across the entire topographic sequence from ridge to gully.

Although there is a policy of not logging within wildlife corridors in the

Figure 12.2 Visual analog of a mathematical model of the Central Place Foraging theory, illustrating the difficulties that some forest vertebrates may encounter if restricted to narrow, long (linear) strips of mature forest (wildlife corridors and wildlife habitat strips) retained in intensively logged production forests. The theory predicts that animals will forage in ways which will maximize energy gain, minimize energy and time expended in foraging, and reduce the risk of predation. For some animals, a linear environment (such as a wildlife corridor) may be too narrow for

southeastern forests, it is not excluded (Recher *et al.*, 1980, 1987; Dobbyns and Ryan, 1983). Unlogged compartments retained along corridors for wildlife would be available for logging after an adjoining logged compartment had matured (Recher *et al.*, 1987) and this pattern is similar to the model proposed by Harris (1984) for the management of Douglas-fir (*Pseudotsuga menziesii*) old growth. The effect was similar to that proposed by Recher *et al.* (1975) who advocated a diversification of the logging cycle throughout the southeast forests for wildlife to provide a mosaic of forest age classes including substantial areas of mature forest. In much of southeastern Australia a forest does not begin to exhibit characteristics necessary for fauna requiring mature trees until more than 150 years following clear-felling (Loyn, 1985).

Corridors and patches of native vegetation retained within plantations are also important for the regional conservation of forest wildlife (Bevege, 1974; Friend, 1980). Retained vegetation is not only important for the conservation of wildlife within plantations but can also be used to link larger areas of native vegetation (Friend, 1980; Recher *et al.*, 1987). Friend (1980) advocated the establishment of corridors of native vegetation in older pine plantations where the original site preparation had cleared all native vegetation. This was necessary to interconnect with other areas of native vegetation. Bevege (1974) argued that up to 30% of the plantation complex should be retained as native vegetation for wildlife conservation and that at least half of this should be within plantation boundaries.

The retention, restoration and management of native vegetation within pine plantations has proven difficult and of limited value. The clearing of forest for plantation establishment exposes corridors and retained patches to wind and increased dryness. Windthrow, particularly of the larger and older trees, is common and surviving trees are affected by the death of terminal vegetation (Recher, personal observation). Eucalypts are also sensitive to the herbicides used to control regrowth in the plantation area and herbicide-killed trees are common in pine plantations in New South Wales (Recher, personal observation). The narrowness of corridors and the small size of retained patches increases the impact of accidental and prescribed fires on wildlife and facilitates colonization by edge species (Recher *et al.*, 1987). These problems could be minimized by the retention or establishment of wider corridors or bigger patches, but this has proven difficult to implement in plantations where the priority is placed on wood production.

---

efficient foraging or lack the full range of resources required. Among the animals likely to be adversely affected by linear environments are those with dispersed or uncommon food resources, those that require resources found along the entire topographic gradient from ridge to gully, and social or colonial animals. (Adapted from Recher *et al.*, 1987.)

Foresters have objected to retaining or adding corridors of native vegetation in pine plantations for two principal reasons: cost (including lost production) and increased risk of fire. Current policies of not clearing native forest for plantation establishment are more effective for wildlife conservation but do not extend to private land, including that purchased by state forest authorities for plantation establishment.

12.6.3   Habitat trees

The large number of vertebrates that are obligate or casual tree-hollow users has led to the identification of hollow trees as an important forest resource for wildlife. Such trees, called 'habitat trees' (Recher *et al.*, 1980), are abundant in eucalypt forests (Table 12.4). Although habitat trees feature prominently in wildlife management prescriptions, guidelines in states other than Tasmania (Forestry Commission, Tasmania, 1993) are vague (Purdie, 1990). However, logging prescriptions in individual management districts often nominate the retention of a specified density of habitat trees. This is essential in clearfelling and Timber Stand Improvement (TSI) operations where habitat trees are cut as pulpwood or culled, but in all instances the numbers retained are much lower than in the original forest and are based on the number of trees and not on the number of hollows. For example, until 1991 in the Eden Management Area of southeastern New South Wales the prescription was five habitat trees per 15 ha (Anon., 1982; Purdie, 1990) but Fanning and Mills (1989) recorded 10 hollow-bearing trees per hectare in a pre-logging survey of one section of forest (Table 12.4). Working in 10–15-year-old regeneration on the south coast of New South Wales, Smith (1985) found an average of five habitat trees per hectare retained along ridges and 12/ha in gullies, compared with 61 and 40/ha respectively in mature forest. In 1991, new guidelines for habitat tree retention were introduced in the Eden Management Area. These guidelines place forest types into three categories of quality – good, moderate and low – for forest fauna dependent on tree hollows for nesting or denning. The recommended retention rate for habitat trees is 5/ha in good quality habitat (e.g. peppermint types), 3/ha in habitat of moderate quality (e.g. messmate types), and 1/ha in low quality habitat (e.g. stringybark types) (Forestry Commission of New South Wales, 1991).

Smith *et al.* (1992) found that the abundance of habitat trees in northern New South Wales differed between forest types, averaging from 4/ha in rainforest to 17/ha in messmate (*E. obliqua*) forest. They recommended a minimum retention of four well-spaced trees per hectare in low site quality forest and six in high site quality forest. Current logging prescriptions in northern New South Wales are variable but most require the retention of fewer than three habitat trees per hectare. Given that the number of hollows used by fauna in an individual tree can vary from one to more than

Table 12.4 Examples of the number of hollows and hollow-bearing trees in Australian forests and woodlands (adapted from Taylor, 1991c)

Density of hollows (No./ha)	Trees with hollows (No./ha)	Location	Forest type
–	3–7	Northeast Victoria	Open forest
16[a]	8.5	Southwest Western Australia	Salmon gum woodland
–	10	Central Victoria	Tall open forest
137	–	Central Victoria	Tall open (ash) forest
–	2.5[b]	Central Victoria	Tall open (ash) forest
–	4.5–5.2[c]	Central Victoria	Tall open (ash) forest
–	1.3–7.3[d]	Central Victoria	Tall open (ash) forest
–	1.4	Central Victoria	Tall open (ash) forest
73	–	West-central Victoria	Open forest
–	18[e]	Southeastern Victoria	Open forest
116[e]	62	Eastern Victoria	Open forest
140[f]	48	Northeastern Tasmania	Dry *E. obliqua* forest
–	10	Southeastern NSW	Open forest
–	61	Southeastern NSW	Open forest
–	40	Southeastern NSW	Tall open forest
–	4	Northeastern NSW	Rainforest
–	17	Northeastern NSW	Open *E. obliqua* forest

[a]Only includes hollows greater than 90 mm diameter.
[b]Area subjected to intense fires in 1939; most hollow trees are stags.
[c]Area subjected to fires and selective logging.
[d]Potential nest trees (≥ 6 m in height and ≥ 0.5 m DHB).
[e]Area subjected to light selective logging.
[f]'Apparent hollows' assessed from the ground; small hollows are included.

30, according to the size of the tree and the number of holes available (Mackowski, 1984), it is clear that prescriptions based on the retention of individual trees are of minimal value. However, Taylor and Haseler (1993) found that the number of hollows prior to logging sclerophyll forest in northeast Tasmania was far greater than required by the number of hollow-nesting birds. Although not all hollow-nesting species nested in streamside areas, the number of hollows retained in unlogged forest along water courses after logging would be sufficient for the number of hollow-nesting bird present before logging.

Some effort is made in New South Wales to retain habitat trees in a cluster with other trees but there are no specific guidelines in logging prescriptions. In Tasmania the procedure is to retain clumps of trees, each of which has several habitat trees. One such clump is retained in every 4–6 ha of forest (Taylor, 1991c).

Because of territorial behaviour, habitat trees that are clustered are less likely to be occupied by hollow-dependent fauna than those with a more regular or random distribution (Ambrose, 1982; Saunders *et al.*, 1982; Lindenmayer *et al.*, 1990). Smith and Lindenmayer (1988) argue for the need to space habitat trees throughout the logging area to avoid the situation in which the available nest or den hollows are contained within a single territory. However, there is also an argument for retaining trees in clumps, as in Tasmania, to provide a habitat cluster within which habitat trees are sheltered by adjacent trees. A cluster of trees also provides a succession of younger trees to mature as habitat trees. Ideally, habitat clusters would be spaced evenly throughout the forest but, to be effective, the spacing must reflect the ecology and behaviour of the hollow-dependent fauna in the region. As noted by Taylor and Haseler (1993), not all hollow-nesting birds nest near streams. Retention of habitat trees along the gradient from ridge to gully therefore appears necessary. Apart from retaining clumps or clusters of trees of different ages, there are no logging prescriptions that identify trees to be retained for maturation as habitat trees. Nor do current guidelines specify the retention of particular types or sizes of hollows or, apart from Tasmania, how these should be spaced throughout the forest. In selective logging operations, as distinct from clearfelling, logging prescriptions are less precise. Defective trees are not felled for reasons of cost and safety and it is argued that adequate numbers for wildlife are therefore retained. Moreover, there is an assumption that defective trees which will develop hollows are inevitable and that special prescriptions are not required to ensure their recruitment (Loyn, 1985).

12.7 HABITAT MANIPULATION AND PREDATOR CONTROL

There have been few attempts to manipulate forest habitats for wildlife in eastern Australia. In Western Australia fire is used to manipulate the density and distribution of shrub vegetation for ground-dwelling mammals and to protect logs for the numbat (*Myrmecobius fasciatus*) (Burrows *et al.*, 1987; Christensen, 1973; Wardell-Johnson and Nichols, 1991). The manipulation of fire regimes is accompanied by poisoning programs to control foxes, the major predator on native mammals in these habitats. Better control or eradication of the European fox in Western Australia would reduce the need to manipulate habitats using fire (Wardell-Johnson, personal communication).

Smith *et al.* (1992) recommend fox control as an adjunct to logging operations in northern New South Wales. They felt that the roading associated with logging may enable foxes to penetrate forest habitats from which they would otherwise be excluded and predation by foxes on ground-dwelling or nesting fauna may be a major reason for the decline of these animals in logging areas. Although they lacked conclusive data, Claridge *et al.* (1991) thought that disturbance associated with logging led to increased predation on the bandicoots *Isoodon obesulus* and *Perameles nasuta* by foxes and dogs in southeastern New South Wales. Smith *et al.* (1992) recommended the closure and revegetation of logging roads to exclude foxes after logging is finished. Claridge *et al.* (1991) recommended the use of fire to increase the area of habitat in logging areas for bandicoots and to include habitat on slopes with the wildlife corridor system. Appropriate fire regimes would provide cover against predators and promote the development of the fungal food resources used by bandicoots.

Artificial nest boxes are used by dependent fauna and may prove useful under special circumstances (e.g. endangered populations), if the cost can be justified (Davey, 1989). However, their use has been limited and largely experimental (Suckling and Macfarlane, 1983; Curry and Nichols, 1986).

12.7.1 Modification of logging regimes

Although not used to modify or create habitat for forest wildlife, logging prescriptions are frequently modified to lessen the impact of logging on fauna. Among the procedures used or recommended are longer logging rotations (Anon., 1988; Dickman, 1991; Loyn, 1985; Recher *et al.*, 1975, 1980, 1987), alternating logged and unlogged coupes (Anon., 1982), dispersing logged areas among unlogged or more mature forest (Dickman, 1991), smaller and/or larger logging coupes (Anon., 1988; Dickman, 1991; Humphreys, 1977; Kavanagh *et al.*, 1985; Recher *et al.*, 1987), selective logging, and retention of a specified proportion of trees or canopy cover (Anon., 1988; Smith *et al.*, 1992). Many of these prescriptions are used for silvicultural reasons (e.g. retention of seed and shelter trees, selective logging), for soil conservation (e.g. small and dispersed logging coupes), or to reduce the long-term economic costs of logging (e.g. alternate large coupes). The recommended length of logging cycles for wildlife management is usually based on the time required for the development of tree hollows and ranges from 80 to 500 year, according to the tree species and the requirements of fauna. Obviously the development of large hollows suitable for large mammals, forest owls and cockatoos requires more time than hollows suitable for the smaller vertebrates. However, this may not always be the case, as is shown by studies of small (< 10 g) insectivorous forest bats in an intensively logged forest. Breeding females fly long distances to roost in large (> 80 cm DBH) old trees in unlogged gullies (Lunney *et al.*, 1985, 1988; Taylor and Savva, 1988). The development of hollows suitable for these small mammals would require as long as those for much larger vertebrates.

Unfortunately, there are few data on the effects of differing levels of canopy retention and none on longer logging rotations. Dunning and Smith (in Smith *et al.*, 1992) found a reduction in greater glider numbers in proportion to the amount of canopy and basal area removed. In a comparison of different levels of canopy retention at Waratah Creek in southeastern New South Wales, arboreal mammals were adversely affected by less than 40% canopy retention (Kavanagh, personal communication). In a study of different sized logging coupes, Kavanagh *et al.* (1985) concluded that coupes of less than 20 ha disadvantaged species of birds with large home ranges or territories. They recommended a minimum area of 50–100 ha to provide sufficient habitat as regeneration proceeded to allow these species to recolonize. Taylor and Haseler (in review) found that the retention of a proportion of older trees can ameliorate the effects of logging on forest birds in Tasmania. However, a few species are adversely affected even if a large proportion of the mature canopy is retained.

Studies on the effects of different fire frequencies are also difficult to interpret and ambiguous. Most workers agree that there is an initial adverse effect on small ground-dwelling mammals and on birds that require understory or shrub vegetation (Christensen and Abbott, 1989; Dickman, 1991; Recher and Christensen, 1981). Populations recover rapidly and there is no evidence of permanent changes to forest fauna from repetitive (~5–6-year intervals) low intensity fires (Christensen and Abbott, 1989). However, some shrub species, especially those that are slow to mature, are killed by fire and regenerate from seed (e.g. heath-leaved banksia, *Banksia ericifolia*), are disadvantaged by frequent fires (Recher and Christensen, 1981). Consequently, frequent prescription burning leads to changes in the composition and structure of the shrub layer. Catting (1991), in a review of the literature, pointed out that low intensity fires reduced forest structure (i.e. number of vegetation layers), while high intensity fires increased forest structure by promoting the germination and growth of understory shrubs. Frequent fires also prevent the build-up of litter and large woody debris with adverse effects on ground-dwelling mammals (e.g. dusky antechinus) (Lunney *et al.*, 1987; Recher and Lunney, unpublished data). Catling (1991) listed 25 species of native forest mammal that are disadvantaged by frequent, low intensity burns (i.e. prescription burns) and 13 species that benefit from such a fire regime. In addition, one introduced mammal is disadvantaged while 12 others, including European fox and feral cat (*Felis domesticus*) benefit.

Unfortunately, all studies of the effects of prescription burning on forest vertebrates have been of limited duration and none has investigated the effects of repetitive fires at the same site. Moreover, all commenced after prescription burning had been practised for some years and may have studied a fauna that had already come to equilibrium with the imposed fire regime. Despite the limitations of the available data, Catling's (1991) recommendation that prescription burning regimes be modified to include the occasional hot, high intensity fire merits consideration.

12.7.2 Restoration and rehabilitation of habitats

Although reafforestation has yet to be significant in Australia, many conservationists advocate the establishment of hardwood (eucalypt) plantations
to reduce the need to log native forests. Extensive eucalypt plantations have
been established in Victoria and Western Australia, but plantations of either
eucalypts or exotic softwoods provide few of the resources required by
fauna dependent on mixed age and old-growth forests and support only a
small proportion of the original forest fauna (Friend, 1980; Recher, 1982;
Woinarski, 1979). The value of eucalypt plantations is further reduced when
exotic species are planted, as in Western Australia where Tasmanian blue
gum (*E. globulus*) is used.

Apart from plantations, forest restoration and rehabilitation are primarily
associated with reafforestation of mine sites. The most significant area of
eucalypt forest affected by mining in Australia is in the northern jarrah
forest of Western Australia where there are extensive deposits of bauxite
(Wardell-Johnson and Nichols, 1991). Mine sites are from 2 to 20 ha in
extent and remain surrounded by forest (Curry and Nichols, 1986). Various
rehabilitation procedures have been followed but the intention has been to
establish species resistant to the pathogenic root fungus *Phytophthora
cinnamoni*. Jarrah and many understory trees and shrubs are killed by
Phytophthora. The opportunity has therefore been taken to introduce commercially useful timber species, usually species of eucalypts from eastern
Australia that are *Phytophthora* resistant. Recolonization and use of the re-
vegetated sites by fauna is rapid, although tree-hole nesters may require the
provision of artificial nest sites (Curry and Nichols, 1986; Nichols *et al.*,
1989; Ward *et al.*, 1990), but the studies published do not provide information on the extent to which exotic eucalypts are used.

12.8 REINTRODUCTIONS

There have been few attempts to reintroduce fauna into forests where they
have become extinct. In Western Australia the numbat has been successfully
reintroduced into several nature reserves after the implementation of fox
control programs (Friend, 1987, 1990). Koalas (*Phascolarctos cinereus*) have
been successfully reintroduced into suitable habitats in Victoria (Martin and
Handasyde, 1990) but efforts to reintroduce parma wallaby (*Macropus
parma*) in New South Wales have failed due to fox predation. All reintroduction programs for ground-dwelling mammals are now founded on the
need to control or eradicate foxes and feral cats before and during reintroductions and a national program has been established to develop a biological control of foxes. For koalas, an arboreal species that spends little time
on the ground, reintroduction programs emphasize habitat suitability and
restoration along with disease control (Lunney *et al.*, 1990).

12.9 CONCLUSIONS

The status of Australia's eucalypt forest vertebrate fauna and the effective-ness of current management prescriptions can only be evaluated in the context that half of the continent's original forest has been cleared and the remainder is fragmented. The greatest part of the remaining forest (> 95%) is affected by logging, mining, roads, power corridors, changed fire regimes, introduced plants and animals, altered hydrological cycles, grazing by domestic animals and recreational activities. Only 19% of eucalypt forest is unlogged, representing less than 10% of the pre-European area of forest, and half of this is low site quality, dry open-forest (Table 12.3).

It also needs to be recognized that these are not recent changes. Writing a century ago, Hamilton (1892) estimated that a third of the forests in New South Wales had been destroyed and that much of the remainder had been degraded in other ways by settlement. Recognition of the need to protect forests occurred as early as 1801 when Governor King issued a General Order prohibiting the felling of cedar (Carron, 1985). The need to protect forests and wildlife was being debated in Parliament by the 1860s (Carron, 1985; Reed, 1991). From 1882 to 1920 forestry departments were progres-sively established by the states and forests reserved for timber production against the objections of agricultural interests (Carron, 1985; Frawley, 1988). The extent of clearing and change to forest environments during the nineteenth century suggests that the impact on forest wildlife was also early and extensive (Lunney and Leary, 1988; Recher and Lim, 1990; Recher *et al.*, 1993). While the impact of Europeans on forest ecosystems commenced early in the history of the colony, it has proceeded through the nineteenth and twentieth centuries with clearing for agriculture continuing unabated (RAC, 1992). At the same time, forestry practices (e.g. logging, plantation establishment, hazard reduction burning) have become increasingly intensive (e.g. the adoption of clearfelling for pulpwood production), as well as more extensive.

Quantitative studies of the ecology of forest fauna and the effects of forest management on forest ecosystems only commenced in the 1960s. Most work is more recent and unlikely to detect changes that occurred a century or more earlier. Changed patterns of abundance and distribution accompanied by the loss of populations and species as a result of human activities occurs much more rapidly than is generally appreciated. Viewing the Australian vertebrate fauna as a whole, Recher and Lim (1990) pointed out that significant changes in the fauna occurred within 20 years of a region being settled and either went unnoticed or were quickly forgotten by settlers. There is no reason to believe that the impact of forest management on forest ecosystems has operated more slowly or that foresters and the timber industry are any more aware of the changes than farmers and graziers were aware of the changes that occurred in agricultural districts.

Studies, such as many of those reviewed by Christensen and Abbott (1989), that do not find an effect of forest management – in this case prescribed burns – may be studying an ecosystem that has already changed and a new equilibrium has been established. Conversely a major factor in forest management is the repetitive nature of many operations. The effects of prescription burns or of logging may not be evident until after a number of cycles. In Western Australia, where dry forests are burnt on a 5–6-year cycle, there have only been seven or eight burns since broad area prescription burning was adopted in the 1950s. Permanent changes in the flora and fauna or in ecosystem processes and functions may not be evident until after a considerably longer period. Until the time scale on which effects occur is determined, forest management should be conservative.

Loyn (1985) identified three strategies of potential value for fauna conservation in forest managed primarily for timber production: the conservation of old trees, long cutting cycles or logging rotations, and the retention of high quality wildlife habitats with little or no harvesting or very long logging rotations, while the remainder of the forest was managed intensively for timber production. He considered the latter to be most attractive as it maximized faunal values with the least cost to production. He contended that the first two options involved substantial losses in production without necessarily providing the requirements of sensitive fauna or those dependent on relatively large areas of mature forest and old growth. Shaw (1983) emphasized a flexible approach to wildlife management and suggested that priorities should be assigned to species. Loyn (1985) agreed that management prescriptions would vary between forest types or to suit the requirements of particular species.

Davey (1989) emphasized the need to make wildlife management an integral part of forest management, while Norton and Lindenmayer (1991) called for 'a coherent strategy to integrate the management of wildlife across all forested lands'. According to Davey, this requires regional and national coordination and a high level of flexibility. His priorities were to maintain viable populations of forest wildlife without significant economic costs or losses in timber production. The procedures used to achieve this would differ according to the requirements of species, forest type, production systems and cost. Davey (1989) contrasted his integrated approach to managing forests 'solely on a zone basis', which he argued is inappropriate. Nonetheless, a system of zones is the prevailing basis for wildlife management in Australian forests.

Current wildlife management prescriptions in production forests emphasize special management zones including wildlife corridors and the preservation of habitat trees. These are supplementary to a conservation reserve system that is intended to sample the biological diversity of forests and provide sufficient habitat to conserve most, if not all, forest fauna. In effect, conservation reserves, such as national parks, are just large 'zones'. As argued by Davey (1989), it is unlikely that zoning forests and allocating a different use to each zone will succeed in conserving forest diversity. The dominant view among Australian biologists is that the conservation reserve system is inadequate and unlikely ever to be expanded to the point where it can meet its nature conservation objectives (Hall, 1988; Kirkpatrick *et al.*, 1990; Lunney and Recher, 1986; Pressey, 1990; Recher, 1976a, 1990, 1994; among others).

I agree with Davey that wildlife conservation needs to be integrated with other forms of forest and land use at regional and national levels (Recher, 1985b). This includes extending wildlife management practices to private forest land. Integration across all land tenures is clearly an objective of the National Forest Policy which states that national goals for the conservation and management of the forest estate 'should be pursued within a regionally based planning framework that integrates environmental and commercial objectives' (CWA, 1992: 5).

Without the cooperation of all authorities and forest owners in planning and implementing management strategies, it is unlikely that national or state goals for the conservation of forest biodiversity can be achieved. The absence to date of appropriate and tangible levels of cooperation between different land management authorities, the timber industry, agricultural interests and private land owners is perhaps the most significant problem facing the conservation and management of forest vertebrates in Australia. The absence of coordinated management across all forested land and the attendant conflict between different users of the forests has significantly hampered the application of wildlife management prescriptions and has prevented rationalization, improvement and extension of the forest conservation reserve system throughout Australia. There is little gain in knowing how to manage forest wildlife in production forests if prescriptions are not implemented or are only partially implemented (see Mooney and Holdsworth, 1991, for examples). A measure of the difficulties is seen in the willingness of state governments to adopt a national forest policy, but being unwilling (or unable) to implement its provisions.

12.9.1 Future directions

If Australia is to conserve its forest fauna, then the management procedures adopted must maintain viable populations through a species distribution and retain or restore the fundamental structure and organization of forest communities within which species exist. As expressed by the Resource Assessment Commission (RAC, 1992):

> Any attempt to maintain the biological and genetic diversity of the forest estate must seek to maintain, through time, all forest types as well as populations of their component species. In addition, populations of each

species must be large enough to encompass genetic diversity within the species and to keep the risk of extinction of each species at acceptable low levels.

Including conservation reserves, management must be flexible and capable of rapid change as new information on the resource requirements of wildlife becomes available. They must also be responsive to medium- and long-term changes in climate as well as catastrophic events. This does not mean that some populations cannot decrease in abundance while others increase or that forest ecosystems are immutable. What is important is that basic ecological processes and functions are maintained. With these caveats, timber harvesting and other commercial activities in native eucalypt forests need not be incompatible with wildlife conservation.

Broad wildlife corridors, management priority areas where logging and fire regimes are adjusted to wildlife requirements, habitat tree management and conservation reserves are the basis of forest conservation and management. As stated in the National Forest Policy, these need to be managed in a regional context across all land tenures irrespective of state or administrative boundaries. There is also a need to accelerate programs to restore and rehabilitate degraded forest and agricultural lands with the objective of expanding the forest estate and increasing national self-sufficiency in wood production without having to log native forests on a non-sustainable basis. Recher (1985b) emphasized the need for integration and flexibility in management, while Smith (1991) argued for the need to broaden forest policy to give adequate attention to non-timber values. Norton and Lindenmayer (1991) emphasized changes to forest management based on a more conservative use of native forest, including protection of remaining old-growth forests on fertile soils. It is clear that a reduction in the amount of timber harvested from eucalypt forests in Australia is required to implement sound wildlife management practices and to place forestry on an ecologically sustainable basis.

12.9.2 Changes in forest management

Dickman (1991) identified four areas where changes in the management of production were required for fauna conservation: extension of logging cycles; wide separation of logging areas to prevent the formation of large uniform stands of regrowth; reserve, or harvest on very long cycle, areas known to be rich in species; and restrict the use of prescribed and post-logging burns to a minimum. His recommendations can be amplified and extended.

Within the system of forest reserves, corridors and management zones, special attention needs to be given to retaining all critical resources. Moreover, reserves and management areas need to have flexible boundaries with long-range contingency planning to allow for short-term catastrophic events, such as wildfire or storm, and long-term climatic change. The importance of habitat trees cannot be disputed but existing prescriptions for habitat tree retention allow for neither the diversity of tree hollows required by forest fauna nor the replacement of trees as they die and fall. The emphasis on habitat trees fails to recognize the importance of other forest resources for fauna (e.g. standing dead wood, logs) that are not necessarily provided by habitat tree retention (Dickman, 1991; Recher, 1991). Rather than adopt a plethora of prescriptions, each designed for a single species or type of resource, it would be beneficial to adopt management strategies that emphasize diversification of the total forest environment. Thus, in addition to habitat tree management, variable logging intensities and fire regimes (including the occasional high intensity burn) that promote the development or retention of a variety of resources may be both cost effective and efficient for wildlife conservation. Some adverse effects of forest management can be mitigated by the elimination of grazing by domestic stock and commercial apiaries and by predator control.

Prescriptions for habitat tree retention need to be modified to include recommendations on the distribution, clustering and retention of trees for maturation as future habitat trees. Probably it is also necessary to establish criteria for habitat tree retention based on the size and position of tree holes, the size and species of habitat trees, the location of habitat trees in relation to ridge and gully, and the number of tree holes per unit area. Such prescriptions will need to be applied in all logging operations regardless of logging intensity and must be tailored to suit the specific requirements of the local fauna. Without question, too few trees and tree holes are retained under current prescriptions (Table 12.4).

Prescriptions for wildlife corridors and habitat strips also need to be revised and studies established to monitor their long-term effectiveness for fauna conservation. It is clear that linear strips restricted to single parts of the topographic sequence (e.g. gully or ridge) are inadequate (Recher *et al.*, 1987; Lindenmayer *et al.*, 1993; Lindenmayer and Nix, 1993). Modifications similar to those recommended by Recher *et al.* (1987) and Claridge *et al.* (1991) need to be adopted urgently and existing corridors and strips extended to include the full range of resources required by wildlife and to sample the entire forest fauna from ridge to gully. Although width may not fully explain all the variations in wildlife abundances in corridors (Lindenmayer *et al.*, 1993; Lindenmayer and Nix, 1993), it is evident that most corridors and strips of mature forest being retained for wildlife are too narrow. Minimum viable width appears to exceed 100 m (Recher *et al.*, 1987; Lindenmayer *et al.*, 1993; Lindenmayer and Nix, 1993) but minimum viable widths need to be established on a regional basis. Logging should be excluded from strips and corridors retained for wildlife or managed specifically to enhance wildlife values on selective and long-term logging cycles (e.g. to maintain a mixed age forest structure). Within corridors and strips, fire regimes need to be managed to ensure the retention of forest structure. This requires occasional high intensity burns.

There is one initiative in the conservation and management of forest fauna that needs to be implemented urgently. Until the long-term effectiveness of wildlife management prescriptions can be demonstrated and the long-term effects of logging are understood, as well as for the cultural and aesthetic reasons set out in the reports by the Forest Working Group (FWG, 1991) and the Resource Assessment Commission (1992) and adopted in the National Forest Policy (CWA, 1992), remnant old-growth forest should not be logged. Recognizing that there are significant qualitative differences between the old growth in reserves and that remaining in production forests (Table 12.3), the environmental cost in terms of lost ecosystems and wildlife as old-growth logging continues will be enormous and irreversible. Retaining old-growth ecosystems may prove of critical importance to the long-term survival of many forest animals, but with the limited and highly fragmented area of old growth remaining in Australia there is little room for error and only a limited time in which to act. Virtually all old growth outside the existing conservation reserve system will be logged or significantly affected by logging within the next 10 to 20 years depending on market demands and economic growth. If the present scale and rate of logging continues, by 2015 there will be no unlogged forest remaining in production areas other than that retained in streamside reserves, special prescription areas, wildlife corridors and habitat strips (Lunney, 1991a; RAC, 1992).

It is not enough to establish a reserve system and manage logging regimes for the needs of forest fauna. The effectiveness of all management prescriptions, as well as the effects on forest fauna of logging, grazing and fire, needs to be monitored and constantly reviewed (Davey, 1989; Dickman, 1991; Macfarlane, 1988). Especially, there is an important need for detailed surveys of forest fauna (including invertebrates) and for long-term studies. Such studies must result in recommendations for the conservation and management of the forest resource (Lunney, 1991a). Whatever prescriptions are adopted, the limited data on the ecology of forest ecosystems and the long-term effects of management means that prescriptions will inevitably change as new information comes to hand. This should not be taken as an excuse for doing nothing, but accepted as a challenge.

12.10 SUMMARY

Australia is a sparsely forested continent and within 200 years of European settlement half of the original forest had been cleared. Most of the forest that remains is fragmented and affected by logging, grazing, changed fire regimes and introduced flora and fauna. The continued clearing of forest for agriculture remains a problem. Although substantial areas of forest, including large areas of old growth and rainforest, are protected as conservation reserves, these do not sample all forest types nor guarantee the survival of all forest vertebrates. Local and regional extinctions have been common and many vertebrate species are threatened by the intensification of forestry practices and continued forest clearing. The conservation of forest vertebrates therefore requires a flexible and integrated approach to management on a national level across all forested lands irrespective of tenure, administrative systems or primary use. The absence to date of appropriate and tangible levels of cooperation between different land management authorities, the timber industry, agricultural interests and private land owners is perhaps the most significant problem facing the conservation and management of forest vertebrates in Australia.

Wildlife management in timber production areas has emphasized the retention of wildlife habitat as fauna management priority zones, the provision of wildlife corridors, and the protection of hollow trees used by fauna for shelter and nesting. These are supplemented by modified logging and fire prescriptions to reduce adverse effects on fauna. Despite these initiatives, further modifications of forestry practices are required to make wildlife management and conservation an integral part of forest management. Pending the determination of long-term effects of logging on forest ecosystems, logging in remnant old-growth forest should be suspended. The boundaries of conservation reserves and management priority areas should be relaxed so that they can be adjusted periodically to changing climatic and environmental conditions and the requirements of fauna. Logging and fire regimes should be further diversified to provide specific resources for wildlife and ensure sufficient areas of mature forest containing large numbers of trees with hollows. Particularly important is the lengthening of cutting cycles in forest managed for pulpwood production and where clear-felling is practised. It will also be necessary to apply wildlife management prescriptions to all forestry operations regardless of land tenure. Such approaches, when combined with the control of introduced predators, could ensure the survival of Australia's forest fauna. However, long-term research and the monitoring of the effects of management are required to ensure rapid response to changing conditions and to continue the refinement of management prescriptions.

12.11   ACKNOWLEDGEMENTS

Michael Andren, Jennifer Gervais, Rod Kavanagh, Dan Lunney, Robert Taylor and Grant Wardell-Johnson provided useful and critical comments on the manuscript. I appreciate their advice, although it has not always been followed. Dan Lunney made available unpublished data on the status of the forest fauna of New South Wales. Myella Recher assisted with the typing and helped greatly by tabulating data on the ecology of forest vertebrates.

12.12 REFERENCES

Abbott, I. and Van Heurck, P. (1985) Response of bird populations in jarrah and marri forest in Western Australia following removal of half the canopy of the jarrah forest. *Australian Forestry* **48**, 227–34.

Ambrose, G.J. (1982) *An Ecological and Behavioural Study of Vertebrates Using Hollows in Eucalypt Branches.* Ph.D. Thesis, La Trobe University, Victoria.

Anon. (1982) *Eden Native Forest Management Plan.* Forestry Commission of NSW, Sydney.

Anon. (1988) *The Road, River and Stream Zone System in the Southern Forest of Western Australia: A Review.* Dept. of Conservation and Land Management, Perth.

Australian Academy of Science (1968) *National Parks and Nature Reserves in Australia.* Report No. 9, Australian Academy of Science, Canberra.

Australian Bureau of Statistics (1990) *Natural Resources and Environmental Accounting in the National Accounts.* National Accounts March Quarter, Canberra.

Bevege, D.I. (1974) Pine plantations: farm or forest? *Proceedings 7th Triennial Conference IFA,* Caloundra. pp. 435–43.

Borg, H., Hordacre, A. and Batini, F. (1988) Effects of logging in stream and river buffers on watercourses and water quality in the southern forest of Western Australia. *Australian Forestry* **51**, 98–105.

Braithwaite, L.W. (1983) Studies on the arboreal marsupial fauna of eucalypt forests being harvested for woodpulp at Eden, NSW. I. The species and distribution of animals. *Australian Wildlife Research* **10**, 219–29.

Braithwaite, L.W., Dudzinski, M.L. and Turner, J. (1983) Studies on the arboreal marsupial fauna of eucalypt forests being harvested for woodpulp at Eden, NSW. II. Relationship between the fauna density, richness and diversity, and measured variables of the habitat. *Australian Wildlife Research* **10**, 231–47.

Braithwaite, L.W., Turner, J. and Kelly, J. (1984) Studies on the arboreal marsupial fauna of eucalypt forests being harvested for woodpulp at Eden, NSW. III. Relationship between the fauna densities, eucalypt occurrence and foliage nutrients, and soil parent materials. *Australian Wildlife Research* **11**, 41–8.

Bridges, R.G. (1983) *Integrated Logging and Regeneration in the Silvertop Ash–Stringybark Forests of the Eden Region.* Research Paper No. 2, Forestry Commission of New South Wales, Sydney.

Bureau of Rural Resources (1990) *Intensive Harvesting of Native Eucalypt Forests in the Temperate Regions of Australia: Environmental Considerations for Sustainable Development.* Submission to the Resource Assessment Commission Inquiry into Australia's Forest and Timber Resources, Department of Primary Industries and Energy, Canberra, ACT.

Burrows, N.D., McCaw, W.L. and Maisey, K.G. (1987) Planning for fire management in Dryandra forest, in *Nature Conservation: The Role of Remnants of Native Vegetation,* (eds D.A. Saunders, G.W. Arnold, A.A. Burbidge and A.J.M. Hopkins), pp. 305–312. Surrey Beatty & Sons, Chipping Norton.

Carron, L.T. (1985) *A History of Forestry in Australia.* Australian National University Press, Canberra.

Catling, P. (1991) Ecological effects of prescribed burning practices on the mammals of southeastern Australia. *Conservation of Australia's Forest Fauna,* (ed. D. Lunney), pp. 353–364. Royal Zoological Society of New South Wales, Mosman.

Christensen, P. (1973) Focus on a new concept in forestry – Fauna Priority Areas. *Forest Focus* 10, 2–10.

Christensen, P. (1980a) *The Biology of* Bettongia penicillata *Gray, 1837, and* Macropus eugenii *Desmarest, 1817, in Relation to Fire.* Forests Department of Western Australia, Bulletin No. 19.

Christensen, P. (1980b) A sad day for native fauna. *Forest Focus* 23, 3–12.

Christensen, P. and Abbott, I. (1989) Impact of fire in the eucalypt forest ecosystem of southern Western Australia: a critical review. *Australian Forestry* 52, 103–121.

Claridge, A., McNee, A., Tanton, M. and Davey, S. (1991) Ecology of bandicoots in undisturbed forest adjacent to recently felled logging coupes: a case study from the Eden Woodchip Agreement Area, *Conservation of Australia's Forest Fauna,* (ed. D. Lunney), pp. 331–346. Royal Zoological Society of New South Wales, Mosman.

Clinnick, P.F. (1985) Buffer strip management in forest operations: a review. *Australian Forestry* 48, 34–45.

Cogger, H.G. (1992) *Reptiles and Amphibians of Australia.* Reed Books, Chatswood. 5th edn.

Commonwealth of Australia (CWA) (1992) *National Forest Policy Statement.* Commonwealth of Australia, Canberra.

Cornish, P.M. (1989) Water quality in unlogged and logged eucalypt forests near Bega, NSW, during a nine year period. *Australian Forestry* 52, 276–285.

Cowley R.D. (1971) Birds and forest management. *Australian Forestry* 35, 234–250.

Curry, P.J. and Nichols, O.G. (1986) Early regrowth in rehabilitated bauxite minesites as breeding habitat for birds in the jarrah forest of sourth-western Australia. *Australian Forestry* 49, 112–114.

Dargavel, J., Johnston, T. and Boutland, A. (1985) Characteristics of the Australian forestry and logging workforces. *Australian Forestry* 48, 235–241.

Date, E.M., Ford, H.A. and Recher, H.F. (1991) Frugivorous pigeons, stepping stones, and weeds in northern New South Wales, *Nature Conservation 2: the Role of Corridors,* (eds D.A. Saunders and R.J. Hobbs), pp. 241–245. Surrey Beatty & Sons, Chipping Norton.

Davey, S.M. (1989) Thoughts towards a forest wildlife management strategy. *Australian Forestry* 52, 56–67.

Dickman, C. (1991) Use of trees by ground-dwelling mammals: implications for management, *Conservation of Australia's Forest Fauna,* (ed. D. Lunney), pp. 125–136. Royal Zoological Society of New South Wales, Mosman.

Disney, H.D.S. and Stokes, A. (1976) Birds in pine and native forests. *Emu* 76, 133–138.

Dobbyns, R. and Ryan, D. (1983) Birds, glider possums and monkey gums: the wildlife reserve system in the Eden district. *Forest and Timber* 19, 12–15.

Drielsma, J.H. (1992) Sustained forest management: past achievement, current changes, and future directions in NSW, in *Sustainable Forest Management,* (ed. M. Rowland), pp. 19–46. Board of Environmental Studies, Occasional Paper No. 18, University of Newcastle, Newcastle.

Fanning, D. and Mills, K. (1989) *Natural Resource Survey of the Southern Portion of Rockton Section Bondi State Forest.* Forest Resource Survey No. 6, Forestry Commission of New South Wales, Sydney.

Floyd, A.G. (1962) *Investigations into the natural regeneration of blackbutt* – E. pilularis. Forestry Commission of NSW, Research Note No. 10.

Ford, H.A. and Recher, H.F. (1991) The dynamics of bird communities of eucalypt forests and woodlands in south-eastern Australia. *Proceedings International Ornithological Congress* **20**, 1470–1479.

Forestry Commission of New South Wales (1992) *Proposed Forestry Operations in the Eden Management Area: Supplementary Environmental Impact Statement.* Forestry Commission of New South Wales, Sydney.

Forestry Commission, Tasmania (1993) *Forest Practices Code.* Forestry Commission, Hobart, Tasmania.

Forest Use Working Group (FWG) (1991) *Ecologically Sustainable Development Working Groups – Final Report, Forest Use.* Australian Government Publishing Service, Canberra.

Frawley, K. (1988) The history of conservation and the national park concept in Australia: a state of knowledge reveiw, in *Australia's Ever Changing Forests,* (eds K.J. Frawley and N. Semple), pp. 395–417. Australian Defence Force Academy, Canberra.

Friend, G.R. (1980) Wildlife conservation and softwood forestry in Australia: some considerations. *Australian Forestry* **43**, 217–225.

Friend, J.A. (1987) Local decline, extinction and recovery: relevance to mammal populations in vegetation remnants, *Nature Conservation: the Role of Remnants of Native Vegetation,* (eds D.A. Saunders, G.W. Arnold, A.A. Burbidge and A.J.M. Hopkins), pp. 53–64. Surrey Beatty & Sons, Chipping Norton.

Friend, J.A. (1990) The numbat, *Myrmecobius fasciatus* (Myrmecobiidae): history of decline and potential for recovery. *Proceedings of the Ecological Society of Australia* **16**, 369–377.

Frith, H.J. (1976) The destruction of the Big Scrub. *Parks and Wildlife* **2**, 7–12.

Frith, H.J. (1982) *Pigeons and Doves of Australia.* Rigby, Sydney.

Garnett, S. (ed.) (1992) *Threatened and Extinct Birds of Australia.* RAOU Report No. 82, Royal Australasian Ornithologists Union, Melbourne.

Goldingay, R. and Kavanagh, R. (1991) The yellow-bellied glider: a review of its ecology, and management considerations, in *Conservation of Australia's Forest Fauna,* (ed. D. Lunney), pp. 365–376. Royal Zoological Society of New South Wales, Mosman.

Hall, C.M. (1988) The 'worthless lands hypothesis' and Australia's national parks and reserves, in *Australia's Ever Changing Forests,* (eds K.J. Frawley and N. Semple), pp. 441–458. Australian Defence Force Academy, Campbell, ACT.

Hamilton, A.G. (1892) On the effect which settlement in Australia has produced upon indigenous vegetation. *Journal of the Royal Society (NSW)* **26**, 178–240.

Harris, L.D. (1984) *The Fragmented Forest.* The University of Chicago Press, Chicago.

Haseler, M. and Taylor, R. (1993) Use of tree hollows by birds in sclerophyll forest in north-eastern Tasmania. *Tasforests* **5**, 51–56.

Hawke, R.J. (1989) *Our Country, Our Future: Statement on the Environment.* Australian Government Printing Service, Canberra.

Henry, N.D. and Florence, R.G. (1966) Establishment and development of regeneration in spotted gum–ironbark forests. *Australian Forestry* **30**, 304–316.

Horne, R., Watts, G. and Robinson, G. (1991) Current forms and extent of retention areas with a selectively logged blackbutt forest in NSW: a case study. *Australian Forestry* **54**, 148–153.

Humphreys, N. (1977) Logging coupe size reduced at Eden. *Australian Forest Industries Journal* **43**, 39–45.

Inions, G.B., Tanton, M.T. and Davey, S.M. (1989) The effect of fire on the availability of hollows in trees used by the Common Brushtail Possum, *Trichosurus vulpecula* Kerr, 1792, and Ringtail Possum *Pseudocheirus peregrinus* Boddaert, 1785. *Australian Wildlife Research* **16**, 449–458.

Jenkins, B.A. and Recher, H.F. (1990) *Conservation in the Eucalypt Forests of the Eden Region in South East New South Wales*. Department of Ecosystem Management, University of New England, Armidale.

Joint Scientific Committee (JSC) (1990) *Biological Conservation of the South-East Forests*. Australian Government Publishing Service, Canberra.

Kavanagh, R. (1987) Forest phenology and its effect on foraging behaviour and selection of habitat by the Yellow-bellied Glider, *Petaurus australis* Shaw. *Australian Wildlife Research* **14**, 371–384.

Kavanagh, R. and Bamkin, K. (1995) Distribution of nocturnal forest birds and mammals in relation to the logging mosaic in south-eastern New South Wales, Australia. *Biological Conservation* **71**:41–53.

Kavanagh, R. and Lambert, M. (1990) Food selection by the Greater Glider, *Petauroides volans:* is foliar nitrogen a determinant of habitat quality? *Australian Wildlife Research* **17**, 285–299.

Kavanagh, R. Shields, J., Recher, H.F. and Rohan-Jones, W. (1985) Bird populations of a logged and unlogged forest mosaic in the Eden woodchip area in *Birds of Eucalypt Forests and Woodlands: Ecology, Conservation, Management,* (eds J. Keast, H.F. Recher, H. Ford and D. Saunders), pp. 273–281. Surrey Beatty & Sons, Chipping Norton.

Kellas, J.D., Jarrett, R.G. and Morgan, B.J.T. (1988) Changes in species composition following recent shelterwood cutting in mixed eucalypt stands in the Wombat Forest, Victoria. *Australian Forestry* **51**, 112–118.

Kennedy, M. (ed.) (1990) *Australia's Endangered Species*. Simon & Schuster, Brookvale.

Kirkpatrick, J. (1991) The magnitude and significance of land clearance in Tasmania in the 1980s. *Tasforests* **3**, 15–24.

Kirkpatrick, J., Meredith, C., Norton, T. *et al.* (1990) *The Ecological Future of Australia's Forests*. Australian Conservation Foundation, Fitzroy.

Lindenmayer, D.B. and Nix, H.A. (1993) Ecological principles for the design of wildlife corridors. *Conservation Biology* **7**, 627–630.

Lindenmayer, D.B., Norton, T.W. and Tanton, M.T. (1990) Differences between wildfire and clearfelling on the structure of montane ash forests of Victoria and their implications for fauna dependent on tree hollows. *Australian Forestry* **53**, 61–68.

Lindenmayer, D.B., Cunningham, R.B., Tanton, M.T., *et al.* (1991) The characteristics of hollow-bearing trees occupied by arboreal marsupials in the montane ash forests of the Central Highlands of Victoria, south-east Australia. *Forest Ecology & Management* **40**, 289–308.

Lindenmayer, D.B., Cunningham, R.B. and Donnelly, C.F. (1993) The conservation of arboreal marsupials in the montane ash forests of the central highlands of

Victoria, south-east Australia, IV. The presence and abundance of arboreal marsupials in retained linear habitats (wildlife corridors) within logged forest. *Biological Conservation* **59**, 207–221.

Loyn, R.H. (1985) Strategies for conserving wildlife in commercially productive eucalypt forest. *Australian Forestry* **48**, 95–101.

Loyn, R.H., Fagg, P.C., Piggin, J.E. *et al.* (1983) Changes in the composition of understorey vegetation after harvesting eucalypts for sawlogs and pulpwood in East Gippsland. *Australian Journal of Ecology* **8**, 43–53.

Lunney, D. (1991a) The future of Australia's forest fauna, *Conservation of Australia's Forest Fauna* (ed. D. Lunney), pp. 1–24. Royal Zoological Society of New South Wales, Mosman.

Lunney, D. (ed.) (1991b) *Conservation of Australia's Forest Fauna.* Royal Zoological Society of New South Wales, Mosman.

Lunney, D. (1993) Editorial. *Australian Zoologist* **29**, 1–3.

Lunney, D. and Ashby, E. (1987) Population changes in *Sminthopsis leucopus* (Gray) (Marsupialia: Dasyuridae) and other small mammal species, in forest regenerating from logging and fire near Bega, NSW. *Australian Wildlife Research* **14**, 275–284.

Lunney, D. and Leary, T. (1988) The impact on native mammals of land-use changes and exotic species in the Bega district. New South Wales, since settlement. *Australian Journal of Ecology* **13**, 67–92.

Lunney, D. and O'Connell, M. (1988) Habitat selection by the Swamp Wallaby, *Wallabia bicolor,* the Red-necked Wallaby, *Macropus rufogriseus,* and the Common Wombat, *Vombatus ursinus,* in logged, burnt forest near Bega, New South Wales. *Australian Wildlife Research* **15**, 695–706.

Lunney, D. and Recher, H.F. (1986) The living landscape: an ecological view of national parks and nature conservation, *A Natural Legacy: Ecology in Australia,* (eds H.F. Recher, D. Lunney and I. Dunn), pp. 294–328. Pergamon Press, Sydney. 2nd edn.

Lunney, D., Barker, J. and Priddel, D. (1985) Movements and day roosts of the Chocolate wattled bat *Chalinolobus morio* (Gray) (Microchiroptera: Vespertilionidae) in a logged forest. *Australian Mammalogy* **11**, 167–169.

Lunney, D., Cullis, B. and Eby, P. (1987) Effects of logging and fire on small mammals in Mumbulla State Forest, near Bega, New South Wales. *Australian Wildlife Research* **14**, 163–181.

Lunney, D., Urquhart, C. and Reed, P. (eds) (1990) *Koala Summit. Managing Koalas in New South Wales.* New South Wales National Parks & Wildlife Service, Sydney.

Lunney, D., Eby, P. and O'Connell, M. (1991) Effects of logging, fire and drought on three species of lizards in Mumbulla State Forest on the south coast of New South Wales. *Australian Journal of Ecology* **16**, 33–46.

Lunney, D., Curtin, A., Ayers, D., *et al.* (ms. a) Identifying the endangered fauna of New South Wales: an ecological approach to a systematic evaluation of the status of all species. (In review.)

Lunney, D., Fisher, D., Curtin, A. and Dickman, C.R. (ms. b) The endangered fauna of New South Wales: identifying their ecological attributes. (In review.)

Macfarlane, M.A. (1988) Mammal populations in mountain ash (*Eucalyptus regnans*) forests of various ages in the Central Highlands of Victoria. *Australian Forestry* **51**, 14–27.

Macfarlane, M.A. and Seebeck, J.H. (1991) Draft management strategies for the conservation of Leadbeater's possum, *Gymnobelideus leadbeateri,* in Victoria. *Arthur Rylah Institute Technical Report Series,* No. 111. Department of Conservation and Environment, Melbourne.

Mackowski, C.M. (1984) The ontogeny of hollows in Blackbutt (*Eucalyptus pilularis*) and its relevance to the management of forests for possums, gliders and timber, in *Possums and Gliders,* (eds A.P. Smith and I. Hume), pp. 353–367. Surrey Beatty & Sons, Chipping Norton.

Majer, J.D., Recher, H.F. and Ganeshanandam, S. (1992) Variation in the foliar nutrients and its relation to arthropod communities on Eucalyptus trees in eastern and Western Australia. *Australian Journal of Ecology* 17, 383–394.

Majer, J.D., Recher, H.F. and Postle, A.C. (in press) Comparison of arthropod species richness in eastern and western Australian canopies: a contribution to the species number debate. *Records of the Queensland Museum.*

Martin, R. and Handasyde, K. (1990) Translocations and the re-establishment of koala populations in Victoria (1944–1988): the implications for NSW, in *Koala Summit,* (eds D. Lunney, C.A. Urquhart and P. Reed), pp. 58–66. New South Wales National Parks and Wildlife Service, Sydney.

McColl, J.G. (1969) Regression models relating soil nutrients and growth of *Eucalyptus gummifera* and *E. maculata* seedlings. *Ecology* 48, 157–159.

McColl, J.G. and Humphreys, F.R. (1967) Relationship between some nutritional factors and the distribution of *Euclayptus gummifera* and *E. maculata. Ecology* 48, 766–771.

McIlroy, J.C. (1978) The effects of forestry practice on wildlife in Australia: a reveiw. *Australian Forestry* 41, 78–94.

Mooney, N. and Holdsworth, M. (1991) The effects of distrubance on nesting wedge-tailed eagles (*Aquila audax fleayi*) in Tasmania. *Tasforests* 3, 25–32.

Mosley, G. and Costin, A. (1992) *World Heritage Values and Their Protection in Far South East New South Wales.* Earth Foundation Australia, Sydney.

Newsome, A.E., McIlroy, J. and Catling, P. (1975) The effects of an extensive wildfire on populations of twenty ground vertebrates in south-east Australia. *Proceedings of the Ecological Society of Australia* 9, 107–123.

Nichols, O., Wykes, B.J. and Majer, J.D. (1989) The return of vertebrate and invertebrate fauna to bauxite mined areas in south western Australia, in *Animals in Primary Succession – the Role of Fauna in Reclaimed Lands,* (ed. J.D. Majer), pp. 397–422. Cambridge University Press, Cambridge.

Norton, T.W. (1988) Australia's arboreal marsupial fauna: past and present, in *Australia's Ever Changing Forests,* (eds K.J. Frawley and N. Semple), pp. 99–118. Australian Defence Force Academy, Campbell, ACT.

Norton, T. and Lindenmayer, D. (1991) Integrated management of forest wildlife: toward a coherent strategy across state borders and land tenures, in *Conservation of Australia's Forest Fauna,* (ed. D. Lunney), pp. 237–244. Royal Zoological Society of New South Wales, Mosman.

Noske, R. (1985) Habitat use by three bark-foragers in eucalypt forests, in *Birds of Eucalypt Forests and Woodlands: Ecology, Conservation, Management,* (eds J. Keast, H.F. Recher, H. Ford and D. Saunders), pp. 193–204. Surrey Beatty & Sons, Chipping Norton.

Paton, D.C. (1980) The importance of manna, honeydew and lerp in the diet of honeyeaters. *Emu* 80, 213–226.

Pizzey, G. (1980) *A Field Guide to the Birds of Australia*. Collins, Sydney.

Pressey, R.L. (1990) Reserve selection in New South Wales: Where to from here? *Australian Zoologist* **26**, 70–75.

Purdie, R.W. (1990) *Forests and the National Estate. Part 4A. Protecting the Natural National Estate Value of Forests. References and Appendices: A Submission to the Resource Assessement Commission Inquiry into Australian Forest and Timber Resources*. Australian Heritage Commission, Canberra.

Pyke, G.H. and O'Connor, P.J. (1991) *Wildlife Conservation in South-east Forests of New South Wales*. Technical Report No. 5, Australian Museum, Sydney.

Recher, H.F. (1976a) An ecologist's view: the failure of our national parks. *Australian Natural History* **18**, 64–68.

Recher, H.F. (1976b) *An Interim Report: The Effects of Woodchipping on Wildlife at Eden*. Department of Environmental Studies, The Australian Museum Technical Report No. 76/3, pp. 1–15.

Recher, H.F. (1982) *Pinus radiata* – a million hectare miscalculation. *Australian Natural History* **20**, 319–325.

Recher, H.F. (1985a) Synthesis: a model of forest and woodland bird communities, in *Birds of Eucalypt Forests and Woodlands: Ecology, Conservation, Management,* (eds J. Keast, H.F. Recher, H. Ford and D. Saunders), pp. 129–135. Surrey Beatty & Sons, Chipping Norton.

Recher, H.F. (1985b) A diminishing resource: mature forest and its role in forest management, in *Wildlife Management in the Forests and Forestry-controlled Lands in the Tropics and the Southern Hemisphere,* (ed. J. Kikkawa), pp. 28-33. IUFRO & The University of Queensland, St Lucia.

Recher, H.F. (1990) Wildlife conservation in Australia: State of the nation. *Australian Zoologist* **26**, 5–10.

Recher, H.F. (1991) The conservation and management of eucalypt forest bires: resources requirements for nesting and foraging, *Conservation of Australia's Forest Fauna,* (ed. D. Lunney), pp. 25–34. Royal Zoological Society of New South Wales, Mosman.

Recher, H.F. (1992) Paradigm and paradox: sustainable forest management, in *Sustainable Forest Management,* (ed. M. Rowland), pp. 7–18. Board of Environmental Studies, Occasional Paper No. 18, University of Newcastle, Newcastle.

Recher, H.F. (1994) Why conservation biology? in *Conservation Biology in Australia and Oceania,* (eds C. Moritz, J. Kikkawa and D. Doley), pp. 1–15. Surrey Beatty & Sons, Chipping Norton.

Recher, H.F. (in preparatrion) The extinction of Australia's birds: The role of forestry and agriculture.

Recher, H.F. and Christensen, P. (1981) Fire and the evolution of the Australian Biota, in *Ecological Biogeography of Australia,* (ed. A. Keast), pp. 137–162. Dr. W. Junk, The Hague.

Recher, H.F. and Lim, L. (1990) A review of the current ideas of the extinction, conservation and management of Australia's terrestrial vertebrate fauna. *Proceedings of the Ecological Society of Australia* **16**, 287–301.

Recher, H.F., Clark, S.S. and Milledge, D. (1975) An assessment of the potential impact of the woodchip industry on ecosystems and wildlife in southeastern Australia, in *A Study of the Environmental, Economic and Sociological Con-*

sequences of the *Wood Chip Operations in Eden, New South Wales*, pp. 108–183. W.D. Scott & Co., Sydney.

Recher, H.F., Rohan-Jones, W. and Smith, P. (1980) *Effects of the Eden woodchip industry on terrestrial vertebrates with recommendations for management*. Forestry Commission of NSW, Research Note No. 42.

Recher, H.F., Gowing, G., Kavanagh, R. *et al.* (1983) Birds, resources and time in a tablelands forest. *Proceedings of the Ecological Society of Australia* 12, 101–123.

Recher, H.F., Holmes, R.T., Schulz, M. *et al.* (1985) Foraging patterns of breeding birds in eucalypt forest and woodland of sourth-eastern Australia. *Australian Journal of Ecology* 10, 399–420.

Recher, H.F., Shields, J., Kavanagh, R. and Webb, G. (1987) Retaining remnant mature forest for nature conservation at Eden, New South Wales, in *Nature Conservation: The Role of Remnants of Native Vegetation*, (eds D.A. Saunders, G.W. Arnold, A.A. Burbidge and A.J.M. Hopkins), pp. 177–194. Surrey Beatty & Sons, Chipping Norton.

Recher, H.F., Kavanagh, R.P., Shields, J.M. and Lind, P. (1991) Ecological association of habitats and bird species during the breeding season in southeastern New South Wales. *Australian Journal of Ecology* 16, 337–352.

Recher, H.F., Hutchings, P.A. and Rosen, S. (1993) The biota of the Hawkesbury–Nepean Catchment: reconstruction and restoration. *Australian Zoologist* 29, 3–42.

Reed, P. (1991) An historical analysis of the changes to the forests and woodlands of New South Wales, in *Conservation of Australia's Forest Fauna*, (ed. D. Lunney), pp. 393–406. Royal Zoological Society of New South Wales, Mosman.

Resource Assessment Commission (RAC) (1992) *Forest and Timber Inquiry Final Report*. Vols 1 and 2, Australian Government Publishing Service, Canberra.

Richards, G. (1991) The conservation of forest bats in Australia: do we really know the problems and solutions? in *Conservation of Australia's Forest Fauna*, (ed. D. Lunney), pp. 81–90. Royal Zoological Society of New South Wales, Mosman.

Robinson, D. (1991) Threatened birds in Victoria: their distributions, ecology and future. *Victorian Naturalist* 108, 67–77.

Saunders, D.A., Smith, G.T. and Rowley, I. (1982) The availability and dimensions of tree hollows that provide nest sites for cockatoos (Psittaciformes) in Western Australia. *Australian Wildlife Research* 6, 205–216.

Scotts, D. (1991) Old-growth forests: their ecological characteristics and value to forest-dependent vertebrate fauna of south-east Australia, in *Conservation of Australia's Forest Fauna*, (ed. D. Lunney), pp. 147–160. Royal Zoological Society of New South Wales, Mosman.

Shaw, W.W. (1983) Integrating wildlife conservation and timber production objectives in Australian forests. *Australian Forestry* 46, 132–135.

Shields, J. (1992) Wildlife management in NSW State Forests: perspective in 1992, in *Sustainable Forest Management*, (ed. M. Rowland), pp. 11–24. Board of Environmental Studies, Occasional Paper No. 18, University of Newcastle, Newcastle.

Smith, A.P. (1991) Forest policy: fostering environmental conflict in the Australian timber industry, in *Conservation of Australia's Forest Fauna*, (ed. D. Lunney), pp. 301–314. Royal Zoological Society of New South Wales, Mosman.

Smith, A.P. and Lindenmayer, D. (1988) Tree hollow requirements of Leadbeater's possum and other possums and gliders in timber production ash forests of the Victorian Central Highlands. *Australian Wildlife Research* 15, 347–362.

Smith, A.P., Moore, D.M. and Andrews, S.P. (1992) *Proposed Forestry Operations in the Glen Innes Forestry Management Area: Fauna Impact Statement.* Austeco Pty Ltd., Armidale, NSW.

Smith, P. (1984) The forest avifauna near Bega, New South Wales. I. Differences between forest types. *Emu* 84, 200–210.

Smith, P. (1985) Effects of intensive logging on birds in eucalypt forest near Bega, New South Wales. *Emu* 85, 15–21.

Specht, R.L., Roe, E.M. and Broughton, V.H. (1974) Conservation of major plant communities in Australia and Papua New Guinea. *Australian Journal of Botany Supplement Series* 7, 1–667.

Strahan, R. (ed.) (1983) *Complete Book of Australian Mammals.* Angus & Robertson, Sydney.

Suckling, G.C. and Macfarlane, M.A. (1983) Introduction of the sugar glider, *Petaurus breviceps*, into re-established forest of the Tower Hill State Game Reserve, Vic. *Australian Wildlife Research* 10, 249–258.

Sutton, P. (1991) Forest management in Victoria beyond the 1990s: a bright green, socially just, high economic growth scenario, in *Conservation of Australia's Forest Fauna*, (ed. D. Lunney), pp. 315–350. Royal Zoological Society of New South Wales, Mosman.

Tacey, W.H. (1979) Landscaping and revegetation practices used in rehabilitation after bauxite mining in Western Australia. *Reclamation Review* 2, 123–132.

Taylor, R. (1991a) Fauna management practices in State Forests in Tasmania, in *Conservation of Australia's Forest Fauna*, (ed. D. Lunney), pp. 259–264. Royal Zoological Society of New South Wales, Mosman.

Taylor, R. (1991b) The role of retained strips for fauna conservation in production forests in Tasmania, *Conservation of Australia's Forest Fauna*, (ed. D. Lunney), pp. 265–270. Royal Zoological Society of New South Wales, Mosman.

Taylor, R. (1991c) *Fauna Conservation in Production Forests in Tasmania.* Forestry Commission, Tasmania, Hobart.

Taylor, R. and Haseler, M. (1993) Occurrence of potential nest trees and their use by birds in sclerophyll forest in north-east Tasmania. *Australian Forestry* 56, 165–171.

Taylor, R. and Haseler, M. (in review) Effects of partial logging systems on bird populations in Tasmania.

Taylor, R. and Savva, N. (1988) Use of roost sites by four species of bats in State Forest in south-eastern Tasmania. *Australian Wildlife Research* 15, 637–645.

Traill, B. (1991) Box–ironbark forests: tree hollows, wildlife and management, in *Conservation of Australia's Forest Fauna*, (ed. D. Lunney), pp. 119–124. Royal Zoological Society of New South Wales, Mosman.

Turner, J. and Lambert, M. (1988) Impact of wildlife grazing on nitrogen turnover in the understorey of a Eucalyptus forest. *Australian Forestry* 51, 222–225.

Tyndale-Biscoe, C.H. and Calaby, J.H. (1975) Eucalypt forests as a refuge for wildlife. *Australian Forestry* 38, 117–133.

Ward, S.C., Koch, J.M. and Nichols, O.G. (1990) Re-establishment of a functional ecosystem following bauxite mining in the Darling Range, WA. *Proceedings of the Ecological Society of Australia* 16, 557–565.

Wardell-Johnson, G. and Nichols, O. (1991) Forest wildlife and habitat management in southwestern Australia: knowledge, research and direction, in *Conservation of*

Australia's Forest Fauna, (ed. D. Lunney), pp. 161–192. Royal Zoological Society of New South Wales, Mosman.

Wardell-Johnson, G., Hewett, P. and Woods, Y. (1991) *Retaining Remnant Mature Forest For Nature Conservation: A Review of the System of Road, River and Stream Zones in the Karri Forest.* Occasional Paper 1/92. Department of Conservation and Land Management, Perth.

Wells, K.F., Wood, N.H. and Laut, P. (1984) *Loss of Forests and Woodlands in Australia: A Summary by State Based on Rural Local Government Areas.* CSIRO Division of Water and Land Resources. Technical Memorandum 84/4, Canberra.

Whitehouse, J. (1990) Conserving what?: the basis for nature conservation reserves in New South Wales. 1967–1989. *Australian Zoologist* **26**, 11–21.

Wilson, B. (1991) Conservation of forest fauna in Victoria, in *Conservation of Australia's Forest Fauna,* (ed. D. Lunney), pp. 281–300. Royal Zoological Society of New South Wales, Mosman.

Woinarski, J.C.Z. (1979) Birds of a *Eucalyptus* plantation and adjacent natural forest. *Australian Forestry* **42**, 243–247.

Woodgate, P. and Black, P. (1988) *Forest Cover Changes in Victoria 1869–87.* Department of Conservation, Forests and Lands, Melbourne.

–13

The importance of forest for the world's migratory bird species

John H. Rappole

13.1 INTRODUCTION

One-quarter to one-third of the migrant bird species of the world are forest-dependent during one or more phases of their life cycle (Rappole, 1995) (Table 13.1). This figure alone should be a cause for some apprehension given the rate at which the world's forests are being altered (World Resources Institute, 1992). However, until quite recently, the connection between forest alteration and migratory bird conservation was not recognized as an issue by many students of migrant ecology because of the apparent flexibility of migrant species in terms of habitat use (Morse, 1971; Karr, 1976). Migrant needs in terms of specific habitat requirements are still a subject of debate (Petit *et al.*, 1993; Rappole and McDonald, 1994) but population declines recorded for 109 species of Nearctic migrants to the Neotropics (DeGraaf and Rappole, 1995) have forced the problem into a different context. In the absence of obvious alternative explanations for most migrant declines, the question of the effects of habitat loss in general and forest loss specifically is no longer strictly academic; for some species, it has become a critical conservation issue (Rappole *et al.*, 1994).

In some cases, the loss of a particular forest type can be linked to species decline with a fair degree of certainty. A classic example is the disappearance of the passenger pigeon (*Ectopistes migratorius*), whose ecology was intimately coupled with mast production (Audubon, 1840–1844). This bird vanished from the wild over a period of 20 years (1880–1900) in apparent response to the loss of the American chestnut (*Castanea dentata*) to disease and the clearing of other mast-producing trees in eastern North America (Schorger, 1955).

Conservation of Faunal Diversity in Forested Landscapes. Edited by
R.M. DeGraaf and R.I. Miller. Published in 1996 by Chapman & Hall. ISBN 0 412 61890 7.

Table 13.1 Estimate of the total number of species as well as forest-related species of migratory birds that winter in a tropical environment and breed in a temperate environment, by continent

Continent	Total migrants	Forest related migrants	Source
Temperate North America	338	112	Rappole, 1995
Tropical North America	338	112	Includes Middle America and West Indies
South America			
Boreal	162	54	Rappole *et al.*, 1993
Austral	230	60	Meyer de Schavensee, 1966; Chesser, 1994
Europe	185	45	Moreau, 1972
Africa			
Boreal	185	13	Moreau, 1972
Austral	?	?	Insufficient data
Temperate Asia	336	107	Rappole, 1995
Tropical Asia			
Boreal	336	107	Rappole, 1995
Austral	?	?	Insufficient data

Similarly, we can relate the extinction of the ivory-billed woodpecker (*Campephilus principalis*) throughout the southeastern United States to the nearly complete harvest of mature bottomland forest (DeGraaf *et al.*, 1991a). Ivory-bills fed on standing, dead trees as described by Wilson 1808–1814, Vol. 1:163) as follows:

> [The Ivory-bill] seeks the most towering trees of the forest; seeming particularly attached to those prodigious cypress swamps, whose crowded giant sons stretch their bare and blasted, or moss-hung, arms midway to the skies.

Giant standing dead individuals of cypress (*Taxodium*) or other tree species typical of the swampy coastal plain of the southeast have long been absent from this habitat, as has the woodpecker.

These instances in which the disappearance of a species can be directly related to the disappearance of a particular forest type are rare. Too often, and especially for migratory species, decline is both difficult to document and even more difficult to assign a probable cause. Nevertheless, it seems clear that the extensive alteration or reduction of forest cover worldwide will have an impact on bird species using those forests. One example of a migrant species in which forest decline does appear related to population decline is the wood thrush (*Hylocichla mustelina*). This species breeds in deciduous and mixed coniferous–deciduous forests of eastern North America and winters in lowland wet tropical forests of Middle America. Population declines of roughly 4% per year have occurred over the past decade, according to National Breeding Bird Survey data (Robbins *et al.*, 1989a). Forest habitat for this species has been severely degraded in large portions of the bird's winter range, resulting in an estimated loss of 73% of its winter habitat in northeastern Costa Rica and a 95% loss in the Tuxtla Mountain region of Mexico (Rappole *et al.*, 1994).

This chapter presents a review of the relationship between migrant birds and forests from both an ecological and a conservation perspective.

13.2 ECOLOGICAL QUESTIONS AND PERSPECTIVES

13.2.1 Breeding grounds

(a) Habitat needs

Forest area in the United States has changed dramatically over the years from the time of European arrival on the continent to the present (Powell and Rappole, 1986). Prior to colonization, there were over one billion acres of forest in the country; by 1872, 0.4 billion acres remained (US Dept Commerce, 1924, 1935). But, at present, the acreage has rebounded to 0.6 billion acres (US Dept Agriculture, 1982). These figures, of course, do not tell the whole story. Nearly all virgin forest is gone, and while it has been replaced by forests that contain many of the same tree species that occurred in the region historically, the individual representatives are quite different from those that were present in pre-colonial America. The trees composing the forests of the United States are younger, shorter and thinner than those in the primeval forests. In addition, these secondary forests are depauperate in understory species composition (Flaccus, 1959; Maclean and Wein, 1977; Brewer; 1980). Duffy and Meier (1992) found that in 50–85-year-old forests of the southern Appalachians, succession of herbaceous understory plants 'resulted in only half the species richness and one-third the total cover measured in primary forests'.

Does it make a difference to the avifauna that oaks in the Appalachian forest are 60–70 years of age instead of 200–300 years of age? Is it important to the breeding bird community if 'mature' sycamores (*Platanus occidentalis*) in Illinois today have a DBH of less than a meter, whereas those of a century ago had a DBH of 4 m or more (Figure 13.1), or if half the understory species are missing? These are difficult, perhaps impossible questions to answer, because in most of the temperate region the primary forests have long been absent as a dominant cover type. What little remains is in the form of tiny remnant patches, insufficient to support viable bird populations. Therefore, we have no true reference point to which we can compare

Figure 13.1   Smithsonian ornithologist Robert Ridgway (left) and his brother seated at the foot of a giant sycamore (*Platanus occidentalis*) in the early 1870s near Mount Carmel, Illinois. (Harris, 1928.)

the extant forest avian community. We can only examine temperate forests in their present form as affected by the types of perturbation that are occurring now. But we should not forget that our use of the term 'forest' refers to quite a different set of communities from what was here prior to European colonization in terms of both structure and species composition.

DeGraaf *et al.* (1991a) provide an excellent review of the various forest types currently used by breeding birds of the United States. Do birds actually need these forest types and, if so, why? The question of why a member of any species is found at a particular place at any given time is not well understood, and the theoretical questions are especially difficult in respect of migratory species. The difficulty is that it is not clear for many migrant species that food resources are limiting during the breeding season (Wiens, 1989; Martin, 1992). If resources are not limiting, then what causes the obvious non-random patterns of breeding habitat use revealed in species' distribution patterns reported by DeGraaf *et al.* (1991a) and all other guides? The number of suitable nesting sites, predation patterns on adults, eggs or young, and factors affecting feeding rates for young in the nest could act as selection factors shaping species-specific responses to the range of breeding habitats available. As an example, the importance of nest site availability in determining the abundance of migrants in breeding habitats has been demonstrated for several species (Martin, 1992:465).

Conversion of the virgin temperate forests to various seral stages associated with European colonization apparently was responsible for the northward extension of the range of several species into the northeastern United States in the early 1900s, e.g. Bewick's wren, willow flycatcher and Bachman's sparrow. As abandoned farms now return to woodlands, these same species have begun to disappear over much of this area (LeGrand and Schneider, 1992). These and other changes in the distributions of birds associated with massive alteration of breeding habitats indicate that the structure of the breeding habitat can be critical, though the specific reasons for its importance are not always evident.

(b)   Effects of fragmentation

Numerous studies of the breeding bird communities in temperate forests of the eastern US have documented an effect on populations of migratory birds from the breaking up of large blocks of habitat into smaller pieces (Askins *et al.*, 1990). Whitcomb and his colleagues were among the first to document a decline in species richness as well as numbers of migrant individuals breeding in fragmented forest tracts when compared with populations breeding in continuous forest of the same type (Whitcomb *et al.*, 1976; Whitcomb, 1977; Whitcomb *et al.* 1977; Lynch and Whitcomb, 1978). Extensive research on this phenomenon has demonstrated that, in general, the smaller the remaining piece of habitat, the lower the breeding bird densities (Robbins *et al.*, 1989b; Askins *et al.*, 1990). Some species, particularly those whose populations winter in the Neotropics, are much more sensitive to this 'area effect' than others.

Several explanations for this phenomenon have been advanced, including:

- increased vulnerability to nest parasites (e.g. brown-headed cowbirds. *Molothrus ater*);
- increased vulnerability to predators;
- alteration or loss of microhabitats;
- increased interspecific competition by resident 'generalist' species (Whitcomb, 1977; Askins *et al.*, 1990).

Part of this 'area effect' may be a sampling effect. Most studies of the effects of habitat fragmentation on birds involve breeding bird density assessments based on counts of singing individuals. For these counts to represent a true reflection of density, it must be assumed that each singing bird represents a mated pair and that size of the fragment does not affect song frequency. Studies have shown, however, that neither of these assumptions is correct. Unmated males sing as much or more than paired individuals in many species (Nice, 1964; Krebs, 1971; Baskett *et al.*, 1978; Rappole and Waggerman, 1986; Morton, 1992) and song rates per individual can be quite different for individuals of the same species holding territory in continuous as opposed to fragmented habitat (McShea *et al.*, in review).

An additional problem with fragmentation studies is that they presume that events observed on the breeding ground are controlled by breeding ground factors, e.g. that observed avian densities in forest fragments of different sizes are the direct result of these differences in size. For migratory species, this presumption is not necessarily correct. Reduced population size in fragments could be the result of reduced population pressure resulting from low overwinter survival rates. If populations of migrants are controlled by winter habitat availability, fewer migrants may return north than the breeding habitat can support. Under these circumstances, only the highest quality breeding habitats would be occupied since there would be little pressure to occupy suboptimal habitats (Rappole and McDonald, 1994).

(c)   Effects of habitat alteration

Habitat alteration can be as sharp and distinct as clearcutting or as subtle. as long-term changes caused by the greenhouse effect. Regardless of the nature of the change, determination of its effects on breeding bird populations is seldom straightforward (Thompson *et al.*, 1992; Rodenhouse, 1992).

In many parts of the world, populations of large herbivores in the absence of significant predators reach high population levels and cause major alterations in their habitats (Laws *et al.*, 1975; Jewell and Holt, 1981; Pastor and Naiman, 1992; Tchamba and Mahamat, 1992; Prins and Van der Jeugd, 1993). For instance, in some parts of eastern North America, white-tailed deer now occur at densities several times those that occurred during pre-Columbian times (Alverson *et al.*, 1988; Warren, 1991). These densities can cause changes in understory structure and species composition, and ultimately affect replacement of dominant tree species (Alverson *et al.*, 1988). Studies of breeding bird communities in the northeastern United States have reported negative effects on forest bird communities associated with high deer densities (Baird, 1990; DeGraaf *et al.*, 1991b). We designed a long-term study to test the effects of deer browsing on breeding species in the Appalachian oak forests of northern Virginia. We established eight 4 ha study sites, four of which are exposed to deer browsing and four of which are not (because of fences 3 m in height surrounding the plots). The study is only in its fifth year, so results are preliminary. However, we have found that predation on artificial ground nests was higher in areas exposed to deer browsing than in those areas protected from browsing (Leimgruber *et al.*, 1994). It remains to be seen whether increased predation or other factors, e.g. reduced food density, cause actual changes in the density of birds on the browsed versus the unbrowsed sites. DeCalesta (in press) compared species richness, abundance and diversity of songbirds on forested sites in western Pennsylvania exposed to different levels of deer browsing. He found that while variations in deer density appeared to have little effect on ground or canopy species, intermediate-canopy nesting birds (e.g. eastern wood pewee, *Contopus virens*; least flycatcher, *Empidonax minimus*; yellow-billed cuckoo, *Coccyzus americanus*) showed significant declines with increasing deer densities.

Complete removal of forest cover (clearcutting) has been a common practice in management of US forests (Smith, 1962), which has obvious negative impacts on breeding populations of forest birds. When forests are converted to open areas, many of the species associated with those forests disappear (Szaro, 1981; Vega and Rappole, 1994). However, clearcutting as a management procedure does not always involve complete removal of forest. Often, blocks of forest comprising various percentages of the habitat are removed in a mosaic, where the size of the clearcut block is determined by the timber rotation. When these blocks are relatively small, their removal may have little immediate effect on the breeding bird community. As an example, Thompson *et al.* (1992) examined the impact of clearcutting on the avifauna of oak forests in southern Missouri. They selected nine sites of 200 ha each where clearcutting had been performed (10% < 10 years of age; 10% 11–20 years; 80% > 20 years). They compared these with nine other 200 ha sites of 'mature forest with no recent timber harvest or other disturbance'. Some forest interior species (scarlet tanager, *Piranga olivacea*; red-eyed vireo, *Vireo olivaceus*; pine warbler, *Dendroica pinus*) showed lower numbers on the sites where clearcutting had been done, while others were at equal or higher population levels on these sites when compared with the control sites.

It is not clear what this kind of study has to say about effects of clearcutting on breeding densities of forest birds. Even though only 10% of the 'treatment' sites were cut for this experiment, all of the sampled sites, both treatment and control, had been cut previously. Thus, both treatment and control already represent mosaics of second growth. As mentioned above, Duffy and Meier (1992) found that such second-growth forests can be quite different from old-growth primary forest, and that recovery to a state comparable with that of primary forest may require centuries. Furthermore, avian density estimates in the study by Thompson *et al.* (1992) were based on transect counts of singing birds made between 16 May and 20 June, though singing bird counts are inappropriate for measuring density even if made at a time when transients are not passing through (Rappole *et al.*, 1993).

13.2.2 The post-nesting period

That portion of the cycle beginning with the fledging of the young and terminating with migration is very poorly known for most migrant species. The presumption is that the birds remain on or near the breeding territory for an indeterminate period and then begin a gradual southward movement (Cooke, 1915; Hahn, 1937; Bent, 1953). This pattern may be true for some bird species. For instance, Pulich (1976) reported that a young golden-cheeked warbler, banded in central Texas in May, was recovered in northern Mexico in July of the same year, clearly indicating a southward movement after fledging. However, detailed studies have indicated that avian movements, behavior and habitat use can be extremely complex during this portion of the life cycle, and a gradual southward drift may not be the norm. In his study of prairie warblers (*Dendroica discolor*) in Indiana, Nolan (1978) found that the adults normally split the brood at fledging, with the male often remaining on or near the territory with his portion of the family while the female foraged separately with her portion, also often on or near the territory. When the young reached independence at about four weeks post-fledging, most of the adult females and offspring remained in the vicinity, although not actually on the breeding territory. Roughly half of the adult males stayed on the breeding territory until the beginning of fall migration in September.

The most intriguing aspect of the post-breeding period from a conservation perspective is the indication that birds of at least some and perhaps many species change habitat during the period of two to three months from when the young fledge until southward migration begins. Rappole and Ballard (1987) mist-netted intensively in riparian forest and old field habitat along the North Oconee River in Athens, Georgia from 17 April to 15 October. Individuals of several species known to breed in forests of the Athens area were captured in the old field habitat during July and August (yellow-billed cuckoo, *Coccyzus americanus*; Acadian flycatcher, *Empidonax virescens*; pine warbler, *Dendroica pinus*). We also captured representatives of species for which the nearest known breeding locality was 150 km north of Athens in the Appalachian forests of the north Georgia mountains (northern parula, *Parula americana*; American redstart, *Setophaga ruticilla*; ovenbird, *Seiurus aurocapillus*; scarlet tanager, *Piranga olivacea*). For both groups, it is unlikely that these were migrants since they were in molt, and showed little subcutaneous fat. Rather, it seems that members of some of these species changed their habitat use patterns during the post-breeding period. The migrant life cycle is normally divided into a breeding period, migration period and winter period. However, these findings raise questions regarding our understanding of migrant ecology. It may be that requirements are sufficiently different during the post-breeding/pre-migratory period to warrant consideration of this phase as a fourth identifiable segment of the migrant life cycle.

13.2.3 Migration

(a) Habitat needs

Do birds that breed in forest habitats require forest habitats during migration? The logical answer to this question is, 'Yes'. But this answer is logical only to the extent that the requirements for a transient are equal to those of a breeding individual. Obviously, there are many differences between the requirements for individuals in these two different phases of the life cycle, so we cannot conclude that a bird that breeds in forest necessarily requires forest during migration. In fact, a number of migrants that breed in forested habitats in Europe migrate through and winter in open habitats (Monkkonen *et al.*, 1992). Research groups headed by Berthold in Germany, Moore in Mississippi and Winker in Minnesota have investigated the phenomenon of habitat choice by transients in detail, and have independently reached the conclusion that transients are selective with regard to habitat during migration (Berthold, 1988; Berthold *et al.*, 1976; Bairlein, 1983, 1992; Moore and Simons, 1992; Winker *et al.*, 1992a,b,c). Rappole and Warner (1976) examined a community of transient birds in hackberry (*Celtis*) riparian forest along the Aransas River in the Texas coastal plain. We found that, while habitat preferences for the majority of transients passing through the region were quite broad, those of individuals apparently stopping to build fat reserves were highly specific. As an example, northern waterthrushes (*Seiurus noveboracensis*), which occur only as transients at the site, were captured in large numbers in several different habitat types. However, recaptures were concentrated along a muddy border surrounding a temporary pond. These birds were color banded and were found to defend feeding territories on which they remained for up to a week. They also showed a mean weight gain of 0.7 g, as compared with birds recaptured away from the temporary pond, which showed a mean weight loss of 0.3 g. These means are different at the 0.05 level of significance ($t = 2.28$, $d.f. = 23$). These findings indicate that transients may exist in two different classes during migration: those in a 'flying state' (*Zugstimmung* of Groebbels, 1928), just passing through, which are not highly selective with regard to habitat needs; and those in a 'feeding state' (*Zudisposition* of Groebbels, 1928), which are likely to be selective with regard to habitat needs since they require specific foraging substrates to rebuild fat reserves successfully. Considerably more research is needed to illuminate the details of transient habitat requirements.

(b) Effects of fragmentation

No studies of the effect of fragmentation on density of transients have as yet been published to my knowledge, although Petit has been investigating the phenomenon (D. Petit, personal communication). In any case, I would not expect fragmentation to have a negative effect on transient densities in the remaining fragments. In fact, breaking a stop-over habitat into pieces could have the effect of causing higher concentrations in the remaining forested sites. Whether such alteration is a benefit or injurious to transient populations will be difficult to test empirically. However, if stop-over habitat is important, as seems likely based on studies of transients cited above, then reduction in such habitat is likely to have a negative effect on transient populations.

(c)   Habitat degradation

As noted above, the specific ecological needs of transients are poorly known and difficult to document. If the work by Rappole and Warner (1976) is correct in concluding that only a small percentage of birds observed at stop-over sites are actually using the site to refurbish fat reserves, then comparison of total numbers of individuals in pristine versus degraded forest stop-over sites is unlikely to reveal significant differences. However, intensive studies of those individuals actually stopping in and using these sites may reveal differences. These are the kinds of study that will be required to determine the effects of such pervasive disturbances of stop-over habitat as browsing and grazing by cattle, perhaps the most common form of stop-over habitat degradation in Texas and northern Mexico (Rappole, personal observation). As in the case of the white-tailed deer discussed above, cattle cause changes in the structure and species composition of grazed sites (Rappole *et al.*, 1986; Diamond and Fulbright, 1990; Orodho *et al.*, 1990). The importance of these changes to migrating individuals has yet to be measured but it seems probable that they are significant.

13.2.4   Wintering grounds

(a)   Habitat needs

The habitat requirements for migrants should, if anything, be more straightforward on the wintering ground than on the breeding ground. There are no complicating factors such as nesting requirements during this portion of the life cycle. Food availability, roost sites and vulnerability to predation should seemingly be the principal factors determining habitat use. However, migrants appear to be less rather than more demanding in terms of habitat specificity during this period. As an example, the ovenbird, which is restricted to large blocks of deciduous forests as a breeding bird in the eastern United States (Robbins *et al.*, 1989b), is found in a variety of forest and second-growth habitats in winter in Belize (Petit *et al.*, 1992). Two explanations have been presented for this type of behavior by migrants:

1. Migratory birds are generalists that can use a wide variety of foods in many different habitat types (Morse, 1971).
2. Migratory birds are forced to use a wide range of habitats on the wintering ground because more birds are produced than can be supported in limited preferred winter habitats (Rappole and McDonald, 1994).

These explanations are not mutually exclusive; both have yet to be thoroughly tested in the field.

(b) Effects of fragmentation

Robbins and his colleagues have investigated the effects of fragmentation on wintering populations of forest-related migrants in the West Indies and middle America (Robbins *et al.*, 1987, 1992). In general, they have found little difference in the migrant communities in patches versus continuous forest of the same type. However, the effects of fragmentation can be expected to be quite different between species depending upon such characteristics as whether or not the species in question is normally a mixed-species flock follower during the winter period. Rappole and Morton (1985) found that mixed species foraging flocks along with two species of migrants associated with those flocks (worm-eating warbler, *Helmitheros vermivorus*; black-and-white warbler, *Mniotilta varia*) disappeared when a continuous rainforest in Veracruz was reduced to a 1 ha block. Askins *et al.* (1992) examined the effects of fragmentation on a community of migrants wintering in moist, lowland forest in the US Virgin Islands. They found that abundance and diversity of migrants were higher in large blocks of forest on St John Island when compared with smaller patches of the same habitat on St Thomas Island.

(c) Forest removal or degradation

The most serious conservation problem for migratory species wintering in tropical forest is actual loss of forest habitat. Conversion of forested habitats to agriculture and pasture is occurring at a rapid rate throughout the tropics (World Resources Institute, 1992). While a few studies have demonstrated the importance of forest to wintering migrants (e.g. Winker *et al.*, 1990), none has shown that loss of winter habitat is directly responsible for population decline. This failure is due principally to the nature of migratory species, which occupy several different habitats separated by large distances during different times of the year. Nevertheless, Rappole and McDonald (1994) have proposed that the characteristics of species' populations controlled by breeding ground factors are likely to be different from those whose populations are controlled by wintering ground factors, and that these characteristics can be tested for experimentally. They propose that most long-distance, forest-related migrant species that breed in the Nearctic are controlled by availability of winter forest habitat and are likely to decline as a result of loss of forested habitat in the tropics.

Other factors besides actual forest removal may affect populations of forest-related migrants in the tropics. For instance, grazing can cause degradation of tropical forest habitats. Unfortunately, aside from Martin (1984), there is little published information on this phenomenon.

Another source of winter habitat degradation is the conversion by tree-crop farmers from orchards that mimic second-growth forest habitats, which can be quite rich in migratory birds, to various forms of intensive management in which the orchards are maintained as monocultures, which are poor for bird use. The effect of this type of treatment is currently being tested by several research groups but little has been published on the topic to date.

13.3   CONSERVATION QUESTIONS AND PERSPECTIVES BY REGION

(a)   North America

There are four principal conservation concerns for forest-related migrant birds in North America: forest loss; forest degradation; forest fragmentation; and remote effects on local populations. At present, there is no way to differentiate between the effects of one of these problems as opposed to any other, except perhaps based on the characteristics of individual populations (Rappole and McDonald, 1994). It is probable that different species suffer from different combinations and to different degrees from these difficulties.

(b)   South America

Forest-related migrant birds in South America suffer from the same series of problems as those in North America plus the added problem of an almost complete lack of knowledge of the actual location of many populations during different times of the year. Very little banding work has been done in South America; as a result, the timing of migratory movements, the routes followed and even many of the species participating in migration have yet to be identified, though Chesser (1994) has identified a remarkable 230 species of austral migrants for South America. Conservation of these species must begin with the basic knowledge of where they go and when.

(c)   Europe

Knowledge of the distribution, timing of migration, route selection and even wintering location is excellent for most European forest-related migrants, due to a long history of thorough banding studies throughout the species' ranges in Europe and Africa (Moreau, 1972). Based on this information, the European forest-related avifauna appears to be depauperate when compared with both the Asian and New World migrant avifauna. Moreau (1972) lists a total of 45 species of migrants that breed regularly in forest in Europe as compared with 112 species for the New World and 107 for the Asian migration system (Rappole, 1995). According to Monkonnen et al. (1992:9), only three of these 45 species winter regularly in forested habitats in Africa. Thus, threats to remote wintering habitats are seemingly less important for European migrants than New World species, though apparently breeding populations of the whitethroat (*Sylvia communis*), an Old World warbler, declined by as much as 77% on their European breeding grounds during the 1960s and early 1970s as an apparent result of prolonged drought in their sub-Saharan wintering range (Morse, 1980). Forest fragmentation does not appear to pose a serious threat to European forest-related migrants (Haila, 1986).

Perhaps the most serious pressures on Europe's forest-related migrants come from the same kinds of broad-based dangers to the environment posed by various atmospheric pollutants that threaten vast regions of North America.

(d) Africa

As is the case for Europe, the movements, routes and wintering locations for most Palearctic migrants are well known, thanks to extensive banding studies. These studies demonstrate that very few migrants from Europe winter in forest in Africa (Monkkonen *et al.*, 1992). As in South America, there are austral migrants, some of which are forest-related. The movements and ecology for most of these species are poorly known.

(e) Asia

The Asian migration system appears to be very similar to that of the New World. Nearly the same number of migrants that breed in temperate Asia are forest-related as breed in temperate North America (107 versus 112), and Asian migrants that breed in forest habitats also winter in forest habitats, as is true for most North American migrants (Rappole, 1995). However, like the South American continent, migration in the vast Asian continent has not been well studied. Aside from the landmark banding operations of McClure (1974), very little has been done to trace the movements of Asian migrants. Thus the threats to Asian migrants are neither well known nor understood. Yet, given the similarities between the Asian and North American migration systems, it seems probable that forest habitat reductions on breeding, stop-over and wintering sites are likely to pose serious threats to Asian forest-related migrant species. Summaries on forest cover change show sharp reductions in forest cover in many countries of the region (World Resources Institute 1992).

13.4 CONCLUSIONS

A significant portion of the world's avifauna is dependent upon forest for survival, including many long-distance migrants. For most of these species, there are insufficient data to determine empirically a cause–effect relationship between forest loss and population decline. For a few species, e.g. the wood thrush (*Hylocichla mustelina*), forest loss during one phase of the cycle has been so severe as to present obvious threats to the continued survival of the species. Conservation efforts should be focused on identifying these gravely threatened species and attempting to identify and preserve the weakest link in terms of habitats required for completion of their annual cycle.

13.5 SUMMARY

Migratory birds are major components of many of the world's forest environments. Over one-quarter of the 1100 or so species of bird that migrate from breeding grounds in temperate regions to wintering grounds in the tropics use forest habitats during one or more phases of the annual cycle. Yet the importance of forest to these species is not altogether clear due to the complex nature of their life history. In North America, populations of a number of forest-related migrant species appear to be declining. Removal, degradation and/or fragmentation of breeding, post-breeding, stop-over and/or wintering habitats may all play a role in these declines. For a few species, e.g. the wood thrush (*Hylocichla mustelina*), habitat loss during one phase of the cycle has been severe, potentially threatening continued survival of the species. Conservation efforts should be focused on identifying these threatened species and attempting to identify and preserve the weakest link in terms of habitats required for completion of their annual cycle.

13.6 REFERENCES

Alverson, W.S., Waller, D.M. and Solheim, S.L. (1988) Forests too deer: edge effects in northern Wisconsin. *Conserv. Biol.* 2:348–358.

Askins, R.A., Lynch, J.F. and Greenberg, R. (1990) Population declines in migratory birds in eastern North America. *Curr. Ornith.* 7:1–57.

Askins, R.A., Ewert, D.N. and Norton, R.L. (1992) Abundance of wintering migrants in fragmented and continuous forests in the US Virgin Islands, in *Ecology and Conservation of Neotropical Migrant Landbirds* (eds J.M. Hagan III and D.W. Johnston), pp. 197–206. Smithson. Inst. Press, Washington, DC.

Audubon, J.J. (1840–1844) *The Birds of America*. 7 vols. Audubon and Chevalier, New York.

Baird, T.H. (1990) *Changes in breeding bird populations between 1930 and 1985 in the Quaker Run Valley of Allegheny State Park, New York*. New York State Museum Bull. 477, Univ. of the State of New York, Albany, New York.

Bairlein, F. (1983) Habitat selection and associations of species in European passerine birds during southward, post-breeding migrations. *Ornis Scand.* 14:239–245.

Bairlein, F. (1992) Morphology–habitat relationships in migrating songbirds, in *Ecology and Conservation of Neotropical Migrant Landbirds* (eds J.M. Hagan III and D.W. Johnston), pp. 356–369. Smithson. Inst. Press, Washington, DC.

Baskett, T.S., Armbruster, M.J. and Sayre, M.W. (1978) Biological perspectives for the Mourning Dove call-count survey. *Trans. N. Amer. Wildl. Nat. Res. Conf.* 37:312–325.

Bent, A.C. (1953) *Life histories of North American Wood Warblers*. Bull. US Natl. Mus., Vol. 203, Washington, DC. 733 pp.

Berthold, P. (1988) The control of migration in European warblers. *Int. Ornith. Congr.* 19:215–249.

Berthold, P., Bairlein, F. and Querner, U. (1976) Über die Verteilung von ziehenden

Kleinvögeln in Rastbiotopen und den Fangerfolg von Fanganlagen. *Vogelwarte* 28:267–273.

Brewer, R. (1980) A half-century of changes in the herb layer of a climax deciduous forest in Michigan. *J. Ecol.* 68:823–832.

Chesser, T. (1994) Migration in South America: an overview of the austral system. *Bird Conserv. Int.* 4:91–107.

Cooke, W.W. (1915) Bird migration. *USDA Bulletin* 185:1–47.

deCalesta, D. (in press) Impact of deer on interior forest songbirds in northwestern Pennsylvania. *J. Wildl. Manage.*

DeGraaf, R.M. and Rappole, J.H. (1995) *Neotropical Migratory Birds.* Cornell Univ. Press, Ithaca, New York.

DeGraaf, R.M., Scott, V.E., Hamre, R.H. *et al.* (1991a) *Forest and Rangeland Birds of the United States: natural history and habitat use.* Agric. Handbook 688, US Dept of Argic., Washington, DC.

DeGraaf, R.M., Healy, W.M. and Brooks, R.T. (1991b) Effects of thinning and deer browsing on breeding birds in New England oak woodlands. *Forest Ecol. and Manage.* 41:179–191.

Diamond, D.D. and Fulbright, T.E. (1990) Contemporary plant communities of upland grasslands of the Coastal Sand Plain, Texas. *Southwest. Natur.* 35:385–392.

Duffy, D.C. and Meier, A.J. (1992) Do appalachian herbaceous understories ever recover from clearcutting? *Conserv. Biol.* 6:196–201.

Flaccus, E. (1959) Revegetation of landslides in the White Mountains of New Hampshire. *Ecology* 40:692–703.

Groebbels, F. (1928) Zur Physiologie des Vogelzuges. *Verh. Ornith. Ges. Bayern* 18:44–74.

Hahn, H.W. (1937) Life history of the Ovenbird in southern Michigan. *Wilson Bull.* 49:145–237.

Haila, Y. (1986) North European land birds in forest fragments: evidence for area effects? in *Wildlife 2000. Habitat Relationships of Terrestrial Vertebrates* (eds J. Verner, M. Morrison and C.J. Ralph), pp. 315–319. Univ. of Wisconsin Press, Madison, Wisconsin.

Harris, H. (1928) Robert Ridgway: with a bibliography of his published writings and fifty illustrations. *Condor* 30:5–118.

Jewell, P.A. and Holt, S. (eds) (1981) *Problems in Management of Locally Abundant Wild Mammals.* Academic Press, London.

Karr, J.R. (1976) On the relative abundance of migrants from the north temperate zone in tropical habitats. *Wilson Bull.* 88:433–458.

Krebs, J.R. (1971) Territory and breeding density in the Great Tit, *Parus major* L. *Ecology* 52:2–22.

Laws, R.M., Parker, I.S.C. and Johnstone, R.C.B. (1975) *Elephants and their Habitats.* Clarendon Press, Oxford.

LeGrand, H.E., Jr and Schneider, K.J. (1992) Bachman's Sparrow, *Aimophila aestivalis,* in *Migratory Nongame Birds of Management Concern in the Northeast,* (eds K.J. Schneider and D.M. Pence), pp. 299–314. Dept Interior, Fish and Wildl. Service, Newton Corner, Massachusetts.

Leimgruber, P., McShea, W.J. and Rappole, J.H. (1994) Predation on artificial nests in large forest blocks. *J. Wildl. Manage.* 58:254–260.

Lynch, J.F. and Whitcomb, R.F. (1968) Effects of the insularization of the eastern deciduous forest on avifaunal diversity and turnover, in *Classification, Inventory, and Evaluation of Fish and Wildlife Habitat* (ed. A. Marmelstein), pp. 461–489. US Fish and Wildl. Serv. Publ. OBS-78176.

MacLean, D.A. and Wein, R.W. (1977) Changes in understory vegetation with increasing stand age in New Brunswick forests: species composition, biomass, and nutrients. *Canadian J. Botany* 55:2818–2831.

Martin, T.E. (1984) Impact of livestock grazing on birds of a Colombian cloud forest. *Tropical Ecol.* 25:158–171.

Martin, T.E. (1992) Breeding productivity considerations: appropriate habitat features for management, in *Ecology and Conservation of Neotropical Migrant Landbirds* (eds J.M. Hagan III and D.W. Johnston), pp. 455–473. Smithsonian Inst. Press, Washington, DC.

McClure, H.E. (1974) *Migration and Survival of the Birds of Asia*. US Army Medical Component SEATO Medical Project, Bangkok, Thailand. 478 pp.

McShea, W.J., Rappole, J.H. and Burford, G. (in review) Changes in vocalization rates for Wood Thrushes and Ovenbirds in continuous versus fragmented forest. *J. Field Ornith.*

Meyer de Schauensee, R. (1966) *A Guide to the Birds of South America*. Livingston Publ. Co., Wynnewood, Pennsylvania, USA.

Monkkonen, M., Helle, P. and Welsh, D. (1992) Perspectives on Palaearctic and Nearctic bird migration; comparisons and overview of life-history and ecology of migrant passerines. *Ibis* 134 (Supp. 1):7–13.

Moore, F.R. and Simons, T.R. (1992) Habitat suitability and stopover ecology of Neotropical landbird migrants, in *Ecology and Conservation of Neotropical Migrant Landbirds* (eds J.M. Hagan III and D.W. Johnston), pp. 345–355. Smithson. Inst. Press, Washington, DC.

Moreau, R.E. (1972) *The Palaearctic–African Bird Migration System*. Academic Press, New York.

Morse, D.H. (1971) The insectivorous bird as an adaptive strategy. *Ann. Rev. Ecol. Syst.* 2:177–200.

Morse, D.H. (1980) Population limitation: breeding or wintering grounds?, in *Migrant Birds in the Neotropics* (eds A. Keast and E.S. Morton), pp. 505–516. Smithson. Inst. Press, Washington, DC.

Morton, E.S. (1992) What do we know about the future of migrant landbirds?, in *Ecology and Conservation of Neotropical Migrant Landbirds* (eds J.M. Hagan III and D.W. Johnston), pp. 579–589. Smithson. Inst. Press, Washington, DC.

Nice, M.M. (1964) *Studies of the Life History of the Song Sparrow. Vol. 2. Behavior.* Dover Publ. Inc., New York. 328 pp.

Nolan, V., Jr (1978) The ecology and behavior of the Prairie Warbler *Dendroica discolor*. *Ornith. Monogr.* 26.

Orodho, A.B., Trlica, M.J. and Bonham, C.D. (1990) Long-term heavy-grazing effects on soil and vegetation in the four corners region. *Southwest. Natur.* 35:9–14.

Pastor, J. and Naiman, R.J. (1992) Selective foraging and ecosystem processes in boreal forests. *Amer. Natur.* 139:690–705.

Petit, D.R., Petit, L.J. and Smith, K.G. (1992) Habitat associations of migratory birds overwintering in Belize, Central America, in *Ecology and Conservation of*

Neotropical Migrant Landbirds (eds J.M. Hagan III and D.W. Johnston), pp. 247–256. Smithson. Inst. Press, Washington, DC.

Petit, D.R., Lynch, J.F., Hutto, R.L. *et al.* (1993) Management and conservation of migratory landbirds overwintering in the Neotropics, in *Status and Management of Neotropical Migratory Birds* (eds D.M. Finch and P.W. Stangel), pp. 70–92. Gen. Tech. Rep. RM-229. US Dept Agriculture, For. Ser., Rocky Mountain Forest and Range Experiment Station, Fort Collins, CO.

Powell, G.V.N. and Rappole, J.H. (1986) The Hooded Warbler, in *Audubon Wildlife Report,* Vol. 3 (ed. R.L. Di Silvestro), pp. 827–853. National Audubon Society, New York.

Prins, H.H.T. and Van der Jeugd, H.P. (1993) Herbivore population crashes and woodland structure in East Africa. *J. Ecol.* 81:305–314.

Pulich, W.M. (1976) *The Golden-cheeked Warbler. A bioecological study.* Texas Parks Wildl. Dept., Austin, Texas.

Rappole, J.H. (1995) *Ecology of Migrant Birds: A Neotropical Perspective.* Smithsonian Inst. Press, Washington, DC.

Rappole, J.H. and Ballard, K. (1987) Passerine post-breeding movements in a Georgia old field community. *Wilson Bull.* 99:475–480.

Rappole, J.H. and McDonald, M.V. (1994) Cause and effect in population declines of migratory birds. *Auk* 111:652–660.

Rappole, J.H. and Morton, E.S. (1985) Effects of habitat alternation on a tropical forest community. *Ornith. Monogr.* 6:1013–1021.

Rappole, J.H. and Waggerman, G. (1986) Calling males as an index of density for breeding White-winged Doves. *Wildl. Soc. Bull.* 14:151–155.

Rappole, J.H. and Warner, D.W. (1976) Relationships between behavior, physiology and weather in avian transients at a migration stopover site. *Oecologia* 26:193–212.

Rappole, J.H., Russell, C.E., Norwine, J.R. and Fulbright, T.E. (1986) Anthropogenic pressures and impacts on marginal, Neotropical, semiarid ecosystems: the case of South Texas. *Sci. Total Environ.* 55:91–99.

Rappole, J.H., McShea, W.J. and Vega, J. (1993) Evaluation of two survey methods in upland avian breeding communities. *J. Field Ornith.* 64:55–65.

Rappole, J.H., Powell, G.V.N. and Sader, S. (1994) Remote sensing assessment of tropical habitat availability for a Nearctic migrant: the Wood Thrush, in *Mapping the Diversity of Nature* (ed. R.I. Miller), pp. 91–104. Chapman & Hall, London.

Robbins, C.S., Dowell, B.A., Dawson, D.K. *et al.* (1987) Comparison of Neotropical winter bird populations in isolated patches versus extensive forest. *Acta Oecologia Generalis* 8:285–292.

Robbins, C.S., Sauer, J.R., Greenberg, R.S. and Droege, S. (1989a) Population declines in North American birds that migrate to the Neotropics. *Proc. Natl. Acad. Sci.* 86:7658–7662.

Robbins, C.S., Dawson, D.K. and Dowell, B.A. (1989b) Habitat area requirements of breeding forest birds of the Middle Atlantic states. *Wildl. Monogr.* 103.

Robbins, C.S., Dowell, B.A., Dawson, D.K. *et al.* (1992) Comparison of Neotropical migrant landbird populations wintering in tropical forest, isolated forest fragments, and agricultural habitats, in *Ecology and Conservation of Neotropical Migrant Landbirds* (eds J. M. Hagan III and D.W. Johnston), pp. 207–220. Smithson. Inst. Press, Washington, DC.

Rodenhouse, N.L. (1992) Potential effects of climatic change on a Neotropical migrant landbird. *Conserv. Biol.* 6:263–272.

Schorger, A.W. (1955) *The Passenger Pigeon.* Wisconsin Univ. Press, Madison, Wisconsin.

Smith, D.M. (1962) *The Practice of Silviculture.* J. Wiley and Sons, Inc., New York.

Szaro, R.C. (1981) Bird population responses to converting chaparral to grassland and riparian habitats. *Southwest. Natur.* 26:251–256.

Tchamba, M.N. and Mahamat, H. (1992) Effects of elephat browsing on the vegetation in Kalamaloue National Park, Cameroon. *Mammalia* 56:533–540.

Thompson, F.R., Dijak, W.D., Kulowiec, T.G. and Hamilton, D.A. (1992) Breeding bird populations in Missouri Ozark forests with and without clearcutting. *J. Wildl. Manage.* 56:23–30.

US Dept of Agriculture (1982) *An analysis of the timber situation in the United States 1952–2030.* US Forest Service, Washington, DC. Forest Service Report No. 23.

US Dept of Commerce (1924) *Statistical abstract of the United States: 1923* (46th edn). Bureau of Foreign and Domestic Commerce, Washington, DC.

US Dept of Commerce (1935) *Statistical abstract of the United States: 1925* (57th edn). Bureau of Foreign and Domestic Commerce, Washington, DC.

Vega, J. and Rappole, J.H. (1994) Effects of scrub mechanical treatment on the nongame bird community in the Rio Grande Plain of Texas. *Wildl. Soc. Bull.* 22:165–171.

Warren, R.J. (1991) Ecological justification for controlling deer populations in eastern national parks. *Trans. N. Amer. Wildl. Natural Resources Conf.* 56:56–66.

Whitcomb, B.L., Whitcomb, R.F. and Bystrak, D. (1977) III. Long-term turnover and effects of selective logging on the avifauna of forest fragments. *American Birds* 31:17–23.

Whitcomb, R.F. (1977) Island biogeography and 'habitat islands' of eastern forest. *Amer. Birds* 31:3–5.

Whitcomb, R.F., Lynch, J.F., Opler, P.A. and Robbins, C.S. (1976) Island biogeography and conservation: strategy and limitations. *Science* 193:1030–1032.

Wiens, J.A. (1989) *The Ecology of Bird Communities.* Cambridge Univ. Press, New York.

Wilson, A. (1808–1814) *American Ornithology.* 9 Vols. Bradford and Inskeep, Philadelphia, Pennsylvania, USA.

Winker, K., Rappole, J.H. and Ramos, M.A. (1990) Population dynamics of the Wood Thrush (*Hylocichla mustelina*) on its wintering grounds in southern Veracruz, Mexico. *Condor* 92:444–460.

Winker, K., Warner, D.W. and Weisbrod, A.R. (1992a) The Northern Waterthrush and Swainson's Thrush as transients at a temperate inland stopover site, in *Ecology and Conservation of Neotropical Migrant Landbirds* (eds J.M. Hagan III and D.W. Johnston), pp. 384–402. Smithson. Inst. Press, Washington, DC.

Winker, K., Warner, D.W. and Weisbrod, A.R. (1992b) Migration of woodland birds at a fragmented inland stopover site. *Wilson Bull.* 104:580–598.

Winker, K., Warner, D.W. and Weisbrod, A.R. (1992c) Daily mass gains among woodland migrants at an inland stopover site. *Auk* 109:853–862.

World Resources Institute (1992) *World Resources: 1992–1993.* Oxford Univ. Press, New York.

–14

Wildlife habitat evaluation in forested ecosystems: some examples from Canada and the United States

Paul A. Gray, Duncan Cameron and Ian Kirkham

14.1 INTRODUCTION

Forested ecosystems cover almost 40% of North America and are a dominant, albeit controversial, feature of life in North America, where they are as central to continental culture as they are to the economy (Caldwall *et al.*, 1994; Canadian Council of Forest Ministers, 1992; Fulford, 1992). Forested ecosystems are home to most wildlife species inhabiting the continent. For example, of the estimated 300 000 species found in Canada, 200 000 meet at least part of their life requirements in forested ecosystems (Natural Resources Canada, 1993:23). From another perspective, 90% of the resident or common migrant vertebrate species in the United States inhabit forested ecosystems at some point in their lives (Flather and Hoekstra, 1989:4).

With continued global human population pressures and the escalating demand for natural assets, people are acutely aware of the need for improved management of the earth's ecosystems. Recognition of the fact that we live in a finite world, with limited space and natural assets, has precipitated considerable support for change in the last two decades – change directed towards the creation of societies that subscribe to, and practise, the art and science of living sustainably. The ways in which we care for forested ecosystems and their wildlife assets are an important part of this change. For example, many government agencies in Canada and the United States, in partnership with the private sector, institutions and the public at large, have begun to explore the use of an ecosystem approach to

Conservation of Faunal Diversity in Forested Landscapes. Edited by R.M. DeGraaf and R.I. Miller. Published in 1996 by Chapman & Hall. ISBN 0 412 61890 7.

care for North America's natural assets (Gerlach and Bengston, 1994; Fowle *et al.*, 1991; Gray *et al.*, 1995; Kaufmann *et al.*, 1994; Robertson, 1992; Thomas, 1994; Thompson and Welsh, 1993).

In response to changing societal values and attitudes, agencies are discarding programs in which forested ecosystems are valued and treated solely as 'fiber factories' in exchange for programs which recognize them as plethoric providers of ecospheric health as well as commodities like recreation, lumber and fiber. In addition to significant perceptual and philosophical change with respect to the way we care for forested ecosystems, the management tools with which we make and turn decisions into action are evolving rapidly as well. Managers are devoting considerable time and effort to develop, adopt and employ tools which, as Odum (1977) suggested, allow us to move away from mere component analyses, wherein ecosystem variables are treated as separate entities, to more holistic approaches designed to identify and understand the properties of ecosystems.

Experience has taught us that successful forest ecosystem management must be supported by adequate data bases and carefully designed data and information management systems upon which to base decisions (Gray and Stelfox, 1991). For example, a number of efficient, cost-effective inventory and monitoring tools are now available to account for wildlife values within forest ecosystem management programs, including portable electronic data collectors, computers and sophisticated software, Landsat and other satellite imagery and digital data, aerial photography, and thematic and integrated maps. Effective application of the data and information we collect depends on how we manage and interpret it, and apply it in support of ecologically-based decisions. Accordingly, natural asset managers rely on a variety of modeling tools and techniques to help them make the right decisions.

Models are tools widely used by industry, organizations, institutions and government agencies to assess the state of forested ecosystems, to predict and compare the results of management actions (including no action), to assist with decision-making, and to provide a framework within which program objectives, policies and strategies can be tested, applied and modified (Kessel, 1979, and others). This chapter focuses on some of the wildlife habitat evaluation models employed in support of programs designed to manage human activities in North America's forested ecosystems.

14.2   THE CONTEXT: AN ECOSYSTEM APPROACH TO MANAGEMENT

Fundamentally, an ecosystem approach to management is based on the idea that if humans subscribe and apply an appropriate set of values and are equipped with the required knowledge and tools, they can protect and maintain ecosystems, derive a quality existence from them and simultaneously, ensure that opportunities for future generations are retained. The traditional approach to management has been sectoral (as exemplified by the separate administration of timber, wildlife, parks and protected areas, fisheries, flood control and aggregate management programs) and utilitarian. While this 'stovepipe' approach has provided some programs that can be employed to implement an ecosystem approach, it has failed to provide the 'integrating' tools necessary to understand and make decisions about 'whole' systems.

How do we define, describe and implement an ecosystem approach to management? Is it one large integrated program, centrally defined and managed, or is it a network of policies and programs coordinated and conducted in integrated unison? What are the key elements of an ecosystem approach? Cortner and Moote (1994), Fowle *et al.* (1991), Gerlach and Bengston (1994), Grumbine (1994), Iverson (1993), Kaufmann *et al.* (1994), Lackey (1994), Maerz (1994), Slocombe (1993) and many others call for significant change in our approach to decision-making through development and implementation of an integrated suite of policies and programs. For example, they suggest that we need to: employ ecologically meaningful spatial frameworks in which to plan and make decisions; describe the systems, their parts and interactions; adopt a comprehensive and holistic approach that recognizes the dynamic nature of ecosystems through application of systems dynamics theory; encourage co-operation and collaboration that includes clients in all aspects of management; integrate programs and actively manage issues; employ flexible research and planning processes; entrench the ecosystem concept in our values, attitudes and behaviors; and understand through knowledge, which includes data and information collection and monitoring, and education, extension and training. But this long list of needs and their associated programs exemplifies a complex suite of ideas and programs requiring rationalization, planning, organization (e.g. linkage and integration) and application.

Gray *et al.* (1995) propose that an ecosystem approach to management is spatially inclusive and accounts for the complex array of ecological, social, cultural and economic conditions and forces, which are managed as complementary components within an integrated framework (Figure 14.1). While recognizing that institutional reformation (e.g. changing roles and responsibilities) will proceed at a rapid pace in the 1990s, they propose that programs be organized and implemented within a framework of interrelated modules – a network of current and new programs coordinated and completed in unison (Figure 14.2). The modules, which permit managers to conduct their work and make decisions within a holistic context, are organized according to three themes: context, enablers and tools.

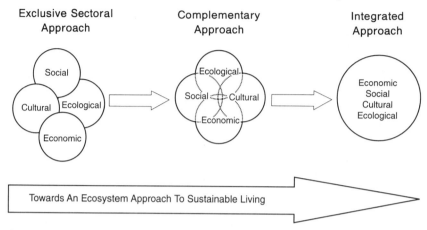

Figure 14.1   An ecosystem approach to management requires society to move from an exclusive, sectoral approach in valuing and using natural assets to an integrated approach. (From Gray *et al.*, 1995.)

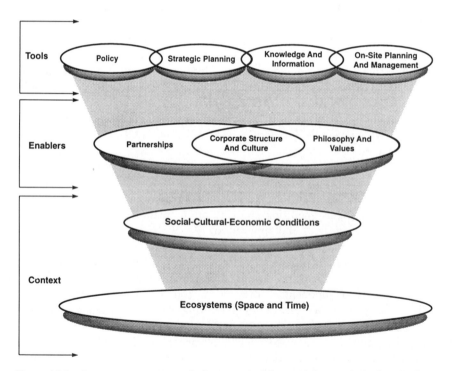

Figure 14.2   An ecosystem approach framework. The modules are linked and often employed in unison to deliver programs. Wildlife habitat evaluation plays an important role in each of the modules in which natural asset managers make decisions. (From Gray *et al.*, 1995.)

- Context:
  - ecosystem definition and description in space and time and the social, cultural and economic forces in them;
  - a philosophy that enables natural asset managers to take account of the range of evolving societal values;
  - corporate or institutional structure and culture designed to provide proactive and integrated programs;
  - inclusive programs involving all sectors in society as partners in decision-making processes;
- Tools:
  - knowledge and information, including science and technology (e.g. inventory, monitoring and assessment) and learning through education, extension and training;
  - strategic thinking and planning to identify, establish and modify short-term and long-term direction;
  - policy, legislation and regulation to guide society in the adoption and attainment of sustainable lifestyles;
  - natural asset planning and program management.

Some of these modules and many of the programs contained in them are not new, nor are they unique to an ecosystem approach to management. Organizations have used them for many years. So, why recycle them? Gray *et al.* (1995) argue that traditional approaches to natural asset management have, in many cases, been based on development and implementation of isolated programs that fail to account for the range of social, cultural, economic and ecological factors and forces. An ecosystem approach provides an all-encompassing framework within which to organize people and their programs, set priorities and make decisions that account for the range of factors and forces.

Given the complex nature of the ecosphere and our limited knowledge of it, it is unlikely that we will be able to 'manage' entire ecosystems in the foreseeable future. Nevertheless, we can employ an 'ecosystem approach' to manage our relationship with the other parts of ecosystems by ensuring that our perceptions, values and behaviors work in support of sustainable living objectives. We employ the modules to manage for the range of social, cultural, economic and ecological forces and factors that ultimately define human–ecosystem relationships. While definition and description of an ecosystem approach to management is not the focus of this chapter, many wildlife habitat evaluation techniques are significant components of the modules.

14.2.1 Ecosystem defined

The earth is home to life or, in the words of Rowe (1991), 'Home Place', and because it is spherical we call this narrow band of land–water–

atmosphere just below, on and above the surface the **ecosphere**. The ecosphere is the earth's largest ecosystem and within it matter and energy combine in infinite ways to create unique, dynamic entities in-space-in-time – geological formations such as hills and valleys, surficial materials and soils, water, climate and organisms. We call these entities 'systems' – connected things or parts that function through the movement of matter and energy (Bertalanffy, 1968; Hall and Fagen, 1956; Miller, 1975; Naveh and Lieberman, 1990:43, and others). Therefore, when we think of the earth as ecosystem, we refer to it as a 'dynamic home of many parts'.

The definition of **ecosystem** has evolved since Arthur Tansley coined the term in 1935; he argued that to understand the natural world we must integrate our knowledge about the 'parts' and conceive them as one system – the ecosystem. Building on the work of Tansley and many others, Gray *et al.* (1995) define ecosystem as a recognizable chunk of earth space in which the flow of energy and the transformation of matter in-space-in-time creates networks of organisms (such as plants and animals, including humans), atmosphere, soil, rock and water, interacting with each other and with other ecosystems.

Although every ecosystem is unique, the concept of ecosystem as an entity comprising many interacting parts of matter and energy allows us to identify ways of describing each system. In this context we refer to ecosystems as having composition, structure and function (King, 1993; Noss, 1990) (Figure 14.3).

(a) Composition

Composition is perhaps the most easily understood of the three. It is a word that denotes matter and energy in various forms – it represents the identity and variety of elements in a collection (Noss, 1990). For example, ecosystem matter includes: surface water, ground water, snow and ice; geological formations, surficial materials and soils; atmospheric gases and particulates; human-made structures; and plants, animals and other organisms. The ecosphere is fuelled by the sun and energy contained in, and occasionally released from, the earth's subsurface lithosphere. Natural-asset managers employ species lists and genetic/species diversity indices, for example, as measures of composition (Noss, 1990).

(b) Structure

Ecosystem structure commonly refers to the distribution and abundance of matter and energy (King, 1993) and is used to describe matter – energy relationships (how the parts fit together at a specific point in time). Structure represents the physical organization of the system as reflected in the variety of shapes, forms and patterns found in it. The study of structure involves measuring components such as wildlife habitat (e.g. corridors and patches) at a variety of map scales (Noss, 1990).

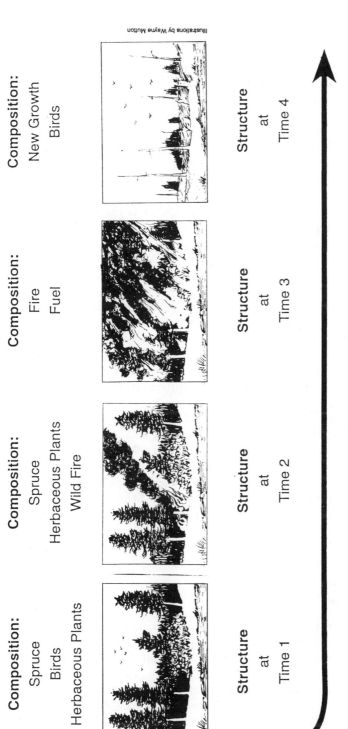

Composition:
Spruce
Birds
Herbaceous Plants

Composition:
Spruce
Herbaceous Plants
Wild Fire

Composition:
Fire
Fuel

Composition:
New Growth
Birds

Illustrations by Wayne Mutton

Structure
at
Time 1

Structure
at
Time 2

Structure
at
Time 3

Structure
at
Time 4

Function

Figure 14.3 Ecosystem composition, structure and function. Composition denotes the elements (matter and energy) that comprise an ecosystem, structure represents the patterns resulting from the interacting elements, and function is composition and structure through time. (From Gray *et al.*, 1995.)

(c) *Function*

As chunks of space-in-time through which energy flows and matter is transformed, ecosystems constantly change. An ecosystem is bound together by a network of food webs: green plants use solar energy to manufacture substances from carbon dioxide in air, from water, and from nutrients in soil. Plants are eaten by herbivores which, in turn, are eaten by carnivores. When an organism dies its body is recycled by decomposer organisms such as bacteria and fungi. Thus, the solar energy originally captured by green plants is gradually consumed as it flows from one level of the food chain to the next.

Ecosystem function, therefore, can be thought of as the flow and dissipation of energy and the transfer and consumption of matter. While composition and structure can be equated to one frame in a movie in which we observe individuals and collections (e.g. patches, corridors and other structural forms) shaped by the movement of energy and matter at a single point in time; function represents all the frames in the movie (Figure 14.3) (Gray *et al.*, 1995). Therefore, what we call function is structural change second by second, hour by hour, day by day, year by year (Noss, 1995; Rowe, 1993). Because structure also can be reduced to its composition, we can define function as composition and structure-through-time (J.S. Rowe, personal communication).

It is critical that we recognize ecosystem productivity (vis-à-vis function) as a limited asset. It is limited by space-in-time, by availability of nutrients, by photosynthetic efficiency and ultimately by the rate of energy input itself (solar flux and lithospheric energy). Ecosystems, therefore, do not − and cannot − grow indefinitely (Fowle *et al.*, 1991; Rees, 1990; and many others). It follows, then, that human use of ecosystems also has limits and we need to know what those limits are. Management agencies examine and employ the concept of function to describe and define relationships and processes as they change through time (e.g. ecological and evolutionary processes, such as gene flow, disturbances and nutrient cycling) and to identify limits (Noss, 1990). Natural asset managers explore many aspects of ecosystem function through wildlife habitat evaluation.

14.2.2 Ecosystems in-space-in-time

An important step to understanding the earth's ecosystems is development of a spatial framework in which to organize information about them. Usually, this is done by arranging them hierarchically. Nested within the largest ecosystem (i.e. the ecosphere) are smaller ecosystems, defined and characterized on the basis of common features which set them apart from other units. These are recognizable because of unique interactions (function) among the components (composition) and the patterns (structure). The hierarchy allows natural asset managers to understand and work with eco-

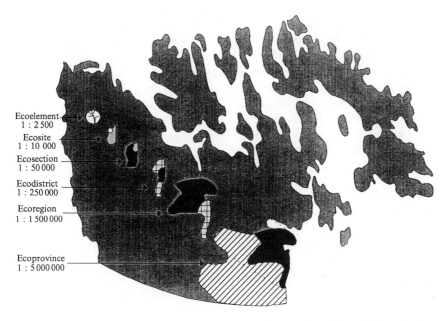

Figure 14.4 Six of the seven levels in the Canadian Ecological Land Classification System hierarchy. In a hierarchical context, an ecosystem mapped and described at a scale of 1:250 000 is seen as both a part of a larger entity (e.g. an ecosystem delineated at a scale of 1:1 000 000) and a whole which can be segmented (e.g. at a scale of 1:50 000).

system composition, structure and function at a variety of map scales. For example, a unit mapped and described at a scale of 1:250 000 is seen as both a part of a larger entity mapped at a scale of 1:1 000 000 and as a whole which can be segmented at a scale of 1:50 000 (Figure 14.4) (Allen and Hoekstra, 1992:52; O'Neill *et al.*, 1986:83; Wiken and Ironside, 1977).

14.2.3 Wildlife defined

Given the comprehensive nature of an ecosystem approach to management, an important integrating concept is the definition and use of the term 'wildlife'. As noted by Thomas *et al.* (1978), many government programs traditionally employed the word in reference to animals of economic and social importance (e.g. hunted and trapped species) and species of special status, like vulnerable, threatened and endangered plants and animals. In Canada, this changed in 1990 when federal, provincial and territorial jurisdictions adopted a comprehensive definition of wildlife in their support of *A Wildlife Policy For Canada* (Andrews *et al.*, 1990) in which 'wildlife' means 'wild mammals, birds, reptiles, amphibians, fishes, invertebrates, plants, fungi, algae, bacteria, and other wild organisms'. In adopting this broad definition, the Ontario Wildlife Working Group (Fowle *et al.*, 1991; Gray *et al.*, 1993) took an additional step by splitting the word (from 'wildlife' to 'wild life') to emphasize 'wild life management' as the management of *all* life in the wild. While we subscribe to this comprehensive definition, most examples in this paper address habitat evaluation of terrestrial vertebrates. Accordingly, it is important to note that wildlife habitat evaluation techniques are available for a spectrum of aquatic and other terrestrial organisms, including plants and micro-organisms, found in forested ecosystems.

14.2.4   Habitat defined

Habitat has been variously described as 'the place', 'the home' or 'the address' of an organism, and consequently confused with the concept of ecosystem as 'Home Place'. Rowe (1994:5) suggests that 'habitat' be used as a more specific term to reference the unique suite of ecosystem parts or components that an organism requires in-space-in-time to survive. Habitat, therefore, is that unique combination of biotic and abiotic ecosystem parts in-space-in-time that provide or detract from the essentials of food, water and cover during an organism's life. It follows that an organism can be understood through its use of many habitats in an ecosystem or different habitats in many ecosystems, sometimes separated by great distances.

14.2.5   Niche defined

Many agree that 'niche' is the functional role, the 'profession', of species in the ecosystem (e.g. Davey and Stockwell, 1991; Forman and Godron, 1986:63). In applied management, it can be defined by how a species uses the ecosystem and how the ecosystem uses it. This is reflected as the organism's 'in-space-in-time' response to existing life requisites (habitat) and interactions with other organisms. Flather and Hoekstra (1985) define niche as a 'bridge' between species–habitat relationships and interspecific interactions (e.g. competition and predation). While 'fundamental niche' is the functional space-in-time that a species can potentially inhabit, the ability to occupy habitat is further determined by forces such as competition and predation. This is called the 'realized niche' (Davey and Stockwell, 1991; Miller, 1967; Whittaker *et al.*, 1973). Historically, many habitat evaluation techniques and programs have not addressed the functional role of organisms and the delineation of niche in the ecosystem. New, evolving modeling techniques, however, show significant potential as tools with which natural asset managers will be able to account for, and make decisions about, the distribution and abundance of wildlife on the basis of niche–habitat dynamics.

14.2.6 Wildlife habitat evaluation

Natural asset managers employ evaluation tools and techniques to measure the distribution and abundance of wildlife, wildlife interactions with each other, the flow of energy associated with wildlife in-space-in-time, and wildlife habitat. Habitat evaluation is the process of assigning value to defined spatial units on the basis of the potential or actual occurrence of wildlife populations (Stelfox, 1991), and measures of habitat quality are the products of evaluation. Flather (1982:20) defines quality as the 'ability' of the habitat to meet an organism's life requisites. While levels of quality have been variously defined in the literature, Davey and Stockwell (1991) employ a simple but useful generic classification to account for population dynamics (as argued for by Hobbs and Hanley, 1990; Lyon *et al.*, 1987; Van Horne, 1983):

- **Optimum habitat:** Habitat that enables fecundity and/or population density of a species to be maximized.
- **Sub-optimal habitat:** Lower quality habitat where species can still reproduce, but with less success.
- **Marginal habitat:** Habitat where a species can exist but not reproduce.

From an applied, temporally sensitive perspective, Stelfox (1991) defines three classes of habitat:

- **Current habitat suitability:** Current habitat suitability is employed to estimate the current ability of a spatial unit to provide the conditions (i.e. food, water and cover) for an organism's survival and perpetuation. It reflects existing conditions, such as vegetation cover (including the current successional stages) determined by natural and anthropogenic disturbance. Because wildlife can be tied to specific habitats, ecosystem components can be employed as a measure of habitat quality (Flather, 1982:11; US Fish and Wildlife Service, 1980a). The Habitat Evaluation Procedures (US Fish and Wildlife Service, 1980a,b,c, 1981) exemplify a number of tools used by decision-makers to compare current habitat suitability (before development) with potential habitat capability (after development) in order to identify the potential impacts on wildlife habitats and populations that would occur under proposed developed scenarios.
- **Potential habitat capability:** Potential habitat capability identifies the future ability of a spatial unit to provide habitat, and is based on the knowledge of potential future conditions of the ecosystem that might result from predictable anthropogenic and natural changes (e.g. vegetation succession) (Stelfox, 1991). For example, in the province of British Columbia, managers employ a classification system in which habitat is rated on the basis of the optimum natural vegetation stage (successional stage) that could be maintained with non-intensive management practices

(Blower, 1973; in Stelfox, 1991), such as prescribed burning, grazing, logging or protection (Demarchi *et al.*, 1983). Potential habitat capability ratings are calculated for wildlife species in-space-in-time in response to proposed management prescriptions.

- **Inherent habitat capability:** Inherent habitat capability is used to evaluate the natural or inherent ability of a spatial unit to provide habitat. It assumes little or no human interference, where natural vegetation cover is representative of a stable climax or disclimax type. The Canada Land Inventory (CLI) maps, for example, emphasize the physical landscape characteristics and the 'natural state' of the land, irrespective of current cover (Perret, 1969; in Stelfox, 1991), and a spatial unit's inherent capability to support ungulates and waterfowl.

Wildlife habitat evaluation is completed through application of a suite of different modeling tools, with which forest ecosystem managers can understand wildlife in the context of ecosystems, and employ that knowledge in decisions about how we humans use our 'dynamic home of many parts'. The habitat evaluation process is normally focused on easily measured biotic and abiotic attributes (variables) in-space-in-time.

14.3   MODELS

Mayr (1982:84) and Merriam (1892) suggest that early humans were keen naturalists because their survival depended on a sophisticated knowledge of nature. It is likely that *Homo sapiens* stepped out of the shadows of evolutionary history with an inherited knowledge of animal behavior, technology and geography. Our predecessor, *Homo erectus*, is believed to have evolved an unprecedented capacity to manipulate the environment with the aid of domestic and killing tools, and a knowledge of dwelling construction techniques (Roberts, 1980:31). Useful survival techniques were passed on, and enhanced, over many thousands of generations. But the tools were facultative as well as physical, and in addition to experimenting and devising specific harvesting strategies for each species of wildlife, early humans learned where and when to find prey – they recognized different ecosystems and habitats with which their prey were associated, and learned to predict where prey might occur at given times (e.g. night and day, summer and winter).

Early humans survived through their ability to describe, to predict an outcome or a consequence of a certain action and to elect a specific course of action. Information collected and passed on over thousands of generations provided the basis for more formal approaches to understanding the earth's ecosystems in concert with the evolution of language and the invention of symbols to catalog cumulative human knowledge. In essence, humans survived by creating and applying models, which are structured, repeatable and selective approximations of the real world (Haggett and Chorley, 1967), developed and employed to gather, synthesize and analyze data and information (Kessel, 1979).

Models are valuable tools because they provide for applied synthesis of the state of the knowledge about natural assets (e.g. species), the expression of linkages between cause and effect, and explicit decision-making rules (Salwasser, 1986). From a pragmatic perspective, they provide a structured process in which natural asset managers can:

- ask questions and test hypotheses;
- rationalize and describe the need and techniques to be employed;
- identify the underlying assumptions;
- define, through written and graphical description and through mathematical formulation, ecological phenomena and wildlife–habitat relationships;
- identify deficiencies in the data and understanding of systems behavior and wildlife ecology (Stelfox, 1991).

The complex nature of ecosystems requires that we use textual, graphical and mathematical tools to approximate the real world (Davey and Stockwell, 1991). There are three classes of models: descriptive, predictive, and decisive. **Descriptive models** characterize various aspects of the real world – they are used to describe a selected set of conditions in-space-in-time. They include maps, drawings, scale models and descriptive equations (Morgan, 1967; Walters, 1971). **Predictive models** represent efforts to delineate and describe what might occur under a given set of conditions (Gélinas, 1986, 1988) and usually require descriptive model data and/or information. Computer simulation models, for example, can be employed to address questions about human-induced impacts on an ecosystem. Decisive or **decision models** identify options about a preferred course of action or response under a specific set of conditions (Jeffers, 1972, 1973; Wyngaarden and Gélinas, 1987). They depend upon the data and information derived from descriptive and/or predictive models. Formal description and application of decisive models in natural asset management, in part, has resulted from significant advances in the field of artificial intelligence, such as 'expert systems'.

The three classes of model are not mutually exclusive – contemporary modeling procedures often contain aspects of all three, and in some cases are employed repeatedly at different stages in the decision-making process (Figure 14.5). Preparation of a descriptive model for ecosystems, for example, requires the natural asset manager to identify the boundary-defining variables, position the boundary on the basis of the variables, test the boundary lines, elect to employ the boundary lines, monitor the boundary and modify it in response to natural and anthropogenic change. The bounded area serves as the descriptive model – as the spatial (and in

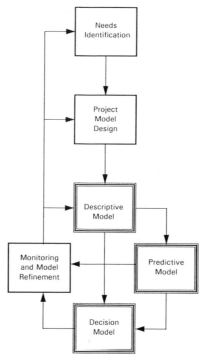

Figure 14.5   An interrelationship diagram of descriptive, predictive and decisive models. Note that the three classes of model are not mutually exclusive.

some cases temporal) context in which to make predictions and decisions and to monitor the accuracy of the predictions and the success of the decisions.

14.4   DESCRIPTIVE MODELS

The earth is a diverse and complex place, and the data and information from which descriptive models are constructed are organized or classified for practical reasons – to create order out of chaos (Bailey *et al.*, 1978), to simplify often bewildering amounts of data and to make sense of our perceptions about the real world and the ideas associated with those perceptions (Rowe, 1979). Classifications are central to the descriptive model.

Detailed accounts of how early civilizations perceived and manipulated the earth's ecosystems are scarce. That the Greeks recognized the concept of the interrelationships between organisms and the ecosystems in which they lived has been amply documented. Anaximander (610–547 BC) is credited with the first attempt to portray the earth as a whole system.

Parmenides (500 BC) developed ideas concerning climatic zonality where he classified the world into two hemispheres with three zones in each hemisphere: a habitable central temperate zone at mid latitudes with significant seasonal differences in temperature, an uninhabitable torrid median (equatorial) zone and a polar zone (Isachenko, 1973:11). The work completed in early Greece heralded the dawn of what Block and Brennan (1993) refer to as the 'Catalogue Era' with the work of Heracleitus and Parmenides in the fifth century BC and later by Aristotle. Aristotle (384–322 BC), often referred to as the founder of biological science and the father of scientific methodology (Allee *et al.*, 1955:14; Mayr, 1982:25), provided the first formal proof of the earth's sphericity (Isachenko, 1973:11), thus lending support to Parmenide's zonal classification, and studied animal–habitat relationships (Allee *et al.*, 1955; Block and Brennan, 1993; Mayr, 1982:88). Perhaps one of the most important benefits of the work completed by the early Greek 'physiologi' (meaning the observers of nature, Wiman, 1990) was that they developed and embraced a holistic approach to classifying and rationalizing the workings of the earth – the 'classical total view' in which the earth and its components were viewed as one large, open system (Wickstrom and Bailey, 1983) and, as taught by Heracleitus (500 BC), a system in which everything is in a constant state of flux (Wiman, 1990).

While scientific progress was by no means a smooth and rapid accumulation of knowledge, people gradually catalogued the earth's landforms, soils, climate and flora and fauna over the centuries that followed the work of the Greeks. However, in the seventeenth century, the 'classical total view' advanced by the Greeks was discarded and science (loosely defined here as the acquisition of knowledge) was divided into narrow, specialized disciplines. Geography, for example, gave rise to geology, geomorphology, climatology and pedology. While these disciplines greatly increased many forms of specific knowledge, an integrated holistic approach to the way we perceive and use the earth's ecosystems was lost (Wickstrom and Bailey, 1983).

As documented observations began to accumulate in each of the disciplines, general descriptions gradually led to serious attempts at classification. The continents and most of the important coastlines were mapped during the eighteenth century, and the continental land masses were explored and documented during the nineteenth century. Most nineteenth century naturalists were the products of the established disciplines who continued to view the world as an orderly grouping of various inorganic and organic categories (Herbertson, 1913). For example, 31 zoologists (zoo-centric) and 25 botanists (phyto-centric) had proposed divisions for North America alone by 1892 (Merriam, 1892). Only since the 1940s have there been serious attempts to classify and regionalize ecosystems (eco-centric) in various countries around the world.

14.4.1   Ecological classification and regionalization

We employ two approaches to organize and help us understand the composition, structure and function of the ecosphere in-space-in-time: classification and regionalization. **Classification** is a means of arranging entities into groups or sets on the basis of their similarities and relationships (Bailey *et al.*, 1978; Cline, 1949; Pfister, 1989; Platts, 1980). **Regionalization** is mapping from above – the partitioning of land- and waterscapes into more or less homogeneous units. It is completed analytically by division and subdivision, where differences between areas are employed to delineate unit boundaries (Rowe, 1984, 1992). Although different, the results of the two approaches can be quite complementary. Ecosystem classification and regionalization are used to organize systems into understandable patterns within a spatio-temporal framework (Bailey, 1982, 1987; Bailey *et al.*, 1978, 1985; Bailey and Hogg, 1986; Demarchi *et al.*, 1990a; Rowe, 1961, 1984). This permits the identification of ecologically founded geographic units on the basis of recognizable patterns of climate, soil, landform, vegetation and other variables (Bailey, 1976, 1978, 1985, 1987; Kansas, 1986:1; Lotspeich and Platts, 1982; Rowe, 1979, Wiken, 1986:2, and many others).

Thousands of classification and regionalization systems have been developed throughout the world; most have been discarded but a few have been retained and continually improved. A number of systems applied by Canadian and American agencies in the last 30 years have served as useful spatial frameworks in which to make management decisions. For example, Rubec (1992) described more than 100 and identified more than 500 ecological survey, classification and regionalization papers, reports and maps completed by Canadian management agencies between 1960 and 1990. Many of these products were completed for forested ecosystems.

14.4.2   Classification – working at large mapping scales

Although it is not possible to describe all forest ecosystem classification systems employed in Canada and the United States, there are several excellent summary papers. Canadian examples include the work of Belangér *et al.* (1992), Bergeron *et al.* (1992), Bowling and Zelazny (1992), Corns (1992), MacKinnon *et al.* (1992), Meades and Roberts (1992), Oswald (1992) and Sims and Uhlig (1992) (Table 14.1). In the United States, classifications have been completed by Arno (1982), Franklin (1980), Steele *et al.* (1983), Pfister (1981, 1989), Wellner (1989) and many others.

Historically, classification of forest ecosystems has been focused on the productivity of trees, silvicultural management, and on the fragility and stability of the systems under use (Rowe, 1984). But vegetation provides only a narrow perspective (Barnes *et al.*, 1982, Rowe, 1984). It cannot, for example, reflect much information about slope stability under different

Table 14.1 Examples of ecosystem classification systems employed by Canadian agencies

Name	Location	Description	Source
Forest Site Classification	Newfoundland and Labrador	Basic unit of classification is the 'forest type', which is defined through a combination of vegetation classification (using a relévé methodology) and soil characteristics of the site.	Meades and Roberts (1992); Damann (1967); Meades and Moores (1989)
Forest Site Classification	New Brunswick	The Province employs this classification as an on-site, pre-harvest assessment tool for use in mature and over-mature natural forest stands. It incorporates vegetation and soil characteristics to classify each stand into a 'Vegetation Type', a 'Soil Type' and a 'Treatment Unit'.	Bowling and Zelazny (1992); Groenwoud and Ruitenberg (1982); Zelazny et al. (1989)
Forest Ecosystem Classification (FEC)	Québec	The Québec FEC program provides a detailed stand classification within a hierarchical ecosystem classification framework. The FEC is used to identify 'forest type' units, define relationships between significant abiotic variables and forest vegetation, and characterize the physical environment.	Bergeron et al. (1992); Saucier (1989); Bélanger et al. (1992)
Forest Ecosystem Classification (FEC)	Ontario	Like Québec, Ontario's FEC program is designed to operate within a hierarchical ecosystem classification framework. It is a stand-level planning tool (a 'key-out' system) where vegetation and soil types are used to delineate each unit.	Sims (1985); Sims and Uhlig (1992); Jones et al. (1983); Sims et al. (1989, 1994); Merchant et al. (1989)
Site Classification System	Alberta	The site classification system was adapted from the biogeoclimatic system developed in British Columbia, and is contained within a hierarchical ecosystem classification system. Sites are classified using a 'key-out' system described in field guides. Vegetation and soils are the principal variables, and the supporting guides employ a 'key-out' system.	Corns (1983, 1991); Annas et al. (1983)
Biogeoclimatic Ecosystem Classification (BEC)	British Columbia	The BEC is a hierarchical system in which ecosystems are organized at three levels: local, regional and chronological. The system employs vegetation and sites (e.g. moisture or nutrient regimes). Site keys are produced for natural asset managers.	MacKinnon et al. (1990, 1992); Pojar et al. (1987); Krajina (1965)

logging regimes. Soil units and vegetation stands are more meaningful when set in the context of landforms whose materials and slopes exert strong influence through climate and substratum moisture regime (Rowe, 1984). In Ontario, managers are currently developing and employing the Forest Ecosystem Classification (FEC) (Sims and Uhlig, 1992) and similar approaches are employed in Alberta (e.g. Corns and Annas, 1986) and British Columbia (e.g. Ecosystems Working Group, 1995; MacKinnon *et al.*, 1992). It is a site-specific, stand-level planning tool that provides descriptive models in which specific types of information about the local forest stand, vegetation in the stand, soil and site (e.g. physiography) conditions are spatially catalogued. Vegetation and soil types form the basic units of the classification system, which is compiled through forest ecosystem 'keys' much like a standard taxonomic textbook (Figure 14.6).

The vegetation data are collected and, through application of computer-assisted statistical programs designed to identify relationships (i.e. analysis of independence), ordinations are completed and matched to soil moisture–nutrient profiles which are displayed in two-dimensional graphs (Figure 14.7). In Ontario the classification is employed in a variety of management programs, including preparation of prescribed burn guidelines, organizing information on competitive species, estimating the impacts of logging on soils, predicting the productivity of selected floral species and describing wildlife habitat (Sims and Uhlig, 1992).

The 'habitat type' classification (not to be confused with habitat as defined in section 14.2.4) developed and refined by Westveld (1951), Daubenmire (1952, 1970), Pfister (1981), Pfister *et al.* (1977), Pfister and Arno (1980) and many others (see Wellner, 1989, for a chronological review of this technique) has been employed widely in the western United States. It is used to classify forest vegetation and sites, but differs from systems that classify current vegetation because it is based on classification of the potential climax vegetation (Pfister, 1981), where climax vegetation is used to reflect the dynamic forces of climate, topography and soils (Figure 14.8). The system contains two parts (Pfister, 1977): a classification of climax vegetation and a classification of sites on which the climax vegetation grows. The 'habitat type' taxonomy provides the basic criteria for mapping land units and biotic potential in many US management programs. Therefore, a given 'habitat type' includes all areas potentially capable of producing similar plant communities at climax. This system provides a permanent land classification that can be applied in research and management in which the site characteristics are summarized in description tables.

14.4.3   Regionalization – working at small mapping scales

The origin of mapping large ecosystems is unclear. Although many nineteenth century authors grasped the concept of ecosystem as a synergistic

Sw/*Viburnum/Aralia*, spruce facies
(white spruce/mooseberry/wild sarsaparilla)

A B C D E
1
2
3
4
5
6
7

Sw/*Viburnum/Aralia*,
aspen facies
(See LBC 5c p. 78-79)

Sw/*Viburnum/Aralia*,
spruce facies
LBC 5a

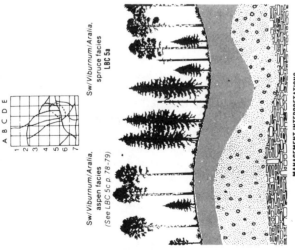

MANAGEMENT INTERPRETATIONS

excess moisture on some sites

Calamagrostis, torbs, shrubs

Sw/*Viburnum/Aralia*, spruce facies

CHARACTERISTIC SPECIES:
(with % cover)

A. *Picea glauca* (51)
   Populus tremuloides (7)
B. *Viburnum edule* (13)
   Rosa acicularis (5)
   Ribes lacustre (1)
   Amelanchier alnifolia (+)[1]
C. *Aralia nudicaulis* (14)
   Cornus canadensis (10)
   Linnaea borealis (9)
   Rubus pubescens (7)
   Mitella nuda (7)
   Mertensia paniculata (4)
   Calamagrostis canadensis (3)
   Maianthemum canadense (1)
   Petasites palmatus (1)
   Galium boreale (1)
   Pyrola asarifolia (1)
   Actaea rubra (+)
   Epilobium angustifolium (+)
   Orthilia secunda (+)
D. *Hylocomium splendens* (20)
   Pleurozium schreberi (12)
   Ptilium crista-castrensis (11)

ENVIRONMENT:

MOISTURE REGIME (MODAL):
Mesic to subhygric (4-5)

pH REGIME (MEAN):
Humus: 3.2-6.0 (5.0)
Mineral: 3.9-7.7 (5.6)

ELEVATION RANGE (MEAN):
575-1060 (831) m

PERCENT SLOPE GRADIENT
(MEAN): 0-37 (4)

ASPECT: Variable (level to
northerly)

SOIL SUBGROUPS: Orthic, Gleyed,
and Brunisolic Gray Luvisols

ASSOCIATED SOIL UNIT/
ASSOCIATION: BKM, DVS, DON,
EDS, LDG, LDM, NMP, SIP, TRC

SOIL DRAINAGE: Moderately well
to imperfectly

LANDFORM: Morainal, lacustrine

PRODUCTIVITY:
MAI (MEAN):
2.1-6.7 (4.8) m³/ha/yr
SI at 70 (MEAN): Sw 14-2? (19) m
                 Pl 19-31 (23) m

NUMBER OF PLOTS: 15

SUCCESSIONAL RELATIONSHIPS

These forests are successionally mature. *Abies* sp. may form a significant component of stands over 100 years. Some of the forests may have succeeded from aspen forests (e.g., Sw/*Viburnum/Aralia*, aspen facies) over a period of 50 to 100 years.

SIMILAR ASSOCIATIONS:

This association is most similar to the Sw/*Viburnum/Rubus pubescens* association, which is somewhat drier. Stands with sparse shrub and herb cover may be transitional to the Sw/*Hylocomium* association. It is similar to the *Picea glauca/Viburnum edule/Rubus pubescens* association of Archibald et al. (1984).

COMMENTS:

When the forest cover is harvested, vegetation cover (especially *Calamagrostis canadensis*) may be expected to increase greatly, competing with tree seedlings.

[1] + indicates less than 0.5% cover

Figure 14.6 A sample page from a typical classification field guide or 'key'. (Reproduced from Corns and Annas, 1986, in Corns, 1992, with permission from *Forestry Chronicle*.)

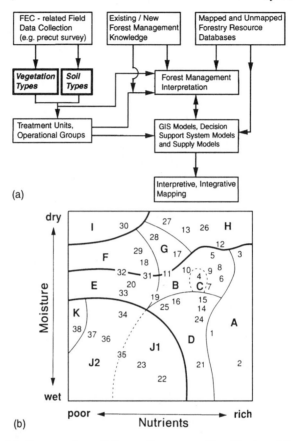

Figure 14.7 (a) Organization of Forest Ecosystem Classification (FEC) Soil and Vegetation Types for ecosystems in Ontario, Canada, used in support of natural asset decision-making; (b) Northwestern Ontario FEC showing Vegetation Types overlaid by Treatment Units and with similar moisture/nutrient gradients. The alphanumerics designate treatment areas characterized by repetitive vegetation patterns on the landscape. The numbers identify the location of 38 vegetation types. A detailed description of the technique is provided by Sims *et al.* (1989). (Reproduced from Sims *et al.*, 1989, in Sims and Uhlig, 1992, with permission from *Forestry Chronicle*.)

system in which all things were interrelated and interdependent (Wickstrom and Bailey, 1983), a holistic entity with many interacting parts, the Russian pedologist Dokuchaev (1899) and the English geographer Herbertson (1905, 1913) were among the first to project their disciplines in the context of the 'total classical view' of the earth's ecosystems. They promoted the concept of a holistic view of the earth's natural assets, a complex network of land, water, air, plant and animal matter in a special relationship constituting

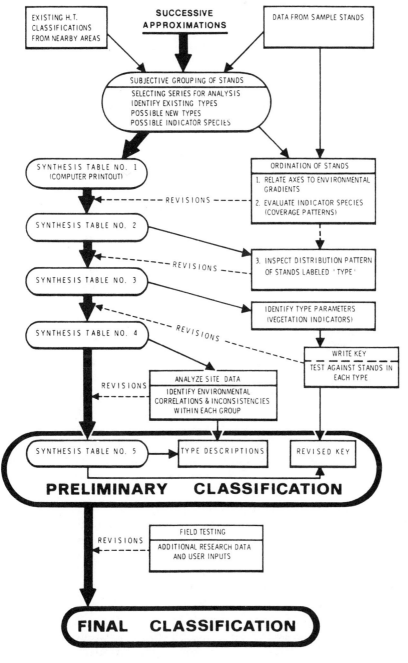

Figure 14.8 Development of a 'Habitat Type' classification. (From Pfister and Arno, 1980.)

defined portions of the earth's surface. Herbertson (1905), and many geographically oriented scientists who followed, asked, for example, what characteristics should be selected to distinguish one spatial unit from another, and how can one determine the different orders of spatial units. On a global scale, Herbertson (1905, 1913) classified the earth into natural regions according to climate and landform.

In Britain, Bourne (1931) developed techniques to make inventories of forests in the British Empire countries and simultaneously completed pioneering work in land classification. He defined a hierarchy of 'sites' nested within 'regions' on the basis of climate, physiography, geology and edaphic factors (Christian and Stewart, 1968; Margules and Scott, 1983). At the same time in the United States, and on the basis of work completed in Michigan, Veatch (1931, 1934) proposed two hierarchical land systems based on natural geographic divisions through the integration of various ecological components such as soil, vegetation and topography. Despite the important work of Herbertson (1905, 1913) and Veatch (1931, 1934), American and Canadian agencies did not begin to search for interdisciplinary classification and regionalization techniques until the 1950s (Bailey *et al.*, 1978; Bailey, 1980, 1982; Hills, 1953; Lotspeich and Platts, 1982).

(a)   United States

In the late 1960s, US Forest Service soil scientists began to examine techniques to rapidly differentiate and classify ecologically significant segments of the land surface at small mapping scales. Wertz and Arnold (1972) delineated spatial units on the basis of a number of ecosystem components, including lithology, climate, soils, geological structure, landform and vegetation (Table 14.2). Within this nine-level hierarchical system, the small-scale upper level units (large ecosystems) were delineated on the basis of climate, geological structure and lithology; the lower level units were determined on the basis of soils, landforms and vegetation.

Around the same time, the US Forest Service initiated the ECOCLASS project (Corliss and Pfister, 1973; Corliss, 1974), which was developed by an interdisciplinary team of specialists who linked existing unidisciplinary (thematic) classifications of important ecosystem components, namely vegetation, land systems (i.e. rock lithology, rock structure, climate, land shape and geomorphology) and aquatic systems (e.g. streams, and upstream land and vegetation characteristics) (Buttery, 1978; Corliss and Pfister, 1973; Corliss, 1974; Pfister, 1977).

Initially, ECOCLASS was designed as a prototype of an integrated national classification system containing aquatic and terrestrial components at comparable levels of regionalization in the hierarchy (Platts, 1980). Despite its potential as a spatial framework in which biological and earth science systems (i.e. terrestrial vegetation, fishery, aquatic and land systems)

Table 14.2 System outline for the landbase portion of the integrated environmental inventory completed in the United States (from Wertz and Arnold, 1972)

Category	Name	Basics for delineation	Size range (square miles)	Principal application
VII	Physiographic Province	Basic elements: structure, lithology and climate. (First order stratification)	1000s	Nationwide or broad regional data summary.
VI	Section	Basic elements: structure lithology and climate. (Second order stratification)	100s to 1000s	Broad regional summary. Basic geologic, climatic, vegetative data for design of individual resource inventories.
V	Subsection	Basic elements: structure lithology and climate. (Third order stratification)	10 to 100s	Strategic management direction and broad area planning.
IV	Landtype Association	Manifest elements: soils, landform and biosphere. (First order stratification)	1 to 10	Summary of resource information and resource allocation.
III	Landtype	Manifest elements: soils, landform and biosphere. (Second order stratification)	1/10 to 1	Comprehensive planning, resource plans, development standards and local zoning.
II	Landtype Phase	Manifest elements: soils, landform and biosphere. (Third order stratification)	1/100 to 1/10	Project development plans.
I	Site	Represents integration of all environmental elements. Units are generally not delineated on map.	Acres or less	Provides precise understanding of ecosystems. Sampling will be for defining broader units, for research, and for detailed on-site project action plans.

Table 14.3 Systems and components of the ECOCLASS method arranged hierarchically: note that the *habitat type* and *community type* designations are inconsistent with the 'Formation' to 'Series' designations, and soils are factored into the Land Subsystem hierarchy at the *landtype* and *landunit* level of mapping (from Buttery, 1978)

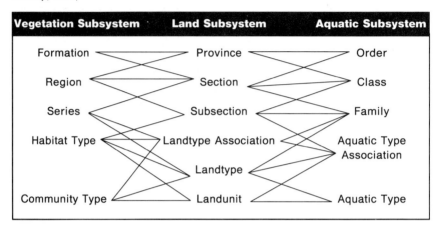

could be integrated, it contained some descriptive inconsistencies and an unworkable mixture of taxonomic and regionalization concepts (Pfister, 1989; Buttery, 1978; Corliss and Pfister, 1973; Corliss, 1974; Platts, 1980) (Table 14.3). Therefore, a second multidisciplinary team was created to develop ECOCLASS II, in which the ECOCLASS land type classification was replaced by separate landform and soil classifications. In addition, the original six levels in the ECOCLASS hierarchy land system (Table 14.3) were replaced with 11 levels depicting landform, a system that closely resembled the hierarchical system proposed by Fairbridge (1968), and a soil classification system was added (Buttery, 1978) (Table 14.4).

The ECOCLASS concept was later expanded and incorporated into ECOSYM to link classification and management needs (Bailey, 1982; Henderson *et al.*, 1978). An attractive feature of ECOSYM is that it contains several component hierarchical classifications, which allow natural asset managers to adopt different approaches (i.e. place emphasis on specific management disciplines) to landscape classification on the basis of its component elements, which can be viewed independently as a series of overlays and integrated by the manager for a specific purpose (Bailey, 1982; Lotspeich and Platts, 1982; Platts, 1980). The ECOCLASS system represents an explicit ecosystem delineation technique with which to identify meaningful ecological boundaries according to the hierarchy proposed by Wertz and Arnold (1972). Wertz and Arnold (1972) recognized the importance of biotic variables in delineating boundaries for the site-specific spatial units (the lower levels in the hierarchy) and ECOCLASS is an attempt to define the

Table 14.4 Systems and components of the modified ECOCLASS method arranged hierarchically (from Buttery, 1978)

Vegetation system	Landform system	Soil system	Aquatic system
Plant Formation	Realm	Order (Level 7)	Order
	Major Division	Suborder (Level 6)	
	Province	Great Group (Level 5)	Class
Plant Region	Section		
	Region	Subgroup (Level 4)	
	District		Family
	Area	Family (Level 3)	
	Zone		
Plant Series	Locale		
	Compartment	Series (Level 2)	Aquatic Type Association
Plant Association	Feature	Type (Level 1)	Aquatic Type

unidisciplinary characteristics in equivalent hierarchical order to enable the natural asset manager to rationalize, with some objectivity, the placement of those boundaries.

The hierarchically defined land systems approach recommended by Wertz and Arnold (1972) also was adopted by Bailey (1976, 1978), who merged the system with an earlier system proposed by Crowley (1967) for Canada. It differed from the ECOCLASS model because Bailey integrated all unidisciplinary characteristics and employed them simultaneously in the boundary delineation process (Table 14.5). This represents an attempt to integrate influential ecosystem properties in defining ecosystem boundaries at various mapping scales – a difficult process because of the trade-offs that have to be made. Bailey's classification involves the delineation, description and evaluation of relatively homogeneous units of land according to their ecological unity at the local and regional scale (Bailey, 1982, 1985, 1987). This approach assumes that not all components may be equally significant at each level in the hierarchy, and that it may not be possible to deal with all components simultaneously. The boundaries at each level in the hierarchy reflect the significance of selected ecosystem variables on the location, size, productivity, structure and function of the ecosystem (Bailey, 1982). For example, differentiating criteria (the variables that control patterns and processes) of large ecosystems are all-encompassing (e.g. climate) while those at the lower levels exhibit a spatially limited sphere of influence (e.g. soil) (Bailey, 1976, 1978).

Table 14.5 A hierarchy of ecosystems proposed by Bailey (1978) (adapted from Crowley, 1967, and Wertz and Arnold, 1972)

Name	Defining criteria
Domain	Subcontinental area of related climates.
Division	Single regional climate at the level of Köppen's types (Trewartha, 1943).
Province	Broad vegetation region with the same type or types of zonal soils.
Section	Climatic climax at the level of Küchler's (1964) potential vegetation types.
District	Part of a section having uniform landforms at the level of Hammond's (1964) land-surface form regions.
Landtype Association	Group of closely related types with recurring pattern of landforms, lithology, soil and vegetative series.
Landtype	Group of neighbouring phases with similar soil series or families with similar plant communities at the level of Daubenbmire's (1968) habitat types.
Landtype Phase	Group of neighbouring sites belonging to the same soil series with closely related habitat types.
Site	Single soil type or phase and a single habitat type or phase.

Bailey and Hogg (1986) and Bailey (1989) employed this approach to map the broad levels in the hierarchy to enhance Udvardy's (1975) map of the **ecoregions** of the world. More recently, and in support of an ecosystem approach to management, the US Forest Service adopted the **ecoregion** for use in management programs; it updates Bailey's map as part of the integrated National Hierarchical Framework and mapping system (Avers *et al.*, 1993; Bailey, 1995; ECOMAP, 1993) (Table 14.6; Figure 14.9).

At the same time, the US Environmental Protection Agency used an integrated approach to locating ecosystem boundaries and delineated the **ecoregions** of the United States on the basis of land-use patterns, land-surface form, potential natural vegetation and soils (Omernik, 1987; Omernik and Gallant, 1988). A number of thematic maps were used to determine the boundaries, including a land-surface form map of the United States, potential natural vegetation, major land uses and various soil maps (Omernik and Gallant, 1988). One of the strengths of this regionalization process is that the **ecoregion** boundaries are sensitive to watershed boundaries. Similarly, Albert *et al.* (1986) developed a regional landscape ecosystem classification for Michigan that expressed the interactive character of landscapes

Table 14.6 USDA Forest Service national hierarchical framework for ecological units (adapted from Avers *et al.*, 1993, and Lachowski *et al*, 1994)

Planning and analysis scale	Ecological unit	Purpose, objectives, and general use	General size range and map scale

Ecoregion:
  Global
  Continental
  Regional
	
	Domain
	Division
	Province
		Broad applicability for modeling and sampling. Strategic planning and assessment, as well as international planning.	1 000 000 to 10 000s mi^2 (1:3 000 000 to 1:1 000 000)

Subregion	Section Subsection	Strategic, multi-forest, statewide and multi-agency analysis and assessment.	1 000s to 10s mi^2 (1:5 000 000) to 1:250 000)
Landscape	Landtype Association	Forest and area-wide planning, and watershed analysis.	1 000s to 100s of acres (1:250 000) to 1:50 000)
Land Unit	Landtype	Project and management area planning and analysis.	100s to < 10 acres (1:50 000 to 1:10 000)
	Landtype Phase		(1:10 000 to 1:2 500)

and their components – climate, geological parent material, physiography (landform and water bodies), soil, plants and animals (Figure 14.10).

(b) Canada

In Canada, a number of post World War II programs were designed to inventory and assess the country's natural assets (Gray and Cameron, 1990). Some classifications were unidisciplinary (i.e. Canada Land Inventory) while others were integrated (e.g. Hills, 1958; Crowley, 1967). Canadian ecosystem classification and regionalization programs evolved rapidly in the 1960s under the auspices of the National Committee on Forest Land, which in 1964 sponsored an interdisciplinary subcommittee to examine techniques and to develop guidelines for a Canadian system (Wiken and Ironside, 1977). The subcommittee employed results obtained previously in Canada (e.g. Hills, 1952, 1953, 1958, 1960, 1961; Krajina, 1965; Rowe, 1961), Australia (e.g. Christian, 1957; Christian and Stewart, 1968, and others), Britain (e.g. Beckett and Webster, 1965; Bourne, 1931) and the former Soviet Union (e.g. Sukachev and Dylis, 1964). It completed its guidelines in 1969 (Lacate, 1969) and developed a methodology to assist ecogeographers in their efforts to identify and map ecologically significant spatial units characterized by

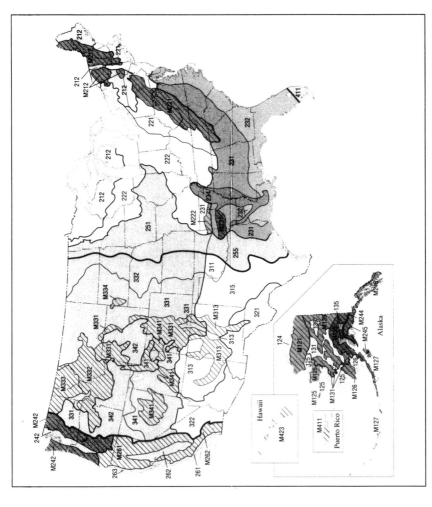

Figure 14.9 USDA Forest Service ecoregion map of the United States. The shades, numbers and alphanumeric designations delineate domains, divisions and provinces. Bailey (1995) provides a description of each unit.

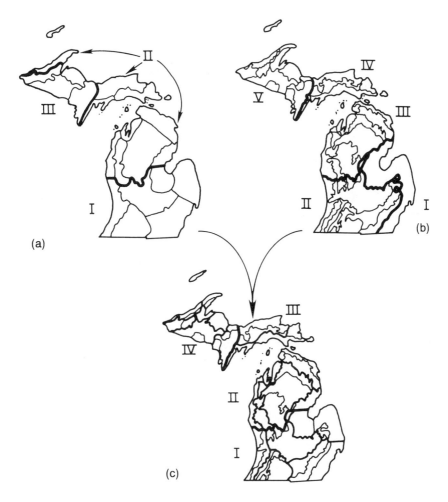

Figure 14.10 Maps illustrating the integration of climate and physiography boundaries in development of the integrated regional ecosystem classification for Michigan: (a) climate regions; (b) physiographic regions; (c) ecosystem regions. (From Albert *et al.*, 1986.)

their inherent biotic and abiotic properties (Wiken, 1980, 1986; Wiken and Ironside, 1977).

A national system, the Canadian Ecological Land Classification System (CELCS) (e.g. Ironside, 1991; Wiken and Ironside, 1977), and many provincial systems were developed and implemented in the 1970s and 1980s (e.g. Demarchi and Lea, 1987; Hills, 1958, 1960; Pojar *et al.*, 1987; Strong and Leggat, 1981). Although these programs are regionalization programs, many were devised in anticipation that site-specific classification programs eventually would be merged with them to complete the hierarchy. For example, many jurisdictions sponsor both regionalization and classification programs, including Newfoundland and Labrador (Meades and Roberts, 1992), Québec (Bergeron *et al.*, 1992), Ontario (Sims and Uhlig, 1992; Sims *et al.*, 1989), Alberta (Corns and Annas, 1986) and British Columbia (MacKinnon *et al.*, 1992).

The CELCS is described here because it is based on many of the general concepts and contains many of the qualities upon which a number of Canadian regionalization mapping systems are based. The CELCS provides a standard framework in which to describe and model ecosystems and ecosystem components. From an anthropogenic perspective, it provides the framework in which natural asset managers can identify, monitor and make decisions about land use. Like many US systems, the classification is hierarchical, where managers work to synthesize biotic and abiotic attributes into ecologically meaningful spatial units at various levels of organization and map scales (Rowe and Sheard, 1981; Wiken, 1980, 1986; Wiken and Ironside, 1977; Wickware and Rubec, 1989; Wiken *et al.*, 1979). The CELCS hierarchy contains seven levels representing mapping scales ranging from 1:20 000 000 to 1:2500 (i.e. ecozone, ecoprovince, ecoregion, ecodistrict, ecosection, ecosite, and ecoelement) (Wiken, 1980; Table 14.7).

Ecozones (Table 14.8; plate K), ecoprovinces, ecoregions and ecodistricts have been mapped for the entire country. This permits natural asset managers to relate their work on small ecosystems (such as ecosites and ecosections) to broader regional, provincial/territorial and national issues and programs. The interdisciplinary techniques applied in Environment Canada's CELCS program offer guidance in managing boundary-setting issues and provide a number of practical advantages (Ironside, 1991; Wiken, 1980):

- They provide a spatial framework within which wildlife managers can address long-term questions and issues and implement long-term inventory and monitoring programs.
- The integrated team approach to boundary delineation minimizes problems arising from discordant boundary lines for descriptions of areas having similar characteristics.
- Data are collected in the context of dynamic ecosystem structure, composition and function, allowing managers to make improved predictions about the response of spatial units to specific management prescriptions.
- They allow natural asset managers to represent a diverse range of ecological data in a common spatio-temporal framework, and to collate reports and maps into a convenient package.
- They minimize the need for redundant data gathering programs – a cost-effective approach to natural asset management whereby the logistical

Table 14.7 The Canadian Ecological Land Classification System (CELCS) hierarchy (from Ironside, 1991, and Wiken, 1980, 1986)

Common map scale	Description	Level of planning
Ecozone (1:20 000 000 to 1:15 000 000)	An area of the earth's surface representative of large and very generalized ecological units characterized by interactive and dynamic abiotic and biotic features, including macro-climatic regimes, plant formations, major soil zones and first-order subcontinental landforms.	Subcontinental International National Provincial
Ecoprovince 1:10 000 000 to 1:5 000 000	An area of the earth's surface characterized by major structural or surface forms, faunal realms, vegetation, hydrological, soil and climatic zones.	International National Provincial
Ecoregion (1:3 000 000 to 1:1 000 000)	Part of an ecoprovince characterized by distinctive ecological responses to climate as expressed by vegetation, soils, water, fauna, etc.	Subprovincial Regional
Ecodistrict (1:500 000 to 1:125 000)	Part of an ecoregion characterized by a distinctive pattern of relief, geology, geomorphology, vegetation, soils, water and fauna	Subregional
Ecosection (1:250 000 to 1:50 000)	Part of an ecodistrict throughout which there is a recurring pattern of terrain, soils, vegetation, water bodies and fauna.	Community
Ecosite (1:50 000 to 1:10 000)	Part of an ecosection having a relatively uniform parent material, soil and hydrology, and a chronosequence of vegetation.	Detailed
Ecoelement (1:10 000 to 1:2 500)	Part of an ecosite displaying uniform soil, topographical, vegetative and hydrological characteristics.	Site-specific

and financial benefits of a single interdisciplinary survey often outweigh those of multidisciplinary and unidisciplinary or thematic surveys.
- The CELCS approach accommodates variation (given that the number, size and descriptive detail of spatial units within a particular study area vary according to the users' needs) through its hierarchical characteristics.

It is important to recognize that no single classification can meet the needs of all users. In some cases natural asset managers require access to more

Table 14.8 Canadian Ecological Land Classification System Ecozone characteristics and land use practices in Canada (from Wiken, 1986)

Ecozone	Defining criteria and land use practices				
	Physiography	Vegetation	Soil order and surface material	Climate	Present use
Tundra Cordillera	Mountainous highlands	Alpine and arctic tundra	Cryosolic and brunisolic; colluvium; moraine and rock	Cold, semi-arid, subarctic	Trapping, hunting, recreation, tourism, mining
Boreal Cordillera	Mountainous highlands with some hills and plains	Boreal with some alpine tundra and open woodland	Brunisolic; collovium; moraine and rock	Moderately cold and moist montane	Hunting, trapping, forestry, recreation, mining
Pacific Maritime	Mountainous highlands with some coastal plains	Coastal western and mountain hemlock	Podzolic; collovium; moraine and rock	Very wet, mild, temperate maritime	Forestry, fishing, urbanization and agriculture
Montane Cordillera	Mountainous highlands and interior plains	Mixed vegetation – conifer stands to sagebrush fields	Luvisolic, brunisolic; moraine, colluvium and rock	Moderately cold and moist to arid montane	Forestry, agriculture, tourism and recreation
Boreal Plains	Plains with some foothills	Conifer and deciduous boreal stands	Luvisolic; moraine, lucustrine materials	Moderately cold, moist boreal	Forestry, agriculture, recreation, trapping
Taiga Plains	Plains with some foothills	Open woodland, including shrublands and wetlands	Cryosolic and brunisolic; organic and moraine	Cold, semi-arid, subarctic to moist boreal	Hunting, trapping, recreation
Prairie	Plains with some foothills	Short and mixed grasslands and aspen parkland	Chernozemic; moraine and lucustrine	Cool, semi-arid	Agriculture, urbanization and recreation

Ecozone	Landform	Vegetation	Soils	Climate	Land use
Taiga Shield	Plains with some interior hills	Open woodlands with some arctic tundra and lichen heath	Cryosolic and brunisolic; moraine and rock	Moist, cold boreal to cold, semi-arid, subarctic	Hunting, trapping and recreation
Boreal Shield	Plains with some interior hills	Conifer and deciduous boreal stands	Brunisolic; moraine, rock and lacustrine materials	Cold, moist boreal	Forestry, mining, recreation and tourism
Hudson Plains	Plains	Wetlands, arctic tundra with some conifer stands	Cryosolic; organic and marine	Cold, semi-arid sub-arctic to cold boreal	Hunting, trapping and recreation
Mixed Wood Plains	Plains with some interior hills	Mixed deciduous and coniferous stands	Luvisolic; moraine, marine and rock	Cold to mild boreal	Agriculture, urbanization and recreation
Atlantic Maritime	Hills and coastal plains	Mixed deciduous and coniferous stands	Brunisolic and luvisolic; moraine, colluvium and marine	Cool, wet temperate maritime	Forestry, agriculture, fishing and tourism
Southern Arctic	Plains with some interior hills	Shrub–herb–heath arctic tundra	Cryosolic; moraine, rock, marine	Cold and dry arctic	Hunting, trapping, recreation and mining
Northern Arctic	Plains and hills	Herb–lichen arctic tundra	Cryosolic; moraine, rock, marine	Very cold and dry arctic	Hunting, trapping, recreation and mining
Arctic Cordillera	Mountainous highlands	Non-vegetated and some shrub–herb arctic tundra	Cryosolic; ice, snow and colluvium	Extremely cold and dry arctic	Hunting

than one regionalization classification system. For example, in British Columbia the provincial government has sponsored development of a hierarchical classification system that permits managers to make decisions about land use in mountainous terrain, characterized by significant changes in elevation over short distances (D. Demarchi, personal communication).

(c)   International mapping projects

Ecosystem boundaries overlap administrative and political boundaries. People of different provinces, states and countries live and work in the same ecosystems. Interprovincial, interstate and international cooperation, therefore, is critical to the successful implementation of an ecosystem approach to management and, by extension, wildlife habitat management.

A number of agencies have begun to explore opportunities for collaborative ecosystem mapping. For example, in response to anticipated requirements under the North American Free Trade Agreement and a general call to action by the United Nations Conference on Environment and Development, Canada, the United States and Mexico have begun to examine joint environmental reporting programs for North America (Ezcurra *et al.*, 1993; Wiken and Lawton, 1995). Ecosystem mapping is an important component of work proposed for this program. Demarchi (1994) produced an ecoprovince map for southwestern Canada, the western United States and northwestern Mexico to enhance current and potential collaborative wildlife management programs (Figure 14.11). In the central part of the continent, Uhlig and Jordan (in press) are exploring joint Canadian–US ecosystem classification and information systems in support of collaborative management programs for the upper Great Lakes.

14.4.4   Wildlife habitat in ecosystem classification and regionalization

Integration of faunal values into ecosystem regionalization/classification systems permits the forest ecosystem manager to develop and apply models in search of significant ecological measures like areas of high faunal diversity, species and ecosystems of socioeconomic importance, and critical habitat (Margules and Scott, 1983; Stelfox, 1988), consistently to employ monitoring and inventory techniques and to identify gaps or weaknesses in protection programs with techniques such as GAP analysis. But until recently (i.e the last 15 years), fauna have received little attention as a component of ecosystem regionalization/classification (Corns, 1991; Kansas, 1991; Margules and Scott, 1983; Miller, 1994).

What separates the ecologically based classification and regionalization from discipline-specific classification and regionalization, such as phytogeographical or pedo-centric classification? What is the significance of this difference and what does it have to do with wildlife habitat evaluation? The answers, in part, can be explained by revisiting the definition of an

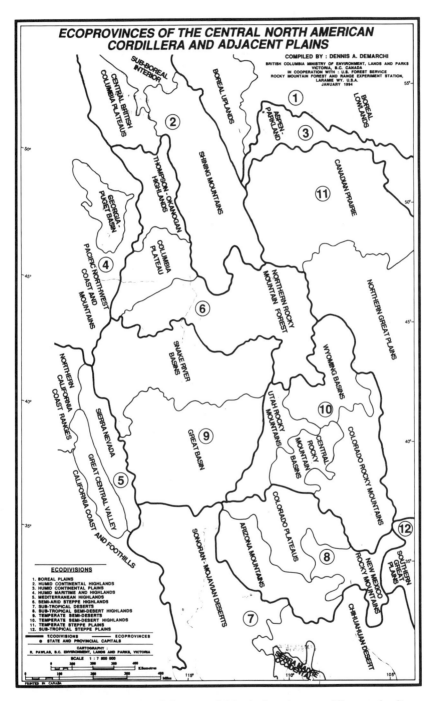

Figure 14.11 Ecoprovinces of the central North American cordillera and adjacent plains. (Map courtesy of D.A. Demarchi.)

ecosystem: 'a network of organisms (such as plants and animals, including humans) and atmosphere, soil, rock, and water, interacting with each other and with other ecosystems'. By definition, an ecosystem is the 'whole' – its description at various levels in the hierarchy is based on the variables that define the boundaries at that level. For example, in the Canadian Ecological Land Classification, the subcontinental ecozones and ecoprovinces are delineated using macroclimatic regimes, plant formations, major soil zones and landforms, where climate and landform are dominant influences in the formation of these spatial units (Table 14.7). Many biotic and abiotic variables are not significant boundary-delineating criteria at this scale of mapping ecosystems. This approach is similar to work completed in the United States by Bailey (ECOMAP, 1993) and Albert *et al.* (1986).

Ecosystem classification and regionalization differ from phytogeographical or zoogeographical classification and regionalization because, in these examples, they replace the bio-centric bias with a more comprehensive eco-centric bias. They employ different criteria at different mapping scales on the basis of the significance of their impact on the ecosystem boundary – they do not dismiss the bio-centric perspective but simply cast it as part of the larger 'whole'. This is important, for example, in distinguishing ecosystem from habitat, where habitat denotes the 'organism-centred' part of the ecosystem (Rowe and Barnes, 1994). This is a critical concept because historically our discipline-specific perspectives significantly limited our ability to cast wildlife and wildlife habitat in a meaningful ecological spatial framework. Discipline-specific shackles caused early naturalists and scientists to look for spatial entities that were not there because the asset of enquiry was simply not a dominant influence in the delineation of boundaries at the mapping scale with which they were working.

(a)   Regionalization: examples

Because most faunal studies historically were conducted without benefit of habitat inventories completed within an ecologically meaningful spatio-temporal framework (Kansas, 1991), natural asset managers relied upon available data and information or on their own unique classification systems (Bailey, 1982). Prescott (1980) examined this problem by completing mammal inventories prior to, during and after ecological inventories in seven national parks in Atlantic Canada. He found that the most effective integration of faunal data and information was made after the ecological inventories were completed because they provided a reliable spatial framework within which to design the surveys.

Descriptive modeling of fauna has been difficult because each species (and in many cases each individual) uses habitat in unique and complex ways, ways often poorly understood by wildlife managers. Many species are highly mobile (East *et al.*, 1979), use a variety of habitats in-space-in-time (Byrne, 1982; East *et al.*, 1979; Kansas, 1991; Mulé, 1982; Noss, 1983), perceive their environment at multiple levels of resolution (East *et al.*, 1979; Holling, 1992), fluctuate widely in total numbers over time (Van Horne, 1983) and are relatively difficult to observe (Kansas, 1991). Wildlife managers attempt to translate their interpretation of an animal's actions into model variables and parameters, which are organized into some type of structured framework such as a classification system (descriptive model), and use them in predictive and decisive (decision/support) models.

Kansas (1986) suggests that the difficult task of integrating faunal values into an ecosystem mapping project might be accomplished by thinking in terms of the spatial unit's 'functional' significance to the species. He argues that, because animals likely react to their environment differently from the way people organize and delineate ecosystems, there is need for a flexible, functional approach. For example, whereas a black bear (*Ursus americanus*) might recognize and respond to changes in vegetation and topography in response to food requirements, it is unlikely to respond as significantly to differences in soil type (on which similar vegetation types occur). Therefore, from a functional perspective, and for the purposes of this type of species-specific question, it is appropriate to aggregate the spatial units into one 'black bear unit' on the basis of their similar habitat suitability. Delineating boundaries in this way makes them responsive and meaningful to the objectives of the project. As Rowe (1990) points out, perhaps the best way to examine and manage faunal habitat is as a collection of spatio-temporal units within a mosaic of ecosystems at different scales in which animals come and go. Therefore, the ecosystem is relatively fixed in space with animals and plants as mobile parts, where spatial units can be aggregated or divided on the basis of defining criteria which describe why the animals come and go.

Some work has been completed in the United States in an effort to determine the relationship between ecoregions (e.g. as delineated by Bailey, 1976, and Omernik and Gallant, 1988) and faunal distribution patterns. Most has focused on aquatic components of ecosystems, particularly in response to Omernik's ecoregions where the boundaries account for both terrestrial and aquatic ecosystems. Omernik's ecoregion boundaries were found to correspond reasonably well to the distribution of fish, macroinvertebrates, water quality and physical habitat in Ohio (Larsen *et al.*, 1986, 1988; Whittier *et al.*, 1987); fish, macroinvertebrates, algae, water quality and physical habitat in Oregon (Hughes *et al.*, 1987; Whittier *et al.*, 1988); fish, water quality and physical habitat in Arkansas (Rohm *et al.*, 1987); fish in Wisconsin (Lyons, 1989); water quality in Minnesota (Heiskary *et al.*, 1987); and phosphorus levels in Minnesota, Wisconsin and Michigan (Omernik *et al.*, 1988). In Canada, Legendre and Legendre (1984) examined the post-glacial dispersal of freshwater fishes on the Québec peninsula and evaluated whether or not climatic, geomorphic and vegetation patterns were more useful in explaining fish distribution than river basin patterns. Although the study was limited because they examined each of the ecological variables separately in relation to fish distribution, they did conclude that the variables were useful in defining the distribution of some species.

From a terrestrial perspective, faunal distribution and abundance values have been matched with regionalization systems in many parts of North America. Rayner and Bennett (1988) developed woodland caribou (*Rangifer tarandus caribou*) habitat suitability indexes within an ecological classification system at the 1:1 000 000 and 1:100 000 mapping scales for Alberta, but they were not validated with distribution, abundance or survival values. Wickware *et al.* (1980) studied breeding Canada geese (*Branta canadensis*) in the Hudson Bay Lowland of northern Ontario and attempted to relate density to habitat capability indices (measured with wetland pattern, percentage of water, hydrologic types such as ponds and lakes, and vegetation physiognomy) developed for ecodistricts (1:500 000–1:250 000; Table 14.2). They established a positive relationship between wildlife density and the ecodistricts. Schmidt (1980) examined the density and distribution of moose (*Alces alces*) in relation to the boundaries of ecological units in northern Manitoba, and Strong and Vriend (1980) examined elk (*Cervus elaphus*) density and distribution in relation to ecoregions in southwestern Alberta. In both studies it was determined that densities could be correlated to the defined spatial units, but neither attempted to model for specific habitat variables. Gray (1993) explored the use of habitat suitability index-type models for winter moose habitat at the ecodistrict level (1:500 000–1:250 000) in the Northwest Territories. In the United States, Inkley and Anderson (1982) identified faunal communities using cluster analysis of the distribution patterns of mammals, birds, reptiles and amphibians, and compared them with Bailey's (1976) ecoregions and ecoregion sections. They found high similarity between the ecoregion sections and the distribution of faunal communities.

Ecosystem classification in British Columbia is multidisciplinary, with parameter identification moving from the very broad, subcontinental level of mapping to site-specific units of land (Demarchi and Chamberlain, 1979; Demarchi and Lea, 1987; Demarchi *et al.*, 1990b) (Table 14.9; Figure 14.12). As part of this program, the government of British Columbia employs a wildlife habitat description project that draws upon data from several disciplines (e.g. surficial geology, soils, plant ecology, wildlife biology and climatology) (Pendergast, 1989) and includes a manual for describing ecosystems in the field (e.g. Luttemerding *et al.*, 1990). All parameters used to describe the site, soils, vegetation, wildlife use and presence, mensuration and humus are provided on standard forms in this manual (D. Demarchi, personal communication). In addition, the government of British Columbia has developed a standardized manual for collecting habitat monitoring information (Habitat Monitoring Committee, 1990). This manual supplements the manual developed by Luttemerding *et al.* (1990) through the

Table 14.9 Physical and biological variables that are considered when defining ecosystems at various scales in British Columbia (from Demarchi and Lea, 1987)

Mapping scales	Physical and biological processes used to determine ecosystem units at various scales					
	Climate	Marine	Landforms	Soils	Vegetation	Fauna
Ecodomain (Global) 1:25 000 000	Broad climatic similarity	Oceanic domains	—	—	Broad physiognomic groups (formations)	Faunal regions
Ecodivision (Continental or Oceanic) 1:12 000 000	Broad climate types	Oceanic currents	Continental landforms	—	Similar physiognomic groups (formation)	Life zones
Ecoprovince (Subcontinental or sub-oceanic) 1:5 000 000	Macro-climatic processes	Specific oceanic processes	Physiographic systems	Soil orders	Assemblages of vegetation regions and/or zones	Potential wildlife population ranges
Ecoregion (Regional) 1:2 000 000	Macro-climatic processes	Regional oceanic processes	Regional physiography	Soil orders	Vegetation regions (assemblages of vegetation zones)	Wildlife biogeography
Ecosection (Sub-regional) 1:500 000	Macro-climatic processes at sub-regional level	Oceanic processes at the sub-regional level	Regional physiography (regional bedrock)	Soil great groups	Vegetation zones	Faunal communities

Table 14.9 Continued

Mapping scales	Physical and biological processes used to determine ecosystem units at various scales					
	Climate	Marine	Landforms	Soils	Vegetation	Fauna
Biogeoclimatic Units (Zonal) 1:250 000	Climatic regimes (macro-climatic parameters)	Specific marine areas	Subdivision of regional physiography to represent groups of local landforms	Soil great groups	Climax communities	Faunal communities with belts of seasonal habitat use by migratory species
Habitat Classes (Local) 1:50 000	Detailed level meso-climatic parameters	Local sea-bed salinity and temperature configurations	Local landforms, topography (slope and aspect) and parent materials	Soil sub-groups	Plant communities (successional stages)	Units of potential and current habitat use by wildlife species
Habitat Classes (Association) 1:20 000	Broad-level micro-climatic parameters	Specific sea-bed salinity and temperature configurations	Specific landforms and materials	Soil series	Plant communities (succession and physiognomy)	The influence of social behavior on distribution and habitat use
Habitat Classes (Site) 1:5000	Micro-climatic parameters	Detailed sea-bed, salinity and temperature parameters	Materials	Soil series phases	Plant community classes	Site-specific habitat use

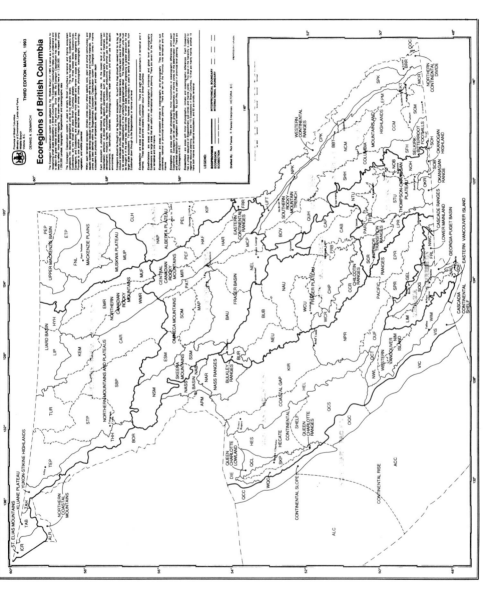

Figure 14.12 Ecoprovinces, ecoregions and ecosections of British Columbia, Canada. (Map courtesy of D.A. Demarchi.)

provision of guidelines and recording sheets for describing habitat manipulation techniques (what was done, why, when and how) and activities such as livestock forage use, prescribed fire, pesticide application and tree-seedling planting. The procedure allows managers to monitor habitat change through time (D. Demarchi, personal communication).

Wildlife habitat rating schemes are being developed for many species in British Columbia. For example, in support of the provincial grizzly bear (*Ursus arctos*) management program, Fuhr and Demarchi (1990) developed a methodology for employing habitat classifications within the hierarchically based provincial ecosystem classification. Grizzly Bear Management Zones, groupings of ecoregion units with similar habitat characteristics that affect grizzly bears, have been delineated and are employed (Figure 14.13). By integrating wildlife management areas into the province's ecosystem classification system, natural asset managers have completed an important step in developing and implementing an ecosystem approach to management. Each biogeoclimatic unit within the ecoregions has been ranked (i.e. high, medium, low, nil) by employing a spatial unit description and rating matrix completed on the basis of the presence/absence of habitat requirements

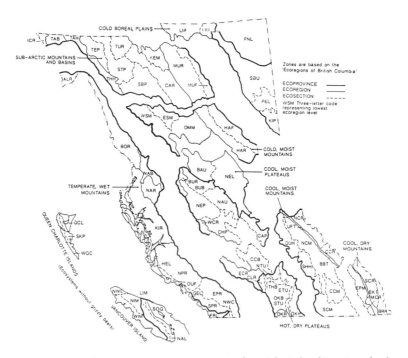

Figure 14.13 Grizzly Bear Management Zones of British Columbia, Canada, based on an aggregation of the province's ecoregions. (From Fuhr and Demarchi, 1990.)

ECOSECTION	BIOGEOCLIMATIC ZONE	DESCRIPTION	RATING
Hart Ranges (HAR)	Alpine Tundra (AT)	rocky, poorly vegetated alpine with deep snow	L
	Engelmann Spruce-Subalpine Fir (ESSF)	dense conifer forests in a wet climate, deep persistent snow, avalanche chutes common	H
	Sub-boreal Spruce (SBS)	dense conifer forests at lower elevations, low gradient floodplains common	H
Nechako Lowlands (NEL)	Sub-boreal Spruce (SBS)	dense conifer forests with a wet climate and a rolling landform	M
	Engelmann Spruce-Subalpine Fir (ESSF)	dense conifer forests in a wet climate, rolling landform	M

(a)

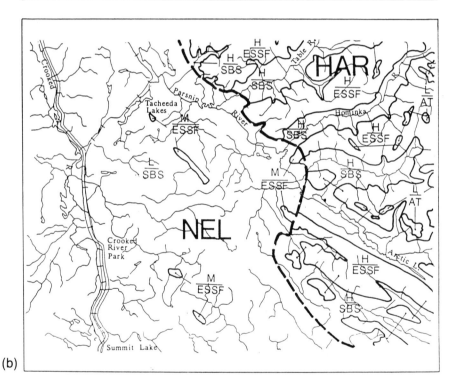

(b)

Figure 14.14 Grizzly bear habitat ratings for ecosections and biogeoclimatic zones (see Table 14.9) in eastern British Columbia, Canada (see Figure 14.13): (a) information table; (b) map. (From Fuhr and Demarchi, 1990.)

(Figure 14.14). Ecoregions provide a method of identifying and stratifying grizzly bear habitat importance at regional and provincial levels of mapping. In another study, Banner *et al.* (1985) worked within the biogeoclimatic classification system to complete a site-specific classification of coastal grizzly bear habitat. Seven climax forest ecosystems and two non-forested ecosystems were described for the 5000 ha study area.

(b)   Classification
As previously defined, classification is the act of arranging objects on the basis of their similarities and relationships. Historically, forest management classification systems employed criteria (variables) that reflected only timber values – descriptions based on tree cover, for example (Thompson and Welsh, 1993; Rowe, 1984; Wellner, 1989). While vegetation is a principal component of habitat, it is not the only factor affecting the distribution and abundance of wildlife. Forest ecosystem classifications must provide a spatio-temporal context for a range of forest ecosystem values for which natural asset managers are, and will continue to be, accountable.

In Ontario, ordination-based interpretations have been completed for woodland caribou (Harris, 1992; Morash and Racey, 1990), moose (Jackson *et al.*, 1991; Racey *et al.*, 1989a) and white-tailed deer (*Odocoileus virginianus*) habitat (Racey *et al.*, 1989b; Sims *et al.*, 1994) using the FEC framework. It is important to note that the FEC system does not necessarily involve mapping. In the white-tailed deer example, a 'vegetation–soil moisture nutrient key' was developed on the basis of applying known habitat requirements to a two-dimensional graph (Figure 14.15).

The key is then used to complete mapping programs as they are required. Figure 14.15 highlights the vegetation types normally capable of producing preferred browse (food) species and winter shelter for deer in areas to be managed for that purpose. Given that winter severity often is a limiting factor to the distribution and abundance of deer in northern regions, tree cover is also evaluated in terms of its ability to provide protection from wind chill and deep winter snow. Because the white-tailed deer has critical energy requirements in winter – and since most energy intake is achieved during the productive, snow-free period – identification, mapping and management of quality summer forage (e.g. grasses, deciduous leaves and herbaceous species) is essential.

In reference to Rowe's (1990) idea that habitat is best perceived and studied as a collection of spatio-temporal units within a mosaic of ecosystems at different scales in which animals come and go, the relatively fixed ecosystem serves as a framework in which to address the functional dynamics of wildlife. In short, organisms have their own inherent spatial and temporal dynamics (Merriam *et al.*, 1991) – they inhabit edges, corridors, patches and other configurations in-space-in-time. Therefore, although the ability to relate habitat descriptions and models to a structured hierarchical system is important, natural asset managers often must employ tools and techniques that provide management solutions for areas within and between ecosystems. The analysis of pattern is an important aspect of

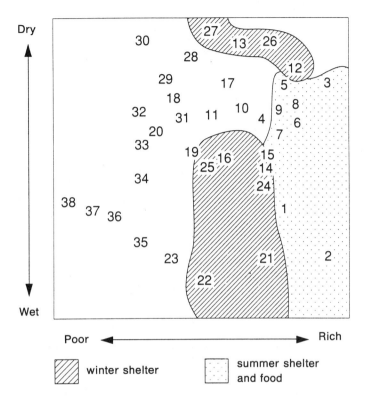

Figure 14.15 Northwestern Ontario Forest Ecosystems Classification vegetation ordination overlaid with winter and summer shelter and food requirements for white-tailed deer. The numbers identify the location of 38 vegetation types. A detailed description of the technique is provided by Sims *et al.* (1989). (Reproduced from Racey *et al.*, 1989a, cited in Sims and Uhlig, 1992, with permission from *Forestry Chronicle*.)

managing wildlife in a spatio-temporal context and within the dynamics of an ecosystem approach to management (Riley and Mohr, 1994:19).

Knowledge of, and modeling for, dispersal (for example) has been examined through development and application of patch dynamics models. Landscapes are divided and linked by corridors and patches, including strip corridors (which contain interior habitat for interior species), narrow line corridors (used by edge species), stream corridors, patches and networks (Forman, 1983; Forman and Godron, 1981, 1986:125; Godron and Forman, 1983). Merriam *et al.* (1991) provide a number of examples of models available to the natural asset manager, including:

- **Corridor models**: Habitats often are connected by corridors; management and protection of these corridors is important.
- **Non-corridor dispersal models**: These models are employed for species whose movements are not restricted to corridors, such as the movement of birds from one habitat patch to another.
- **Pool models**: Some organisms use a variety of unconnected habitat patches.
- **Mixed pool/corridor models**: These models are used for species that use a 'matrix' of connected and unconnected habitats.

Many tools and techniques to address habitat pattern and scale in wildlife habitat management are available. The reader is referred to Forman and Godron (1986), Levin and Buttel (1986), Merriam and Lanone (1990), Middleton and Merriam (1981), Turner (1987, 1989) and Turner *et al.* (1989).

14.4.5   Classification versus regionalization: which is best?

Successful ecosystem management requires that natural asset managers be able to make decisions at all levels in the hierarchy of ecosystems, from site-specific, stand-level spatial units to large regional (landscape) areas. For example, what a natural asset manager might protect and manage effectively at the 'stand' level (e.g. ecosite or ecosection levels of mapping in Canada or the landtype phase or landtype level of mapping in the United States) could be compromised by the activities of others who significantly impact regional landscapes (e.g. ecodistricts, ecoregions and ecozones in Canada or provinces, divisions and domains in the United States). The aphorism that 'everything is connected to everything else' is a relevant paradigm at this point in our history. With continued resource use pressures at any level, management of spatial units in isolation from smaller, adjacent or the larger ecosystems of which they are a part likely will fail to make a significant contribution to ecospheric protection objectives in the long term, and may compromise the values for which the wildlife habitat was managed initially (Gray 1993). Effective forest ecosystem management is, in part, founded on the basis of tools which allow managers to work at all levels in the spatial hierarchy. Classification and regionalization allow managers to describe and map all levels in the space-in-time hierarchy of ecosystems. For example, Ontario's FEC program, a classification system, is being linked to the province's regionalization program (Sims and Uhlig, 1992). In British Columbia wildlife managers must work at all levels in the ecological hierarchy to implement the programs for which they are responsible (Table 14.10). An integrated classification and regionalization system permits managers:

- To make statements about the local, regional, provincial/state and national status of wildlife (Eccles and Stelfox, 1985) and other assets-in-

Table 14.10 Examples of mapping themes and associated scales used by the Habitat Inventory and the Wildlife Conservation Sections of the British Columbia Wildlife Branch (table kindly provided by D. Demarchi)

Common input and output scales	Special requests	Program planning	Ecoregions	Biogeoclimatic	Ecosystem classes	Broad vegetation cover	Special habitats	Terrain	Terrain hazard	Plot locations	Capability for animals	Suitability for animals	Enhanceability/animals	Regional wildlife	Wildlife aerial survey	Administration boundaries	Guide outfitters	Trappers	Wildlife distribution	Wildlife property	Protected areas	Wildlife reserves	Endangered species
	Habitat inventory section																*Wildlife conservation section*						
1:2 000 000	■	■	■	■	■												■	■		■		■	
1:600 000	■																■						
1:500 000	■	■	■	■	■												■	■					
1:250 000	■	■	■	■	■	■	■	■	■	■	■	■	■	■	■	■	■	■			■		■
1:50 000	■	■		■	■	■	■	■	■	■	■	■	■	■	■	■					■	■	■
1:20 000	■	■		■	■	■	■	■	■	■	■	■	■	■							■	■	■
>1:20 000	■			■		■	■	■	■	■	■	■	■	■									

time (Lachowski *et al.*, 1994). For example, Maus *et al.* (1992) employed the USDA Forest Service ecological classification system to monitor vegetation change in the Mark Twain Forest, Missouri. Monkkennen and Welsh (1991) and Thompson and Welsh (1993) reported that changes in forest structure at the landscape level are having deleterious effects on year-round resident forest bird species in Canada.

- To identify and relate local, regional, provincial/state and national planning priorities (Albert *et al.*, 1986:1, and others). The effects of habitat alteration and loss are insidious because changes in habitat supply are cumulative over large areas and the general effects may not be noticeable in ecological time by those working at site-specific scales (Thompson and Welsh, 1993).

- To determine and defend budgets at all levels of program planning and implementation.

- To identify present and future wildlife supply and demand trends (Eccles and Stelfox, 1985). For example, Gray (1993:183) developed habitat suitability index-type models at the ecodistrict level (1:500 000–1:250 000) of mapping, a scale that allows managers to delineate moose 'functional units' and to identify areas of concern which might be affected by extensive regional industrial development, to give priority to areas requiring detailed inventory and monitoring and to reduce costs through design of efficient aerial survey programs. Some species have large home ranges and cannot be managed any other way.

14.4.6   Data and information management

Planning and development of comprehensive and consistent data and information management systems is one of the most difficult and pressing problems facing natural asset managers (Davis, 1980). Any type of inventory and assessment project must ultimately be part of an information management system designed to acquire and organize knowledge in a manner which assists decision-making and action planning (Gray and Stelfox, 1991) (Figure 14.16). Baseline inventory programs must be structured on the premise that they will form an essential basis from which reasoned action can be taken.

Information systems can be as simple or as complex as necessary. It is important to design evaluation programs that can be integrated upward with broader decision-making activities and downward with more site-specific (and perhaps more detail) decision-making activities. An effective data and information management system must have appropriate data storage capacity and be capable of data and information flow through an interpretative framework. This need has been recognized by agencies and companies around the world. In the United States, many federal and state agencies began to develop information management systems in the 1970s

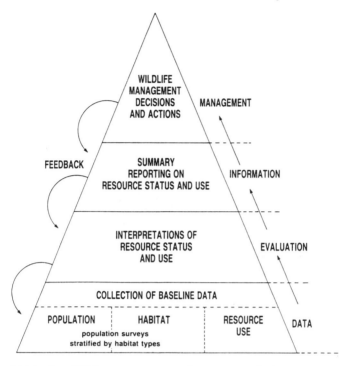

Figure 14.16 A wildlife management information and decision-making system, illustrating the systematic collection of large volumes of baseline data and conversion of them into smaller volumes of useful information tailored to focused management decision-making. (From Gray and Stelfox, 1991.)

(e.g. Naveh and Lieberman, 1990:148). In Canada, the government of Alberta has developed and implemented such a system, called the Integrated Resource Planning (IRP) system, and the government of Ontario is currently developing the Integrated Natural Resource Inventory System (INRIS). Like many of the other systems currently used in North America, INRIS will enable managers to gain automated access to data and information on the quality, quantity and characteristics of natural assets within an ecogeographical context – it will permit managers to address a variety of natural asset issues, including GAP analysis.

(a) On-site inventories
Habitat evaluation that supports implementation of an ecosystem approach requires natural asset managers to examine habitat data and information management needs during the initial planning of the data and information collection program. The level(s) of planning and/or management will determine the mapping scale, and will require that project staff understand the

types of data and information which can be employed at each level in the ecosystem hierarchy.

In an applied context, some of the important ecosystem composition, structure and function variables required to measure values – such as habitat suitability, habitat capability and carrying capacity – normally are not collected under the auspices of other programs (Laymon and Barrett, 1986), such as forest or geomorphological inventory programs. This has been identified as a limiting factor by Bonar (1991), Gray (1993:185) and others. For example, Bonar (1991) found that habitat variables required for some of the 30 faunal species models being developed in west-central Alberta were not available in existing inventory data bases. Therefore, wildlife managers must, during the early planning stages, identify those variables required for the types of models they know, or anticipate, will be required to manage the spatial unit. Corns (1991) suggests that a few minutes of additional time per sample plot by the vegetation ecologist describing attributes not normally identified, such as snags, may significantly enhance the quality and usefulness of the data base and improve the manager's ability to integrate faunal values into ecosystem assessments. Many agencies have begun to initiate integrated inventory programs that account for a range of forest values. In British Columbia, field teams comprise a surficial geologist/pedologist, a plant ecologist and a wildlife biologist (Demarchi and Lea, 1987). In Ontario, the Ministry of Natural Resources (MNR) collects data on wildlife species and habitat (e.g. mast production, density and type of cavity trees, and modified stocking/structure for featured species such as the red-shouldered hawk (*Buteo lineatus*)) during forest inventories (Anderson and Rice, 1993:117).

(b)   *Remotely sensed data*

Remote sensing is '*the acquisition of knowledge by unattached means*' and includes aerial photography, satellite imagery, radar, thermal imagery and side-looking radar (Anderson *et al.*, 1976; Lunt, 1989a; Mayer, 1984; Naveh and Lieberman, 1990:114). It is an important data capture tool, and is employed to inventory and monitor natural assets (including the flow of energy and matter in the ecosystem), to quantify ecosystem variables and to evaluate change and management strategies (Quattrochi and Pelletier, 1991).

Forest managers have a long history of developing and employing remotely sensed data in Canada and the United States (Colwell, 1985; Leckie, 1990; Lunt, 1989a), beginning with early experimentation and use of aerial photography in the 1920s and 1930s. In the 1950s and 1960s, forest ecosystem managers developed and implemented large area management inventories through use of aerial photography. Remotely sensed data provides the basis for development of descriptive models at all mapping scales in ecological classification/regionalization hierarchies (Table 14.11). Use of digitally remotely sensed data for forestry and the application of

Table 14.11 Relationship between remote sensing systems and ecological land survey mapping scales (from Rubec, 1983)

Remote sensing system	Spatial units of the Canadian Ecological Land Classification System (CELC)						
	Scale I	Scale II	Scale III	Scale IV	Scale V	Scale VI	Scale VII
Satellite imagery	Ecozone (1:20 000 000 to 1:15 000 000)	Ecoprovince (1:10 000 000 to 1:5 000 000)	Ecoregion (1:3 000 000 to 1:1 000 000)				
High altitude spacecraft or aircraft photography				Ecodistrict (1:500 000 to 1:1:125 000)			
Moderately high altitude aircraft photography					Ecosection (1:250 000 to 1:50 000)		
Low altitude aircraft photography						Ecosite (1:50 000 to 1:10 000)	
Low altitude aircraft and ground photography							Ecoelement (1:10 000 to 1:2500)

these data for wildlife habitat mapping began in the early 1970s with the construction of Landsat satellite reception and processing facilities (Anderson *et al.*, 1976; Leckie, 1990; Naveh and Lieberman, 1990:114).

The multispectral capabilities of satellite imagery such as Landsat permit natural asset managers to observe and measure many ecosystem variables at multiple scales and at different times (Quattrochi and Pelletier, 1991). There are seven Thematic Mapper spectral bands with which natural asset managers can identify and model a variety of assets, including coastal areas, vegetation discrimination and vigour assessment, soil moisture and discrimination of mineral and rock types. For example, Wiken *et al.* (1980) employed Landsat to map Canada's ecodistricts.

During the 1970s and 1980s, many application projects, including wildlife habitat evaluation, were implemented in support of enhanced land management in both countries (Table 14.12). For a general summary of tools and techniques, consult Kerr (1986), Leckie (1990), Mayer (1984) and Quattrochi and Pelletier (1991). Two important advantages of remote sensing for wildlife habitat evaluation are that global–local measurements can be completed repeatedly and that a wide variety of spectral ranges and sensors are available to provide remotely sensed data (Lulla and Mausel, 1983; in Quattrochi and Pelletier, 1991).

(c)   Geographical information systems (GIS)

As technology advances and new issues emerge, forest ecosystem managers are adapting their approaches to data and information collection and management to provide the most meaningful information to user groups (e.g. McAllister *et al.*, 1994). Prior to the 1980s, most of the ecogeographic studies were represented on 'hard copy' maps and in detailed, often complex, legends. Therefore modeling capabilities were limited. For example, a detailed mapping program which required five years to complete in the 1960s was considered efficient and useful, but in the 1990s such maps would be antiquated at the time of publication and of little use to those attempting to make meaningful decisions. Through the combined use of remote sensing technologies and GIS, maps can be produced and revised quickly, accurately and at reasonable cost.

In general, a GIS contains four fundamental components: a data input subsystem to receive and process spatial and associated descriptive data (e.g. from maps, photographs, digital images and on-site surveys); a data base management subsystem to store and retrieve data; an analysis subsystem to interpret within and among data themes; and a reporting subsystem to display maps and reports (Coulson *et al.*, 1991) (Figure 14.17). Continuing development in GIS technology and the increased number of available data bases provide natural asset managers with a tool that can be used to simulate real-world processes spatially and temporally (Aronoff, 1989:2; Ball, 1994; Lunt, 1989b). Although initially created as a data base

Table 14.12 Examples of Landsat imagery applications used in support of wildlife habitat evaluation programs in Canada and the United States

Description	Source
Application of remote sensing techniques to wildlife habitat evaluation programs	Mayer (1984)
Lesser prairie chicken (*Tympanuchus pallidicinctus*) habitat in western Oklahoma	Cannon *et al.* (1982)
Caribou habitat in northern Manitoba	Dixon and Horn (1981)
Moose habitat in Alaska	Laperriere *et al.* (1980)
Moose habitat in north-central Manitoba	Dixon *et al.* (1984)
Spatial relations of wildlife habitat in the Great Dismal Swamp	Mead *et al.* (1981)
California condor (*Gymnogyps californianus*) habitat in California	Scepan and Blum (1987)
Barren-ground caribou (*Rangifer tarandus groenlandicus*) winter range in the boreal forest of the southern Northwest Territories, Canada	Peterson (1987)
Identify, measure and analyze forest patches in southern Ontario	Pearce (1992)
Ground cover mapping in northern Ontario	Piirvee and Braun (1978)
Analysis of black-tailed deer (*Odocoileus hemionus columbianus*) habitat in California	Stenback *et al* (1987)
Analysis of forested habitat in southern Ontario to assess habitat suitability of area-sensitive and forest-interior bird species	Hounsell (1992)
Elk habitat evaluation in Oregon	Eby and Bright (1985)
Spotted owl (*Strix occidentalis*) habitat evaluation in the old growth forests of Washington	Young *et al.* (1987)
Evaluation of habitat of the black-capped vireo (*Vireo atricapillus*) and the golden-cheeked warbler (*Dendroica chrysoparia*) in central Texas	Shaw and Atkinson (1990)
Nesting habitats of kestrel falcons (*Falco sparverius*) located using Landsat computer-classified data and a nesting habitat model	Lyon (1983)
Forest ecosystem classification in northern Québec	Beaubien (1994)

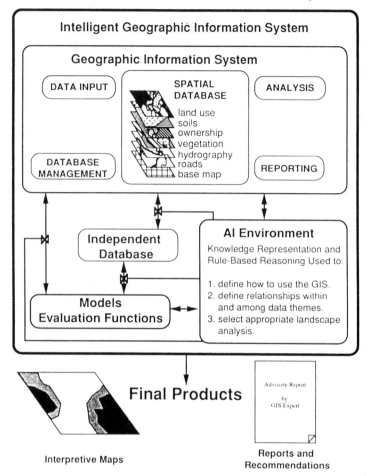

Figure 14.17 Basic elements of a Geographic Information System (GIS). All GISs have facilities for data input, data base management, analysis and reporting. An Intelligent GIS (IGIS) includes an artificial intelligence (AI) environment for knowledge representation and rule-based reasoning. (Reproduced from Coulson *et al.*, 1991, with permission from Springer-Verlag.)

query tool which allowed users to identify the spatial location of a set of values or the location in which a set of criteria could be met (Ball, 1994), GIS systems have been developed into powerful modeling tools which, when combined with natural asset models, permit users to model efficiently and iteratively.

Although GIS applications related to wildlife are in their infancy, a number of projects have been completed in North America (Table 14.13). For example, the British Columbia Ministry of Environment and Parks and the Ministry of Forests and Lands have sponsored a number of GIS projects (Kansas and Usher, 1988:10) for regional planning and forest management (e.g. McNay and Page, 1988). In response to forest habitat loss and fragmentation in the southern part of the province, Ontario Hydro is conducting an evaluation of landscape patterns and their conservation in the remaining forested areas. Using satellite imagery, the digital data are analyzed at spatial scales ranging from 1:250 000 to 1:10 000 using a GIS. Through analysis of forest patch pattern, size, shape, percentage cover and other measures, managers are assessing the relative conservation value of forested landscapes for supporting populations of forest interior and area-sensitive wildlife species (Hounsell, 1992). Similarly, Pearce (1992) completed a pattern analysis of forest cover in southwestern Ontario using remote sensing and GIS technology.

Many agencies in the United States employ GIS as a management tool. For example, Hart *et al.* (1985) employed GIS to manage natural assets in Flathead National Forest, Montana. The data base comprised digital terrain data, vegetation associations derived from Landsat satellite data, land types and ownership, precipitation, drainage and timber harvest. Maus *et al.* (1992) employed data collected remotely and GIS to monitor vegetation change in the Mark Twain National Forest, Missouri. With recent advances in artificial intelligence (AI) (section 14.6), scientists currently are exploring a variety of expanded modeling techniques, including integrated AI and GIS systems. Coulson *et al.* (1987, 1991), for example, developed an 'intelligent GIS' (IGIS) designed to enhance the manager's ability to automate interpretation of relationships within and among ecosystem data themes, select analytical solutions in response to the questions and issues being addressed and seek guidance in use of the system (Coulson *et al.*, 1991).

14.5 PREDICTIVE MODELS

With increased land-use pressures, and known and potential habitat loss followed by decline or elimination of wildlife species in the last half of this century, natural asset managers increasingly are being asked to provide information on the known and potential impacts of resource use on wildlife habitat and populations. Despite the fact that North American biologists had accumulated a basic knowledge of the habitat requirements of many species by the 1950s, they experienced great difficulty in objectively describing and predicting species occurrence, and in some cases abundance. Predictive models assist natural asset managers in estimating habitat quantity and quality through time. They allow the manager to create various scenarios that describe the impacts of decisions and actions on ecosystems (Ball, 1994) and, by extension, wildlife habitats.

Over the last 50 years, and in association with improved technology like computers and remote sensing tools, scientists have developed many sophis-

Table 14.13   Examples of GIS applications in wildlife management

Software	Description	Source
PAMP (Planning and Assessment Map Product) and ARC-Info	Black-tailed deer and forest management on Vancouver Island, British Columbia.	McNay and Page (1988)
CAPAMP (Computer Assisted Planning and Assessment Map Product)	Classification, mapping and evaluation of coastal grizzly bear habitat in British Columbia.	Banner *et al.* (1985); Hamilton *et al.* (1987)
MPSIII/pc (version 1.1 or later), PAMP, SPS (Stand Projection System)	MAGIS (Multi-Resource Analysis and Geographic Information System) used to determine location and timing of land management activities and road construction in watershed-level planning areas.	H. Zuuring, G. Jones and W. Wood, cited in Schuster *et al.* (1993:231)
MOSS (Map Overlay and Statistical System)	Natural asset planning in the Nicolet National Forest, Wisconsin.	Watry (1989); Lachowski (1989)
MOSS and Oracle 6.0	TROPPS (Treatment Opportunity Model) identifies potential treatment of stands; employed as a 'scoping' tool.	D. Gregson and L. Smit, cited in Schuster *et al.* (1993:203)
MOSS and Oracle	SERAL (Seral Stage Analysis Module) identifies the seral stage of each stand; employed to quantify change in biodiversity and other wildlife attributes.	D. Gregson, cited in Schuster *et al.* (1993:95)
MOSS	To examine the cumulative effects of land-use activities on grizzly bears in Greater Yellowstone Area.	Key (1989)
ARC/Info and HEP	Used GIS to integrate wildlife habitat attributes (HEP models) into ecological land classification system in Clay Brook watershed, central Vermont.	Herbig (1988)
ARC/Info and PC-FORPLAN	In GIS/FORPLAN, the linear programming model is applied within the GIS to provide a spatial context for decision making.	D. Norris, cited in Schuster *et al.* (1993:226)
ARC/Info and Oracle	ARCFOREST contains forest management decision support tools.	ESRI Canada and Ontario Ministry of Natural Resources, cited in Schuster *et al.* (1993:215)
ARC/Info	Spotted owl habitat evaluation in Washington.	Young *et al.* (1987)

Table 14.13 Continued

Software	Description	Source
ARC/Info	The SDSS (Spatial Decision Support System for Timber Sale Planning) is a GIS-based decision support system employed in the Jefferson National Forest, Virginia.	D.P. Kenney and T.W. Reisinger, cited in Schuster *et al.* (1993:111)
ARC/Info	GAP (Gap Analysis Project) is a geographic approach to quantifying biodiversity indicators in areas that are managed primarily for the long-term maintenance of native species and natural ecosystems.	B. Butterfield, cited in Schuster *et al.* (1993:81)
ARC/Info	SE ALASKA HABCAP (Habitat Capability Models for the management of Indicator Species in Southeastern Alaska) models are created using habitat variables accessed through a GIS. Habitat capability estimates can be calculated for 13 indicator species in the Tongrass National Forest.	L.H. Suring, cited in Schuster *et al.* (1993:189)
ARC/Info, but not specific to any GIS software.	The HSG Wood Supply Model is a spatial timber management simulation model to which managers can link wildlife habitat and biodiversity information. It operates on large data sets and over long time periods (e.g. 50–200 years).	T. Moore and C. Lockwood, cited in Schuster *et al* (1993:229)
ARC/Info	Employed to investigate summer habitat selection by mule deer in Arizona.	Haywood (1989)
ERAS (Earth Resources Data Analysis System)	Used GIS to identify and describe remaining potential habitat for the California condor.	Scepan and Blum (1987)
Tydac SPANS and DBase III +	Used GIS with Habitat Evaluation Models for moose in Nahanni National Park, Northwest Territories, Canada.	Gray (1993)
Tydac SPANS	Described Canada's avifaunal and mammalian diversity with a species richness index. In total, 386 bird species and 126 mammal species are represented at the ecoregion level of mapping.	Desrochers and Lash (1990)
MIRIS (Michigan Resource Information System)	Calculates and displays Habitat Suitability Indices for wildlife in Michigan.	Donovan *et al.* (1987)
ELAS digital image processing software	Evaluates black-tailed deer habitat in California.	Stenback *et al.* (1987)

ticated techniques with which to explore species–habitat relations. Following World War II, ecologically oriented scientists (e.g. MacArthur, 1958; Whittaker, 1962, 1967a,b; Greig-Smith, 1964) began to adopt, develop and apply quantitative techniques to the measurement of wildlife populations and habitats. By the mid 1970s managers were organizing these diffuse and diverse data bases in support of land-use planning (Thomas, 1986). Thousands of species–habitat predictive models have been developed, ranging from matrices and simple linear and curvilinear regression equations to complex multiple regressions and GIS models (Lyon, 1989). They are used to formulate and evaluate alternative management strategies and plans, and to estimate compensation requirements in relation to specific land-use actions (Farmer, 1978; Schamberger and Farmer, 1978; Schamberger and O'Neill, 1986).

14.5.1   Statistical models

Given that habitat comprises a complex suite of ecosystem properties in-space-in-time, its study often requires simultaneous examination of a number of interacting variables (Gauch, 1982:1; Jones, 1986; Shaw and Wheeler, 1985:229; ter Braak and Looman, 1987). The adoption and application of multivariate statistical analyses in the last 30 years has provided the manager with a range of quantitative tools with which to explore wildlife–habitat relationships, to treat multivariate data as a whole, and a technique to explore structural relationships (Gauch, 1982:1).

Some multivariate techniques allow managers to simplify and summarize large data sets with many variables (often referred to as analysis of interdependence), including ordination (factor analysis and principal component analysis) and cluster analysis, while others, such as multiple regression and discriminant analysis techniques, are employed to examine relationships between variables (analysis of dependence) (Figure 14.18). Thousands of studies employing multivariate statistical procedures have been completed in Canada and the United States. The reader is referred to Capen (1981), Verner et al. (1986) and Szaro et al. (1988) for examples of papers outlining statistical applications to wildlife habitat management, while Ruggiero et al. (1991) is recommended to those interested in work focused on wildlife in a specific area, the Douglas fir forests of the Pacific Northwest. Despite this large body of work, comparatively few studies have been completed within a hierarchical ecological classification or regionalization system.

While it is not the purpose of this chapter to describe each statistical procedure, a few studies have been selected that employ a relatively new statistical application: multiple logistic regression. Logistic regression has emerged as an important modeling tool because it is useful when the manager's goal is to examine the relationship between a dichotomous dependent

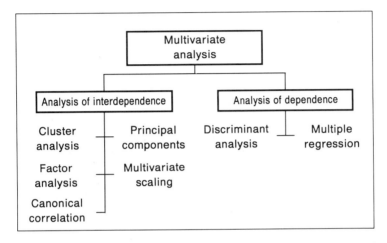

Figure 14.18 A typology of multivariate statistical techniques. (Adapted from Shaw and Wheeler, 1985:229, courtesy of Wiley, Chichester.)

variable (presence/absence) and independent habitat variables which cannot be assumed to satisfy the stringent assumptions of discriminant analysis (Ludeke *et al.*, 1990). Logistic regression does not require, for example, the assumption of multivariate normality (Seber, 1984; Thomasma *et al.*, 1991). Austin (1984), Austin and Cunningham (1981) and Austin *et al.* (1983) used categorical data in multiple logistic regression equations to examine the altitudinal distribution of several eucalypt species in relation to ecological parameters (i.e. rainfall, radiation and geology) in Australia. Similarly, Ludeke *et al.* (1990) used logistic regression analysis to determine variables most closely associated with deforestation in Honduras. In North America, Smith and Connors (1986) employed logistic regression to build predictive models of occurrence for avifauna in Alaska, but they did not employ multiple logistic regression techniques; single ecological variables were used to examine avifaunal distribution, which limits their usefulness as a decision-aid. Gore (1988) employed logistic regression to examine the influence of habitat structure on the distribution of small mammals in New Hampshire old-growth northern hardwood forests. Unfortunately, the model failed to predict the presence of small mammals in the independent study area. Robbins *et al.* (1989) used logistic regression to model the habitat area requirements of breeding forest birds of the middle US Atlantic states. Gray (1993) employed multiple logistic regression to identify habitat variables employed in winter moose habitat evaluation models at a regional mapping scale (ecodistrict level of mapping; 1:250 000 to 1:500 000).

14.5.2   Matrix models

Matrix or wildlife–habitat association models are simple and traditionally have been used in environmental impact statements to describe the current state of wildlife which may be impacted by a proposed land-use activity. Matrix models denote a suite of procedures with which managers organize and describe wildlife–habitat relationships in tabular format. Although not a powerful predictive tool, the matrix model permits the natural asset manager to organize data and information and to address preliminary planning and assessment questions. In some cases, matrix models are used to represent the results of statistical analyses of organism–habitat relationships (e.g. Ruggiero *et al.*, 1991). There are a number of variations of the matrix model.

(a)   Life-form

In the United States, matrix models were developed and employed in support of natural asset management planning following the work completed by Haapanen (1965) for avifauna inhabiting Finnish forests. He employed the life-form (a group of species whose habitat requirements are satisfied by similar successional stages within plant communities). Thomas *et al.* (1978) and Thomas (1979) adapted this technique, developed a wildlife–habitat association model and applied it to forest management planning in the Blue Mountains of Oregon and Washington by displaying information describing animal response to proposed land uses – 379 terrestrial vertebrates were grouped according to 16 life-forms correlated with the successional stages of each plant community. Managers then provided detailed life history information in a user-friendly matrix format, explained the relationship of animal life-forms to plant communities and the relationship of species (within life-forms) to plant communities, and subjectively estimated the relative vulnerability of each species to the proposed land uses. In Ontario, Baker (1988) described the habitat preferences of birds, mammals, amphibians and reptiles breeding in the forested ecosystems of the province. Habitat preferences were divided into eight habitat types and each species was assigned to a life-form (after Thomas, 1979). On the basis of a literature review, species were then allocated to one or more of the habitat types. The matrices are used as a preliminary impact assessment and evaluation tool.

(b)   Guilds

'Guilding' is similar to the life-form approach through the clustering of species with similar habitat requirements for feeding and reproduction, and can be employed to estimate the response of species to land-use activities

(Berry, 1986). Guilds are defined according to the habitat layers a species uses for breeding and feeding sites (Cooperrider, 1986a). Species are then assigned to appropriate matrix cells based on an evaluation of habitat requirements. Species occupying the same combination of cells are lumped into guilds. Using such models, a natural asset manager can predict the guilds (and the species that constitute the guilds) that would be impacted in the event that the habitat is lost or adversely affected by land use (Cooperrider, 1986a). Short (1984), for example, developed guild and layer models to evaluate the structural complexity of desert and riparian habitats in Arizona. He argued that the vertical dimension of vegetation is important in determining the number of wildlife species that inhabit it, and identified eight habitat layers on the basis of vegetation height, density and the presence of water. Species were allocated to cells in the species–habitat matrix on the basis of foraging and nesting locations (requirements).

In Alberta, Green *et al.* (1986) employed guilding to examine wetland drainage scenarios for increasing agricultural production and evaluated the potential impacts to wildlife. Because it was not possible to examine and model each species individually, and given that many of the species demonstrated similar habitat requirements, the guilding tool was employed to group species with strong similarities in their use of various wetland zones. A representative species was then selected from each guild and a habitat assessment model was developed for that species (Stelfox, 1991). Bonar *et al.* (1990) employed a similar approach as a data reduction technique to identify groups of wildlife with similar habitat requirements and selected indicator species for which habitat supply models were created in west-central Alberta. In all, 30 species were used to represent 284 terrestrial species identified in the study unit. The reader is referred to the work of Root (1967), Short (1983, 1984), Short and Burnham (1982), Sveringhaus (1981) and Verner (1984) for descriptions and examples of guilding techniques.

(c) Single species habitat matrices
Species–habitat matrix descriptions have been employed widely as planning and assessment tools in Canada and the United States. For example, Verner and Boss (1980) developed matrices for wildlife inhabiting forested and non-forested habitats according to successional stage and tree canopy cover classes in California. The matrix cells provided information on seasonal use (spring, summer, fall and winter) and habitat function (breeding, feeding, resting and season of occurrence), and were employed to allocate the species to one of optimum, suitable and marginal habitat classes. DeGraaf and Rudis (1986) employed matrix models to describe natural history profiles (e.g. breeding season, breeding and feeding, and winter feeding) of New England wildlife species and their associations with forested and non-forested habitats. They also related the forested habitats to Bailey's (1980) ecoregions map of the United States. In a second study, DeGraaf *et al.*

(1992) used species occurrence and utilization matrices for forested and non-forested habitats and provided silvicultural treatments for six major forest cover types in New England (Figure 14.19). Sidle and Suring (1986) developed a species–habitat matrix and a species status matrix for Alaskan wildlife. Hoover and Wills (1987) employed matrices to provide forest ecosystem managers with the necessary information and outlined techniques to use the matrices in development of management plans that address options for bringing together and analyzing a wide variety of information about wildlife habitat needs and habitat responses to various silvicultural treatments in Colorado.

The matrix model is a two-dimensional, often static representation of the real world. It is a simple tabular technique used to convey information about a natural asset, such as wildlife and habitat. Although simple, and more often than not limited to presence or absence, or a ranked population status (e.g. rare, common, abundant), matrices do reflect considerable science. The work completed by DeGraaf and Rudis (1986), for example, was based on an extensive review of available literature. Ruggiero *et al.* (1991) compiled matrices for plant and animal species (and guilds for birds as well) in the Douglas-fir forests of the Pacific Northwest on the basis of 24 quantitative studies involving the use of numerous multivariate statistical procedures.

14.5.3   Index models

Many index models have been developed and applied since the mid 1970s (Table 14.14). The Habitat Evaluation Procedures (HEP) and the Pattern Recognition (PATREC) models are employed throughout North America, and exemplify this suite of modeling techniques.

(a)   *Habitat evaluation procedures (HEPs)*
In 1973 the US Fish and Wildlife Service commissioned a Task Force to identify and examine evaluation systems that could be employed in water planning programs in the United States (Daniel and Bachant, 1980). The Task Force recommended that the system developed and employed by the Missouri Department of Conservation and the US Soil Conservation Service be adopted (Daniel and Lemaire, 1974) because it was community-oriented, achieved an acceptable degree of resolution and sensitivity (for site design and calibration), could be made available to the environmental community and could be modified to meet national requirements for a standardized evaluation system (Daniel and Bachant, 1980; Farmer, 1978). This recommendation led to the creation of the Habitat Evaluation Procedures (HEPs).

HEPs permit managers to develop models which employ an Habitat Suitability Index (HSI). The HSI is determined by using variables known or perceived to be important to the species. The technique is simply an

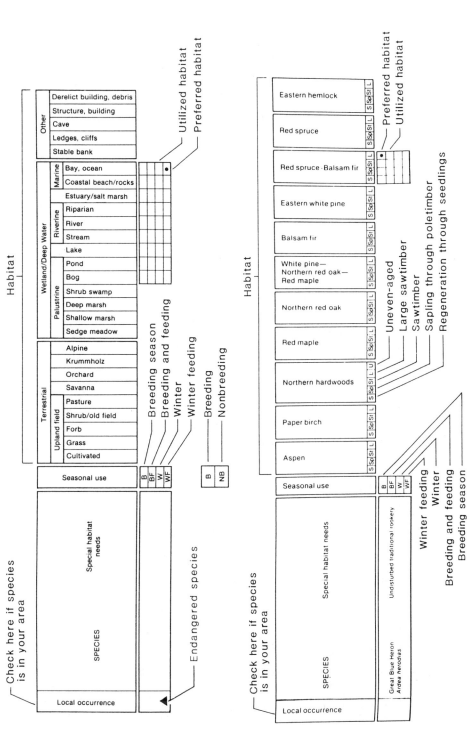

Figure 14.19 The key to elements of the species–habitat matrices developed for New England. (From DeGraaf *et al.*, 1992.)

Table 14.14 Examples of wildlife habitat index models

Name	Location	Description	Source
Habitat Evaluation Procedures (HEP)	United States and Canada	Employs a Habitat Suitability Index (HSI). HSI is determined by using variables known or perceived to be important to a species. An accounting procedure in which variables are rationalized, ranked and sometimes weighted and linked to other variables in an additive or multiplicative equation.	US Fish and Wildlife Service (1981); Allen et al. (1987, 1991); Schroeder (1986); Schroeder and Allen (1983)
Current Habitat Suitability (CHS) Models	Alberta	A ranking and weighting procedure similar to the HEP model. These models have been developed within an ecosystem classification system.	Eng and Stelfox (1988); Green et al. (1986)
Pattern Recognition Models (PATREC)	United States	The PATREC model employs habitat variables to predict the expected capability of habitat to support a species. It is based on probability, where the model predicts the likelihood that a habitat with a given set of characteristics will support varying densities of animal populations.	Williams et al. (1977); Kling (1980); Wilson (1983); Evans (1983); Grubb (1988)
Habitat Evaluation System (HES)	Lower Mississippi Valley	These multi-species models were developed to assess wildlife communities in aquatic and terrestrial habitats. The model is constructed within a spatial framework of four plant community types, each described with eight variables (e.g. percentage ground cover). The variable ratings (0–1) are weighted and summed to calculate a Habitat Quality Index (HQI).	US Army Corps of Engineers (1980), cited in Schroeder (1987)
Wildlife Habitat Appraisal Procedure (WHAP)	Texas	This is a multi-species model in which the Biological Habitat Component (BHC) is calculated for species inhabiting plant community types. The model output score (0–1) is derived from the sum of weighted values for habitat variables. The model measures alpha diversity.	Frye (1984), cited in Schroeder (1987)

Model	Location	Description	Reference
Saskatchewan Forest Habitat Project (SFHP)	Saskatchewan	Employs HEP models for six species selected as indicators of habitat change.	Yurach et al. (1991)
Stream Corridor Inventory and Evaluation System (SCIES)	United States	Multi-species model used to assign a habitat value to stream corridors. Model variables are used to calculate a score on the basis of the habitat's ability to support species richness and density. The terrestrial habitats are divided into guild units using vegetative and topographic descriptors, and the variables for each guild are rated, weighted and summed.	Garcia et al. (1984), cited in Schroeder (1987)
Wildlife Wetland Evaluation Model (WWEM)	Northeastern United States	The WWEM evaluates wildlife values in wetlands. It provides an output score ranging between 36 and 108 as a measure of diversity and productivity. The index is calculated by summing the weighted ranks of the habitat variables.	Golet (1976), cited in Schroder (1987)
Wildlife Species Richness Model for Wetlands	United States	Based on a 'key' that provides a qualitative (i.e. low, medium and high) species richness designation. The model employs 25 variables, and the wetland is evaluated on the basis of whether or not the variables represent the desired condition. It employs the wetland classification developed by Cowardin et al. (1979).	Adamus (1983), cited in Schroeder (1987).
Habitat Evaluation Index (HEI)	Illinois	The model's spatial units are classified according to 23 plant communities, and the model output provides a measure of species richness of the habitat being evaluated.	Graber and Graber (1976), cited in Schroeder (1987)

accounting procedure employing indices, whereby each variable is rationalized according to its importance to the species, ranked (and in some cases weighted) and linked to other variables in an additive or multiplicative equation (US Fish and Wildlife Service, 1981) (Figure 14.20). Hundreds of habitat suitability models, or models of similar design, have been developed in the United States and Canada using tailored software specifically designed for the application of HEPs (Hays, 1985, 1987, 1989; Mangus, 1990), using Dbase III + and BASIC (e.g. Eccles and Stelfox, 1985; Eng and Stelfox, 1988; Gray, 1993; Green *et al.*, 1986; O'Leary, 1990; Stelfox, 1988, 1991; Rayner and Bennett, 1988) and spreadsheet packages (Gray and Keith, 1988).

Many of the models have been developed for managers working with habitats in ecosystems at the lower end of the ecological classification system hierarchy (i.e. at large mapping scales). For example, the HEP model for moose in the Lake Superior Region developed and tested by Allen *et al.* (1987, 1991) is based on vegetation data characteristics (e.g. browse biomass, browse diversity, canopy cover, species composition of the trees and distance between forage and cover). The vegetation is mapped at a scale of 1:15 840 and the distribution and abundance of moose at a scale of 1:24 000. Application of this model is therefore intended for use at the site-specific forest stand level, which equates to the ecoelement, ecosite and in some cases the ecosection mapping scale in the Canadian Ecological Land Classification System hierarchy and the landtype association, landtype and landtype phase mapping scales in the USDA Forest Service classification system.

(b) Pattern recognition models (PATREC)

Williams *et al.* (1977) developed PATtern RECognition (PATREC) models to evaluate wildlife habitat, based on pattern recognition theory and medical diagnostic techniques. They approached habitat evaluation from the perspective that it could be expressed as a problem of conditional probability. PATREC models are similar to Habitat Suitability Index models in that they employ habitat variables to predict the expected capability of habitat to support an animal species, but differ from HSI models because they are based on a formal statistical procedure, probability (Cooperrider, 1986a; Grubb, 1988). For example, Williams *et al.* (1977) employed PATREC to predict the probabilities that a habitat with a given set of characteristics will support varying densities of pronghorn (*Antilocapra americana*) populations. PATREC is based on the calculation of likelihood or probability that a spatial unit could provide habitat for a species as determined by examining the unit's ecological attributes (Flather, 1982:67; Flather and Hoekstra, 1985; Grubb, 1988; Kling, 1980:8) (Figure 14.21). Williams *et al.* (1977) employed Bayesian theory as a primary constituent of the evaluation technique. Three steps are involved: estimation of the initial probability distribution that some condition or set of conditions exists;

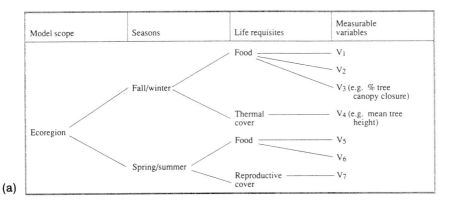

(a)

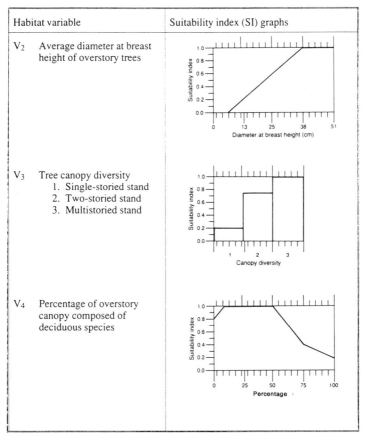

(b)

Figure 14.20 Example of the HEP model. (a) The model is defined by spatial unit (e.g. ecoregion), season, life requisites and measurable variables. Multivariate statistical analysis can be used in support of expert opinion to verify the significance of the variables to the species. (b) Habitat Suitability Indices are prepared whereby the variables are ranked according to their importance to the species. The scores are than calculated for each spatial unit using additive of multiplicative equations. (From Stelfox, 1991.)

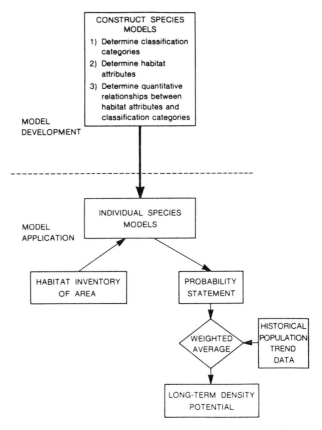

Figure 14.21 The PATREC modeling process. (From Flather and Hoekstra, 1985, courtesy of The Wildlife Society.)

collection of the sample data; and application of the sample results to revise the initial probability estimates.

Kling (1980) tested PATREC models for mule deer (*Odocoileus hemionus*), pronghorn antelope, sharp-tailed grouse (*Pediocetes phasianellus*), sage grouse (*Centrocercus urophasianus*), golden eagle (*Aquila chrysaetos*) and Brewer's sparrow (*Spizella breweri*) inhabiting southeast Montana and northeast Wyoming. Wilson (1983) employed PATREC to develop and evaluate habitat evaluation models for mule deer, elk, moose, ruffed grouse (*Bonasa umbellus*), prairie falcon (*Falco mexicanus*) and sandhill crane (*Grus canadensis*) for sagebrush-grasslands and deciduous and coniferous forests in southeastern Idaho. The models were tested and modified by comparing model output to estimated population densities developed by the Idaho Department of Fish and Game. Evans (1983) employed PATREC models to predict the impact of phosphate mining on large mammal species (mule deer, elk and moose) in southeast Idaho and explored options for incorporating the models into mitigation planning procedures. PATREC has not been employed as widely as the HEPs but has considerable potential as a natural asset planning and management tool.

(c) Limitations of index models

As outlined by Schroeder (1987) and others, limitations of many index models produced to date are:

- The variables used to represent a species habitat are selected subjectively and often not well rationalized.
- The assignment of variable ranks and weights is subjective and not well rationalized.
- The final scores or indices which purport to indicate multiple conditions are ambiguous.
- The models are static point-in-time representations of wildlife habitat.
- The methodology is poorly described.
- The assumptions are not explicit.
- The habitat data bases, often compiled to meet non-wildlife objectives, do not include suitable habitat variables. and/or adequate parameters for model construction and the generation of meaningful results.
- The model has not been, or cannot be, tested against the dependent variable on control sites.

Although subjectivity will always be a factor in natural asset models, techniques to maximize model objectivity are now available and being employed. For example, the HEP procedures can be quantified at various stages in the development and testing of the model through application of statistical analyses to, for example, select and test habitat variables. In addition, dynamic variables (e.g. fire) can be incorporated into the models, particularly if the model is tested against the dependent variable (e.g. presence and absence of a species through time). In response to some management issues, modelers can elect to use index model results in other models, such as forest ecosystem succession models.

14.5.4 Dynamic, linked models

Historically, land-use planning and environmental impact assessment and management programs have been limited through application of static models. Modeling for change is important because real world simulation requires that a model accounts for the passage of time in one or more variables (Ball, 1994). Static models are flawed because they ignore dynamic ecosystem composition, structure and function – they do not reflect the dynamic life history of wildlife and their habitat.

Ecosystems constantly change, and as dynamic 'wholes' their interacting parts are 'drivers' and/or 'receptors' of change. Different dynamic models are used to explore various aspects of ecosystem function through time, but vegetation growth, succession and climate models are examined here because of their obvious relationship to wildlife habitat. In many cases, wildlife habitat models have been incorporated into dynamic models (e.g. succession) as sub-models. For example, the DYNAST model combines resource sub-models (e.g. wildlife) with natural succession sub-models (Holthausen, 1986) and HABSIM links wildlife habitat carrying capacity estimates to a succession model (Raedeke and Lehmkuhl, 1986). The reader is referred to Ferguson *et al.* (1989), Means (1982), Shugart (1984), Shugart and West (1980) and Shugart *et al.* (1988) for descriptions and review of forest succession models. One of the finest collections is contained in the Wildlife 2000 Symposium edited by Verner *et al.* (1986), where numerous papers link wildlife habitat evaluation models with dynamic forest management models (Table 14.15).

(a) *Succession and growth*

Succession results from natural (e.g. windstorms and fire) and anthropogenic disturbance (e.g. logging) (Primack, 1993:35). Since the early work of Clements (1916), Gleason (1926) and other botanists, succession theory and management has received considerable attention because the ability to predict future vegetative conditions provides considerable insight into the functioning of ecosystems, which in turn has significant social and economic implications. Growth models are employed to evaluate forest growth and the viability of different timber harvesting scenarios while succession models predict changes in forest stand composition and structure in-space-in-time (Host *et al.*, 1992). Succession models include statistical or empirical models and mechanistic or process models.

As with most statistically based models, **empirical succession models** are based on calculations of the derived relationship between independent variables (e.g. site attributes and disturbance) and dependent variables (e.g. species composition). Modeling tools include: application of regression techniques designed to predict the presence of one or more species (e.g. Keane 1987); Markov chain models (managers calculate the probability that a given configuration of vegetation will change to a new configuration after a given time interval); and transition rate analyses based on the application of differential equations (Roberts and Morgan, 1989; Shugart *et al.*, 1988). For example, Keane (1989) developed a classification system to identify successional vegetation communities and stages within a 'habitat type' (as per Pfister *et al.*, 1977; Pfister and Arno, 1980) in west-central Montana, linked it to forest management strategies ('treatments') and developed a computer model (FORSUM) to predict coverage of plant species in 'treatment' and 'pre-disturbance' forest stands (Figure 14.22).

Empirical model users benefit from easy access to statistical techniques and high powered statistical software packages, such as SAS, that provide efficient and relatively objective analyses. Given the need for statistical rigor, sample size often is a problem, particularly for models that employ a large number of variables. Sklar and Costanza (1991) provide an excellent review of empirical models.

Mechanistic models (or 'stand' or 'gap' models) are employed to examine the relationship between tree species and ecosystem variables in forest stand development (Sklar and Costanza, 1991). They have considerable promise in climate change and wildlife habitat modeling as well. There are many types of stand model which incorporate a variety of ecosystem variables, such as light and moisture availability, nutrient availability and fire response (Roberts and Morgan, 1989). These models simulate succession by calculating the annual change in the diameter of each tree in a small spatial unit and can be used to predict the sequence of species replacement through time (Shugart *et al.*, 1988). For example, Botkin *et al.* (1972) pioneered the JABOWA model to simulate birth, growth and death of individual trees in a stand (Roberts and Morgan, 1989; Shugart, 1990). The JABOWA and related models (e.g. the FORET model developed by Shugart and West, 1977) explicitly incorporate the effects of light and moisture availability on photosynthetic response and growth rate. Growth, reproduction and death also are modeled as a function of variables such as soil type, nitrogen availability, shade tolerance, slope, climate, flooding regime and disturbance frequency (Sklar and Costanza, 1991). The models are data intensive and require quantitative estimates of a species response to the selected independent ecosystem variables.

(b) Climate change

Through significant and continuing modification of the earth's atmosphere, at least three major ecospheric impacts have been identified: wet and dry deposition of acidic, toxic and other undesirable substances in terrestrial and aquatic habitats; depletion of stratospheric ozone; and significant global climate change (Anderson, 1989). While all three are time sensitive and significant factors in the management of wildlife habitat, this chapter deals only with global climate change.

Given the social and economic impacts of climate on human populations, agencies traditionally have devoted significant resources to programs designed to understand the role of climate in ecosystem composition, structure and function. In the last 15 years the challenges of climate change (e.g. global warming) and the need to understand through models has emerged as an important area of endeavour. At all mapping scales, climate as a modifier, or a variable that is modified by other natural forces, significantly influences availability and access to wildlife habitat. Therefore, climate change models have significant potential as wildlife habitat evaluation tools.

There is reasonable consensus that increased greenhouse gas concentra-

Table 14.15 Examples of dynamic, linked models that can be employed in wildlife habitat evaluation programs

Name	Description	Source
PROGNOSIS–COVER	PROGNOSIS simulates forest stand development through growth of individual trees. Historically, it has been used to summarize stand conditions and to predict future patterns of stand growth and the potential impacts of alternative management programs on stand development. Moeur (1985, 1986) extended the use of PROGNOSIS by integrating it with COVER, a model that permits managers to represent tree crown and understory vegetation structure, which can be used to evaluate changes in wildlife habitat.	Stage (1973); Moeur (1985, 1986); Monserud and Haight (1990); Crookston (1990)
PPE (Parallel Processing Extension of the Prognosis Model	Multi-stand simulation model that combines the PROGNOSIS model and a technique for specifying management policies through rules and activity schedules. PPE simulates and displays management alternatives using decision trees.	Crookston and Stage (1991); Crookston, cited in Schuster et al. (1993:89)
DYNAST (Dynamically Analytic Silviculture Technique)	Developed for comparing multiple-use scenarios under different management options. Combines resource models (e.g. wildlife and timber production) with natural succession models. The simulation model operates by moving the size of each successional stage within each forested land type through time. Analyzes the cumulative effects of multiple resource management by tracking the consequences of a proposed management action. Simulates vegetation dynamics using a large, coarse-resolution, deterministic approach.	Boyce (1977, 1978, 1980); Biesterfeldt and Boyce (1978); Benson and Laudenslayer (1986); Sweeney (1986); Holthausen (1986)
FORPLAN	Linear optimization model, initially developed and evolved to provide timber harvesting scheduling and sustained-yield programs. Permits managers to identify the combination of activities and outputs that will maximize or minimize the desired objective.	Holthausen (1986); K. Sleavan, cited in Schuster et al. (1993:222)
FORSOM (Forest Simulation Optimization Model)	Spreadsheet model, allows managers to examine the implications of various harvesting scheduling strategies. Components are based on cover types (e.g. jack pine, aspen, red pine, etc.) and wildlife models (i.e. PATREC).	L. Leefers, cited in Schuster et al. (1993:224)

HABSIM	Links wildlife habitat carrying capacity with a habitat succession model and allows the manager to employ data that are generally available. It is cost-effective and can simulate the effects of alternative management practices. HABSIM output provides a habitat-succession matrix for the time period covered by the simulation. Raedeke and Lehmkuhl (1986) employed it to model responses of black-tailed deer (*Odocoileus hemionus columbianus*) and elk populations to forest management in the northwestern United States.	Raedeke and Lehmkuhl (1986)
RO3 WILD	Measures habitat capability of indicator species where vegetation structural stage is the variable and forage and cover are the coefficients.	B. Rickle, cited in Schuster *et al.* (1993:178)
HABCAP	Quantifies the capability of an area to support wildlife populations using vegetation cover types. Used as a monitoring tool to detect habitat quality decline.	W.C. Aney, cited in Schuster *et al.* (1993:180); J. Fenwood, cited in Schuster *et al.* (1993:181)
STEMS (Stand and Tree Evaluation Modeling System) and TWIGS	TWIGS is the microcomputer version of STEMS. It is a forest growth projection model based on individual trees, and employed to assess management practice impacts on the forest overstory. For example, Brand *et al.* (1986) integrated TWIGS with a gray squirrel (*Sciurus carolinensis*) HSI model.	Brand *et al.* (1986); Belcher *et al.* (1982); Belcher (1982)
FORHAB (Forest Habitat simulation model)	Modified version of FORET. Simulates annual change of forest stand by calculating growth increment of each tree on the stand. For example, Smith (1986) employed FORHAB to predict the impact of selected forest management decisions on the availability of breeding habitat for a bird community in Tennessee.	Smith *et al.* (1981a, 1981b); Shugart and West (1977); Smith (1986)
SAMM (Southeast Alaska Multi-resource Model)	Interactive microcomputer program that allows managers to explore relations among different assets and the effects of logging on those assets. Framework comprised of a number of sub-models (e.g. timber, wildlife, hydrology and fisheries) which are linked as required in response to natural assessment management requirements and issues.	McNamee *et al.* (1986); D. Gegson and L. Smith, cited in Schuster *et al.* (1993:94)
SPOTTED OWL HC (Northern Spotted Owl Habitat Capability Estimator)	Estimates current and future habitat capability (number of potential pairing sites) for owls in western Washington, western Oregon and northwestern California. Does not predict population size but provides a relative gauge to potential habitat capability.	B. G. Marcot, cited in Schuster *et al.* (1993:195)

Table 14.15 Continued

Name	Description	Source
FIRESUM (FIRE Succession Model)	Simulates effect of different fire regimes on tree composition, stand structure and fuel loading in forests of inland portion of northwestern United States. Deterministic model with stochastic properties: tree growth, woody fuel accumulation and litterfall are simulated deterministically, whereas tree establishment and mortality are stochastic algorithms.	Keane *et al.* (1989); Shugart and West (1980); Keane, cited in Schuster *et al.* (1993:80)
CLIMACS (Computer Linked Integrative Model for Assessing Community Structure)	Simulates forest succession for western Oregon and Washington. Tracks characteristics of individual trees growing in small forest opening. 'Gap' model that simulates characteristics of individual trees in small portion of forest; particularly useful for evaluation of long-term and large-scale changes and effects of those changes on forest succession.	Shugart and West (1980); Dale and Hemstrom (1984)
SRS (Snag Recruitment Simulator)	Identifies snag densities according to size and decay class, projects snag densities over time within the forest stand, and assists managers to develop and employ stand management prescriptions.	B.G. Marcot, cited in Schuster *et al.* (1993:199)
UNITPLAN	Snag-dynamics model employed for snag retention and creation in forest management units.	M. Hunter, cited in Schuster *et al* (1993:208)
Habitat Supply Analysis (HSA)	Employs wood supply models (FORMAN + 1 and FORMAN + 2) to forecast availability of forest stands (habitat) through time. Forest projection models are matched with wildlife species habitat requirements models to predict future availability of habitat.	Patch (1987); Sullivan (1989); New Brunswick Department of Natural Resources (1992)
HEICALC/HEIWEST (Elk Habitat Effectiveness Index)	These two simulation models allow managers to calculate a habitat effectiveness index (HEI) for elk habitat quality. HEICALC evaluates spatial proximity of forage, marginal and satisfactory cover, and harvest treatment areas for habitat in the Blue Mountains of Eastern Oregon and Washington. HEIWEST evaluates spatial proximity of forage, cover (optimal, hiding, thermal) and effects of silvicultural treatments and fertilization on forest production. Elk habitat indices are calculated from these spatial relationships.	Ager and Hitchcock (1992); Thomas *et al.* (1988); Wisdom *et al.* (1986); A. Ager and M. Hitchcock, cited in Schuster *et al.* (1993:150)

ELK COVER	Simulation model calculates security of forest stand cover for elk. Data derived from stand inventory data are PROGNOSIS simulation models.	R. Teck, cited in Schuster et al. (1993:135)
CALWHRS (California Wildlife Habitat Relationships System)	Simulation tool completes word-level Habitat Evaluation Procedures (HEP) for terrestrial vertebrates in California.	California Department of Fish and Game, cited in Schuster et al. (1993:127)
BOISED (Boise and Payette National Forest Sediment Yield Model)	Sediment yield model employed in Boise and Payette National Forests that estimates on-site erosion, delivery to stream channels and routing of sediment downstream to critical reaches where impacts to fisheries are interpreted. The model is applied to the watershed level of mapping.	R. Beverage, cited in Schuster et al. (1993:124)
SLAVES (Stand Layer Analysis and Vegetation System)	Uses current and potential plot inventory data to determine vegetative structure within and across forest stands. For example, Wallowa-Whitman National Forest and Umatilla National Forest staff employ the vegetation data base to map habitat.	N. Cimon, cited in Schuster et al. (1993:96)
HIDE2	HIDE2 predicts elk hiding cover with tree density and diameter data.	J. Lyon, cited in Schuster et al. (1993:151)
HIDE2X	Variation of HIDE2. Calculates hiding cover values for any species. Has been used in Winema National Forest to calculate hiding cover values for every stand represented in FORPLAN timber tables.	J. Haugen, cited in Schuster et al. (1993:152)
NICOLET HABCAP (Nicolet National Forest Wildlife Habitat Capability Model)	Generalized habitat capability model written in FORTRAN. Can be used with the US Forest Service Oracle data base to employ vegetation data in model development.	A. Albee, cited in Schuster et al. (1993:165)
THINX	Demonstration models for calculating and displaying the effects of oil and gas drilling and timber harvest on soils, water quality, stream sedimentation, forest vegetation and some types of wildlife habitat.	P. Case, cited in Schuster et al. (1993:202)
WATSED (Water and Sediment Yields)	Can be locally calibrated to reflect impacts from specific management activities. Predicts changes in water yield and sediment from roads, fire, logging and site preparation activities.	R. Nygaard, cited in Schuster et al. (1993:211)

Table 14.15 Continued

Name	Description	Source
ATLAS	Multiple rotation, spatially explicit, block scheduling and road network analysis program for detailed analysis at watershed level of planning.	J. Nelson, cited in Schuster *et al.* (1993:245)
RM SPATIAL ANALYSIS	Statistical modelling program, analyzes spatial characteristics of mapped attributes (e.g. land types, vegetation types and soil types, etc.) and can be used for analysis of biodiversity and landscape ecology (i.e. landscape patterns and processes).	C.H. Flather, cited in Schuster *et al.* (1993:248)
SEEDFLO	Empirical tree species dispersal model designed to address impact of landscape structure and forest fragmentation on dynamics of spatially extensive areas.	Hanson *et al.* (1990)

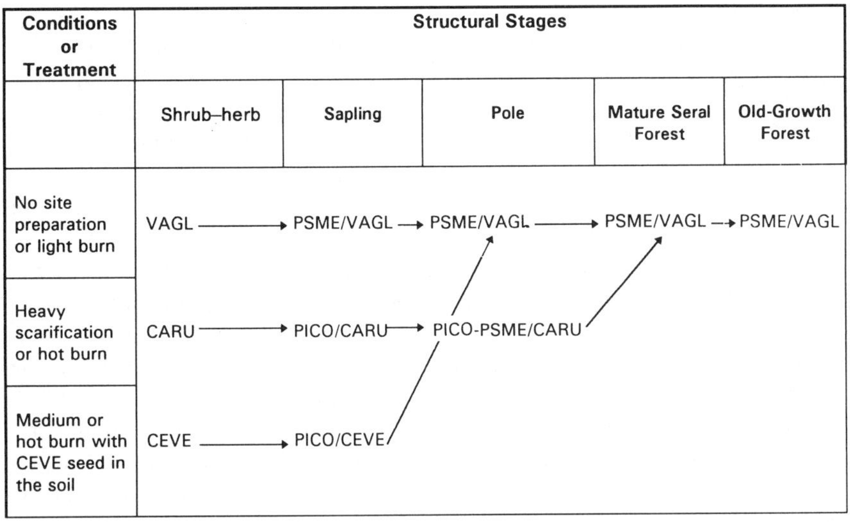

Conditions or Treatment	Structural Stages				
	Shrub–herb	Sapling	Pole	Mature Seral Forest	Old-Growth Forest
No site preparation or light burn	VAGL ⟶	PSME/VAGL ⟶	PSME/VAGL ⟶	PSME/VAGL ⟶	PSME/VAGL
Heavy scarification or hot burn	CARU ⟶	PICO/CARU ⟶	PICO-PSME/CARU		
Medium or hot burn with CEVE seed in the soil	CEVE ⟶	PICO/CEVE			

Variables	Stand Characteristics				
Tree Canopy Cover %	0–15	15–90	50–90	50–80	50–70
DBH of Important Trees (inches)	0–1	2–5	6–10	11–15	16–25
Basal Area (ft/acre)	0–1	2–100	100–250	100–250	50–250
Age (years)	5–15	15–30	30–100	100–200	200–300

Figure 14.22 Forest stand descriptions, succession pathway keys and conditions or treatment. VAGL = *Vaccinium globulare*; CARU = *Calamagrostis rubescens*; CEVE = *Ceanothus velutinus*; PSME = *Pseudotsuga menziesii*; PICO = *Pinus contorta*. For example, the CEVE community type results after moderate to high severity broadcast burn or wildfire in a stand containing CEVE seeds in the soil. (From Kean, 1989.)

tions in the atmosphere will change global climate but the accuracy of the rate and magnitude of the estimates are at issue (Smith, 1990). It is anticipated that the planet will warm several degrees during the next 100 years and that there will be widespread changes in moisture regimes, increased atmospheric carbon dioxide concentrations, increased ocean levels (up to 1.5 m) and altered soil chemistry regimes (Peters, 1988). It is possible that global warming will result in fundamental changes to the way in which the earth's ecosystems currently function, including changes in the locations of the boundaries and the establishment of new ecosystems not currently in evidence (Shugart, 1990). The impact of global warming will affect wildlife species differently, where some species will benefit and others will not (Topping and Bond, 1988).

Climate is influenced by the amount of energy received from the sun, the shape of the earth's orbit, the wobble and tilt of its axis, the placement of the oceans, continents and landforms on those continents, the gaseous composition of the atmosphere, cloudiness, the chemistry of the oceans and the variation of landforms and vegetative cover (Rowe and Rizzo, 1990). These factors and forces are important elements in climate change models. Considerable effort is now being expended to model climate change in Canada and the United States and to evaluate the impacts of climate change on the earth's natural assets, including wildlife. Development of computer simulation models to predict (or with which to generate scenarios of) future change currently is attracting considerable attention (Malanson, 1993). Three types of model have been explored: empirical models (transfer function models), process models ('stand' or 'gap' models) and physiological models.

Transfer function models are statistical models that employ mathematically derived functions which link climatic parameters and indices (independent variables) to the properties (e.g. vegetation) of specific spatial units (dependent variables) (Rowe and Rizzo, 1990). The climatic variables are generated from General Circulation Models (GCMs), which are mathematical representations of the earth's climate (Anderson, 1989; Smith, 1990; Stocks, 1993) and are used to predict temperature, precipitation, clouds, winds and the exchange of energy in the ecosphere through time. There are primary and secondary effects models.

Primary effects models relate present vegetation to present climate and are used to project the future distribution of vegetation (Malanson, 1993). For example, Shugart (1990) reports that GCM predictions based on increased carbon dioxide using correlation analysis result in a change of 30% of the earth's terrestrial surface. Such change would be characterized by a significant shift in vegetation – from an area occupied by one broad vegetation type to one being occupied by another. In Canada, Rizzo and Wiken (1992), Wheaton *et al.* (1987) and Zoltai (1988) developed empirically based impact models that provide scenarios for the future. Rizzo (1989) and Rizzo and Wiken (1992) employed discriminant analysis to model nine climatic variables using the ecoclimatic regions of Canada (Ecoregions Working Group, 1989). The transfer functions derived from the analysis were then applied to the GCMs, and scenarios were developed and mapped (Figures 14.23, 14.24). In secondary effects models, variables such

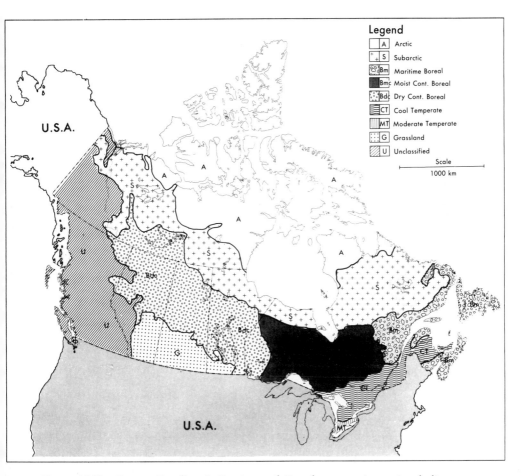

Figure 14.23 Current Ecoclimatic Provinces of Canada, expressing regional climate through the development of vegetation and soil conditions. As climate changes over time, it should be possible to anticipate and predict related changes to vegetation and soils. (From Rizzo, 1989.)

as land-use disturbance are incorporated into the climate–vegetation relationships model (e.g. Baker *et al.*, 1991, cited in Malanson, 1993).

While these models provide useful scenarios that describe the magnitude of potential change (Malanson, 1993) to large ecosystems, transfer function models are limited because they assume equal and spontaneous dispersal of all species in the ecosystem. In other words, they assume an equilibrium response to change. This assumption is unrealistic, given that the diversity of organisms and their differential dispersal rates (in response to a variety

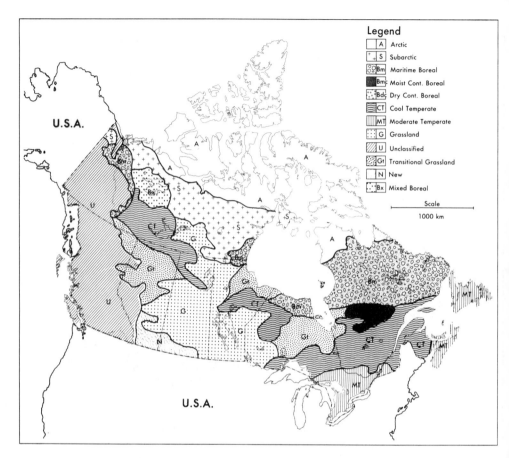

Figure 14.24 Predicted Ecoclimatic Provinces of Canada, based on climatic change resulting from a doubling of carbon dioxide in the atmosphere. There are significant transformations affecting the location of Canada's ecosystems. For example, in Ontario the Moist Continental Boreal Ecoclimatic Province depicted in Figure 14.23 will disappear and be replaced by Transitional Grassland, Grassland and Cool Temperate Ecoclimatic Provinces. These predictive models suggest that if climatic change progresses unchecked, there will be substantial impacts on Canada's natural assets, including wildlife. (From Rizzo, 1989.)

of forces like competition and predation) are not modeled (Malanson, 1993; Shugart, 1990).

As described in Section 14.5.4(a), **process or stand models** simulate the birth, growth and death of the individual plant. While these 'mechanistic' models traditionally have been employed for areas less than a hectare in size, they provide a truer indication of the actual changes that might occur in the ecosystem being modeled (Malanson, 1993). These models can address questions relating to biodiversity, can simulate non-equilib-rium responses as necessary, and can include secondary impacts of climate change (Malanson, 1993). While limited by their spatial scale and large data requirements, they likely will emerge as important components of larger models designed to simulate change in large ecosystems in the future (Shugart, 1990).

Physiological models employ information on the response of individual species to changes in climate and hydrology. They are being developed at three generic mapping scales: plant, stand-to-region and continental (Malanson, 1993). The plant scale models simulate plant development, phy-siological processes and the influence of external environmental factors for single species (e.g. Isebrands *et al.*, 1990, in Malanson, 1993). The stand-to-region scale models employ information extrapolated from sites that do not have an explicit spatial cell size (Malanson, 1993). For example, the FOREST-BGC model developed by Running and Coughlan (1988) projects response to climate change by calculating net primary productivity (and other related factors) on a daily basis and litterfall and decomposition on an annual basis. Working at a continental scale of mapping, Neilson *et al.* (1992) developed a rule-based model in which secondary effects, such as change resulting from disturbance patterns, can be predicted. Malanson (1993) suggests that this type of model operates like a complex transfer function model, because it relies on accurate calculation of water balance and a series of decision rules (like a dichotomous key), made on the basis of climate and physiognomic types.

Like the stand models, physiological models can provide a truer impres-sion of change than the transfer function models. Despite their potential, physiological models require large data sets and traditionally have focused on few species, which limits their ability to address biodiversity issues (Malanson, 1993; Rowe and Rizzo, 1990).

14.5.5 Predictive model variables

Models are only as good as the data and information used to construct them. Natural asset managers are constantly faced with difficult decisions respecting the selection and use of independent and dependent model vari-ables.

(a) Independent variables

In response to a general lack of ecological data at various levels in ecosys-tem classification and regionalization hierarchies, natural asset managers traditionally have relied on study results completed elsewhere. Although comparative analysis of the individual variables (i.e. slope, topography and vegetation type) important to a species can assist the manager in a pre-liminary scoping of the variables and parameters, and in understanding wildlife ecology dynamics in general, the results of other studies must be used cautiously. For example, the composition, structure and function of the respective ecosystems invariably are different. Selection of the independent model variables is, therefore, a significant consideration in model construction. For example, a principal limitation of HEP and many other index models traditionally has been the selection and subsequent testing of the independent variables.

Although Irwin and Cook (1985) are correct in their assertion that analysis of a large number of variables allows the manager to develop a greater understanding of the limitations and performance of a model, use of many variables can mask the effect of the highly significant ones, reduce model success and quality and potentially reduce the significance of the recommendations advanced by the natural asset management. Van Horne (1983) suggests that it is useful to identify *a priori* those habitat variables most likely to be significant. *A priori* statistical analyses of potentially important habitat variables could minimize time and cost, and significantly reduce subjectivity in the selection of variables.

Statistical techniques, such as factor analysis and multivariate regression analysis, assist managers to manage large, often unwieldy data sets. Despite the reservations of Peters (1991:184), who correctly argues that multivariate techniques like regression analysis are opaque and hide important information about habitat variable performance in the model, they do provide a tool with which to scope and relate the habitat variables to the distribution and, in some cases, abundance of a species.

(b)   *Dependent variables*

It seems logical to assume that natural selection will favor animals which select better habitats and therefore population density should be higher in optimal habitat than in sub-optimal and marginal habitat. Models often are tested by relating population density and habitat use. Density is difficult to measure accurately – observer bias resulting from variable weather conditions, observer competence and experience, the terrain and vegetation over which the survey is being completed, the type of survey and survey equipment (e.g. the aircraft employed) all affect density estimates. As a result of the uncertainty surrounding the accuracy and consistency of the density estimates, a model could well be invalidated or validated by questionable population data (Clark and Lewis, 1983).

Hobbs and Hanley (1990), Lyon *et al.* (1987) and Van Horne (1983) argue that density alone can be a misleading indicator of habitat quality. For example, a density estimate, derived from a survey involving only a few hours of a season which can extend over a period of many months, is a poor indicator of the number of animals using the habitat. They recommend a cautious approach to the use of animal census data, particularly in view of the fact that populations can fluctuate widely over years and decades as a result of factors not accounted for in the model defining the relationship. In rejecting density as sole indicator of habitat quality, Van Horne (1983, 1986) suggests that habitat quality can be measured more accurately as a function of an equation which accounts for density, survival and reproduction rates. This suggestion also increases the utility of the model because it addresses factors affecting niche as well. Few habitat evaluation models developed to date account for all of these factors (Schamberger and O'Neill, 1986; Van Horne, 1983, 1986, and others) but with the rapidly increasing power of computer systems and sophistication of models (e.g. dynamic process models), development and testing in response to a variety of dependent variables will improve.

It should be noted, however, that there are examples of studies where density did appear to reflect habitat quality. Smith *et al.* (1988) delineated four strata (best, moderate, poor, inadequate) on the basis of habitat quality (using vegetation types and access) in support of a moose aerial inventory program in central Alberta. Hammill and Moran (1986) used ruffed grouse densities in an HEP validation procedure in Michigan, and Latka and Yahnke (1986) employed densities for a sandhill crane model in Nebraska. Lancia and Adams (1985) validated HEP models for pine warblers (*Dendroica pinus*) and prairie warblers (*D. discolor*) found in North Carolina using densities (call counts) as well, although the validation procedure did not work for three other species tested in the same model.

14.5.6 Model validation

Validation determines if a model produces acceptable and consistent results (Gaudette and Stauffer, 1988) and serves two primary purposes: to provide information about model performance and reliability under specific applications, and to provide data and information in support of model development (Schamberger and O'Neill, 1986). Although models cannot be proven true, they can be invalidated (Shugart and West, 1980; Stelfox, 1991), which provides the natural asset manager with some idea about the reliability and utility of the model (Shugart and West, 1980).

Habitat evaluation model validation is difficult because there are no standard methods for defining and measuring habitat quality. Further, the quantitative data needed to develop the model often are lacking. In many cases, existing data are not available in a consistent format and reside in disparate data bases, collected years apart by different people. In addition, natural asset managers are plagued with small sample sizes which limit or prevent the application of some validation procedures, including the creation of experimental and control study sites. Despite these limitations, validation with historical data does improve the manager's confidence in a model as a decision-aid tool (Hoekstra and Joyce, 1988). Numerous validation techniques are available and it is important that they be examined and selected as part of the project planning and data collection process to ensure that the appropriate types and quantity of data are available.

As a case in point, although Habitat Suitability Index models for over 160 species have been developed in the United States, few have been tested and validated. Some of the statistical procedures employed to date include studies by Allen *et al.* (1991), Bart *et al.* (1984), Clark and Lewis (1983), Cole and Smith (1983), Cook and Irwin (1985), Hammill and Moran (1986), Lancia *et al.* (1982) and Thomasma *et al.* (1991). In addition, Mulé (1982) and Byrne (1982) used subjective Delphi techniques to assess and invalidate HEP models developed for a number of birds and mammals in Alaska.

The clapper rail (*Rallus longirostris*) HEP model developed by Lewis and Garrison (1983) was invalidated by Clark and Lewis (1983) using density as the dependent variable and correlation analysis with the Habitat Suitability Index (HSI) ratings. Irwin and Cook (1985) used simple correlation and multiple regression to examine the relationship of habitat and non-habitat variables with pronghorn densities and fawn:doe ratios on winter range in an effort to validate the use of selected variables for a model developed by Allen and Armbruster (1982). They found that of five variables assumed to be important in the initial model, three were important. Thomasma *et al.* (1991) tested the HEP model for the fisher using a preference index and logistic regression.

These examples illustrate the importance of identifying model limitations, the factors which might influence the validation of results and the overall performance of the final model in various management applications when creating a validation procedure. The limitations can include imperfect life history data and information about the modeled species, geographic resolution, the presence and influence of predators and the availability and accuracy of the habitat data and information (Laymon and Barrett, 1986).

14.6   DECISIVE (DECISION/SUPPORT) MODELS

While most decisions are tempered by uncertainty (Saveland *et al.*, 1988), a primary goal of effective natural asset management is to eliminate as much uncertainty as possible. This requires good data and sound application of the data in decision-making. Many of the descriptive and predictive models described in previous sections are empirically based, mechanistic approximations of the real world where inputs (i.e. matter and energy) are processed, transformed and converted into outputs (different forms of energy and matter) (Davey and Stockwell, 1991). Although useful and instructive, these models are limited because most mathematical representations of the real world are intractable and plagued with many, often implicit assumptions.

As Kourtz (1990) aptly states, there is a 'perennial dream of human beings . . . to develop a thinking machine that can mimic human behaviour in the process of carrying out complex tasks'. In the 1940s, a few scientists began to develop a language that allowed computers to mimic human thought processes, and by the mid 1960s Edward Feigenbaum and his associates reached the stage in their research where they were able to encode human knowledge in symbolic nomenclature (usually English words) (Harmon and King, 1985; Kourtz, 1990). Feigenbaum *et al.* (1971) recognized that these symbols could be used as the basis of a language to solve real-world problems. This language has come to be known as **artificial intelligence** (AI). An exact definition of intelligence is elusive but generally centres around the concept of knowledge, its acquisition and use (Kourtz, 1990).

AI involves flexible responses, an ability to make sense out of uncertain, incomplete or contradictory messages, and the ability to recognize the importance of the different elements of a situation (Hofstadler, 1980:26). AI blends cognitive psychology and computer science to make computers 'smarter' and more useful, and capable of doing things that, at present, humans can do more efficiently (Gray *et al.*, 1991; Hofstadler, 1980; Rich, 1983). A number of sub-disciplines have evolved within the AI research field, including: the development of natural language processing, speech recognition, learning and computer vision; robotics; and expert systems (Harmon and King, 1985:1; Stock, 1987; White *et al.*, 1985).

In the realm of natural language processing research, for example, scientists are building machines that can read, speak and understand language, the goal being to develop smart machines that can engage in spontaneous discourse in support of decision-making. Robotics focuses on the development of smart robots, while expert systems research involves the development of programs which employ symbolically (as opposed to numerically) represented knowledge to simulate and mimic human experts (Harmon and King, 1985:7; Stock, 1987). For example, through expert system programming, axioms are manipulated with rules of inference to identify the consequences of a proposed action. Although all AI disciplines have existing and potential significant implications to wildlife habitat management, this chapter is focused on the application of expert systems. **Expert systems** emerged as a dominant concept in the computer software industry in the 1980s (Olson and Hansory, 1988); they are one of the fastest growing computer technologies in the world (Stock, 1987) and will, in all likelihood, play a major role in computer systems development into the next century.

Decisive or knowledge-based models can be employed to represent human beliefs (values) which in part stem from the knowledge derived from descriptive and predictive models, and are developed and applied to identify specific options about a preferred course of action or response to a given set of conditions (Coulson *et al.*, 1991; Jeffers, 1972, 1973; Robinson *et al.*, 1987). A decisive model, therefore, while relying on data and information derived from descriptive and/or predictive models, is a flexible, dynamic approximation of the real world that explains the consequences of a proposed action. Inference rules are integrated with the output from descriptive and predictive models and employed to interpret the numerically based simulation or statistical model (Davey and Stockwell, 1991). While not numerically based quantitative analogs that describe and predict ecosystem composition, structure and function, they do provide a logical 'grounding' through which managers can challenge model assumptions and identify the consequences of employing selected beliefs and values (Davey and Stockwell, 1991).

Generally, the power and usefulness of expert systems is not derived from clever programming; their power is the knowledge base they contain and the flexible, intelligent ways in which that knowledge is employed (White *et al.*, 1985). They are user-friendly, easy to modify and highly interactive (Harmon and King 1985:8; White *et al.*, 1985). They allow natural asset managers to install computer-based, human-derived knowledge as facts which are applied to the heuristics developed for a specific task or problem solving exercise (Kourtz, 1990).

Through expert systems research, significant emphasis has been placed on the creation and application of formal 'knowledge system' models designed to reduce the impact of the laws of chance – the uncertainty (Saveland *et al.*, 1988). For example, wildlife habitat evaluation is an uncertain proposition, particularly when attempting to account for random events affecting the distribution and habitat of wildlife (Davey and Stockwell, 1991). Accordingly, heuristics are used to reduce the complex task of assessing probabilities and predicting values to make judgements and decisions (Saveland *et al.*, 1988). Conventional computer programs cannot do this; they are employed to process large volumes of data in programmed step-by-step procedures to complete a pre-determined task or meet an objective (Harmon and King, 1985) (Table 14.16). Thus, this knowledge can be used to analyze unique situations and to generate recommendations or solutions comparable to the advice that might be provided by a human expert.

An expert system has four principle parts: a knowledge base, a user interface with supporting computer operating systems (including data base structure and graphic tools) and an 'inference engine' (Kourtz, 1990; Rauch-Hindin, 1983; Stock, 1987; White *et al.*, 1985) (Table 14.17). Three tools are used to develop and employ expert systems: computer programming language, expert shells and 'tool kits' (Kourtz, 1990). Traditional computer programming language was the first and only technique available until the mid 1970s; therefore, most expert systems of that era were developed in BASIC, FORTRAN and PASCAL (Kourtz, 1990). Because of their limited capacity to accommodate the use of symbols, AI researchers developed

Table 14.16 Comparison of conventional programming with expert system programming (adapted from Gray *et al.*, 1991; Harmon and King, 1984; Stock, 1987; White *et al.*, 1985)

Conventional Programming	Expert system
Task previously performed by humans, commonly with little expertise, but according to the instructions of experts.	Task previously performed by a human expert.
General and broad applications.	Normally a narrow subject area.
Maintained by specialized programmers.	Maintained by knowledge engineers and experts.
Often difficult to read and modify.	Knowledge base is readable and easy to modify.
Sequential batch processing.	Highly interactive processing, and can explain the rationale that led to the solution.
Numerically structured data base.	Symbolically structured knowledge base.
Relies on algorithms for overall structure.	Relies on heuristics for structure; therefore, problems which have algorithmic or mathematical process are not suitable for expert systems.
Cannot manage uncertainty.	Includes techniques to handle uncertainty.
Tables of numbers and graphs as output.	Concrete advice as output.

special languages such as LISP (LISt Programming, used principally in the United States) and PROLOG (PROgramming LOGic, used primarily in the rest of the world) (Koutz, 1990). Since its invention, a number of dialects have been developed for LISP, including MACLISP (and its successor, Common LISP), INTERLISP, FranzLISP and SCHEME (Olson and Hanson, 1988). PROLOG and LISP are best suited for symbolic language processing and not well suited for numerical analysis (Olson and Hanson, 1988).

Expert shells are frameworks (written for LISP, PROLOG, C or other languages) that permit the user to add appropriate facts, rules and the user interface (Kourtz, 1990). They lie anywhere along a continuum from simple language interpreters to elaborate systems development environments (Citrenbaum *et al.*, 1987). With significant advancement of microcomputers and associated software in the 1980s, a variety of expert shells were developed (Citrenbaum *et al.*, 1987). The tool kits represent a compromise

Table 14.17 Components of an expert system (after Gray *et al.*, 1991; Harmon and King, 1985; Kourtz, 1990; Rauch-Hindin, 1983; Stock, 1987; White *et al.*, 1985)

Component	Description
Knowledge base	Contains facts and rules of operation. It is encoded in symbolic form such as 'if-then' clauses.
User interface	Allows user and expert system to communicate where questions posed to, or received from, operating software through user interface. Interface usually includes features such as color graphics, pulldown menus, raster pictures, maps, voice output and cursor control (e.g. mouse).
Interface with supporting operations	Allows access to existing supporting operations data bases.
Inference engine	Automatically carries out search of knowledge base for appropriate facts and rules to link together to form new facts and rules that lead to decision or recommendation. Inference engine interprets the knowledge base and performs logical deduction and certain knowledge base modifications in support of a decision.

between the language-based systems and the expert shells, and are used on LISP machines, mini-computers, work stations and the newest micro-computers (Kourtz, 1990). They comprise utilities and building components which can be configured to meet the specific needs of the user (Kourtz, 1990), such as customization of the inference engine.

Expert systems can be used for many tasks, including diagnostic evaluations, monitoring, prediction, planning and training (Stefik *et al.*, 1982; White *et al.*, 1985). Some of the first diagnostic expert systems were developed for physicians. For example, MYCIN was created to assist physicians in the diagnosis and treatment of meningitis and bacteremia infections through administration of the appropriate drugs (Harmon and King, 1985:16; Kourtz, 1990). In natural asset management, diagnostic programs can be used for forest ecosystem pest management, genetic mapping and wildlife parasite and disease management (Table 14.18). Planning of wildlife population and habitat survey programs is a logical expert system task as well. They can be employed to develop specific recommendations about the actions most likely to succeed –- they can be used to determine pesticide application strategies or the most appropriate timber harvesting techniques to enhance production of wildlife habitat. Kourtz (1990) suggests that design-oriented expert systems can be used for forest inventory survey designs, regeneration survey techniques and forest stand treatment programs. Wildlife habitat evaluation also is a logical objective.

They are being developed and applied to GISs for map design, geographic feature extraction and geographic decision support (Coulson *et al.*, 1991). Some of the most promising applications of integrated GIS-expert systems for wildlife habitat evaluation include development of intelligent user interfaces, efficient spatial database search techniques, improved image classification (e.g. Landsat imagery) and the production of high quality cartographic products (Robinson *et al.*, 1987).

Expert systems have demonstrated significant potential as a land-use conflict resolution tool. For example, in British Columbia, natural asset managers tested two expert systems ('The Deciding Factor' and 'Prospector II') in support of efforts to resolve conflicts resulting from the decision to harvest timber and the need to manage black-tailed deer (McNay *et al.*, 1987). McNay *et al.* (1987) used Prospector II to develop winter habitat suitability ratings for black-tailed deer in four habitats (Figure 14.25).

Agencies in British Columbia also employ the HAP (Habitat Assessment Planning) tool to assist in the management of old-growth forest stands protected from cutting because of their significance as winter range for black-tailed deer and elk. The HAP tool permits managers to assess the implications of decisions at regional or watershed levels of management. HAP contains a series of microcomputer-based models that allow managers to incorporate the spatial and temporal aspects of wildlife habitat into operational forestry plans (Eng *et al.*, 1991), whereby managers can:

- assess wildlife habitat suitability and the impacts of proposed forestry plans and activities on that habitat;
- develop plans that minimize negative impacts and increase the benefit of forestry to wildlife management;
- identify the risk of uncertain management actions and where possible the data required to reduce uncertainty;
- assess priorities for, and cost effectiveness of, habitat management projects;
- document the rationale used to make decisions that affect wildlife habitat.

There are three components to HAP: a regional priorities model, a watershed assessment model and a management options model. The **regional priorities** model is used to rank planning in terms of the need for habitat management. For each priority planning unit, the **watershed assessment** model provides an evaluation of the proposed forestry scenarios and assists with the specification of habitat requirements. **Management options** are then developed and assessed interactively with a GIS.

The advantages of expert systems include a knowledge base derived from more than one expert, reliability (e.g. no sick days or periods of low productivity), total recall (White *et al.*, 1985) and easy user access to the rationale behind the predictions being made (Hurley, 1986). They are

Table 14.18 Examples of expert systems developed for applied decision-making

Software	Description	Source
Not described	The knowledge-based land information manager and simulator (KBLIMS) is a GIS compatible system for managing spatio-temporal ecological simulations. Users can explore spatial patterns of forest productivity and responses of watersheds to climate change.	Mackay *et al.* (1993)
LISP machine and expert system development kits	Moose management expert system with which specific harvesting strategies can be examined to predict impact of moose populations on conifer regeneration. Integrated with a GIS, rules to simulate hourly behavior of moose and forest growth and harvesting models.	Kourtz (1990)
Not described	Aquaculture decision and fish disease diagnostic expert system. Constructed with a commercial shell.	Kourtz (1990)
'The Deciding Factor'	Black-tailed deer/forest relationships model, develops a heuristic framework for the decision-maker through an editor.	McNay *et al.* (1987)
PROSPECTOR II	Black-tailed deer/forest relationships model.	McNay *et al.* (1987)
Symbolics 3640 computer and prototype in LISP	Moose/forest dynamics. Employs an expert shell (KEE).	Saarenmaa *et al.* (1988)
Not described	SYTEREP predicts ecological effects of various site preparation methods using expert system technology in British Columbia.	M. Johnson, cited in Schuster *et al.* (1993:98)
CHAMPS	Silvicultural advisor expert system for ruffed grouse and white-tailed deer.	Buech (1990); Buech *et al.* (1990)
DBVISTA III for DOS; INFORMIX for UNIX	Grazing Land Applications (GLA) is a decision support system for planning on rangelands, woodlands, pastureland, cropland and hayland. Emphasizes forage and animal inventories, wildlife/livestock relationships and nutritional management.	J.W. Stuth, J.R. Conner and W.T. Hamilton, cited in Schuster *et al.* (1993:141)

Data General Goldworks for Mvs.	DIAGNOSIS is a knowledge-based system used to identify treatment requirements for forest stands through comparison between the current and target conditions. The target stand is characterized by desired vegetative conditions for the given land use.	J. Chew, cited in Schuster et al. (1993:132)
Not described	The Northeast Decision (NED) Model is a silvicultural prescription tool in support of multiple management goals.	M. Twery, cited in Schuster (1993:163)
MS Windows; PC Oracle, ARC-Info and MOSS system	Integrated Resource Management Automation (IRMA) is a PC-based tool employed to link spatial GIS data and other data for the same area.	D. Loh, cited in Schuster et al. (1993:159)
BRUSH is written in BASIC for use on an IBM PC and GUILD is written in LISP (dialect ALISP) on a CDC Cyber 170/172 mainframe.	Two expert systems to predict the presence of bird species in brushfield habitat following clear-cut logging in a coastal forest in California.	Marcot (1986)
GIS (MOSS [Map Overlay and Statistical System]) and the integrated Resource Management Automation (IRMA)	Employed in support of variety of natural asset management programs in Nicolet National Forest.	Watry (1989)
Oracle 5 or 6; MOSS G.	INtegrated FOrest Resource Management System – Data General Version (INFORM-DG) is a decision support system that relies on simulation models such as growth and yield (PROGNOSIS), fish production, and elk. The system was developed for parts of Oregon and Montana.	USDA Forest Service, cited in Schuster et al. (1993:156)
Common LISP	Fire Effects Information System (FEIS) is a knowledge processor that accesses state-of-the-knowledge about the effects of fire on plant species, plant communities, and animal species.	USDA Forest Service, cited in Schuster et al. (1993:78)
Not described	Parameter Selection System for Streams in Forested Areas (PASSSFA) is an expert system that assists managers to select appropriate parameters for monitoring the effects of different management activities on streams in forested areas. It was developed for the Pacific Northwest and Alaska.	L. MacDonald, cited in Schuster et al. (1993:117)

Figure 14.25 Winter habitat suitability ratings obtained from Prospector II for black-tailed deer in British Columbia, Canada. (From McNay *et al.*, 1987, courtesy of the Wildlife Management Institute.)

particularly useful for natural asset management disciplines where biological knowledge is often symbolic, as opposed to numeric (Olson and Hanson, 1988). Limitations include generation of incorrect decisions and questionable advice requiring implementation of a comprehensive audit process, and the lack of knowledge (particularly about ecological systems) (White *et al.*, 1985).

14.7 MONITORING

Monitoring is employed to detect change in ecosystem health (Freedman *et al.*, 1992). For example, habitat monitoring serves to identify changes in the capability of a spatial unit to support wildlife (Cooperrider, 1986b). It is the act of repeatedly measuring ecosystem part(s) over time. In the case of wildlife habitat, it is employed to:

- determine wildlife use of habitat;
- evaluate the effects of land-use practices on species and habitats and the correctness of the land-use decision;
- detect species and habitat change and determine the cause of that change, evaluate mitigation tools and techniques to protect wildlife and habitat, and identify additional habitat improvement needs to benefit a species or habitat;
- update descriptive models;
- determine the accuracy of predictive and decisive models, which is a fundamental component of adaptive management (e.g. Barrett and Salwasser, 1982; Holling, 1978; Ontario Forest Policy Panel, 1993);
- collect and enhance natural asset data bases.

In addition to measuring habitat condition and change, an effective monitoring program is supported by research designed to develop measurement techniques. Long-term monitoring is contingent upon a long-term commitment by government and partners to provide programs designed to monitor slow processes, rare events, subtle changes and multifactor processes (Anderson *et al.*, 1992). Successful adoption of an ecosystem approach to management, and by extension wildlife habitat management, is contingent upon a comprehensive monitoring program (characterized by an interdisciplinary, systems-oriented approach) and knowledge of the processes which link system components and partnerships.

14.8 GUIDELINES FOR HABITAT EVALUATION MODEL DEVELOPMENT

Salwasser (1986) suggests that models should be easy to understand and communicate, provide information that contributes to the resolution of the problem or issue, and be reliable in proportion to the risks and values involved in the decisions. The following guidelines, while not exhaustive, are provided to assist managers in their efforts to select, develop and/or apply models that meet the standards proposed by Salwasser (1986) and others:

- Identify the goals and objectives of the model prior to its development, and decide what the model will be used to measure, predict and/or decide.
- Predictive models should be developed within a standard descriptive model or spatial framework, such as the Canadian Ecological Land Classification System or the USDA Forest Service ecosystem classification hierarchies. The spatial units in the classification system, or the new units derived from them, must be conducive to the development of practical and useful wildlife models.
- The composition, structure and function of ecosystems, and by extension wildlife habitat, can be perceived and understood differently at a variety of scales. Scale, therefore, is critical to program planning and design (Allen and Hoekstra, 1988; Quattrochi and Pelletier, 1991). Natural asset managers must recognize scale in management decisions and ensure that the inventory and monitoring data are appropriate; they must permit sufficient discrimination for the level of decisions being made. It is important that managers do not attempt to work, for example, at site-specific mapping scales when the supporting data and information do not lend themselves to decisions at these scales. The scale of the model must be sensitive to the scale of the habitat and niche variables employed, and the kinds of decisions which must be made with support from the model. Although some overlap of mapping scale does occur in the CELCS hierarchy, wildlife managers cannot, for example, use ecodistrict-level mapping units (1:500 000–1:250 000) to make decisions about wildlife distribution at the ecosite-level (1:50 000–1:10 000) of mapping.
- Account for seasonal shifts in habitat use. Modelers should be careful in their use of data and information from published studies conducted outside the study area. In addition, managers should be wary of season definitions which can be based on a number of different parameters for each study area, such as temperature and plant phenology. For example, early winter conditions at sea level in southern Alaska are different from early winter conditions in the southwestern Northwest Territories at 1500 m asl.
- Annual shifts in habitat use should be accounted for. Shifts could potentially result from predation, survival, productivity, annual change in climate patterns such as snow depth, wild fire dynamics and other seasonal influences. Some spatial units, for example, may be more important only during years when snow depths are limiting; therefore, a model may be limited if it does not account for a variable which only may be important once in 10 years. Short-term monitoring will not detect the importance of such variables. The results of many studies are weakened because they fail to address long-term habitat requirements.

- Because of the wide variation in potential responses of an organism to the ecosystem(s) in which it lives, all models should be field tested and validated. The sample size must be large enough to create and test the model, and to avoid bias from anomalies such as aberrant behavior by a few animals in the sample. Where possible, models should be constructed using part of the data (i.e. the experimental data) and tested with the remaining data (i.e. the control data set, which for this purpose must be treated as a completely independent data set).

- The model should be validated at steps where data translation occurs, particularly if real numbers are being translated into abstract number sets such as indices or ranks. Modelers should use more than one validation technique at each step if possible.

- To the greatest degree possible, survey techniques should be applied consistently within and between study periods (e.g. hours, days, months, years or seasons) and study sites.

- Dependent variables traditionally used as a measure of habitat quality, such as density, can provide misleading and in some cases erroneous results. Density estimates should be used cautiously. As a general rule, and if possible, habitat quality should be measured against density, survival and reproduction rates. Simple presence/absence data may be the most appropriate dependent variable at small mapping scales (e.g. 1:5 000 000) depicting large areas where the goal is simply to detect the presence of habitat or patterns of use by wildlife, or change in-space-in-time in response to natural and anthropogenic forces. Density is a more reliable indicator of habitat quality for habitat specialists. In addition, behavioral patterns may be used to account for the distribution and abundance of some species as well. It is important to note that the measures used will depend on the logistics, time and funding available, and that in some cases trade-offs will have to be made. These trade-offs should be reflected in how the model results are represented to the decision-makers. For example, in some studies, model accuracy and precision is reduced by aerial survey accuracy and coverage, and little or no data on survival and reproduction rates.

- The literature will not always provide an indication of what a species' optimum habitat conditions are, particularly in view of the fact that adaptations to ecosystems by local, regional and ecogeographically isolated demes and populations can be significant. Comparison and use of results from published studies must be treated with caution because it is difficult to translate and extrapolate the results of other studies conducted at different spatio-temporal scales. If the models are scoped and developed using the literature as a principal source of information, the natural asset manager must ensure that a comprehensive literature review is completed, preferably by a species expert.

- All model limitations should be documented. The following list of potential limitations may be relevant:
  - small sample size, which may limit statistical analyses, including validations;
  - problems with resolution between the data used in the predictive model and data and information used in the descriptive model (mapping scale);
  - temporal scale (e.g. hours, days, seasons, years and decades);
  - inadequate life history information, including lack of knowledge about predators and/or prey;
  - the quality of the data used to describe independent variables, such as vegetation and climate;
  - sampling to ensure that a diverse array of ecosystem quality estimates can be made for the purposes of comparison and demonstrations;
  - the quality and accuracy of the dependent variable data (e.g. population estimates, survival, productivity and behavioral data);
  - inappropriate combinations of ecological variables;
  - improper weighting and ranking of independent habitat variables;
  - limitations of the methods used to interpret the relationships between independent habitat variables;
  - a study area which does not represent the range of habitats in which the species is found;
  - the position of the population estimates in short-term and long-term population cycles (e.g. was the density estimate derived during the high or low period in the cycle?);
- Many government agencies across Canada and the United States employ standard techniques for surveying wildlife. To the greatest extent possible, these standard techniques should be employed. In addition, techniques should be updated continually and integrated into the data and information collection system(s). It is important that management staff be formally trained in survey techniques. Whenever possible, avoid multi-species aerial inventory programs. Visual search patterns employed by the observer to detect animals are different for each species. Switching search patterns on transect may reduce observation rates.
- Develop wildlife model guidelines (e.g. Schroeder and Haire, 1993) and handbooks. Currently, model handbooks are used widely in the United States (e.g. Baskett *et al.*, 1980; Byrne, 1982; Flood, 1977; Sparrowe and Sparrowe, 1978; US Fish and Wildlife Service, 1980a; Urich and Graham, 1984; Urich *et al.*, 1983) and in some jurisdictions in Canada (e.g. Habitat Monitoring Committee, 1990; IEC Beak, 1984; Eccles *et al.*, 1988; Green *et al.*, 1986; Luttmerding *et al.*, 1990; Province of British Columbia, 1994). Development and use of handbooks results in consistent precision, reduced variability and increased repeatability (Mulé, 1982:125).

14.9 CONCLUSIONS

Modern forest ecosystem managers function in a decision-oriented environment where the stakes are high and where the wrong decision could result in significant impacts on wildlife values. Habitat evaluation denotes a suite of tools that enhances the manager's ability to understand forested ecosystems at all levels in the ecological hierarchy (local–global) – they help us identify the 'natural capital' that must be maintained if periodic 'interest' is to be drawn off in the form of water, wildlife, recreational activities and other less tangible benefits.

Habitat is a principal limiting factor affecting wildlife. This is reflected in a survey completed by Flather and Hoekstra (1989:97) in which habitat loss and habitat degradation consistently topped the list as the most important management issues facing forest ecosystem managers in the United States. It is an important issue in Canada as well. Although habitat modeling is a relatively new field of endeavor, significant progress has been made to improve and enhance modeling techniques during the last 20 years. But in the long term, natural asset management agencies, organizations, institutions and industry must address a number of issues that continue to limit the usefulness of habitat modeling techniques:

- Multi-scale modeling: Human impacts range from local (e.g. small-scale quarrying and logging) to regional (e.g. depletion of aquifers) and global (e.g. global warming) (Environment Canada, 1991; Flather and Hoekstra, 1989:111). Historically, most habitat modeling has been focused on site-specific areas, but because decisions and impacts also are being made at regional, national and international levels (Flather and Hoekstra, 1989:111; Thompson and Welsh, 1993), habitat evaluation must be completed at comparable mapping scales. Although a relatively new tool, the conceptual frameworks for regional modeling have been developed. Applications research is, however, required.
- Species–habitat relationship information: Given that this information is basic to any management program, continued intensive and extensive research on the life history of organisms is required.
- Reduce model subjectivity: Despite the fact that considerable effort is being devoted to the development of objective, quantitative wildlife habitat models (Flather and Hoekstra, 1989:110; Verner *et al.*, 1986), continued research is required.
- Data base management: Habitat and population data bases compiled through inventory and monitoring programs often are a limiting factor in model development because model data requirements often outstrip availability of the data. For example, model validation requires better wildlife census techniques than are currently available for the majority of species, and monitoring techniques that can be employed at a variety of map

scales are required (Anderson *et al.*, 1992; Flather and Hoekstra, 1989:111; Lyon, 1989).

• Model validation: Many models lack clearly defined and testable output (Schroeder, 1987:24). Model development has exceeded validation and the testing of basic assumptions. The scientific community must explore the basic underlying assumptions and test model performance (Fausch *et al.*, 1988; Flather and Hoekstra, 1989:111; Lyon, 1989). For example, many index models make assumptions about the importance of a variable to a wildlife species or a group of species. These assumptions require validation.

• Update and refine spatial descriptive models: Natural asset management agencies must ensure that spatial models are kept current, through sponsorship of research, inventory and monitoring focused on improving integrated boundary delineation techniques.

• Integrate US and Canadian ecosystem classification systems: Canada and the United States should integrate their respective national ecosystem classification hierarchies (Uhlig and Jordan, in press). This will result in a continental spatial framework in which to collaborate in the development and implementation of bi-national ecosystem management programs.

• Education and extension: Transferring knowledge is fundamental to the successful development and implementation of a model. It is also a technique with which natural asset managers can market models and garner support for their development and application.

Wildlife habitat evaluation models are employed to examine and evaluate forest ecosystems with greater precision, at a far greater speed than ever before (Lyon, 1989), and generate a wide variety of decision scenarios. While they represent a suite of dynamic, rapidly evolving tools, it is important to remember that models are only a means to an end – they are not the end in and of themselves (Thomas, 1986). A model is simply a formalized decision-aid.

14.10 SUMMARY

Wildlife habitat evaluation is an important component in the development and implementation of an ecosystem approach to management. We have organized our review of selected wildlife habitat evaluation techniques according to three generic classes of models: descriptive, predictive and decisive. The models described in each class are not mutually exclusive, and often are linked in management programs. In addition, it is important to note that this review is far from comprehensive – a comprehensive review would fill many volumes.

A comprehensive spatio-temporal context in which to address natural asset management issues and programs, such as wildlife habitat evaluation, is critical to implementation of an ecosystem approach to management. Natural asset managers describe ecosystems and their components through classification and regionalization to organize data and information collected for application in real-world decision-making. These models often are expressed as maps, images, photographs and drawings. The development and advancement of data collection techniques (such as remote sensing) and data management tools (such as geographic information systems) have catapulted the natural asset manager into a new age of interactive and dynamic descriptive modeling.

With predictive models, natural asset managers develop and employ a variety of management scenarios in support of the decision-making process. Predictive tools range from simple matrices to sophisticated, mathematically based computer models designed to address questions and issues at a variety of spatio-temporal scales. New technologies now permit natural asset managers to integrate 'smart' systems into the modeling and decision-making processes. Potentially, these technologies can serve to reduce the level of uncertainty, a principal limitation of natural asset management, through consistent application of rules in the development of decisions.

The guidelines for habitat evaluation model development propose that natural asset managers adopt a systematic approach to program design and development, beginning with the identification of clear goals and objectives, establishing the spatio-temporal context for the project and concluding with a validated model and a model performance monitoring program.

14.11 ACKNOWLEDGEMENTS

We thank J. Stan Rowe (Professor Emeritus, University of Saskatchewan, Saskatoon) and Jim MacLean (Director, Research, Science and Technology Branch, Ministry of Natural Resources, Maple, Ontario) for their support. We also appreciate the time spent by Dennis Demarchi (Wildlife Branch, Ministry of the Environment, British Columbia) and Peter Uhlig (Ministry of Natural Resources, Ontario) in reviewing parts of this chapter.

14.12 REFERENCES

Adamus, P.R. (1983) *A method for wetland functional assessment.* Volume II. US Department of Transportation, Federal Highway Administration, Report No. FHWA-IP-82-24. 134 pp.

Ager, A. and Hitchcock, M. (1992) *Microcomputer software for calculating the Western Oregon elk habitat effectiveness index.* US Department of Agriculture, Forest Service, Pacific Northwest Research Station, PNW-GTR-303, Portland, Oregon. 12 pp.

Albert, D.A., Denton, S.R. and Barnes, B.V. (1986) *Regional landscape ecosystems of Michigan.* School of Natural Resources, The University of Michigan, Ann Arbour, Michigan. 32 pp.

Allee, W.C., Emerson, A.E., Park, O. *et al.* (1955) *Principles of Animal Ecology.* W.B. Saunders Company, Philadelphia. 837 pp.

Allen, A.W. (1983) *Habitat suitability index models: Fisher.* US Department of the Interior, Fish and Wildlife Service, FWS/OBS-82/10.45, Washington, DC. 19 pp.

Allen, A.W. and Armbruster, M.J. (1982) Preliminary evaluation of a habitat suitability model for pronghorn. *Proceedings of the Biennial Pronghorn Antelope Workshop* 19:93–105.

Allen, A.W., Jordan, P.A. and Terrell, J.W. (1987) *Habitat suitability index models: Moose, Lake Superior Region.* US Department of the Interior, Fish and Wildlife Service, Research and Development, Biological Report 82(10.155), Washington, DC. 48 pp.

Allen, A.W., Terrell, J.W., Mangus, W.L. and Lindquist, E.L. (1991) Application and partial validation of a habitat model for moose in the Lake Superior region. *Alces* 27:50–64.

Allen, T.F.H. and Hoekstra, T.W. (1988) The critical role of scaling in land modelling, in *Perspectives On Land Modelling*, (eds R. Gélinas, D. Bond and B. Smit), pp. 9–13. Polyscience Publications Inc., Montreal, Québec, Canada. 230 pp.

Allen, T.F.H. and Hoekstra, T.W. (1992) *Toward a Unified Ecology.* Columbia University Press, New York. 384 pp.

Anderson, H.W. and Rice, J.A. (1993) *A tree-marking guide for the tolerant hardwoods working group in Ontario.* Queen's Printer for Ontario, Toronto, Ontario, Canada. 227 pp.

Anderson, J.C. (1989) *Climate change and the Canadian Wildlife Service: A discussion paper.* Canadian Wildlife Service, Ottawa, Ontario, Canada. 15 pp. (Unpublished.)

Anderson, J.R., Hardy, E.E., Roach, J.J. and Witmer, R.E. (1976) *A land use and land cover classification system for use with remote sensor data.* US Geological Survey Professional Paper 964, US Government Printing Office, Washington, DC. 28 pp.

Anderson, J., Kurvits, T. and Wiken, E. (1992) A national monitoring and assessment network: The concept, in *State of the Environment Reporting: Proceedings of the National Ecological Monitoring and Research Workshop, Toronto, Ontario, Canada, May 5–8, 1992*, pp. 3–12. Atmospheric Environment Service and State of the Environment Reporting, Environment Canada, Occasional Paper Series No. 1, Ottawa, Canada. 77 pp.

Andrews, R., Beck, T., Blundell, G. *et al.* (1990) *A Wildlife Policy for Canada.* Canadian Wildlife Service, Environment Canada, Catalogue No. CW66-59/1990E, Ottawa, Canada. 29 pp.

Annas, R.J., Robertson, S.L., Bentz, J.A. and Nemeth, Z.J. (1983) *An ecosystem classification of the Boreal Cordilleran Ecozone: Sundre area.* Forest Research Branch, Alberta Forest Service and Resource Evaluation Branch, REAP. Internal Report. Volumes I and II, Edmonton, Alberta, Canada.

Arno, S.F. (1982) Classifying forest succession on four habitat types in western Montana, in *Proceedings of the Symposium on Forest Succession and Stand Development Research in the Northwest, March 26, 1981, Corvalis, Oregon*, pp. 54–62. Forest Research Laboratory, Oregon State University, Corvalis, Oregon.

Aronoff, S. (1989) *Geographic Information Systems: A management perspective.* WDL Publications, Ottawa, Canada. 294 pp.

Austin, M.P. (1984) Problems of vegetation analysis for nature conservation, in *Survey Methods for Nature Conservation*, (eds K. Myers, C.R. Margules and I. Musto), pp. 101–130. Volume I, Commonwealth Scientific and Industrial Research Organization, Division of Land and Water Resources, Australia. 399 pp.

Austin, M.P. and Cunningham, R.B. (1981) Observational analysis of environmental gradients. *Proceedings of the Ecological Society of Australia* 11:109–119.

Austin, M.P., Cunningham, R.B. and Good, R.B. (1983) Altitudinal distribution of several eucalypt species in relation to other environmental factors in southern New South Wales. *Australian Journal of Ecology* 8:169–180.

Avers, P.E., Cleland, D.T., McNab, W.H. *et al.* (1993) *Summary, national hierarchical framework of ecological units.* Ecomap, US Department of Agriculture, Washington, DC. 10 pp.

Bailey, R.G. (1976) *Ecoregions of the United States* (map). US Department of Agriculture, Forest Service, Intermountain Region, Ogden, Utah.

Bailey, R.G. (1978) A new map of the ecosystem regions of the United States, in *Classification, Inventory, and Analysis of Fish and Wildlife Habitat*, (ed. A. Marmelstein), pp. 121–127. US Department of the Interior, Fish and Wildlife Service, Office of Biological Services, FWS/OBS-78/76, Washington, DC. 604 pp.

Bailey, R.G. (1980) *Ecoregions of the United States*. US Department of Agriculture, Miscellaneous Publication 1391, Washington, DC. 77 pp.

Bailey, R.G. (1982) Classification systems for habitat and ecosystems, in *Research on Fish and Wildlife Habitat*, (ed. W.T. Mason), pp. 16–26. US Environmental Protection Agency, Office of Research and Development, Washington, DC.

Bailey, R.G. (1985) The factor of scale in ecosystem mapping. *Environmental Management* 9:271–276.

Bailey, R.G. (1987) Suggested hierarchy of criteria for multiscale ecosystem mapping. *Landscape and Urban Planning* 14:313–319.

Bailey, R.G. (1989) Explanatory supplement to ecoregions map of the continents. *Environmental Conservation* 16(4):307–309.

Bailey, R.G. (1995) *Description of the Ecoregions of the United States*. US Department of Agriculture, Forest Service, Miscellaneous Publication 1391, Washington, DC. 108 pp.

Bailey, R.G. and Hogg, H.C. (1986) A world ecoregions map for resource reporting. *Environmental Conservation* 13(3):195–201.

Bailey, R.G., Pfister, R.D. and Henderson, J.A. (1978) Nature of land and resource classification: A review. *Journal of Forestry* 76:650–655.

Bailey, R.G., Zoltai, S.C. and Wiken, E.B. (1985) Ecological regionalization in Canada and the United States. *Geoforum* 16(3):265–275.

Baker, J.A. (1988) *The classification of habitat of terrestrial vertebrates within forest management units of Ontario*. Ontario Ministry of Natural Resources, Toronto, Ontario, Canada. Unpublished manuscript.

Baker, W.L., Egbert, S.L. and Frazier, G.F. (1991) A spatial model for studying the effects of climate change on the structure of landscapes subjected to large disturbances. *Ecological Modelling* 56:109–125.

Ball, G.L. (1994) Ecosystem modeling with GIS. *Environmental Management* 18(3):345–349.

Banner, A., Pojar, J., Trowbridge, R. and Hamilton, A. (1985) Grizzly bear habitat in the Kimsquit River Valley, coastal British Columbia, in *Proceedings – Grizzly*

Bear Habitat Symposium, (G.P. Contreras and K.E. Evans, compilers), pp. 36–49. US Department of Agriculture, Forest Service, Intermountain Research Station, Ogden, Utah.

Barnes, B.V., Pregitzer, K.S., Spies, T.A. and Spooner, V.H. (1982) Ecological forest site classification. *Journal of Forestry* 80:493–498.

Barrett, R.H. and Salwasser, H. (1982) Adaptive management of timber and wildlife habitat using DYNAST and wildlife–habitat relationship models. *Western Association of Fish and Wildlife Agencies Proceedings* 62:350–365.

Bart, J., Petit, D.R. and Linscombe, G. (1984) Field evaluation of two models developed following habitat evaluation procedures. *Transactions of the North American Wildlife and Natural Resources Conference* 49:489–499.

Baskett, T.S., Darrow, D.A., Hallett, D.L. *et al.* (eds) (1980) *A Handbook for Terrestrial Habitat Evaluation in Central Missouri*. US Department of the Interior, Fish and Wildlife Service, Resource Publication No. 133, Washington, DC. 155 pp.

Beaubien, J. (1994) Landsat TM satellite images of forests: From enhancement to classification. *Canadian Journal of Remote Sensing* 20(1):17–26.

Beckett, P.H. and Webster, R. (1965) *A Classification System for Terrain*. Military Engineering Experiment Establishment Report No. 872, Hampshire, England. 248 pp.

Bélanger, L., Bergeron, T. and Camire, C. (1992) Ecological land survey in Québec. *Forestry Chronicle* 68(1):42–52.

Belcher, D.M. (1982) TWIGS: The woodsman's ideal growth projection system, in *Microcomputers: A New Tool for Foresters*, (ed. J.W. Moser, Jr), pp. 70–95. Purdue University Press, West Lafayette, Indiana.

Belcher, D.M., Holdaway, M.R. and Brand, G.J. (1982) *A Description of STEMS – The Stand and Tree Evaluation Modeling System*. US Department of Agriculture, Forest Service, North Central Forest Experiment Station, Forest Service General Technical Report NC-79, St. Paul, Minnesota. 18 pp.

Benson, G.L. and Laudenslayer, W.F. Jr, (1986) DYNAST: Simulating wildlife responses to forest management strategies, in *Wildlife 2000: Modeling Habitat Relationships of Terrestrial Vertebrates*, (eds J. Verner, M.L. Morrison and C.J. Ralph), pp. 351–355. The University of Wisconsin Press, Madison, Wisconsin. 470 pp.

Bergeron, J.F., Saucier, J.P., Robitaille, A. and Robert, D. (1992) Quebec forest ecological classification program. *Forestry Chronicle* 68(1): 53–63.

Berry, K.H. (1986) Introduction: Development, testing, and application of wildlife–habitat models, in *Wildlife 2000: Modeling Habitat Relationships of Terrestrial Vertebrates*, (eds J. Verner, M.L. Morrison and C.J. Ralph), pp. 3–4. The University of Wisconsin Press, Madison, Wisconsin. 470 pp.

Bertalanffy, L. von (1968) *General Systems Theory, Foundations, Development, and Application*. George Braziller, New York.

Biesterfeldt, R.C. and Boyce, S.G. (1978) Systemic approach to multiple-use management. *Journal of Forestry* 76:342–345.

Block, W.M. and Brennan, L.A. (1993) The habitat concept in ornithology: Theory and applications, in *Current Ornithology*, (ed. D.M. Power), pp. 35–91. Volume 11, Plenum Press, New York, New York.

Blower, D. (1973) *Methodology: Land Capability for Ungulates in British Columbia*.

British Columbia Wildlife Division, Victoria, British Columbia, Canada. 24 pp. (Unpublished.)

Bonar, R.L. (1991) The Weldwood Hinton timber–wildlife integrated management pocket, in *WILDFOR91, Wildlife and Forestry: Towards a Working Partnership*, (ed. R.L. Bonar), pp. E25–E28. Canadian Society of Environmental Biologists and Canadian Pulp and Paper Association, Montreal, Québec, Canada. 133 pp.

Bonar, R., Quinlan, R., Sikora, T. *et al.* (1990) *Integrated Management of Timber and Wildlife Resources on the Weldwood-Hinton Forest Management Agreement Area.* Weldwood of Canada Ltd, Hinton Division, and the Alberta Forests, Lands, and Wildlife Division, Alberta Government, Edmonton, Alberta, Canada. 44 pp.

Botkin, D.B., Janak, J.F. and Wallis, J.R. (1972) Some ecological consequences of a computer model of forest growth. *Journal of Ecology* **60**:849–873.

Bourne, R. (1931) Regional survey and its relation to stocktaking of the agricultural and forest resources of the British Empire, in *Oxford Forestry Memoirs 13*, pp. 7–62. Clarendon Press, Oxford, England. 169 pp.

Bowling, C. and Zelazny, V. (1992) Forest site classification in New Brunswick. *Forestry Chronicle* **68**(1):34–41.

Boyce, S.G. (1977) *Management of eastern hardwood forests for multiple benefits (DYNAST-MB).* US Department of Agriculture, Forest Service, Southeastern Forest Experiment Station, Asherville, North Carolina, Research Paper SE-168. 116 pp.

Boyce, S.G. (1978) *Management of forest for timber and related benefits (DYNAST-TM).* US Department of Agriculture, Forest Service, Southeastern Forest Experiment Station, Research Paper SE-184, Asherville, North Carolina. 140 pp.

Boyce, S.G. (1980) *Management of forests for optimal benefits (DYNAST-OB).* US Department of Agriculture, Southeastern Forest Experiment Station, Research Paper NC-204, Asherville, North Carolina. 92 pp.

Brand, G.J., Shifley, S.R. and Ohmann, L.F. (1986) Linking wildlife and vegetation models to forecast the effects of management, in *Wildlife 2000: Modeling Habitat Relationships of Terrestrial Vertebrates*, (eds J. Verner, M.L. Morrison and C.J. Ralph), pp. 383–387. The University of Wisconsin Press, Madison, Wisconsin. 470 pp.

Buech, R.R. (1990) Silvicultural prescription advisors for deer and grouse: An application of expert systems. *Canada Committee On Ecological Land Classification, Wildlife Working Group Newsletter* **10**:8–9. Sustainable Development and State of the Environment Reporting Branch, Environment Canada, Ottawa, Canada.

Buech, R.R., Martin, C.J. and Rauscher, H.M. (1990) An application of expert systems for wildlife: Identifying forest stand prescriptions for deer. *AI Applications in Natural Resource Management* **4**:1–8.

Buttery, R.F. (1978) Modified ECOCLASS – A Forest Service method for classifying ecosystems, in *Integrated Inventories of Renewable Natural Resources: Proceedings of the Workshop*, (H.G. Lund, V.J. La Bau, P.F. Ffolliot and D.W. Robinson, technical co-ordinators), pp. 157–168. US Department of Agriculture, Forest Service, Rocky Mountain Forest and Range Experiment Station, General Technical Report M-55, Fort Collins, Colorado. 482 pp.

Byrne, L.C. (1982) Field testing the habitat evaluation procedures for Alaska. MSc Thesis, University of Alaska, Fairbanks, Alaska. 114 pp.

Caldwell, L.K., Wilkinson, C.F. and Shannon, M.A. (1994) Making ecosystem policy: Three decades of change. *Journal of Forestry* 92(4):7–10.

Canadian Council of Forest Ministers (1992) *Sustainable forests: A Canadian commitment*. Canadian Council of Forest Ministers, Hull, Québec, Canada. 51 pp.

Cannon, R.W.F., Knopf, F.L. and Pettinger, L.R. (1982) Use of Landsat data to evaluate lesser prairie chicken habitats in western Oklahoma. *Journal of Wildlife Management* 46(4):915–922.

Capen, D.E. (ed.) (1981) *The use of multivariate statistics in studies of wildlife habitat*. United States Department of Agriculture, Forest Service, General Technical Report RM-87.

Christian, C.S. (1957) The concept of land units and land systems. *Proceedings of the Ninth Pacific Science Congress* 20:74–81.

Christian, C.S. and Stewart, G.A. (1968) Methodology of integrated surveys. *Conference on Aerial Survey and Integrated Studies, Natural Resources Research* 6:233–280.

Citrenbaum, R., Geisman, J.R. Jr and Schultz, R. (1987) Selecting a shell. *AI Applications in Natural Resource Management* 2(1):3–13.

Clark, J.D. and Lewis, J.C. (1983) A validity test of a habitat suitability index model for clapper rail. *Proceedings of the Annual Conference of the Southeast Association of Fish and Wildlife Agencies* 37:95–102.

Clements, F.E. (1916) Plant succession: An analysis of the development of vegetation, in *Ecological Succession*, (ed. F.B. Golley) (1977). Dowden, Hutchinson, and Ross, Inc., Stroudsburg, Pennsylvania. 373 pp. (Reprinted from the Carnegie Institute, Washington Publication No. 242:1–512).

Cline, M.G. (1949) Basic principles of soil classification. *Soil Science* 67:81–91.

Cole, C.A. and Smith, R.L. (1983) Habitat suitability indices for monitoring wildlife populations – An evaluation. *Transactions of the North American Wildlife and Natural Resources Conference* 48:367–375.

Colwell, R.N. (1985) Has remote sensing come of age? Volume III. *Remote Sensing Aided Assessment of Natural Resources, A Syllabus*. Thirteenth Alberta Remote Sensing Course, University of Alberta, Edmonton, Alberta, Canada. 32 pp.

Cook, J.G. and Irwin, L.L. (1985) Validation and modification of a habitat suitability model for pronghorns. *Wildlife Society Bulletin* 13:440–448.

Cooperrider, A.Y. (1986a) Habitat evaluation systems, in *Inventory and Monitoring of Wildlife Habitat*, (eds A.Y. Cooperrider, R.J. Boyd and H.R. Stuart), pp. 757–776. US Department of the Interior, Bureau of Land Management, Service Centre, Denver, Colorado. 858 pp.

Cooperrider, A.Y. (1986b) Introduction, in *Inventory and Monitoring of Wildlife Habitat*, (eds A.Y. Cooperrider, R.J. Boyd and H.R. Stuart), pp. xvii–xviii. US Department of the Interior, Bureau of Land Management, Service Centre, Denver, Colorado. 858 pp.

Corliss, J.F. (1974) ECOCLASS: A method for classifying ecosystems, in *Foresters in Land Use Planning. Proceedings of the National Convention of the Society of American Foresters*, September 23–27, 1973, Portland, Oregon, pp. 264–271. 275 pp.

Corliss, J.C. and Pfister, R.D. (1973) *ECOCLASS – A Method for Classifying Ecosystems*. US Department of Agriculture, Forest Service, Missoula, Montana. 52 pp.

Corns, I.G.W. (1983) Forest community types of west-central Alberta in relation to selected environmental variables. *Canadian Journal of Forest Research* 13:995–1010.

Corns, I.G.W. (1991) Ecological inventories and applications for wildlife habitat, in *WILDFOR91, Wildlife and Forestry: Towards a Working Partnership*, (ed. R.L. Bonar), pp. E29–E33. The Canadian Society of Environmental Biologists and the Canadian Pulp and Paper Association, Montreal, Québec. 133 pp.

Corns, I.G.W. (1992) Forest site classification in Alberta: Its evolution and present status. *Forestry Chronicle* 68(1):85–93.

Corns, I.G.W. and Annas, R.M. (1986) *Field guide to forest ecosystems of west-central Alberta*. Canadian Forest Service, Northern Forest Centre, Edmonton, Alberta.

Cortner, H.J. and Moote, M.A. (1994) Trends and issues in land and water resources management: Setting the agenda for change. *Environmental Management* 18(2):167–173.

Coulson, R.N., False, L.J. and Loh, D.K. (1987) Artificial intelligence and natural resource management. *Science* 237:262–267.

Coulson, R.N., Lovelady, C.N., Flamm, R.O. *et al.* (1991) Intelligent geographic information systems for natural resource management, in *Quantitative Methods in Landscape Ecology*, (eds. M.G. Turner and R.H. Gardner), pp. 153–172. Ecological Studies, Volume 82, Springer-Verlag, New York, New York. 536 pp.

Cowardin, L.M., Carter, V., Golet, F.C. and LaRoe, E.T. (1979) *Classification of wetlands and deepwater habitats of the United States*. US Department of the Interior, Fish and Wildlife Service, FWS/OBS-79/31, Washington, DC. 102 pp.

Crookston, N.K. (1990) *User's Guide to the Event Monitor: Part of PROGNOSIS model, Version 6*. US Department of Agriculture, Forest Service, Intermountain Research Station, General Technical Report INT-275, Ogden, Utah, 21 pp.

Crookston, N.L. and Stage, A.R. (1991) *User's Guide to the Parallel Processing Extension of the PROGNOSIS Model*. US Department of Agriculture, Forest Service, Intermountain Research Station, General Technical Report INT-281, Ogden, Utah. 88 pp.

Crowley, J.M. (1967) Biogeography. *Canadian Geographer* 11:312–326.

Dale, V.H. and Hemstrom, M. (1984) *CLIMACS: A computer simulation model of forest stand development for western Oregon and Washington*. US Department of Agriculture, Forest Service, Pacific Northwest Forest and Range Experiment Station, Research Paper PNW-327, Portland, Oregon. 60 pp.

Damann, A.W.H. (1967) The forest vegetation of western Newfoundland and site degradation associated with vegetation change. PhD Thesis, University of Michigan. 319 pp.

Daniel, C. and Bachant, J.P. (1980) Foreword, in *A Handbook for Terrestrial Habitat Evaluation in Central Missouri*, (eds T.S. Basket, D.A. Darrow, D.L. Halley *et al.*), pp. v–vi. US Department of the Interior, Fish and Wildlife Service, Resource Publication No. 133, Washington, DC. 155 pp.

Daniel, C. and Lemaire, R. (1974) Evaluating effects of water resource developments on wildlife habitat. *Wildlife Society Bulletin* 2:114–118.

Daubenmire, R. (1952) Forest vegetation of northern Idaho and adjacent Washington, and its bearing on concepts of vegetation classification. *Ecological Monographs* 22:301–330.

Daubenmire, R. (1968) *Plant Communities: A Textbook of Plant Synecology.* Harper and Row, New York, New York. 300 pp.

Daubenmire, R. (1970) *Steppe Vegetation of Washington.* Technical Bulletin 62, Agricultural Experiment Station, College of Agriculture, Washington State University, Pullman, Washington. 131 pp.

Davey, S.M. and Stockwell, D.R.B. (1991) Incorporating wildlife habitat into an AI environment: Concepts, theory, and practicalities. *AI Applications In Natural Resource Management* 5(2):59–104.

Davis, L.S. (1980) Strategy for building a location-specific multi-purpose information system for wildland management. *Journal of Forestry* (July) 1980:402–406, 408.

DeGraaf, R.M. and Rudis, D.D. (1986) *New England Wildlife: Habitat, Natural History, and Distribution.* US Department of Agriculture, Forest Service, Northeastern Forest Experiment Station, General Technical Report NE-108, Radnor, Pennsylvania. 491 pp.

DeGraaf, R.M., Yamasaki, M., Leak, W.B. and Lanier, J.W. (1992) *New England Wildlife: Management of Forested Habitats.* United States Department of Agriculture, Forest Service, Northeastern Forest Experiment Station, General Technical report NE-144, Radnor, Pennsylvania. 271 pp.

Demarchi, D. (1994) Ecoprovinces of the central North American Cordillera and adjacent plains. Appendix A in *The Scientific Basis for Conserving Forest Carnivores: American Marten, Fisher, Lynx, and Wolverine in the Western United States,* (eds L.F. Ruggiero *et al.*), pp. 153–167 plus map (1:7 500 000), US Department of Agriculture, Forest Service, Rocky Mountain Forest and Range Experiment Station, General Technical Report RM-254, Fort Collins, Colorado. 184 pp.

Demarchi, D.A. and Chamberlain, T.W. (1979) The Canadian experience: An approach toward biophysical interpretation, in *Classification, Inventory, and Analysis of Fish and Wildlife Habitat,* (ed. A. Marmelstein), pp. 145–155. US Department of The Interior, Fish and Wildlife Service, Office of Biological Services, FWS/OBS-78/76, Washington, DC. 604 pp.

Demarchi, D.A. and Lea, T. (1987) Biophysical habitat classification in British Columbia: An interdisciplinary approach to ecosystem evaluation, in *Proceedings – Land Classifications Based on Vegetation: Applications for Resource Management,* (D.E. Ferguson, P. Morgon and F.D. Johnson, compilers), pp. 275–278. US Department of Agriculture, Forest Service, Intermountain Research Station, General Technical Report INT-257, Moscow, Idaho. 315 pp.

Demarchi, D.A., Fuhr, B., Pendergast, B.A. and Stewart, A.C. (1983) *Wildlife Capability Classification for British Columbia: An ecological (biophysical) approach for ungulates.* British Columbia Ministry of the Environment, Manual No. 4, Victoria, British Columbia, Canada. 56 pp.

Demarchi, D.A., Clark, D. and Lea, T. (1990a) Ecological (biophysical) inventory, classification and mapping conducted by the British Columbia Ministry of Environment, in *Proceedings of the Wildlife/Forestry Symposium: A Workshop on Resource Integration for Wildlife and Forest Managers, 7–8 March, 1990,* (ed. A. Chambers), pp. 107–114, Prince George, British Columbia, 182 pp.

Demarchi, D.A., Marsh, R.D., Harcombe, A.P. and Lea, E.C. (1990b) The environment: A regional ecosystem outline of British Columbia, in *The Birds of British Columbia* Vol. 1, (eds R. Campbell, N.K. Dawe, I.M. Cowan *et al.*), pp. 55-143. Royal British Columbia Museum, Victoria, British Columbia, Canada. 514 pp.

Desrochers, B. and Lash, T. (1990) Assessing Canada's biological diversity. Environment Canada Sustainable Development and State of the Environment Reporting Branch, Ottawa, Canada Committee on Ecological land Classification, *Wildlife Working Group Newsletter* 10:5–6.

Dixon, R.J. and Horn, L. (1981) Digital colour enhancement of Landsat data for mapping vegetation of barren-ground caribou winter range in northern Manitoba, in *Proceedings of the Seventh Canadian Symposium on Remote Sensing*, pp. 197–205, September 8–11, 1981, Winnipeg, Manitoba, Canada.

Dixon, R., Bowles, L. and Knudsen, B. (1984) Moose habitat analysis in north-central Manitoba from Landsat data. *Canadian Symposium on Remote Sensing* 8:623–629.

Dokuchaev, V.V. (1899) *On the theory of natural zones.* Sochineniya (collected works) Volume 6, Academy of Sciences of the USSR, Moscow-Leningrad. 1951.

Donovan, M.L., Rabe, D.L. and Olson, C.E. Jr (1987) Use of geographic information systems to develop habitat suitability models. *Wildlife Society Bulletin* 15:574–579.

East, K.M., Day, D.L., Le Sauteur, D. *et al.* (1979) Parks Canada application of biophysical land classification for resources management, in *Applications of Ecological (Biophysical) Land Classification in Canada*, (ed. C.D.A. Rubec), pp. 209–220. Environment Canada, Lands Directorate, Ottawa, Ecological Land Classification Series No. 7, Ottawa, Canada. 396 pp.

Eby, J.R. and Bright, L.R. (1985) A digital GIS based on Landsat and other data for elk habitat effectiveness analysis, in *Proceedings of the 19th International Symposium on Remote Sensing of the Environment*, pp. 855–863, Ann Arbour, Michigan.

Eccles, T.R. and Stelfox, H.A. (1985) Habitat evaluation techniques in Alberta. Paper presented at the *Alberta Society of Professional Biologists Symposium: Fish and Wildlife Management in Alberta: Current Practice – Future Strategies*, April 16–17, Edmonton, Alberta, Canada. 16 pp.

Eccles, T.R., Salter, R.E. and Green, J.E. (1988) *A proposed evaluation system for wildlife habitat reclamation in the mountains and foothills biomes of Alberta: Proposed methodology and assessment handbook.* The Delta Environment Management Group Ltd., Report No. RRTC 88-1, Calgary, Alberta, Canada. 101 pp.

ECOMAP (1993) *National hierarchical framework of ecological units.* US Department of Agriculture, Forest Service, Washington, DC. 14 pp.

Ecoregions Working Group (1989) *Ecoclimatic Regions of Canada. First approximation.* Ecoregions Working Group of the Canada Committee on Ecological land Classification, Ecological Land Classification Series, No. 23. Sustainable Development Branch, Canadian Wildlife Service, Conservation and Protection, Environment Canada, Ottawa, Canada. 119 pp.

Ecosystems Working Group (1995) *Standards for Terrestrial Ecosystems Mapping for British Columbia* (1:100000+ scale). Terrestrial Ecosystems Task Force, Resource Inventory Committee, Ministry of the Environment, Victoria, British Columbia, 99 pp. (Review Draft: 6 February, 1995.)

Eng, M. and Stelfox, J.A. (1988) A prototype assessment of wildlife resource status of the Rocky Mountain House (83B) NTS map sheet using an ecological land classification approach, in *Land/Wildlife Integration No. 3*, pp. 49–57. Environment Canada, Canadian Wildlife Service, Land Conservation Branch, Ecological Land Classification Series No. 22, Ottawa, Canada. 215 pp.

Eng, M.A., McNay, R.S. and Page, R.E. (1991) Integrated management of forestry and wildlife habitat with the aid of a GIS-based habitat assessment and planning tool, *GIS Applications in Natural Resources*, (eds M. Heit and A. Shortreid), pp. 331–336. GIS World Inc., Fort Collins, Colorado.

Environment Canada (1991) *The State of Canada's environment*. Government of Canada, Ottawa.

Evans, L.C. (1983) Impact assessment and mitigation planning with habitat evaluation models. MS Thesis, Colorado State University, Fort Collins, Colorado. 282 pp.

Ezcurra, E., Rump, P. and Ross, P. (compilers) (1993) *Proceedings of the North American Workshop on Environmental Information, Mexico City, October 19–22, 1993*. State of the Environment Directorate, Environment Canada, Ottawa. 125 pp.

Fairbridge, R.W. (ed.) (1968) *The Encyclopedia of Geomorphology*. Reinhold Book Corp., New York, New York. 1295 pp.

Farmer, A.H. (1978) The habitat evaluation procedures, in *Classification, Inventory, and Analysis of Fish and Wildlife Habitat*, (ed. A. Marmelstein), pp. 407–419. US Department of the Interior, Fish and Wildlife Service, Office of Biological Services, Washington, DC, FWS/OBS-78/76. 604 pp.

Fausch, K.D., Hawkes, C.L. and Parsons, M.G. (1988) *Models that predict standing crop of stream fish from habitat variables: 1950–1985*. Us Department of Agriculture, Forest Service, Pacific Northwest Research Station, General Technical report PNW-213, Portland, Oregon. 52 pp.

Feigenbaum, E.A., Buchanan, B.G. and Lederberg, J. (1971) On generality and problem solving: A case study using the DENDRAL program. *Machine Intelligence* 6:165–190.

Ferguson, D.E., Morgan, P. and Johnson, F.D. (compilers) (1989) *Proceedings – Land Classifications Based on Vegetation: Applications for resource management*. US Department of Agriculture, Forest Service, Intermountain Research Station, General Technical Report INT-257, Ogden, Utah. 315 pp.

Flather, C.H. (1982) Use of ecological theory to evaluate pattern recognition: Implications to wildlife assessments. MS Thesis, Colorado State University, Fort Collins, Colorado. 109 pp.

Flather, C.H. and Hoekstra, T.W. (1985) Evaluating population–habitat models using ecological theory. *Wildlife Society Bulletin* 13:121–130.

Flather, C.H. and Hoekstra, T.W. (1989) *An Analysis of the Wildlife and Fish Situation in the United States. 1989–2000: A technical document supporting the 1989 USDA Forest Service RPA Assessment*. US Department of Agriculture, Forest Service, Rocky Mountain and Forest and Range Experiment Station, General Technical report RM-178, Fort Collins, Colorado. 146 pp.

Flood, B.S. (1977) Development and testing of a handbook for habitat evaluation procedures. MS Thesis, University of Missouri, Columbia, Missouri. 149 pp.

Forman, R.T.T. (1983) Corridors in a landscape: Their ecological structure and function. *Ekologia* (CSSR) 2(4):375–387.

Forman, R.T.T. and Godron, M. (1981) Patches and structural components for a landscape ecology. *Bioscience* 31(10):733–740.

Formann, R.T.T. and Godron, M. (1986) *Landscape Ecology*. John Wiley and Sons, Inc., New York, New York. 619 pp.

Fowle, C.D., James, R., McAfee-Ryan, S. *et al.* (1991) *Looking Ahead: A Wild Life Strategy for Ontario*. Queen's Printer for Ontario, Toronto, Ontario, Canada. 172 pp.

Franklin, J.F. (1980) Ecological site classification activities in Oregon and Washington. *Forestry Chronicle* 56(2):68–70.

Freedman, B., Staicer, C. and Shackell, N. (1992) A framework for a national ecological monitoring program, in *State of the Environment Reporting*, pp. 21–25. Proceedings of the National Ecological Monitoring and Research Workshop, Toronto, Ontario, May 5–8, 1992, Environment Canada, Occasional Paper Series No. 1, Ottawa, Canada. 77 pp.

Frye, R.G. (1984) *Wildlife Habitat Evaluation Procedure*. Texas Parks and Wildlife Department, Texas. 23 pp.

Fuhr, B.L. and Demarchi, D.A. (1990) *A Methodology for Grizzly Bear Habitat Assessment in British Columbia*. Habitat Inventory Section, Wildlife Branch, Ministry of Environment, Wildlife Bulletin No. B-67, Victoria, British Columbia, Canada. 36 pp.

Fulford, R. (1992) The forest and Canadian culture. *Forestry Chronicle* 68(1): 17–20.

Garcia, J., Pratt, J., Ahlborn, G. *et al.* (1984) *A Method for Assessing the Value of Stream Corridors to Fish and Wildlife Resources*. Volumes I and II. Biosystems Analysis, Sausalito, California.

Gauch, H.G., Jr (1982) *Multivariate Analysis in Community Ecology*. Cambridge University Press, New York, New York. 298 pp.

Gaudette, M.T. and Stauffer, D.F. (1988) Assessing habitat of white-tailed deer in southwestern Virginia. *Wildlife Society Bulletin* 16:284–290.

Gélinas, R. (1986) *Modelling: The terms and their use in ecological land evaluation*. Canada Committee on Ecological Land Classification, Lands Directorate, Environment Canada, Ottawa. 3 pp (Mimeo).

Gélinas, R. (1988) Definition, characteristics, and types of 'models'. Environment Canada, Lands Directorate, Ottawa, *Canada Committee On Ecological Land Classification Newsletter* 16:1–2.

Gerlach, L.P. and Bengston, D.N. (1994) If ecosystem management is the solution, what's the problem? *Journal of Forestry*, August (1994):18–21.

Gleason, H.A. (1926) The individualistic concept of the plant association. *Torrey Botanical Club Bulletin* 53:7–26.

Godron, M. and Forman, R.T.T. (1983) Landscape modification and changing ecological characteristics, in *Disturbance and Ecosystems: Components of response*, (eds H.A. Mooney and M. Godron), pp. 12–28. Springer-Verlag, New York, New York.

Golet, F.C. (1976) Wildlife wetland evaluation model, in *Models for Assessment of Freshwater Wetlands*, (ed. J.S. Larson), pp. 13–34. Water Resources Research Centre, University of Massachusetts, Amherst, Massachusetts.

Gore, J.A. (1988) Habitat structure and the distribution of small mammals in a northern hardwood forest, in *Management of Amphibians, Reptiles, and Small Mammals in North America*, (R.C. Szaro, K.E. Severson and D.R. Patton, technical co-ordinators), pp. 319–327. US Department of Agriculture, Forest Service,

Pacific Northwest Research Station, General Technical Report RM-166, Portland, Oregon. 458 pp.

Graber, J.W. and Graber, R.R. (1976) *Environmental evaluations using birds and their habitats.* Illinois Natural History Survey, Biological Note 97. 39 pp.

Gray, P.A. (1993) Ecosystem and wild life integration: Descriptive and predictive models for managing moose (*Alces alces*) in Nahanni National Park Reserve, Northwest Territories, Canada. PhD Thesis, York University, North York, Ontario. 363 pp.

Gray, P.A. and Cameron, D. (1990) A review of some wildlife habitat inventory and evaluation programs applied in Canada, in *International Union of Forest Research Organizations, XIX World Congress*, Division 1, Volume 2, pp. 173–182. Montreal, Québec, Canada.

Gray, P.A. and Keith, J. (1988) The application of a spreadsheet formatting proce- dure in the development and implementation of a habitat evaluation model for the Northwest Territories, in *Perspectives On Land Modelling*, (eds R. Gélinas, D. Bond and B. Smit), pp. 195–210. Polyscience Publications Inc., Montreal, Québec, Canada. 203 pp.

Gray, P.A. and Stelfox, H.A. (1991) Supporting wildlife and habitat management, in *Guidelines for the Integration of Wildlife and Habitat Evaluations with Ecological Land Survey*, (eds H.A. Stelfox, G.R. Ironside and J. Kansas), pp. 47–56. The Wildlife Working Group, Canada Committee on Ecological Land Classification and Wildlife Habitat Canada, Ottawa, Canada. 107 pp.

Gray, P.A., Kansas, J.L. and Cameron, D. (1991) Automated data and information management, in *Guidelines for the Integration of Wildlife and Habitat Evalua- tions with Ecological Land Survey*, (eds H.A. Stelfox, G.R. Ironside and J. Kansas), pp. 59–76. The Wildlife Working Group, Canada Committee on Ecological Land Classification and Wildlife Habitat Canada, Ottawa, Canada. 107 pp.

Gray, P.A., Kirkham, I., Fowle, D. *et al.* (1993) *A proposed action plan for 'Looking Ahead: A wild life strategy for Ontario'.* Prepared by the Wild Life Strategy Action Plan Ad Hoc Committee on behalf of the Wild Life Forums for the Minis- ter of Natural Resources, Ministry of Natural Resources, Toronto, Ontario, Canada. 239 pp.

Gray, P.A., Demal, L., Hogg, D. *et al.* (1995) *An Ecosystem Approach to Living Sus- tainably: A Perspective for the Ministry of Natural Resources.* Ministry of Natural Resources, Terrestrial Ecosystems Branch, Box 7000, Peterborough, Ontario. 77 pp.

Green, J.E., Salter, R.E. and Cooper, C.R. (1986) *Habitat Assessment Models for Wetland-associated Wildlife in the Agricultural Areas of Alberta.* LGL Ltd, Calgary, Alberta for Alberta Agricultural and Alberta Environment Interdepart- mental Steering Committee on Drainage, Edmonton, Alberta, Canada. 246 pp.

Greig-Smith, P. (1964) *Quantitative Plant Ecology.* Butterworths, London.

Groenwoud, H. van and Ruitenberg, A.A. (1982) *A productivity oriented forest site classification for New Brunswick, Canada.* Department of the Environment, Canadian Forestry Service, Maritimes Forest Research Centre, Information Report M-X-136. 9 pp.

Grubb, T.G. (1988) *Pattern recognition – A simple model for evaluating wildlife habitat.* US Department of Agriculture, Forest Service, Rocky Mountain Forest and Range Experiment Station, Research Note RM-487, Fort Collins, Colorado. 5 pp.

Grumbine, E. (1994) What is ecosystem management? *Conservation Biology* 8(1): 27–38.

Haapanen, A. (1965) Bird fauna of the Finnish forests in relation to forest succession. I. *Annals of Zoologica Fennica* 2:153–196.

Habitat Monitoring Committee (1990) *Procedures for Environmental Monitoring in Range and Wildlife Habitat Management* (Draft version 4.1). Ministry of the Environment, Wildlife Branch, Victoria, British Columbia. 196 pp.

Haggett, P. and Chorley, R.J. (1967) Models, paradigms and the new geography, *Socio-economic Models in Geography*, (eds R.J. Chorley and P. Haggett), pp. 19–41. Methuen and Company Ltd, London. 468 pp.

Hall, A.D. and Fagen, R.E. (1956) Definition of a system, in *General Systems*, (eds L. von Bertalanffy and A. Rapport), pp. 18–28. Volume I. Ann Arbour, Michigan.

Hamilton, A.N., Bryden, C.A. and Lofroth, E.C. (1987) *Developing interpretive products using CAPAMP: Classification, mapping, and evaluation of coastal grizzly bear habitat for forestry interpretations*. British Columbia Ministry of Environment and Parks, Wildlife Branch, Victoria, British Columbia, Canada. 15 pp.

Hamill, J.H. and Moran, R.J. (1986) A habitat model for ruffed grouse in Michigan, in *Wildlife 2000: Modeling Habitat Relationships of Terrestrial Vertebrates*, (eds J. Verner, M.L. Morrison and C.J. Ralph), pp. 15–18. University of Wisconsin Press, Madison, Wisconsin. 470 pp.

Hammond, E.H. (1964) *Classes of land-surface form in the forty eight States*. Association of American Geographers 54, Map Supplement 4, Scale 1:5 000 000.

Hanson, J.S., Malanson, G.P. and Armstrong, M.P. (1990) Landscape fragmentation and dispersal in a model of riparian forest dynamics. *Ecological Modeling* 49: 277–296.

Harmon, P. and King, D. (1985) *Expert Systems. Artificial Intelligence in Business.* John Wiley and Sons Inc., Toronto. 283 pp.

Harris, A. (1992) *Post-logging regeneration of reindeer lichen (*Cladina *spp.) as related to woodland caribou winter habitat*. Ontario Ministry of Natural Resources, Northwestern Ontario Forest Technology Development Unit, Technical Report 69, Thunder Bay, Ontario, Canada. 33 pp.

Hart, J.A., Wherry, D.B. and Bain, S. (1985) An operational GIS for Flathead National Forest, in *Proceedings of Autocarto* 7, pp. 244–253. American Society for Photogrammetry and Remote Sensing, Falls Church, Virginia.

Hays, R.L. (1985) *A Users Manual for HEP Accounting Software for Microcomputers: Version 2*. US Department of the Interior, Fish and Wildlife Service, Fort Collins, Colorado.

Hays, R.L. (1987) *A Users Manual for Micro-HSI: Habitat suitability index modelling software for microcomputers: Version 2*. US Department of the Interior, Fish and Wildlife Service, Fort Collins, Colorado.

Hays, R.L. (1989) *Micro-HSI: Master model library, cover type list, and variable lexicon reference manual: Version 2*. US Department of the Interior, Fish and Wildlife Service, Fort Collins, Colorado.

Haywood, D.D. (1989) Application of GIS toward evaluation of habitat selection by Kaibab mule deer, in *Proceedings from the Geographic Information Systems Aware-*

518 Wildlife habitat evaluation: N. American examples

ness Seminar, Salt Lake City, Utah, May 16–19, 1988, pp. 32–36. US Department of Agriculture, Forest Service, Intermountain Region, Ogden, Utah. 84 pp.

Heiskary, S.A., Wilson, C.B. and Larsen, D.P. (1987) Analysis of regional patterns in lake water quality: Using ecoregions for lake management in Minnesota. *Lake Reservoir Management* 3:337–344.

Henderson, J.A., Davis, L.S. and Ryberg, E.M. (1978) *ECOSYM, a classification and information system for wildland management.* Department of Forest Resources, Utah State University, Logan, Utah. 30 pp.

Herbertson, A.J. (1905) The major natural regions: An essay in systematic geography. *Geographical Journal* 25:300–312.

Herbertson, A.J. (1913) The higher units. *Scientia* 14:199–212.

Herbig, R.A. (1988) Application of GIS technology to the HEP habitat layers model, *Perspectives On Land Modelling*, (eds R. Gélinas, D. Bond and B. Smit), pp. 211–223. Polyscience Publications Inc., Montreal, Québec, Canada. 230 pp.

Hills, A.G. (1952) *The classification and evaluation of site for forestry.* Ontario Department of Lands and Forests, Division of Research, Research Report No. 24, Toronto, Ontario, Canada. 41 pp.

Hills, A.G. (1953) The use of site in forest management. *Forestry Chronicle* 29:128–136.

Hills, A.G. (1958) Soil–forest relationships in the site regions of Ontario, in *Proceedings of the First North American Forest Soils Conference, September 8–11, 1958*, pp. 190–212. Michigan State University, East Lansing, Michigan.

Hills, A.G. (1960) Regional site research. *Forestry Chronicle* 36:401–423.

Hills, A.G. (1961) *The Ecological Basis for Land Use Planning.* Ontario Department of Lands and Forests, Research Branch, Research Report 46, Toronto, Ontario, Canada. 244 pp.

Hobbs, N.T. and Hanley, T.A. (1990) Habitat evaluation: Do use/availability data reflect carrying capacity? *Journal of Wildlife Management* 54:515–522.

Hoekstra, T.W. and Joyce, L.A. (1988) An overview of the southern United States resource modelling, in *Perspectives On Land Modelling*, (eds R. Gélinas, D. Bond and B. Smith), pp. 17–25. Polyscience Publications Inc., Montreal, Québec, Canada. 230 pp.

Hofstadler, D.R. (1980) *Gödel, Escher, and Bach: An eternal golden braid.* Vintage Books, Random House, New York, New York. 777 pp.

Holling, C.S. (ed.) (1978) *Adaptive Environmental Assessment and Management.* Volume 3. International Series on Applied Systems Analysis. John Wiley and Sons, New York, New York. 377 pp.

Holling, C.S. (1992) Cross-scale morphology, geometry, and dynamics of ecosystems. *Ecological Monographs* 62(4):447–502.

Holthausen, R.S. (1986) Use of vegetation projection models for management problems, in *Wildlife 2000: Modeling Habitat Relationships of Terrestrial Vertebrates*, (eds J. Verner, M.L. Morrison and C.J. Ralph), pp. 371–375. University of Wisconsin Press, Madison, Wisconsin. 470 pp.

Hoover, R.L. and Wills, D.L. (eds) (1987) *Managing Forested Lands for Wildlife.* Colorado Division of Wildlife in Co-operation with the US Department of Agriculture, Forest Service, Rocky Mountain Region, Denver, Colorado. 459 pp.

Host, G.E., Rauscher, H.M. and Schmoldt, D. (1992) SYLVATICA: An integrated framework for forest landscape simulation. *Landscape and Urban Planning* 21:281–284.

Hounsell, S.W. (1992) Relationships between forest-interior birds and woodland pattern in southern Ontario Landscapes. *Midwest Fish and Wildlife Conference* 54:136–137. (Abstract).

Hughes, R.M., Rexstad, E. and Bond, C.E. (1987) The relationship of aquatic ecoregions, river basins, and physiographic provinces to the icthyogeographic regions of Oregon. *Copeia* 1987:423–432.

Hurley, J.F. (1986) Summary: Development, testing, and application of wildlife-habitat models – The managers viewpoint, in *Wildlife 2000: Modeling Habitat Relationships of Terrestrial Vertebrates*, (eds J. Verner, M.L. Morrison and C.J. Ralph), pp. 151–153. University of Wisconsin Press, Madison. 470 pp.

IEC Beak (1984) *Species–habitat relationship model for moose*. Prepared for the Alberta Fish and Wildlife Division, Alberta Energy and Natural Resources, by IEC Beak Consultants Ltd, Report 5107.1, Calgary, Alberta, Canada. 63 pp.

Inkley, D.B. and Anderson, S.H. (1982) Wildlife communities and land classification systems. *Transactions of the North American Natural Resources and Wildlife Conference* 47:73–81.

Ironside, G.R. (1991) Ecological land survey: Background and general approach, in *Guidelines for the Integration of Wildlife and Habitat Evaluations with Ecological Land Survey*, (eds H.A. Stelfox, G.R. Ironside and J. Kansas), pp. 3–9. The Wildlife Working Group, Canada Committee on Ecological Land Classification and Wildlife Habitat Canada, Ottawa, Canada. 107 pp.

Irwin, L.L. and Cook, J.G. (1985) Determining appropriate variables for a habitat suitability model for pronghorns. *Wildlife Society Bulletin* 13:434–440.

Isachenko, A.G. (1973) *Principles of Landscape Science and Physical–geographic Regionalization*. Melbourne University Press, Melbourne, Australia.

Isebrands, J.G., Rauscher, H.M., Crow, T.R. and Dickmann, D.I. (1990) Whole-tree growth process models based on structural–functional relationships, in *Process Modeling of Forest Growth Responses to Environmental Stress*, (eds R.K. Dixon, R.S. Meldahl, G.A. Ruark and W.G. Warren), pp. 96–112. Timber Press, Portland, Oregon.

Iverson, D.C. (1993) A shared approach to ecosystem management, in *Silvicultural Systems, Societal Needs, and Policy Development*, pp. 129–134. Society of American Foresters National Convention Proceedings.

Jackson, G.L., Racey, G.D., McNicol, J.G. and Godwin, L.A. (1991) *Moose Habitat Interpretation in Ontario*. Ontario Ministry of Natural Resources, Northwestern Ontario Forest Technology Development Unit, Technical Report 52, Thunder Bay, Ontario, Canada. 74 pp.

Jeffers, J.N. (1972) The statistician's role in environmental sciences. *Statistician* 31:3.

Jeffers, J.N. (1973) Systems modelling and analysis in resource management. *Journal of Environmental Management* 1:13–28.

Jones, K.B. (1986) Data types, in *Inventory and Monitoring of Wildlife Habitat*, (eds A.Y. Cooperrider, R.J. Boyd and H.R. Stuart), pp. 11–28. US Department of the Interior, Bureau of Land Management, Service Centre, Denver, Colorado. 858 pp.

Jones, R.K., Pierpoint, G., Wickware, G.M. *et al.* (1983) *Field Guide to Forest Eco-*

system Classification for the Clay Belt, Site Region 3e. Ministry of Natural Resources, Government of Ontario, Toronto, Ontario, Canada. 123 pp.

Kansas, J.L. (1986) The 'functional unit' concept: Applications for evaluating the importance of ecological map units to wildlife. Canada Committee on Ecological Land Classification, Wildlife Working Group, Environment Canada, Ottawa, *Wildlife Working Group Newsletter* 7:7–10.

Kansas, J.L. (1991) Wildlife-enhanced survey proposal and ecological land classification, in *Guidelines for the Interaction of Wildlife and Habitat Evaluations with Ecological Land Survey*, (eds H.A. Stelfox, G.R. Ironside and J. Kansas), pp. 11–28. Wildlife Working Group, Canada Committee on Ecological Land Classification and Wildlife Habitat Canada, Ottawa, Canada. 107 pp.

Kansas, J.L. and Usher, R. (1988) *A review of applications of geographical information systems (GIS) for wildlife planning and management.* Recreation, Parks and Wildlife Foundation, Edmonton, Alberta, Canada. 36 pp.

Kaufmann, M.R., Graham, R.T., Boyce, D.A. Jr, *et al.* (1994) *An ecological basis for ecosystem management.* US Department of Agriculture, Forest Service, Rocky Mountain Forest and Range Experiment Station, General Technical Report RM-246, Fort Collins, Colorado. 22 pp.

Keane, R.E. (1987) *Forest succession in western Montana – A computer model designed for resource managers.* US Department of Agriculture, Forest Service, Intermountain Research Station, Research Note INT-376, Ogden, Utah. 8 pp.

Keane, R.E. (1989) Classification and prediction of successional plant communities, in *Proceedings – Land Classifications Based On Vegetation: Applications for Resource Management, Moscow, Idaho, November 17–19, 1987.* US Department of Agriculture, Forest Service, Intermountain Research Station, General Technical Report INT-257, Ogden, Utah. 315 pp.

Keane, R.E. Arno, S.F. and Brown, J.K. (1989) *FIRESUM – An ecological model for fire succession in western conifer forests.* US Department of Agriculture, Forest Service, Intermountain Research Station, General Technical Report INT-266, Ogden, Utah. 76 pp.

Kerr, R.M. (1986) Habitat mapping, in *Inventory and Monitoring of Wildlife Habitat*, (eds A.Y. Cooperrider, R.J. Boyd and H.R. Stuart), pp. 49–69. US Department of the Interior, Bureau of Land Management, Service Centre, Denver, Colorado. 858 pp.

Kessel, S.R. (1979) *Gradient Modeling: Resource and Fire Management.* Springer-Verlag, New York. 432 pp.

Key, P.F. (1989) Grizzly bears, cumulative effects, and geographical information systems – A ranger district application, in *Proceedings from the Geographic Information System Awareness Seminar, Salt Lake City, Utah, May 16–19, 1988*, (eds D.S. Winn, R.E. Beverly and G.W. Moore), pp. 81–83. US Department of Agriculture, Forest Service, Intermountain Region, Ogden, Utah. 84 pp.

King, A.W. (1993) Considerations of scale and hierarchy, in *Ecological Integrity and the Management of Ecosystems*, (eds S. Woodley, J. Kay and G. Francis), pp. 19–45. St. Lucie Press. 220 pp.

Kling, C.L. (1980) Pattern recognition for habitat evaluation. MS Thesis, Colorado State University, Fort Collins, Colorado. 244 pp.

Kourtz, P. (1990) Artificial intelligence: A new tool for forest management. *Canadian Journal of Forest Research* 20:428–437.

Krajina, V.J. (1965) Biogeoclimatic zones and classification of British Columbia. *Ecology of Western North America* 1:1–17.

Küchler, A.W. (1964) *Potential Natural Vegetation of Conterminous United States.* American Geographical Society, Special Publication 36. 116 pp.

Lacate, D.S. (1969) *Guidelines for biophysical land classification.* Department of Fisheries and Forests, Canadian Forest Service, Ottawa, Publication No. 1624. 61 pp.

Lachowski, H.M. (1989) The Forest Service approach to implementation of Geographic Information Systems, in *Proceedings from the Geographic Information System Awareness Seminar, Salt Lake City, Utah, May 16–19, 1988,* (eds D.S. Winn, R.E. Beverly and G.W. Moore), pp. 8–12. US Department of Agriculture, Forest Service, Intermountain Region, Ogden, Utah. 84 pp.

Lachowski, H.M., Wirth, T., Maus, P. and Avers, P. (1994) Remote sensing and GIS: their role in ecosystem management. *Journal of Forestry,* August (1994):39–40.

Lackey, R.T. (1994) *The Seven Pillars of Ecosystem Management* (Draft Report). US Environmental Protection Agency, Corvallis, Oregon. 22 pp.

Lancia, R.A. and Adams, D.A. (1985) A test of habitat suitability index models for five bird species. *Proceedings of the Annual Southeast Association of Fish and Wildlife Agencies* 39:412–419.

Lancia, R.A., Miller, S.D., Adams, D.A. and Hazel, D.W. (1982) Validating habitat quality assessment: an example. *Transactions of the North American Wildlife and Natural Resources Conference* 47:96–100.

Laperriere, A.J., Lent, P.C., Gassaway, W.C. and Nodler, F.A. (1980) Use of Landsat data for moose habitat analyses in Alaska. *Journal of Wildlife Management* 44(4):881–887.

Larsen, D.P., Omernik, J.M., Hughes, R.M. *et al.* (1986) Correspondence between spatial patterns in fish assemblages in Ohio streams and aquatic ecoregions. *Environmental Management* 10:816–828.

Larsen, D.P., Dudley, D.R. and Hughes, R.M. (1988) A regional approach for assessing attainable surface water quality: an Ohio case study. *Journal of Soil and Water Management* 43:171–176.

Latka, D.C. and Yahnke, J.W. (1986) Simulating the roosting habitat of sandhill cranes and validating suitability-of-use indices, in *Wildlife 2000: Modeling Habitat Relationships of Terrestrial Vertebrates,* (eds J. Verner, M.L. Morrison and C.J. Ralph), pp. 19–22. University of Wisconsin Press, Madison, Wisconsin. 470 pp.

Laymon, S.A. and Barrett, R.H. (1986) Developing and testing habitat-capability models: Pitfalls and recommendations, in *Wildlife 2000: Modeling Habitat Relationships of Terrestrial Vertebrates,* (eds J. Verner, M.L. Morrison and C.J. Ralph), pp. 87–91. University of Wisconsin Press, Madison, Wisconsin. 470 pp.

Leckie, D.G. (1990) Advances in remote sensing technologies for forest surveys and management. *Canadian Journal of Forest Research* 20:464–483.

Legendre, P. and Legendre, V. (1984) Post glacial dispersal of freshwater fishes in the Québec peninsula. *Canadian Journal of Fisheries and Aquatic Science* 41:1781–1802.

Levin, S.A. and Buttel, L. (1986) *Measures of Patchiness in Ecological Systems.* Ecosystem Research Center, Publication No. ERC-130, Cornell University, Ithaca, New York, New York.

Lewis, J.C. and Garrison, R.L. (1983) *Habitat suitability index models: Clapper rail.*

US Department of the Interior, Fish and Wildlife Service, FWS/OBS-83/10 Washington, DC. 15 pp.

Lotspeich, F.B. and Platts, W.S. (1982) An integrated land–aquatic classification system. *North American Journal of Fisheries Management* 2:138–149.

Ludeke, A.K., Maggio, R.C. and Reid, L.M. (1990) An analysis of anthropogenic deforestation using logistic regression. *Journal of Environmental Management* 31:247–259.

Lulla, K. and Mausel, P. (1983) Ecological applications of remotely sensed multi-spectral data, in *Introduction to Remote Sensing of the Environment*, (ed. B.F. Richason, Jr), pp. 354–377. 2nd edn. Kendall/Hunt, Dubuque, Iowa.

Lunt, M. (1989a) Remote sensing and the GIS, in *Proceedings from the Geographic Information Systems Awareness Seminar, Salt Lake City, Utah, May 16–19, 1988*, (eds D.S. Winn, R.E. Beverly and G.W. Moore), pp. 37–42. US Department of Agriculture, Forest Service, Intermountain Region, Ogden, Utah. 84 pp.

Lunt, M. (1989b) Geographic information system – what is it? *Proceedings from the Geographic Information Systems Awareness Seminar, Salt Lake City, Utah, May 16–19, 1988;* (eds D.S. Winn, R.E. Beverly and G.W. Moore), pp. 1–7. US Department of Agriculture, Forest Service, Intermountain Region, Ogden, Utah. 84 pp.

Luttermerding, H.A., Demarchi, D.A., Lea, E.C. *et al.* (1990) *Describing Ecosystems in the Field* (2nd edn), MOE Manual II. Ministry of Environment in cooperation with Ministry of Forests, Victoria, British Columbia. 213 pp.

Lyon, J.G. (1983) Landsat-derived land-cover classifications for locating potential kestrel nesting habitat. *Photogrammetric Engineering and Remote Sensing* 49(2):245–250.

Lyon, J.G., Heinen, J.T., Mead, R.A. and Roller, N.E.G. (1987) Spatial data for modelling wildlife habitat. *Journal of Surveying Engineering* 113(2):88–100.

Lyon, L.J. (1989) Succession modeling and wildlife habitat management, in *Proceedings – Land Classifications Based On Vegetation: Applications for Resource Management, Moscow, Idaho, November 17–19, 1987.* US Department of Agriculture, Forest Service, Intermountain Research Station, General Technical Report INT-257, Ogden, Utah. 315 pp.

Lyons, J. (1989) Correspondence between the distribution of fish assemblages in Wisconsin's streams and Omernik's ecoregions. *American Midland Naturalist* 122:163–182.

MacArthur, F.H. (1958) Population ecology of some warblers of northeastern coniferous forests. *Ecology* 39:599–619.

Mackay, D.S., Robinson, V.B. and Band, L.E. (1993) An integrated knowledge-based system for managing spatio-temporal ecological simulations. *AI Applications in Natural Resource Management* 7(1):29–36.

MacKinnon, A., Delong, C. and Meidinger, D. (1990) *A Field Guide for Identification and Interpretation of Ecosystems of the Northwest Portion of the Prince George Forest Region.* Land Management handbook No. 21, British Columbia Ministry of Forests, Victoria, British Columbia, Canada. 116 pp.

MacKinnon, A., Meidinger, D. and Klinka, K. (1992) Use of the biogeoclimatic ecosystem classification system in British Columbia. *Forestry Chronicle* 68(1):100–120.

Maerz, J.C. (1994) Ecosystem management: a summary of the Ecosystem Management Round Table of 19 July 1993. *Bulletin of the Ecological Society of America*, June 1994:93–95.

Malanson, G.P. (1993) Comment on modeling ecological response to climate change. *Climate Change* 23:95–109.

Mangus, W. (1990) Habitat evaluation and management modelling software. Environment Canada, Sustainable Development and State of the Environment Reporting Branch, Ottawa, Canada Committee on Ecological Land Classification, *Wildlife Working Group Newsletter* 10:6–7.

Marcot, B.G. (1986) Use of expert systems in wildlife-habitat modeling, in *Wildlife 2000: Modeling Habitat Relationships of Terrestrial Vertebrates*, (eds J. Verner, M. Morrison and C.J. Ralph), pp. 145–150. University of Wisconsin Press, Madison, Wisconsin. 470 pp.

Margules, C.R. and Scott, R.M. (1983) Review and evaluation of integrated surveys for conservation, *Survey Methods For Nature Conservation*, (eds K. Myers, C.R. Margules and I. Musto), pp. 1–17. Volume 1. Commonwealth Scientific and Industrial Research Organization, Division of Water and Land Resources, Canberra, Australia. 399 pp.

Maus, P., Landrum, V., Johnson, J. *et al.* (1992) Using satellite data and GIS to map land cover changes. *GIS 1992 Symposium Proceedings, Canada and the Province of British Columbia*, Vancouver, British Columbia, Canada.

Mayer, K.E. (1984) A review of selected remote sensing and computer technologies applied to wildlife habitat inventories. *California Fish and Game* 70(2):102–112.

Mayr, E. (1982) *The growth of Biological Thought: Diversity, Evolution, and Inheritance.* Belknap Press of Harvard University, Cambridge, Massachusetts. 974 pp.

McAllister, D.E., Schueler, F.W., Roberts, C.M. and Hawkins, J.P. (1994) Mapping and GIS analysis of the global distribution of coral reef fishes on an equal-area grid, in *Mapping the Diversity of Nature*, (ed. R.I. Miller), pp. 155–175. Chapman & Hall.

McNamee, P.J., Bunnell, P. and Sonntag, N.C. (1986) Dealing with wicked problems: a case study of models and resource management in southeast Alaska, in *Wildlife 2000: Modeling Habitat Relationships of Terrestrial Vertebrates*, (eds J. Verner, M.L. Morrison and C.J. Ralph), pp. 395–399. University of Wisconsin Press, Madison, Wisconsin. 470 pp.

McNay, R.S. and Page, R.E. (1988) Integrating wildlife and forest management using GIS. *Proceedings of the Twelfth Land Resources Science Workshop*, University of British Columbia, Vancouver, British Columbia, Canada.

McNay, R.S., Page, R.E. and Campbell, A. (1987) Application of an expert-based decision model to promote integrated management of forests and deer. *Transactions of the North American Wildlife and Natural Resources Conference* 52:82–91.

Mead, R.A., Sharik, T.L., Prisley, S.P. and Heinen, J.T. (1981) A computerized spatial analysis system for assessing wildlife habitat from vegetation maps. *Canadian Journal of Remote Sensing* 7(1):34–40.

Meades, W.J. and Moores, L. (1989) *Forest Site Classification Manual: A Field Guide to the Damman Forest Types of Newfoundland.* Forest Resources Development Agreement, FRDA Report 003, St. Johns, Newfoundland, Canada.

Meades, W.J. and Roberts, B.A. (1992) A review of forest site classification activities in Newfoundland and Labrator. *Forestry Chronicle* 68(1):25–33.

Means, J.E. (ed.) (1982) *Forest Succession and Stand Development Research in the*

Northwest. Proceedings of the Symposium, March 26, 1981, Covalis, Oregon. Oregon State University, Forest Research Laboratory, Corvallis, Oregon. 170 pp.

Merchant, B.G., Baldwin, R.D., Taylor, E.P. *et al.* (1989) *Field guide to a productivity oriented pine forest ecosystem classification for Algonquin Region, Site Region 5E, first approximation.* Ontario Ministry of Natural Resources, Toronto, Ontario, Canada.

Merriam, C.H. (1892) The geographic distribution of life in North America. *Proceedings of the Biological Society, Washington* 7:1–64.

Merriam, G. and Lanone, A. (1990) Corridor use by small mammals: field measurements for three experimental types of *Peromyscus leucopus*. *Landscape Ecology* 4:123–131.

Merriam, G., Henein, K. and Stuart-Smith, K. (1991) Landscape dynamics models, in *Quantitative Methods in Landscape Ecology*, (eds M.G. Turner and R.H. Gardner), pp. 399–416. Ecological Studies, Volume 82, Springer-Verlag, New York, New York. 536 pp.

Middleton, J.D. and Merriam, G. (1981) Woodland mice in a farmland mosaic. *Journal of Applied Ecology* 18:703–710.

Miller, J.G. (1975) The nature of living systems. *Behavioural Science* 20:343–365.

Miller, R.I. (1994) *Mapping the Diversity of Nature.* Chapman & Hall, London. 218 pp.

Miller, R.S. (1967) Pattern and process in competition, in *Advances in Ecological Research*, Vol. 4 (ed. J.B. Cragg), pp. 1–74. Academic Press, London.

Moeur, M. (1985) *COVER: A user's guide to the CANOPY and SHRUBS extension of the stand PROGNOSIS model.* US Department of Agriculture, Forest Service, Intermountain Research Station, General Technical Report INT-190, Ogden, Utah. 49 pp.

Moeur, M. (1986) Predicting canopy cover and shrub cover with the PROGNOSIS-Cover model, in *Wildlife 2000: Modeling Habitat Relationships of Terrestrial Vertebrates*, (eds J. Verner, M.L. Morrison and C.J. Ralph), pp. 339–345. University of Wisconsin Press, Madison, Wisconsin. 470 pp.

Monkkönen, M. and Welsh, D.A. (1991) Predicting the effects of human caused landscape changes on European and North American forest birds. *World Congress of Landscape Ecology*, July, 1991, Ottawa, Ontario, Canada.

Monserud, R.A. and Haight, R.G. (1990) *A programmer's guide to the PROGNOSIS optimization model.* US Department of Agriculture, Forest Service, Intermountain Research Station, General Technical Report INT-269, Ogden, Utah. 24 pp.

Morash, P.R. and Racey, G.D. (1990) *The northwestern Ontario forest ecosystem classification as a descriptor of woodland caribou* (Rangifer tarandus caribou) *range.* Ontario Ministry of Natural Resources, Northwestern Ontario Forest Technology Development Unit, Thunder Bay, Ontario, Technical Report No. 55. 22 pp.

Morgan, M.A. (1967) Hardware and models in geography, in *Models In Geography*, (eds R.J. Chorley and P. Haggett), pp. 727–774. Methuen and Co. Ltd, London. 468 pp.

Mulé, R.S. (1982) An assessment of a wildlife habitat evaluation methodology for Alaska. MSc. Thesis, University of Alaska, Fairbanks. 215 pp.

Natural Resources Canada (1993) *The State of Canada's forests, 1993.* Canadian Forest Service, Catalogue Fo1-6/1994-E, Ottawa, Canada. 112 pp.

Naveh, Z. and Lieberman, A.S. (1990) *Landscape Ecology: Theory and Application.* Springer-Verlag, New York, New York. 356 pp.

Neilson, R.P., King, G.A. and Koerper, G. (1992) Towards a rule-based biome model. *Lanscape Ecology* 7:27–43.

New Brunswick Department of Natural Resources and Energy (1992) *Forest land habitat management in New Brunswick: Progress report (October 1, 1992).* Department of Natural Resources and Energy, Fredericton, New Brunswick, and Wildlife Habitat Canada, Ottawa, Canada. 28 pp.

Noss, R. (1983) A regional landscape approach to maintain diversity. *Bioscience* 33:700–706.

Noss, R. (1990) Indicators for monitoring biodiversity, a hierarchical approach. *Conservation Biology* 4(4):355–364.

Noss, R. (1995) *Maintaining Ecological Integrity in Representative Reserve Networks.* World Wildlife Fund Canada/World Wildlife Fund United States Discussion Paper, Toronto, Ontario. 77 pp.

Odum, E.P. (1977) The emergence of ecology as a new integrative discipline. *Science* 195 (4284):1289–1293.

O'Leary, D.J. (1990) Wildlife habitat assessment models for integrated resource planning in Alberta. Environment Canada, Sustainable Development and State of the Environment Reporting Branch, Ottawa, Canada Committee on Ecological Land Classification, *Wildlife Working Group Newsletter* 10:1–2.

Olson, R.L. and Hanson, J.D. (1988) AI tools and techniques: a biological perspective. *AI Applications in Natural Resource Management* 2:31–38.

Omernik, J.M. (1987) Ecoregions of the conterminous United States. *Annals of the Association of American Geographers* 77:118–125.

Omernik, J.M. and Gallant, A.L. (1988) *Ecoregions of the upper midwest states.* US Environmental Protection Agency, EPA/600/3-88/037. 56 pp.

Omernik, J.M., Larsen, D.P., Rohm, C.M. and Clarke, S.E. (1988) Summer total phosphorous in lakes: A map of Minnesota, Wisconsin, and Michigan, USA. *Environmental Management* 12:815–825.

O'Neill, R.V., De Angelis, D.L., Wade, J.B. and Allen, T.F.H. (1986) *A Hierarchical Concept of Ecosystems.* Princeton University Press, Monographs in Population Ecology 23, Princeton, New Jersey. 253 pp.

Ontario Forest Policy Panel (1993) *Diversity: Forests, People, Communities: A comprehensive forest policy framework for Ontario.* Queen's Printer for Ontario, Toronto, Ontario. 147 pp.

Oswald, E.T. (1992) Forest site classification activities in northern Canada. *Forestry Chronicle* 68(1):94–99.

Patch, J. (1987) *Habitat supply forecasting and forest management in New Brunswick: Final report.* Fish and Wildlife Branch, Department of Natural Resources and Energy, Fredericton, New Brunswick, Canada. 46 pp.

Pearce, C.M. (1992) Pattern analysis of forest cover in southwestern Ontario. *East Lakes Geographer* 27:65–76.

Pendergast, B.A. (1989) Ecological habitat mapping in British Columbia, in *Proceedings of the Western Association of Fish and Wildlife Agencies and the Western Division of the American Fisheries Society*, pp. 134–138. American Fisheries Society, Seattle, Washington.

Perret, N.G. (1969) *Land classification for wildlife.* Environment Canada, Lands Directorate, Canada Land Inventory Report No. 7, Ottawa, Canada. 30 pp.

Peters, R.L. (1988) The effect of global climate change on natural communities, in

Biodiversity, (ed. E.O. Wilson), pp. 450–461. National Academy Press, Washington, DC. 521 pp.

Peters, R.H. (1991) *A Critique for Ecology*. Cambridge University Press, Cambridge. 366 pp.

Peterson, G.H. (1987) Ground cover mapping on the winter range of the Beverly barren-ground caribou herd using remote sensing techniques: An aid to management. MED Thesis, University of Calgary, Calgary, Alberta, Canada. 111 pp.

Pfister, R.D. (1977) Ecological classification of forest land in Idaho and Montana, *Proceedings Ecological Classification of Forest Land in Canada and Northwestern USA*, pp. 329–358. University of British Columbia, Vancouver, British Columbia, Canada.

Pfister, R.D. (1981) Habitat type classification for managing western watersheds, in *Interior West Watershed Management, Proceedings of a Conference held April 8–10, 1980, Spokane, Washington*, (ed. D.M. Baumgartner), pp. 59–67. Washington State University, Pullman, Washington.

Pfister, R.D. (1989) Basic concepts of using vegetation to build a site classification system, in *Proceedings – Land Classifications Based On Vegetation: Applications for Resource Management, Moscow, Idaho, November 17–19, 1987*, (D.E. Ferguson, P. Morgan and F.D. Johnson, compilers), pp. 22–28. US Department of Agriculture, Forest Service, Intermountain Research Station, General Technical Report INT-257, Ogden, Utah. 315 pp.

Pfister, R.D. and Arno, S.F. (1980) Classifying forest habitat types based on potential climax vegetation. *Forest Science* 26:52–70.

Pfister, R.D., Kovalchik, B.L., Arno, S.F. and Presby, R.C. (1977) *Forest Habitat Types of Montana*. US Department of Agriculture, Forest Service, Intermountain Forest and Range Experiment Station, General Technical Report INT-34, Ogden, Utah. 174 pp.

Piirvee, R. and Braun, K.N. (1978) An application of the ARIES system to ground vegetation mapping for forestry. *Canadian Symposium on Remote Sensing* 5:79–85.

Platts, W.S. (1980) A plea for fishery habitat classification. *Fisheries* 5:2–6.

Pojar, J., Klinka, K. and Meidinger, D. (1987) Biogeoclimatic ecosystem classification in British Columbia. *Forest Ecology and Management* 22:119–154.

Prescott, W.H. (1980) Integrating wildlife and biophysical land inventories in the National Parks of Atlantic Canada, in *Land/Wildlife Integration No. 1*, pp. 61–73. Environment Canada, Lands Directorate, Ottawa, Ecological Land Classification Series No. 11, 160 pp.

Primack, R.B. (1993) *Essentials of Conservation Biology*. Sinauer Associates, Inc., Sutherland, Massachusetts. 564 pp.

Province of British Columbia (1994) *British Columbia Forest Practices Code: Standards and Revised Rules and Field Guide References*. Government of British Columbia, Victoria, British Columbia. 216 pp.

Quattrochi, D.A. and Pelletier, R.E. (1991) Remote sensing for analysis of landscapes: an introduction, in *Quantitative Methods in Landscape Ecology*, (eds M.G. Turner and R.H. Gardner), pp. 399–416. Ecological Studies, Volume 82, Springer-Verlag, New York, New York. 536 pp.

Racey, G.D., Whitfield, T.S. and Sims, R.A. (1989a) *Northwestern Ontario Forest Ecosystem Interpretations*. Ontario Ministry of Natural Resources, Toronto, Ontario, Canada. 160 pp.

Racey, G.D., McNicol, J. and Timmermann, H.R. (1989b) Application of the moose and deer habitat guidelines: Impact of investment, in *Forest Investment: a Critical Look*, (eds R.F. Calvert, B. Payandeh, M.F. Squires and W.D. Baker), pp. 119–132. Forestry Canada, Ontario Region, Information Report No. O-P-17, Sault Ste. Marie, Ontario, Canada. 216 pp.

Raedeke, K.J. and Lehmkuhl, J.F. (1986) A simulation procedure for modeling the relationship between wildlife and forest management, in *Wildlife 2000: Modeling Habitat Relationships of Terrestrial Vertebrates*, (eds J. Verner, M.L. Morrison and C.J. Ralph), pp. 377–381. University of Wisconsin Press, Madison, Wisconsin. 470 pp.

Rauch-Hindin, W. (1983) Artificial intelligence: a solution whose time has come. *Systems and Software* 1:150–177.

Rayner, M.R. and Bennett, K.J. (1988) Ecological land classification based wildlife habitat evaluation in west-central Alberta: Woodland caribou winter habitat suitability, *Land/Wildlife Integration No. 3*, (H.A. Stelfox andd G.R. Ironside, compilers), pp. 33–40. Environment Canada, Canadian Wildlife Service, Land Conservation Branch, Canada Ecological Land Classification Series No. 22, Ottawa, Canada. 215 pp.

Rees, W.E. (1990) The ecology of sutainable development. *Ecologist* 20(1):18–23.

Rich, E. (1983) *Artificial Intelligence*. McGraw Hill, New York, New York.

Riley, J.L. and Mohr, P. (1994) *The natural heritage of Southern Ontario's settled landscapes*. Ontario Ministry of Natural Resources, Southern Region, Science and Technology Transfer Unit, Technical Report TR-001, Aurora, Ontario, Canada. 78 pp.

Rizzo, B. (1989) The sensitivity of Canada's ecosystems to climate change. *Canada Committee on Ecological Land Classification Newsletter* 17:10–12. Environment Canada, Lands Directorate, Ottawa.

Rizzo, B. and Wiken, E. (1992) Assessing the sensitivity of Canada's ecosystems to climatic change. *Journal of Climatic Change* 21:37–55.

Robbins, C.S., Dawson, D.K. and Dowell, B.A. (1989) Habitat area requirements of breeding forest birds of the middle Atlantic states. Wildlife Monographs 103. 34 pp.

Roberts, D.W. and Morgan, P. (1989) Classification and models of succession, in *Proceedings – Land Classifications Based On Vegetation: Applications for Resource Management, Moscow, Idaho, November 17–19, 1987*, (D.E. Ferguson, P. Morgan and F.D. Johnson, compilers), pp. 49–53. US Department of Agriculture, Forest Service, Intermountain Research Station, General Technical Report INT-257, Ogden, Utah. 315 pp.

Roberts, J.M. (1980) *The Pelican History of the World*. Penguin Books, New York, New York. 1052 pp.

Robertson, F.D. (1992) *Ecosystem Management of the National Forests and Grasslands*. US Department of Agriculture, Forest Service, Washington, DC.

Robinson, V.B., Frank, A.U. and Karimi, H.A. (1987) Expert systems for geographic information systems in resource management. *AI Applications in Natural Resource Management* 1(1):47–57.

Rohm, C.M., Giese, J.W. and Bennett, C.C. (1987) Test of an aquatic ecoregion classification of streams in Arkansas. *Journal of Freshwater Ecology* 4:127–140.

Root, R.B. (1967) The niche exploration pattern of the blue-gray gnatcatcher. *Ecological Monographs* 37:317–350.

Rowe, J.S. (1961) The level-of-integration concept and ecology. *Ecology* 42:420–427.

Rowe, J.S. (1979) Revised working paper on methodology/philosophy of ecological land classification in Canada, in *Applications of Ecological (Biophysical) Land Classification*, (ed. C.D.A. Rubec), pp. 23–30. Ecological Land Classification Series No. 7, Environment Canada, Lands Directorate, Ottawa, Canada. 396 pp.

Rowe, J.S. (1984) Forestland classification: Limitations of the use of vegetation, in *Forestland Classification: Experience, Problems, Perspectives*, (ed. J.G. Bockheim), pp. 132–147. US Department of Agriculture, Soil Service, Washington, DC.

Rowe, J.S. (1990) Letter to Dr. C.D. Fowle, November 15, 1990, Toronto, Ontario, Canada. 12 pp.

Rowe, J.S. (1991) *Home Place: Essays on Ecology*. NeWest Publishers Ltd., Edmonton, Alberta, Canada. 253 pp.

Rowe, J.S. (1992) Forest site classification in Canada: A current perspective: prologue. *Forestry Chronicle* 68(1):22–24.

Rowe, J.S. (1994) *Ecosystems and intellectual muddles*. Ontario Ministry of Natural Resources, Research, Science and Technology Report No. 2, Maple, Ontario, Canada. 8 pp.

Rowe, J.S. (1993) Eco-diversity, the key to biodiversity, in *World Wildlife Fund, Endangered Spaces Campaign: A Protected Areas GAP Analysis Methodology: Planning for the Conservation of Biodiversity*. World Wildlife Fund Canada, Toronto, Ontario. 68 pp.

Rowe, J.S. and Barnes, B.V. (1994) Geo-ecosystems and bio-ecosystems. *Bulletin of the Ecological Society of America* 75:40–41.

Rowe, J.S. and Rizzo, B. (1990) How climate affects ecosystems – a Canadian perspective on what we know, in *Climate Change: Implications for Water and Ecological Resources*, (eds G. Wall and M. Sanderson), pp. 61–68. Proceedings of an International Symposium/Workshop, Department of Geography Publication Series, Occasional Paper No. 11, University of Waterloo, Waterloo, Ontario, Canada. 342 pp.

Rowe, J.S. and Sheard, R. (1981) Ecological land classification: a survey approach. *Environmental Management* 5:451–464.

Rubec, C.D.A. (1983) Applications of remote sensing in ecological land survey in Canada. *Canadian Journal of Remote Sensing* 9(1):19–30.

Rubec, C.D.A. (1992) Thirty years of ecological land surveys in Canada, from 1960 to 1990, in *Landscape Approaches to Wildlife and Ecosystem Management. Proceedings of the Second Symposium of the Canadian Society for Landscape Ecology and Management, University of British Columbia, May 1990*, (eds G.B. Ingram and M.R. Moss), pp. 61–66. PolyScience Publications Inc., Morin Heights, Québec, Canada. 267 pp.

Ruggiero, L.F., Aubry, K.B., Carey, A.B. and Huff, M.H. (1991) *Wildlife and Vegetation of Unmanaged Douglas-fir Forests*. US Department of Agriculture, Forest Service, Pacific Northwest Research Station, General Technical Report PNW-GTR-285, Portland, Oregon. 533 pp.

Running, S.W. and Coughlan, J.C. (1988) A general model of forest ecosystem processes for regional applications, I. Hydrologic balance, canopy gas exchange and primary production processes. *Ecological Modelling* 42:152–154.

Saarenmaa, H., Stone, N.D., Folse, L.J. *et al.* (1988) An artificial intelligence modelling approach to simulating animal/habitat interactions. *Ecological Modelling* 44:125–141.

Salwasser, H. (1986) Modeling habitat relationships of terrestrial vertebrates – the manager's viewpoint, in *Wildlife 2000: Modeling Habitat Relationships of Terrestrial Vertebrates*, (eds J. Verner, M.L. Morrison and C.J. Ralph), pp. 419–424. University of Wisconsin Press, Madison, Wisconsin. 470 pp.

Saucier, J.P. (1989) *La végétation forestière de la région écologique 5a Basses et Moyennes Appalaches*. Minist. Energie Ressour., Service Inventaire Forest, Division Ecologique, Québec City, Québec. 196 pp.

Saveland, J.M., Stock, M. and Cleaves, D.A. (1988) Decision-making bias in wilderness fire management: Implications for expert system development. *AI Applications in Natural Resource Management* 2(1):17–29.

Scepan, J. and Blum, L.L. (1987) A geographical information system for managing California condor habitat, in *GIS '87, San Francisco. Second Annual International Conference, Exhibits and Workshops on GIS*, pp. 476–486. American Society of Photogrammetry and Remote Sensing, and the American Congress on Surveying and Mapping, 1987, Falls Church, Virginia. 386 pp.

Schamberger, M.L. and Farmer, A. (1978) The habitat evaluation procedures: their application in project planning and impact evaluation. *Transactions of the North American Wildlife and Natural Resources Conference* 43:274–283.

Schamberger, M.L. and O'Neill, L.J. (1986) Concepts and constraints of habitat-model testing, in *Wildlife 2000: Modeling Habitat Relationships of Terrestrial Vertebrates*, (eds J. Verner, M.L. Morrison and C.J. Ralph), pp. 5–10. University of Wisconsin Press, Madison, Wisconsin. 470 pp.

Schmidt, R.K. (1980) Wildlife distribution in the Haye's River Area, Manitoba, in *Land/Wildlife Integration No. 1*, (ed. D.G. Taylor), pp. 75–91. Environment Canada, Lands Directorate, Ecological Land Classification Series No. 11, Ottawa, Canada. 160 pp.

Schroeder, R.L. (1986) *Habitat suitability index models: wildlife species richness in shelter belts*. US Department of the Interior, Fish and Wildlife Service, Biological Report 82 (10.128), Washington, DC. 17 pp.

Schroeder, R.L. (1987) *Community models for wildlife impact assessment: a review of concepts and approaches*. US Department of the Interior, Fish and Wildlife Service, Biological Report 87(2), Washington, DC. 41 pp.

Schroeder, R.L. and Allen, A.W. (1992) *Assessment of habitat of wildlife communities on the Snake River, Jackson, Wyoming*. US Department of the Interior, Fish and Wildlife Service, Resource Publication 190, Washington, DC. 21 pp.

Schroeder, R.L. and Haire, S.L. (1993) *Guidelines for the development of community-level habitat evaluation models*. US Department of the Interior, Fish and Wildlife Service, Biological Report 8, February, 1993, Washington, DC 8 pp.

Schuster, E.G., Leefers, L.A. and Thompson, J.E. (1993) *A Guide to Computer-based Analytical Tools for Implementing National Forest Plans*. US Department of Agriculture, Forest Service, Intermountain Research Station, General Technical report INT-296, Ogden, Utah. 269 pp.

Seber, G.A.F. (1984) *Multivariate Observations*. John Wiley and Sons Inc., New York, New York. 678 pp.

Shaw, D.M. and Atkinson, S.F. (1990) An introduction to the use of geographical information systems for ornithological research. *Condor* 92:564–570.

Shaw, G. and Wheeler, D. (1985) *Statistical Techniques in Geographical Analysis*. John Wiley and Sons, New York, New York. 364 pp.

Short, H.L. (1983) *Wildlife Guilds in Arizona Desert Habitats*. US Department of the Interior, Bureau of Land Management, Technical Note 362. 258 pp.

Short, H.L. (1984) *Habitat suitability index models: the Arizona guild layers of habitat models*. US Department of the Interior, Fish and Wildlife Service, FWS/OBS-82/10.70. 37 pp.

Short, H.L. and Burnham, K.P. (1982) *Techniques for structuring wildlife guilds to evaluate impacts on wildlife communities*. US Department of the Interior, Special Science Report – Wildlife 244. 34 pp.

Shugart, H.H. (1984) *A Theory of Forest Dynamics: the Ecological Implications of Forest Succession Models*. Springer-Verlag, New York, New York. 278 pp.

Shugart, H.H. (1990) Using ecosystem models to assess potential consequences of global climate change. *Trends in Ecological Evolution* 5(9):303–307.

Shugart, H.H. and West, D.C. (1977) Development of an Appalachian deciduous forest model and its application to assessment and impact of the Chestnut Blight. *Journal of Environmental Management* 5:161–179.

Shugart, H.H. and West, D.C. (1980) Forest succession models. *Bioscience* 30(5):308–313.

Shugart, H.H., Michaels, P.J., Smith, T.M. *et al.* (1988) Simulation models of forest succession, in *Scales and Global Change: Spatial and Temporal Variability in Biospheric and Geospheric Processes. Scope 35*, (eds T. Rosswall, R.G. Woodmansee and P.G. Risser), pp. 125–151. John Wiley and Sons, New York, New York. 355 pp.

Sidle, W.B. and Suring, L.H. (1986) *Management Indicator Species for the National Forest Lands of Alaska*. US Department of Agriculture, Forest Service, Alaska Region, Technical Publication R10-TP-2. 62 pp.

Sims, R.A. (1985) Developing a forest ecosystem classification for the northcentral region of Ontario, in *Proceedings of IUFRO Workshop, Working Party S1.02.06, October 7–10, 1985*, pp. 101–113. Fredericton, New Brunswick, Canada.

Sims, R.A. and Uhlig, P. (1992) The current status of forest site classification in Ontario. *Forestry Chronicle* 68(1):64–77.

Sims, R.A., Towill, W.D., Baldwin, K.A. and Wickware, G.M. (1989) *Field Guide to the Forest Ecosystem Classification for Northwestern Ontario*. Ontario Ministry of Natural Resources, Toronto, Ontario, Canada. 191 pp.

Sims, R.A., Mackey, B.G. and Baldwin, K.A. (1994) Stand and landscape level applications of a forest ecosystem classification for northwestern Ontario, Canada. *Proceedings, International Union of Forest Research Organizations, 1.02–06 Site Classification and Evaluation research Group Workshop, 19–22 October, 1993*, Clemont-Ferrand, France.

Sklar, F.H. and Costanza, R. (1991) The development of dynamic spatial models for landscape ecology: a review and prognosis, in *Quantitative Methods in Landscape Ecology*, (eds M.G. Turner and R.H. Gardner), pp. 239–288. Ecological Studies, Volume 82, Springer-Verlag, New York, New York. 536 pp.

Slocombe, D.S. (1993) Implementing ecosystem-based management. *Bioscience* 43(9):612–622.

Smith, J.B. (1990) The potential effects of global climate change on the United States, in *Climate Change: Implications for Water and Ecological Resources*, (eds

G. Wall and M. Sanderson), pp. 79–123. Proceedings of an International Symposium/Workshop, Department of Geography Publication Series, Occasional Paper No. 11, University of Waterloo, Waterloo, Ontario, Canada. 342 pp.

Smith, K.G. and Connors, P.G. (1986) Building predictive models of species occurrence from total count transect data and habitat measurements, in *Wildlife 2000: Modeling Habitat Relationships of Terrestrial Vertebrates*, (eds J. Verner, M.L. Morrison and C.J. Ralph), pp. 389–393. University of Wisconsin Press, Madison, Wisconsin. 470 pp.

Smith, K., Edmonds, J. and Stelfox, H. (1988) Draft: WMU 346 habitat stratified moose survey, February, 1988. Alberta Forestry, Lands and Wildlife, Fish and Wildlife, Edson, Alberta, Canada. 20pp + maps. (Unpublished.)

Smith, T.M. (1986) Habitat-simulation models: integrating habitat classification and forest simulation models, in *Wildlife 2000: Modeling Habitat Relationships of Terrestrial Vertebrates*, (eds J. Verner, M.L. Morrison and C.J. Ralph), pp. 389–393. University of Wisconsin Press, Madison, Wisconsin. 470 pp.

Smith, T.M., Shugart, H.H. Jr and West, D.C. (1981a) FORHAB: A forest simulation model to predict habitat structure for nongame bird species, in *The Use of Multivariate Statistics in Sutides of Wildlife Habitat*, (ed. D.E. Capen), pp. 114–123. US Department of Agriculture, Forest Service, Rocky Mountain Forest and Range Experiment Station, General Technical Report RM-87, Fort Collins, Colorado.

Smith, T.M., Shugart, H.H. and West, D.C. (1981b) Use of forest simulation models to integrate timber harvest and nongame bird management. *Transactions North American Wildlife and Natural Resources Conference* 46:501–510.

Sparrowe, R.O. and Sparrowe, B.F. (1978) Use of critical parameters for evaluating wildlife habitat, in *Classification, Inventory, and Analysis of Fish and Wildlife Habitat*, (ed. A. Marmelstein), pp. 385–405. US Department of the Interior, Fish and Wildlife Service, FWS/OBS-78/76, Washington, DC. 604 pp.

Stage, A.P. (1973) *PROGNOSIS model for stand development*. US Department of Agriculture, Forest Service, Intermountain Forest and Range Experiment Station, Research Paper INT-137, Ogden, Utah. 32 pp.

Steele, R., Cooper, S.V., Ondov, D.M. *et al.* (1983) *Forest Habitat Types of Eastern Idaho–Western Wyoming*. US Department of Agriculture, Forest Service, Intermountain Forest and Range Experiment Station, General Technical Report INT-144, Ogden, Utah. 122 pp.

Stefik, M., Aikons, J., Balzer, R. *et al.* (1982) *The Organization of Expert Systems: a perspective tutorial*. XEROX, Palo Alto Resources Centre, Palo Alto, California.

Stelfox, H.A. (1988) Wildlife resource evaluation and land/wildlife relationship models, in *Land/Wildlife Integration No. 3*, (H.A. Stelfox and G.R. Ironside, compilers), pp. 59–68. Environment Canada, Canadian Wildlife Service, Land Conservation Branch, Ottawa, Ecological Land Classification Series No. 22, Ottawa, Canada. 215 pp.

Stelfox, H.A. (1991) Assessing land/wildlife relationships via the ecological land evaluation, in *Guidelines for the Integration of Wildlife and Habitat Evaluations with Ecological Land Survey*, (eds H.A. Stelfox, G.R. Ironside and J. Kansas), pp. 29–46. Wildlife Working Group, Canada Committee on Ecological Land Classification and Wildlife Habitat Canada, Ottawa, Canada. 107 pp.

Stenback, J.M., Travlos, C.B., Barrett, R.H. and Congalton, R.G. (1987) Application of remotely sensed digital data and a GIS in evaluating deer habitat suitability on the Tehama deer winter range, in *GIS '87, San Francisco. Second Annual International Conference, Exhibits and Workshops on GIS*, pp. 440–445. American Society of Photogrammetry and Remote Sensing, and the American Congress on Surveying and Mapping, 1987, Falls Church, Virginia. 386 pp.

Stock, M. (1987) AI and expert systems: An overview. *AI Applications in Natural Resource Management* 1(1):9–17.

Stocks, B.J. (1993) Global warming and forest fires in Canada. *Forestry Chronicle* 69(3):290–293.

Strong, W.L. and Leggat, K.R. (1981) *Ecoregions of Alberta*. Alberta Energy and Natural Resources, Resource Evaluation and Planning Division, Edmonton, Alberta, Canada.

Strong, W.L. and Vriend, H.G. (1980) Ecological land classification hierarchies and distribution in southwestern Alberta, in *Land/Wildlife Integration No. 1*, (ed. D.G. Taylor), pp. 99–106. Environment Canada, Lands Directorate, Ecological Land Classification Series No. 11, Ottawa, Canada. 160 pp.

Sukachev, V.N. and Dylis, N. (1964) *Fundamentals of Forest Biogeocoenology*. (Translated by J.M. MacLennan.) Oliver and Bond, London. 672 pp.

Sullivan, M. (1989) *New Brunswick forest land habitat project: Progress report, October 1989*. Wildlife Habitat Canada and New Brunswick Department of Natural Resources, Fredericton, New Brunswick, Canada. 18 pp.

Sveringhaus, W.D. (1981) Guild theory development as a mechanism for assessing environmental impact. *Environmental Management* 5:187–190.

Sweeney, J.M. (1986) Refinement of DYNAST's forest structure simulation, in *Wildlife 2000: Modeling Habitat Relationships of Terrestrial Vertebrates*, (eds J. Verner, M.L. Morrison and C.J. Ralph), pp. 357–360. University of Wisconsin Press, Madison, Wisconsin. 470 pp.

Szaro, R.C., Severson, K.E. and Patton, D.R. (technical co-ordinators) (1988) *Management of Amphibians, Reptiles, and Small Mammals in North America*. US Department of Agriculture, Forest Service, Pacific Northwest Research Station, General Technical Report RM-166, Portland, Oregon. 458 pp.

Tansley, A. (1935) The use and misuse of vegetation terms and concepts. *Ecology* 16:284–307.

ter Braak, C.J.F. and Looman, C.W.N. (1987) Regression, in *Data Analysis in Community and Landscape*, (eds R.H.G. Jongman, C.J.F. ter Braak and O.F.R. van Tongeren), pp. 29–77. Pudoc, Wageningen, The Netherlands. 299 pp.

Thomas, J.W. (1979) *Wildlife Habitats in Managed Forests: the Blue Mountains of Oregon and Washington*. US Department of Agriculture, Handbook No. 553, Washington, DC. 512 pp.

Thomas, J.W. (1986) Wildlife-habitat modeling – cheers, fears, and introspection, in *Wildlife 2000: Modeling Habitat Relationships of Terrestrial Vertebrates*, (eds J. Verner, M.L. Morrison and C.J. Ralph), pp. xix–xxv. University of Wisconsin Press, Madison, Wisconsin. 470 pp.

Thomas, J.W. (1994) Forest ecosystem management assessment team: objectives, process and options. *Journal of Forestry* 92(4):12–17, 19.

Thomas, J.W., Miller, R., Maser, C. *et al.* (1978) The relationship of terrestrial vertebrates to plant communities and their successional stages, in *Classification, Inventory, and Analysis of Fish and Wildlife Habitat*, (ed. A. Marmelstein), pp.

281–303. US Department of the Interior, Fish and Wildlife Service, FWS/OBS-78/76, Washington, DC. 604 pp.

Thomas, J.W., Leckenby, D.A., Henjum, M. *et al.* (1988) *Habitat effectiveness index for elk on Blue Mt. winter range ranges.* US Department of Agriculture, Forest Service, Pacific Northwest Experiment Station, PNW-GTR 218, Portland, Oregon. 28 pp.

Thomasma, L.E., Drummer, T.D. and Peterson, R.O. (1991) Testing the habitat suitability index model for fisher. *Wildlife Society Bulletin* **19**:291–297.

Thompson, I.D. and Welsh, D.A. (1993) Integrated resource management in boreal forest ecosystems – impediments and solutions. *Forestry Chronicle* **69**(1):32–39.

Topping, J.C., Jr and Bond, J.P. (1988) *The potential impact of climate change in fisheries and wildlife in North America.* Report of the Climate Institute to the US Environmental Protection Agency, Washington, DC.

Trewartha, G.T. (1943) *An Introduction to Weather and Climate.* 2nd edn. McGraw-Hill, New York, New York.

Turner, M.G. (ed.) (1987) *Landscape Heterogeneity and Disturbance.* Springer-Verlag, New York, New York.

Turner, M.G. (1989) Landscape ecology: the effect of pattern on process. *Annual Review of Ecology and Systematics* **20**:171–197.

Turner, M.G., Costanza, R. and Sklar, F.H. (1989) Methods to compare spatial patterns for landscape modeling and analysis. *Ecological Modelling* **48**:1–18.

Udvardy, M.D.F. (1975) *A classification of the biogeographical provinces of the world.* International Union for the Conservation of nature and Natural resources, Occasional Paper 18, Morges, Switzerland. 48 pp.

Uhlig, P.W.C. and Jordan, J.K. (in press) Hierarchical framework for ecosystem management: evolution of ecosystem classification and information systems in the upper Great Lakes Region. *Journal of Environmental Monitoring and Assessment.*

US Army Corps of Engineers (1980) *A habitat evaluation system for water resources planning.* US Army Corps of Engineers, Lower Mississippi Valley Division, Vicksburg, Mississippi.

US Fish and Wildlife Service (1980a) *Habitat as a Basis for Environmental Assessment.* US Department of the Interior, Fish and Wildlife Service, Division of Ecological Services, Ecological Services Manual 101, Washington, DC.

US Fish and Wildlife Service (1980b) *Habitat Evaluation Procedures (HEP): 102 ESM.* US Department of the Interior, Fish and Wildlife Service, Division of Ecological Services, Ecological Services Manual 102, Washington, DC.

US Fish and Wildlife Service (1980c) *Human Use and Economic Evaluation (HUEE): 104 ESM.* US Department of the Interior, Fish and Wildlife Service, Division of Ecological Services, Ecological Services Manual 103, Washington, DC.

US Fish and Wildlife Service (1981) *Standards for the Development of Habitat Suitability Index Models.* US Department of the Interior, Fish and Wildlife Service, Division of Ecological Services, Ecological Services Manual 103, Washington, DC. 169 pp.

Urich, D.L. and Graham, J.P. (1984) *A Handbook for Wetland Habitat Evaluation in Missouri.* Missouri Department of Conservation and the US Soil Conservation Service, Columbia, Missouri. 59 pp.

Urich, D.L., Graham, J.P. and Cook, C.C. (1983) *A Handbook for Habitat Evalua-*

tion in Missouri. Missouri Department of Conservation and the US Soil Conservation Service, Columbia, Missouri. 148 pp.

Van Horne, B. (1983) Density as a misleading indicator of habitat quality. *Journal of Wildlife Management* 47:893–901.

Van Horne, B. (1986) Summary: When habitats fail as predictors – the researcher's viewpoint, in *Wildlife 2000: Modeling Habitat Relationships of Terrestrial Vertebrates*, (eds J. Verner, M.L. Morrison and C.J. Ralph), pp. 257–258. University of Wisconsin Press, Madison, Wisconsin. 470 pp.

Veatch, J.O. (1931) Natural geographic divisions of land. Papers of the Michigan Academy of Science, Arts and Letters, University of Michigan, Ann Arbor, Volume XIV: 417–432.

Veatch, J.O. (1934) Classification of land on a geographical basis. Papers of the Michigan Academy of Science, Arts and Letters, University of Michigan, Ann Arbor, Volume XIX:359–365.

Verner, J. (1984) The guide concept applied to management of bird populations. *Environmental Management* 8(1):1–13.

Verner, J. and Boss, A.S. (eds) (1980) *California Wildlife and their Habitats: Western Sierra Nevada.* US Department of Agriculture, Forest Service, General Technical Report PSW-37. 439 pp.

Verner, J., Morrison, M.L. and Ralph, C.J. (eds) (1986) *Wildlife 2000: Modeling Habitat Relationships of Terrestrial Vertebrates.* University of Wisconsin Press, Madison, Wisconsin. 470 pp.

Walters, C.J. (1971) Systems ecology: the systems approach and mathematical models in ecology, in *Fundamentals of Ecology*, (ed. E.P. Odum), pp. 276–292. W.P. Saunders Co., Philadelphia. 574 pp.

Watry, M.M. (1989) Utilizing expert systems and geographic information systems in forest planning, in *Proceedings from the Geographic Information Systems Awareness Seminar, Salt Lake City, Utah, May 16–19, 1988*, (eds D.S. Winn, R.E. Beverly and G.W. Moore), pp. 23–25. US Department of Agriculture, Forest Service, Intermountain Region, Ogden, Utah. 84 pp.

Wellner, C.A. (1989) Classification of habitat types in the western United States, in *Proceedings – Land Classifications Based on Vegetation: Applications for Resource Management, Moscow, Idaho, November 17–19, 1987*, (D.E. Ferguson, P. Morgan and F.D. Johnson, compilers), pp. 7–21. US Department of Agriculture, Forest Service, Intermountain Research Station, General Technical Report Int-257, Ogden, Utah. 315 pp.

Wertz, W.A. and Arnold, J.F. (1972) *Land systems inventory.* US Department of Agriculture, Forest Service, Intermountain Region, Ogden, Utah. 12 pp.

Westveld, M. (1951) Vegetation mapping as a guide to better silviculture. *Ecology* 32:508–517.

Wheaton, E.E., Singh, T., Dempster, R. *et al.* (1987) *An exploration and assessment of the implications of climatic change for the boreal forest and forestry economics of the prairie provinces and the Northwest Territories: Phase one.* Saskatchewan Research Council Publication E-906-36-B-87, SRC Technical Report 211, Saskatoon, Saskatchewan, Canada. 18 pp.

White, G.C., Carpenter, L.H. and Anderson, D.R. (1985) Application of expert systems in wildlife management. *Transactions of the North American Wildlife and Natural Resources Conference* 50:363–366.

Whittier, T.R., Larsen, D.P., Hughes, R.M. *et al.* (1987) *The Ohio Stream Regionalization Project: a compendium of results.* US Environmental Protection Agency, EPA/600/3-87/025, Corvallis, Oregon. 66 pp.

Whittier, T.R., Hughes, R.M. and Larsen, D.P. (1988) Correspondence between ecoregions and spatial patterns in stream ecosystems in Oregon. *Canadian Journal of Fisheries and Aquatic Sciences* 45:1264–1278.

Whittaker, R.H. (1962) Classification of natural communities. *Botanical Review* 28:1–239.

Whittaker, R.H. (1967a) Gradient analysis of vegetation. *Biological Review* 42:207–264.

Whittaker, R.H. (1967b) *Classification of Plant Communities.* Dr. W. Junk bv Publishers, The Hague, Boston. 408 pp.

Whittaker, R.H., Levin, S.A. and Root, R.B. (1973) Niche, habitat, and ecotope. *American Naturalist* 107(955):321–338.

Wickware, G. and Rubec, C.D.A. (1989) *Ecoregions of Ontario.* Environment Canada, Sustainable Development Branch, Ecological Land Classification Series No. 26, Ottawa, Canada. 37 pp.

Wickware, G., Cowell, D., Ross, K. and Sims, R.A. (1980) Utilization of ecological land classification data for the study and management of waterfowl resources in the Hudson Bay Lowland, in *Land/Wildlife Integration No. 1*, (ed. D.G. Taylor), pp. 45–50. Environment Canada, Lands Directorate, Ecological Land Classification Series No. 11, Ottawa, Canada. 160 pp.

Wiken, E.B. (1980) Rationale and methods of ecological land surveys: an overview of Canadian approaches, in *Land/Wildlife Integration No. 1*, (ed. D.G. Taylor), pp. 11–18. Environment Canada, Lands Directorate, Ecological Land Classification Series No. 11, Ottawa, Canada. 160 pp.

Wiken, E.B. (1986) *Terrestrial ecozones of Canada.* Environment Canada, Lands Directorate, Ecological Land Classification Series No. 19, Ottawa, Canada. 19 pp.

Wiken, E.B. and Ironside, G.R. (1977) The development of ecological (biophysical) land classification in Canada. *Landscape Planning* 4:273–282.

Wiken, E.B. and Lawton, K. (1995) North American protected areas: an ecological approach to reporting and analysis. *The George Wright Forum: A Journal of Cultural and Natural Parks and Reserves* 12(1):25–34.

Wiken, E.B., Welsh, D.M., Ironside, G. and Taylor, D.G. (1979) Ecological land survey of the Northern Yukon, in *Applications of Ecological (Biophysical) Land Classification in Canada*, (ed. C.D.A. Rubec), pp. 361–372. Environment Canada, Lands Directorate, Ecological Land Classification Series No. 7, Ottawa, Canada. 396 pp.

Wiken, E.B., Pierce, T.W. and Ironside, G.R. (1980) Multistage remote sensing in exploratory ecodistrict land classification. *Canadian Symposium on Remote Sensing* 6:63–71.

Wikstrom, J.H. and Bailey, R.G. (1983) Land classification – possible or impossible? *Renewable Resources Journal* 1(4):13–18.

Williams, G.L., Russell, K.R. and Seitz, W.K. (1977) Pattern recognition as a tool in the ecological analysis of habitat, in *Classification, Inventory and Analysis of Fish and Wildlife Habitat*, (ed. A. Marmelstein). US Department of the Interior, Fish and Wildlife Service, FWS/OBS-78/76, Washington, DC. 604 pp.

Wilson, J.A.J. (1983) Pattern recognition habitat evaluation models for southern Idaho. MS Thesis, Colorado State University, Fort Collins, Colorado. 104 pp.

Wiman, I.M.B. (1990) Expecting the unexpected: some ancient roots to current perceptions of nature. *AMBIO* **19**(2):62–69.

Wisdom, M.J., Bright, L.R. and Carey, C.G. (1986) *A model to evaluate elk habitat in western Oregon.* US Department of Agriculture, Forest Service, Pacific Northwest Region, R6-F&WL-216-1986, Portland, Oregon. 36 pp.

Wyngaarden, R. van and Gélinas, R. (1987) *Environmental modelling scenarios for the MacKenzie Mountains study region, Yukon and Northwest Territories.* Environment Canada, Lands Directorate, Ottawa, Canada. 23 pp. (Unpublished.)

Young, T.N., Eby, J.R., Allen, H.L. *et al.* (1987) Wildlife habitat analysis using Landsat and radiotelemetry in a GIS with application to spotted owl preference for old growth, in *GIS '87, San Francisco. Second Annual International Conference, Exhibits and Workshops on GIS*, pp. 595–600. American Society of Photogrammetry and Remote Sensing, and the American Congress on Surveying and Mapping, 1987, Falls Church, Virginia. 386 pp.

Yurach, K., Rock, T., Kowal, E. *et al.* (1991) *Saskatchewan Forest Habitat Project, Annual Report, April 1991.* Saskatchewan Parks and Renewable Resources, Wildlife Branch, Prince Albert, Saskatchewan, Canada. 101 pp.

Zelazny, V.F., Ng, T.T.M., Hayter, M.G. *et al.* (1989) *Field Guide to Forest Site Classifiction in New Brunswick.* Canada–New Brunswick Forest Subsidy Agreement Publication. New Brunswick Department of Natural Resources and Energy, Fredericton, New Brunswick, Canada.

Zoltai, S.C. (1988) Ecoclimatic provinces of Canada and man induced climatic change. Environment Canada, Lands Directorate, Ottawa. *Canada Committee on Ecological Land Classification Newsletter* **17**:12–15.

—15

The 'spatial solution' to conserving biodiversity in landscapes and regions

Richard T.T. Forman and Sharon K. Collinge

15.1 INTRODUCTION

Land planners, conservationists and natural resource managers are inherently optimists, yet few would deny that the land continues to degrade at an alarming rate. Human population growth curves all imply continued or accelerated land degradation in the present decade and subsequent decades (Ehrlich, 1988; Worldwatch Institute, 1992).

This is reminiscent of the 'paradox of management' (Forman, 1995). The probability of having a significant effect is greater in small areas, whereas the probability of successful long-term management is greater in large areas. We can see the results of someone's work protecting a rare butterfly in a local grassland, but few persons will impact a continent or the planet. At the same time, over human generations the chance of finding the butterfly at that spot is low, whereas the globe is likely to continue in similar form.

Our land use and conservation efforts are focused at too fine a scale. We must move them to the landscape and regional scales. Clearly both global and local efforts are important but the emphasis must be in the broad middle ground.

Our basic thesis is that a spatial solution has now emerged that addresses key environmental and land-use issues in any large land area. The ecology of landscapes and regions is at the root of this solution (Forman and Godron, 1986; Saunders et al., 1987; Saunders and Hobbs, 1991; Turner and Gardner, 1991; Forman, 1995). The focus here applies to species richness and animal movement but the principles apply broadly to ecological issues including wind, water and soil erosion.

Conservation of Faunal Diversity in Forested Landscapes. Edited by
R.M. DeGraaf and R.I. Miller. Published in 1996 by Chapman & Hall. ISBN 0 412 61890 7.

The broad objective of this chapter is to highlight the major components of the spatial solution in a changing land. The specific goals addressed are as follows.

1. To outline indispensable spatial patterns, as well as some propitious ones (section 15.2).
2. To pinpoint some patterns of disturbance spread and animal movement through landscape mosaics that have land planning significance (section 15.3).
3. To illustrate how fragmentation fits into the spatial processes that occur during land transformation (section 15.4).
4. To identify an ecologically optimum sequence for changing land (section 15.5).
5. To illustrate the use of some principles from the preceding analyses in a world of shrinking natural vegetation (section 15.6).
6. To outline a range of approaches for planning and conserving whole landscapes and regions (section 15.7).

15.2   THE SPATIAL SOLUTION

Some agreement has emerged in recent years that effective conservation of nature and natural processes requires inventories of the components to be conserved. Thus in a landscape biological surveys, and in some areas taxonomic work, are considered critical for conserving biodiversity and species of special conservation importance (Wilson, 1988, 1992; Raven, 1988). Population biology and demographic studies are needed to protect rare species (Ehrlich *et al.*, 1975; Ehrlich, 1983). Soil surveys and mapping are keys to identifying erosion rates and susceptible areas (Hole and Campbell, 1985). Mineral nutrient cycles need to be measured in the diverse ecosystem types present to prevent scarce nutrient loss, or eutrophication (Likens, 1985; Binford and Buchenau, 1993). Hydrological monitoring and water quality or fish community assays are essential to protect aquatic systems receiving inputs from the land (Dunne and Leopold, 1978; Karr, 1991). Land use and conservation decisions then can be based on the solid foundations of these extensive data sets.

This chapter will highlight an alternative approach, essentially a 'spatial solution'. We hypothesize that there are spatial land-use patterns that make good ecological sense and that will conserve the bulk of nature and natural processes in any landscape or major portion thereof. These patterns will not protect every species, every soil particle, every nutrient concentration or every portion of water bodies, but the spatial patterns will conserve the bulk of the attributes, as well as the most important ones.

At first glance this is merely a provocative hypothesis. Yet important lines of intersecting evidence have appeared in landscape ecology, as well as conservation biology, forestry and other fields, in the past few years (Harris, 1984; Franklin, 1993; Soule, 1987; Meffe and Carroll, 1994). These coalescing lines of evidence are pinpointed in three categories below (Forman, 1995): first, the 'indispensable patterns' are those for which no known or reasonable alternative exists to provide their benefits; second, an 'aggregate-with-outliers' principle provides an especially effective arrangement for fitting different land uses together in a landscape; and finally, 'strategic points' in a landscape are an important component of a spatial solution.

Most of the indispensables are individual building blocks or components of the overall spatial pattern. A few are considered here; future research will doubless discover more. In contrast, the aggregate-with-outliers pattern fits individual components together in a mosaic, and this will emerge as an equally important frontier. Only then will the full form of the spatial solution crystallize.

Rather than eliminating the need for inventories, the spatial solution is a higher, more urgent priority for conservation and land-use protection. Detailed inventories should be pursued concurrently and can be expected to lead to supplementary protection.

15.2.1 Indispensable patterns

(a) A few large patches

A few large patches of natural vegetation are required to protect the species richness of a landscape (Forman *et al.*, 1976; Game and Peterken, 1984; Forman, 1995). A large natural-vegetation patch is especially needed for interior species and species with large home ranges. The number of such patches required is unknown, but probably ranges from about two to five. If the percentage of the total species pool of the landscape in one large patch is very high (say >95%), two or three large patches may be sufficient to accomplish the ecological objectives. If the percentage of the species pool in one patch is lower, perhaps four or five may be sufficient. No substitute for a few large natural-vegetation patches is known.

(b) Vegetated corridors along major streams and rivers

Stream and river corridors of natural vegetation (Figure 15.1) provide a wide range of ecological benefits otherwise not accomplished (Binford and Buchenau, 1993; Malanson, 1993). These include:

- controlling streambank erosion;
- reducing mineral nutrient and other substance inputs to streams;
- providing shade and fallen branches for fish habitat;
- providing leaves and branches as the base of aquatic food chains;
- protecting conduits for species moving along the dendritic system.

Stream corridors also provide hydrologic and biodiversity benefits. Width

Figure 15.1 Indispensible spatial patterns illustrated in a portion of an agricultural landscape in Iowa, USA, with snow-covered fields and darker forest. River corridor in center, and stream corridor in upper right. A few large patches of natural vegeta-

and connectivity appear to be key variables determining the effectiveness of stream corridors. The minimum width required differs markedly from an intermittent or first-order stream to a high-order river (Forman, 1995). Rather little is known about the effect of connectivity of stream corridors, except that a gap in the vegetated strip can have a major effect on stream temperature and trout populations for a few kilometers downstream (Barton *et al.*, 1985).

(c) Connectivity between large patches

Wide connected corridors or, in some locations, a scatter of small patches are keys to effective movement of species between large natural-vegetation patches (Figure 15.1) (Harris and Gallagher, 1989; Saunders and Hobbs, 1991; Noss, 1993a; Forman, 1995). Other mechanisms such as a row of stepping stones or a broken corridor may be significantly less effective. The movement of species among large patches enhances recolonization following local extinction, reduces the probability of inbreeding depression and decreased genetic variability, and spreads risk in case of severe disturbance to a patch.

(d) Bits of nature across the matrix

Where the surrounding matrix is ecologically unsuitable or less suitable than the large patches, small patches and/or corridors scattered over the area provide several key ecological benefits (Figure 15.1) (Forman, 1995). These include:

- protecting rare habitats and species outside the large patches;
- providing stepping stones for species movement in all directions;
- breaking up extensive stretches of matrix to reduce wind or water erosion;
- providing heterogeneous conditions throughout the landscape as a hedge against disturbance.

15.2.2 Aggregate-with-outliers guideline

Most landscapes with a significant human imprint contain a few major land uses, each covering at least 10% of the area. For instance, consider a large area composed of woodland, pasture and houses. Is there an ecologically optimum spatial arrangement of the three types? At one extreme all

---

tion assumed to be below and to left of image, and attached to river corridor. Connectivity between large patches (stepping stones or disconnected corridors provide only partial connectivity). Bits of nature scattered across the matrix. (Photograph by R. Forman.)

woodland, all pastures and all houses could each be aggregated, so that the landscape contains only three different huge patches. At the other extreme each little woodlot, pasture and house could be randomly or regularly intermixed over the landscape. Both scenarios contain significant ecological shortcomings. Numerous intermediate options exist, of which the 'aggregate-with-outliers' principle appears to be the best as yet detected (Forman, 1995).

The principle states that one should aggregate land uses, yet maintain corridors and small patches of nature throughout developed areas, as well as outliers of human activity spatially arranged along major boundaries. To illustrate the spatial pattern produced, find a blackboard and use woodland, pasture and houses (or the three major land uses in your landscape) to make a sequential diagram of the following steps.

1. Aggregate the land uses. To spread risk a few large patches of woodland, pasture and houses is better than a single patch of each (Figure 15.2).
2. Add connectivity between the large woodland patches using corridors and/or scatters of small wooded patches.

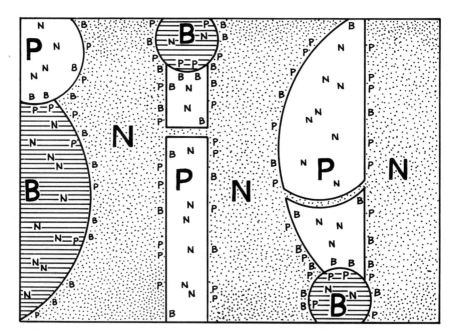

Figure 15.2 Aggregate-with-outliers guideline for ecologically meshing land uses; N = natural vegetation; P = pasture; B = built area. Large letters indicate a large patch; small letters indicate locations of small patches. (Drawing courtesy of Kimberly Hill.)

3. Distribute small patches or bits of nature over the developed pastureland and built areas.
4. Align small pastures on each side of the woodland–houses border.
5. Move the small pastures so that near a large pasture area small pastures are close to one another, but small pastures distant from the large area are isolated from each other.
6. Align small clusters of houses (or individual houses) on each side of the woodland–pasture border.
7. Move the house-clusters so they are more isolated from one another with increasing distance from a large built area.

The resulting spatial model (Figure 15.2) illustrates the aggregate-with-outliers principle, which in turn accomplishes the following major ecological objectives.

- **Large patches** of natural vegetation are present that can protect populations of interior species, large-home-range species, aquifers, low-order stream networks and natural disturbance regimes.
- The landscape is **coarse grained,** but with fine-grained portions present. The coarse grain provides not only the preceding ecological attributes but also efficiency in livestock production and social linkages and specialization in the human community. The fine-grained areas provide efficiency for resource use by multihabitat species, including humans.
- **Risk spreading,** i.e. not putting 'all eggs in one basket', is manifest in the few large patches of each type, as well as in the many small distributed patches.
- **Genetic diversity** associated with reproduction in separated patches provides a buffer against the effect of disturbance. Hermits and town dwellers are both provided for, and genetic strains in isolated pastures may be used for repopulation after massive diseases of grass or livestock in large patches.
- **Boundary zones** between major land-use patches are fine-scale areas useful for species that use two or more habitat types. By concentrating the small patches of human development (pastures and house clusters) along these boundaries, the large aggregated land-use patches gain protection.
- **Small patches** of natural vegetation provide stepping stones for species movement throughout the developed areas, and additional ecological benefits introduced in the previous section.
- **Corridors** of natural vegetation provide for movement between large natural-vegetation patches. In addition, the major boundary zones are effective corridors for transportation and movement between large built areas and pastureland. Distinct areas of natural vegetation distributed along a boundary zone prevent continuous ribbon or strip development and enhance the identity of neighborhoods or towns.

In effect, the aggregate-with-outliers principle is a flexible spatial model applicable for fitting land uses together in any landscape, from grassland to forest and from agriculture to suburbia.

15.2.3   Strategic points in a landscape

It is likely that the spatial solution to land planning and conservation must include strategic points in a landscape, although the subject is little explored (Forman, 1995). These are locations whose ecological importance is exceptional and long term.

One type of strategic point emphasizes the content of a location. Thus an occasional city, a large natural-vegetation patch or the only two mountains present could be strategic as a habitat, source or sink for many ecological phenomena. A second type of strategic location may be a 'flux center', where movements or flows are concentrated, such as a migration route, an erosion surface and an intersection in a network. A third type focuses on change, such as an especially sensitive habitat or a location that, if altered, has a long recovery time.

A different approach to identifying strategic locations that warrant protection in a landscape ignores the site characteristics and focuses on position relative to the size, and especially the shape, of a landscape. For example, a mathematical model predicted that the population of an uncommon songbird (Bachman's sparrow, *Aimophila aestivalis*) in South Carolina would be 12% greater in a mature wood in the center of a landscape than if the wood was in a randomly located position (Liu *et al.*, 1994). The population was predicted to be 21% greater in a central wood than for one in the corner of a landscape.

Such a positional approach should probably at least consider access to, protection from, escape from and control over certain terrain. This pinpointing of strategic points focuses on the interaction between a location and its surroundings. The surroundings may be entirely within the landscape, as for a central location, or may be mainly an adjacent landscape, as for a point near the tip of a lobe.

In conclusion, a spatial solution in land planning, conservation, management and policy permits ecologically based decisions to be made long before most of the ecological details in a landscape are known. The indispensable patterns should be widely incorporated now. The aggregate-with-outliers, strategic points and other spatial patterns to be developed enhance the spatial repertoire of the land planner. In view of the rapid rates of human population growth, agricultural intensification and spreading land development, waiting to document the richness of ecological detail in a landscape is no longer ecologically warranted. We believe that leadership and history demand the use of a spatial solution.

15.3 ANIMAL MOVEMENT AND SPREAD OF DISTURBANCE IN LAND MOSAICS

The preceding patterns result mainly from land use planning and human activities without land planning. Natural disturbance also creates patterns in the landscape. To understand the implications of landscape pattern for the conservation of biological diversity, it is important to consider how functional characteristics, i.e. movement and flow, relate to spatial structure. Therefore this section introduces some key aspects of natural disturbance and of faunal responses to pattern at the landscape scale. A full review of these subjects is beyond the scope of this chapter but they are significant in their own right, as well as foundations for subsequent sections. The material in this section is presented in three parts: natural disturbance; faunal movement response to patterns; and the black bear as a case study.

15.3.1 Natural disturbance

The patterns produced by natural disturbance, and the gap dynamic and successional processes following disturbance, are extensively studied (Pickett and White, 1985). Here we use some North American studies to identify several characteristic spatial patterns produced by fire and hurricane that are of ecological interest at the landscape scale. Then the focus is broadened to consider the effect of disturbance on overall heterogeneity, the spread of disturbance and the minimum dynamic area required for conservation (Forman, 1995).

(a) Fire patterns

Burned areas in Yellowstone Park had an average length-to-width ratio of *c.* 2.5:1 (range 1.1–3.7:1), with wavy borders and about three major lobes (Romme and Knight, 1981; Romme, 1982). In southeastern Labrador *Betula papyrifera* areas up to several square kilometers have sharp borders with the adjoining conifer forest, and are restricted to steep slopes that recently burned (Foster, 1983). On a barrier island in Georgia fire starts in open habitats, moves into wooded areas and is a major determinant of landscape heterogeneity (Turner and Bratton, 1987). In the New Jersey Pine Barrens natural fire disturbance produced a coarse-grained mosaic of variable-sized patches (average 36 ha), on which a fine-scale mosaic of 6 ha patches was superimposed after the introduction of fire control (Forman and Boerner, 1981). In severe droughts approximately every 20–25 years, a landscape-homogenizing effect appears when extensive hot fires kill stream-corridor vegetation. In northwestern Ontario the highest landscape heterogeneity is associated with intermediate levels of forest fire (Suffling *et al.*, 1988). In Quebec and Labrador average natural-fire sizes were recorded as 12 710 and

7 764 ha, respectively, whereas internal areas dissected by lakes and rivers had average fire sizes of 322 and 770 ha, respectively (Hunter, 1993).

(b)   Hurricane patterns

Hurricane patterns are illustrated by a 1938 storm that produced severe tree damage over 25%, and moderate damage over 50%, of a mainly forested area in southern New England (Foster, 1988a, 1988b; Foster and Boose, 1992). Few trees were flattened in agricultural and young forest patches. South-facing and east-facing slopes suffered heavy forest damage, while northwest slopes were protected. Trees blew down at the northwest ends of ponds and lakes but survived on south and east shores. Tall stands, conifer stands and canopies of fast-growing pioneer species were severely damaged. Most blowdown patches were small (< 10 ha) though they varied up to 3.25 km². Within these patches damage intensity varied markedly, so most relatively homogeneous blowdowns were < 2 ha. In this landscape natural disturbance increased heterogeneity. Also landscape structure strongly influenced the spread of disturbance.

(c)   Heterogeneity and disturbance spread

It appears that some natural disturbances decrease landscape heterogeneity, but the bulk of them create more heterogeneity and structure in the mosaic (Arno, 1980; Forman and Boerner, 1981; Suffling *et al.*, 1988; Foster and Boose, 1992). In considering the spread of disturbance, certain local eco-systems are especially susceptible to the initiation of disturbance, whereas others are resistant. Furthermore, some spatial elements enhance or channel disturbance spread, whereas others are barriers or filters that reduce its spread.

Large disturbances are important natural processes, with which certain species evolved and which are required for maintenance of many species. Where humans cause fragmentation of habitat, large natural disturbances become rare. Reuniting the habitats and/or recreating former disturbance regimes may be important planning and management objectives. In other areas where human disturbance predominates, reducing its spread may be the goal. In this situation three approaches focused on landscape pattern are suggested. The first is to minimize the sites that are especially susceptible to the initiation of disturbance. The second is to increase mosaic heterogeneity (Turner, 1987; Forman, 1987). The third is to increase the barriers or filters that inhibit spread, especially the corridors and patch boundaries at right angles to disturbance movement (Romme, 1982; Forman and Godron, 1986; Foster, 1988a).

(d)   Minimum dynamic area

One final dimension of natural disturbance is of particular importance in later sections. Species–area curves with minimum area points have been used in planning, conservation and management (Lovejoy *et al.*, 1984; Meffe *et al.*, 1994). But much larger areas of natural vegetation are required to sustain species richness over time. The minimum dynamic area is the patch space required so that a natural disturbance regime will not eliminate species (Pickett and Thompson, 1978). For example, a fire or hurricane blowdown usually covers only a portion of a large patch, and hence species can repopulate the disturbed area from within the patch. Therefore the effective sizes of large patches should be scaled to the sizes and frequencies of disturbances (Pickett and White, 1985; Baker, 1993). Natural vegetation patches effective for faunal conservation are much larger than the average size of a disturbance (Forman, 1995).

15.3.2 Faunal movement response to patterns

Animals respond to spatial patterns in diverse ways, ranging from metapopulation dynamics to dispersal and population growth to genetic differentiation. This section focuses on one of the important functional characteristics, animal movement, in response to habitat spatial configuration. Do animals respond to the spatial configuration of landscape elements? If so, are these responses predictable? How do such patterns vary among organisms which perceive the structure of the landscape at different spatial scales? We will primarily illustrate animal movement patterns related to spatial configurations outlined in the preceding section.

(a) Corridors that enhance

Diverse types of animal respond to both remnant and human-introduced vegetated corridors by using them as conduits of movement. For example, cougars (*Felis concolor*) in southern California use a canyon as a corridor connecting two disjunct, large natural areas in the Santa Ana mountain range (Beier, 1993). Juvenile red squirrels (*Sciurus vulgaris*) in Belgium use tree rows and hedgerows as dispersal corridors, and adults use them to move between feeding patches (Wauters *et al.*, 1994). This connectivity is crucial to habitat use, as squirrels avoid small woodlots if not connected to other woodlots. Other small mammals and birds use fences and hedgerows in agricultural landscapes to move between remnant wooded patches (Wegner and Merriam, 1979; Johnson and Adkisson, 1985; Henderson *et al.*, 1985; Merriam and Lanoue, 1990). Road verges and fencelines containing native vegetation in Western Australia are used for movement between remnant woodland patches by two species of kangaroo (Arnold *et al.*, 1991). Use of road verges and fencelines near small patches is greater than by large patches, suggesting that corridors may be more critical for connecting small remnants than large remnants.

The key role of animal movement along vegetated corridors in maintaining viable populations is only beginning to be more clearly understood.

Vegetated corridors enhance the dispersal rates of meadow voles (*Microtus pennsylvanicus*) between old-field patches, and vole population densities are higher in connected than in isolated patches (La Polla and Barrett, 1993).

Vegetated corridors may also enable geographic range expansions of particular animal species. *Microtus pennsylvanicus* has extended its geographic range southward in Illinois due to the establishment of dense, grassy vegetation along interstate highways (Getz *et al.*, 1978). The voles did not use grassy areas along two-lane roads or railroad corridors, presumably because grassy areas were interrupted when encountering villages and small towns. The western harvest ant (*Pogonomyrmex occidentalis*) has similarly expanded its range using roadside ditches (DeMers, 1993).

(b)   Corridors that inhibit

Certain types of corridor, particularly roads, may serve as barriers to animal movement and habitat use. Grizzly bears (*Ursus horribilis*) in Montana use roadside habitat significantly less than expected, based on its percentage of the total available habitat (McLellan and Shackleton, 1988), and therefore are displaced by roads. Black bears (*Ursus americanus*) in North Carolina avoid residential areas and appear to need large, inaccessible (roadless) areas for escape and denning (Landers *et al.*, 1979). Breeding male willow warblers (*Phylloscopus trochilus*) in the Netherlands move away from the heavy traffic on a highway to wooded areas further from the road (Foppen and Reijnen, 1994).

In the same landscape four groups of species were compared for their ability to cross highways and rivers/canals (Knaapen *et al.*, 1992). Large mammals readily crossed, butterflies and small mammals were intermittent, and forest songbirds were most inhibited by these corridors.

Rivers appear to serve as barriers to movement for some Amazonian primate species (Ayres and Clutton-Brock, 1992). Thus in areas with high river density it is predicted that forest fragmentation will lead to greater separation of primate populations than in areas with low river densities.

(c)   Boundary shape

The shape of a particular patch may enhance the rate of animal movement across boundaries. A high perimeter-to-area ratio of habitat patches is expected to affect positively the emigration of small mammals from source patches into sink habitat (Buechner, 1989). Male mink (*Mustela vison*) radio-tracked in Manitoba, Canada, are more frequently found in large wetlands with irregular shorelines than near smoother boundaries (Arnold and Fritzell, 1990). Deer and elk tend to move across a woodland–grassland boundary where those boundaries are curvilinear rather than straight (Forman *et al.*, in preparation).

(d) Landscape heterogeneity

Both the direction and distance moved by animals may vary according to the spatial heterogeneity of the habitat and the scale at which heterogeneity exists. For example, bank voles in a forested landscape travel further in heterogeneous (habitat mosaic including mixed-wood forest, alder wood and crop field) than in homogeneous (large mixed-wood forest) landscapes (Kozakiewicz *et al.*, 1993). Black-tailed deer (*Odocoileus hemionus columbianus*) on Vancouver Island, Canada, show little response to edge where the habitat is a fine-grained mosaic of forage and cover areas (Kremsater and Bunnell, 1992). Where these two habitat types occur in a more distinct configuration of larger patches, responses to edge are detectable and positive. Experiments with insects in short-grass prairie have shown that some species move further where the habitat is more spatially heterogeneous (Wiens and Milne, 1989; With, 1994). A simulation model of ungulates foraging in Yellowstone National Park (United States) suggests that animals have greater feeding efficiency, and higher probability of survival, in a landscape with aggregated rather than dispersed (random) patches of forage (Turner *et al.*, 1993).

(e) Habitat juxtaposition

Where habitats are not continuous, a cluster of 'stepping stones' may facilitate movement between large patches. Carnaby's cockatoo (*Calyptorhynchus funereus latirostris*) in Western Australia is hypothesized to be able to persist in a mosaic of remnant patches of native vegetation which comprise visually contiguous stepping stones (Saunders and Ingram, 1987). Wild turkeys (*Meleagris gallopavo*) need clusters of habitat patches for continuous movement, since they avoid traversing open areas > 180 m across (Gustafson *et al.*, 1994). Herbivorous insects colonize clover patches close to adjacent meadows in greater numbers than patches more distant from adjacent meadows (Kruess and Tscharntke, 1994). Bird species composition within cornfields bordered by herbaceous habitat differs from that in cornfields adjacent to woodland habitat (Best *et al.*, 1990).

The animal movement patterns summarized above suggest that animals do respond to landscape spatial structure, and that these movements may enhance or inhibit the persistence of populations. Continuous vegetated corridors are probably the best solution to maintaining animal movement among large remnant patches. Clusters of small patches may be sufficient for movement of some species. Where a continuous corridor is not feasible, stepping stone clusters may be necessary for species movement. Some animals distinctly avoid roads and residential developments, thus emphasizing the need for protecting large areas with little human impact. More information is needed on how different animals respond to particular spatial arrangements, especially patch shape and the proximity of other similar and different patch types, and how these movement patterns influence species persistence.

15.3.3   Black bear as a case study

Black bear (*Ursus americanus*) is chosen because it is widely distributed, is a multihabitat species that moves across the landscape, is sensitive to spatial arrangement and, where present, is a species that influences planning, conservation and management. In the Smoky Mountains of North Carolina and Tennessee, black bears use different habitats in summer and autumn (Garshelis and Pelton, 1981). The summer and autumn ranges overlap somewhat for some bears but are separate for about half of the individuals studied. The intervening space between seasonal ranges is barely used. The center of summer activity moves about 1.1 km from year to year. On the other hand, the distance between centers of summer and autumn activity is 2.8 km for animals with overlapping seasonal ranges and 10.0 km for bears with separate seasonal ranges. The boundaries of summer ranges often correlate with drainage basins and are relatively constant from year to year. The length-to-width ratio of summer ranges averages about 2:1. Width is relatively constant, compared with the more variable length of summer ranges. Seasonal ranges are commonly less than half the total area of the respective annual home ranges.

For this same multihabitat species in the coastal plain 400 km eastward, spatial patterns provide additional insight (Landers *et al.*, 1979). Here the matrix is usually a hardwood swampland. Bears generally avoid five patch types that cover a quarter of the area (residential, lake, most farmland, pine plantation and sparsely vegetated sand ridge). But three ecosystem types are important for bears. Firstly, scattered ridges with moderately dense vegetation provide food in late autumn. Secondly, dense woodlands (Carolina bays) provide both winter denning habitat and rich spring–summer foraging habitat. However, these woodland patches used for winter dens are relatively small and hence the bears are subject to disturbance.

To escape disturbance by high water, hunting dogs and hunters with radio communication, some bears relocate dens to hollow trees or dry spots in the third important ecosystem type: the swampland matrix. If either the ridges or the dense woodlands disappeared due to human activity, a few bears could probably survive. But the swampland represents escape cover, the most critical habitat to maintain the bear population.

In the Adirondack Mountain region of New York, black bear populations decrease with higher human density, higher road density or less forest cover (Brocke *et al.*, 1990). However, road density in mainly forested areas is the best predictor of bear density. Rather than road length per se inhibiting the species (Rost and Bailey, 1979; Mech *et al.*, 1988), tiny 'first-order' roads provide access to remote areas for hunters.

The bear example is instructive in planning, conservation and managing for multihabitat species in general (Forman, 1995). Such species tend to be generalists and hence can utilize a range of foods and find suitable spots for denning or nesting in different habitats. Two or more habitats must be protected. These are used at different seasons for different objectives and a minimum size of each is required. The area between the habitats may be little used. Hence a corridor system may be sufficient to connect the habitats. Optimal corridor width varies by season, e.g. wider when bear cubs must move in spring. Escape cover is essential for each of the patch and corridor steps along the route. Minimizing road density is an effective management tool, both in the planning stage and in closing networks of tiny roads.

The black bear example illustrates faunal movement response to several of the landscape spatial patterns already mentioned. Doubtless all of the structural patterns presented, plus others that could be added, are important in conserving the diverse species composing whole faunas.

15.4 FRAGMENTATION AND SPATIAL PROCESSES IN LAND TRANSFORMATION

The spatial solution described in section 15.2 focuses on pattern or structure of a landscape, and section 15.3 (animal movement) highlights the functioning of a landscape. Most planning, conservation and management is designed around this concept of structure and functioning. Yet over time landscape structure changes. Aerial photographs or maps show a different mosaic each year, each decade and each century. Should planning, conservation and management be tailored to the existing trajectory of mosaic change? Is there an 'ecologically optimum' sequence of mosaics for a landscape that should form the basis of decision-making? We believe the answers to both questions are 'yes', and will outline a spatial or **mosaic sequence** model that appears to be the ecologically optimum manner of transforming a landscape from one type to another (Forman, 1995).

Habitat fragmentation has become of major conservation concern because of small isolated populations, elevated extinction rates and loss of rare species. Fragmentation has sometimes been considered as habitat loss plus isolation (Lovejoy *et al.*, 1984; Wilcox and Murphy, 1985; Wilcove *et al.*, 1986). However, many types of land change produce habitat loss and/or isolation without fragmentation. We prefer to use fragmentation in a sense somewhat similar to that of the dictionary, i.e. breaking an object into pieces. Obviously habitats rarely move horizontally like the pieces of a dropped plate. However, aerial photographs, taken say at decade intervals, often reveal a net increase or decrease in fragmentation, habitat loss and/or isolation. In this manner a habitat type may become more fragmented, yet with no habitat loss or isolation. In short, habitat loss and isolation are useful, but different and broader, concepts than fragmentation.

This provides a framework within which to consider fragmentation along with other spatial processes in the context of land transformation or landscape change. To simplify, we assume that an all-green landscape (surrounded by a brown area) is progressively transformed to become all brown. The change in the landscape can be described or modeled as a sequence of mosaics (Franklin and Forman, 1987; Hansen *et al.*, 1992; Li *et al.*, 1993). We assume that the initial green land type is more ecologically suitable than the less suitable brown type (the opposite assumption could be made in land restoration). We also assume that, once a spot turns brown, it remains brown.

The all-green landscape may initially be **perforated** by patches, e.g. logged clearings or mine sites (Forman, 1995). Alternatively, it may be **dissected**, for example, by roads or railroads slicing through it. As agriculture or other new land uses spread out from the roads, the landscape becomes **fragmented** into separated pieces. Some of the green pieces then **shrink** in size, and indeed green pieces disappear by **attrition**. These five spatial processes (perforation, dissection, fragmentation, shrinkage and attrition) produce the dynamic patterns in a mosaic sequence.

The curves for perforation and dissection peak early in this general land-transformation model. Fragmentation and then shrinkage are most prominent in the middle portion of the sequence. Attrition peaks near the end. Overall, the five spatial processes appear broadly to overlap during land transformation. Data from Cadiz Township, Wisconsin, USA (Sharpe *et al.*, 1987; Dunn *et al.*, 1991) and the Miyaka River Basin, Chiba, Japan (Ohsawa and Liang-Jun, 1987) illustrate the model during the phase when green land drops from *c.* 35% to 10%. In these landscapes both the relative heights and the overlapping of curves for the spatial processes are consistent with the general model (Forman, 1995).

Unusual land-change processes in a landscape cause distinctive shapes in the five curves of the land-transformation model. For example, the cutting of paths (rides) in wooded patches in England in the eighteenth and nineteenth centuries (Rackham, 1986) produced a secondary peak of dissection late in the transformation sequence.

15.5 DETECTING AN ECOLOGICALLY OPTIMUM SEQUENCE IN CHANGING LAND

With an understanding of the spatial processes in land transformation, we now can use the mosaic sequence model in an attempt to identify an ecologically optimum land transformation. In a survey of more than two dozen types of land transformation worldwide, Forman (1995) and G.F. Peterken (personal communication) noticed that most were described by five mosaic sequences. An **edge** sequence starts at one border and the landscape progressively changes from green to brown using parallel strips across the

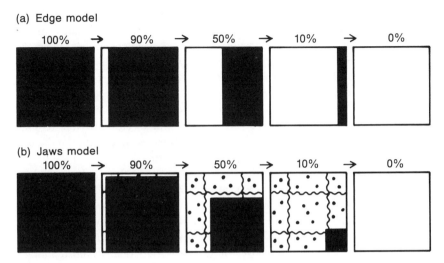

Figure 15.3 Edge model and jaws model of land transformation. Black indicates the initial, more ecologically suitable land type, which is replaced by the less suitable type (white). Narrow corridors and small patches are present in the jaws model of landscape change. (Drawing by Kimberly Hill.)

landscape (Figure 15.3a). A **corridor** sequence starts with a brown strip through the middle, and brown spreads in opposite directions from this strip. A **nucleus** sequence has brown spreading in concentric rings from an initial spot. A **nuclei** sequence has concentric rings spreading from a few initial brown spots. A **dispersed** sequence has brown patches progressively scattered over the landscape, such that the variance in inter-patch distance is minimal (Franklin and Forman, 1987). A recent study evaluated the consequences of these different spatial sequences for insects in native grassland habitat (Collinge and Forman, in preparation).

The edge sequence was determined to be the ecologically best of the five options, based on calculating many spatial attributes for each time interval for each model, and comparing many ecological characteristics known from the literature to correlate with spatial attributes. However, the edge sequence has significant shortcomings. For example, near the end when the land is 90% brown, the green remaining is only a strip along a border, with little or no interior habitat (Figure 15.3a). Also the huge brown area is continuous, with no scattered greenery to enhance species movement or inhibit wind or water erosion.

To reduce these shortcomings, a **jaws** sequence or model was proposed (Forman, 1995). In this case the initial brown is a strip along two adjacent borders (Figure 15.3b). Such L-shaped brown strips progressively spread across the landscape to the opposite corner, much like a fish moving from one corner to the opposite corner with its jaws open. Thus the large green patch is a shrinking square, an ecological improvement, especially in the latter portion of land transformation. The other problem addressed in the jaws sequence is that bits of scattered greenery are skipped as each L-shaped brown area is added. These tiny patches and/or corridors left behind as the brown area spreads might constitute only a small percentage of the area. They remain to become the last green removed before the land is all brown. The bits of nature protect isolated rare habitats and species, provide stepping stones or connectivity for movement of some species across the landscape, and inhibit wind and water erosion in the brown area.

The jaws sequence can be further improved ecologically in what might be called a **jaws-and-chunks** sequence or model. In the preceding jaws model the early phase is fine, and appears much like a whole green landscape (Figure 15.3b). The late phase is also ecologically appropriate, where one large patch retains as many interior species, large-home-range species, aquifer protection, etc. as possible. The middle phase, though, has two related shortcomings. First, as noted in section 15.2, a few large patches contain somewhat more species in the landscape than one large patch. Second, as described above for risk spreading, escape from disturbance is more likely if more than one patch is present. The middle phase of the jaws sequence manifests no risk spreading and no species-richness advantage offered by two or more large patches.

To incorporate the few-large-patches concept in the jaws sequence, we make the simplifying assumption that a green square covering approximately 10% of the landscape is both the area covered by most large disturbances and the minimum area needed to accomplish most ecological values of large patches. In this fashion we relate or scale the size of large patches to the size of the landscape. Large-patch size is also related to disturbance and ecological benefits.

A few large patches or 'chunks' of green in the middle phases can be incorporated (Figure 15.4). For example, three green squares each of 10% and one square of approximately 20% of the landscape area would accomplish this. The four large patches are separated by brown strips wide enough to inhibit the spread of most disturbances. As the jaws-and-chunks sequence proceeds, the chunks of green sequentially decrease in number rather than concurrently shrinking in size.

In effect, the model metaphor is of open jaws moving forward, gripping a huge chunk, followed by a few large chunks, followed by a single large chunk, and finally scattered bits of green just before the end. This hypothesis awaits empirical evaluation. Where an ecologically more suitable habitat type (green) is replaced by a less suitable type (brown), the jaws-and-chunks model appears to be an ecologically optimum mosaic sequence for land transformation. Land planning, conservation and management that recognize a changing landscape can use such a model to pinpoint the best, and the worst, locations for any proposed change.

Jaws and chunks model

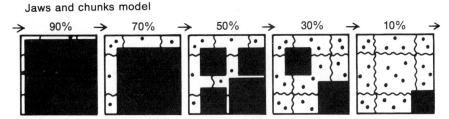

Figure 15.4 Jaws-and-chunks model of land transformation (see caption in Figure 15.3). The model begins all black and ends all white; note that the points illustrated along the sequence differ from those of Figure 15.3. (Drawing by Kimberly Hill.)

15.6 CONSERVATION PLANNING FOR LESS NATURAL VEGETATION

Planners, conservationists and managers often assume that the existing amount of natural vegetation will remain the same, or increase somewhat. Their expertise then focuses on how to arrange the existing ecosystems better and where increased protection should best take place.

This inherent optimism is in shocking contrast to the history of civilizations. Human population growth is typically exponential, a pattern replicated in nation after nation around the globe today. Technological development in humankind's ability to alter the land likewise grows exponentially. The same is evident for the extraction of non-renewable resources. Harvesting of renewable resources follows the same pattern. Agricultural intensification is widespread. The spread of development continues unabated, that is, the replacement of natural vegetation by varied land uses associated with population growth. All signs point to these processes and trends continuing for at least the coming decades.

Land planners, conservationists and managers will be relicts unless they can effectively deal with the basic fact that natural vegetation and ecological resources will significantly decrease during the next few decades.

Which are the 'first removals' – the least ecologically significant natural-vegetation areas prime for elimination? Which are the 'last stands' – the most valuable ecological areas to be protected longest? Last stands must be considered not in the present landscape context but in a nearly all-brown landscape. For most landscapes these are key questions, and few people are providing answers. The preceding sections permit us to address these questions in a preliminary manner, using the green-changing-to-brown metaphor.

Several first removals are evident. When a green matrix surrounds brown patches, major green lobes and corners can be removed with little loss.

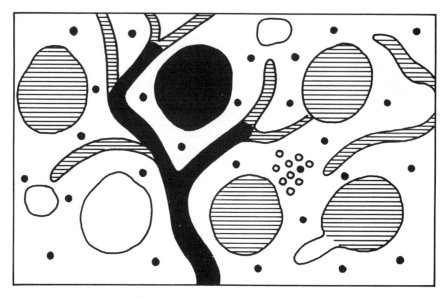

Figure 15.5 Landscape diagram indicating the optimal order of natural-vegetation loss. First removals are white (patches and corridors only); last stands are black. Shaded areas are removed during the in-between phase. (Drawing by Kimberly Hill.)

Large patches and small patches are much studied, and their differing ecological significance has been outlined in section 15.2.2. No special ecological importance is known for medium-sized patches, and hence these are prime candidates for shrinkage or elimination (Figure 15.5). Major lobes of large patches may be readily truncated. If there are more than a few large patches, the excess can be eliminated. Removing small patches does not gain much brown area, but in areas where they are dense, most can be readily obliterated.

The last stands are also reasonably evident for green patches. These include patterns described above with the 'indispensables' and in the 'jaws-and-chunks' model. Two, and then one, large green patches should remain (Figure 15.5). Bits of nature scattered over the brown matrix are last stands (tiny corridors are especially valuable in reducing erosion, while tiny patches may better protect rare isolated habitats or species). Vegetated corridors along major streams warrant long-term (last stand) protection.

This approach does not justify removing any natural vegetation. Rather, identifying first removals and last stands emphasizes planning and the setting of priorities. Two circumstances render inappropriate the loss of a site identified as a first removal. Firstly, the types of situation identified as first removals are generic, for landscapes in general. In contrast, any specific location so identified is a specific case. There are always special ecological attributes of a specific location that require evaluation to determine whether removal is appropriate.

Secondly, the appropriateness of eliminating a first removal depends on the percentage of natural vegetation present, i.e. where a landscape stands on the gradient of land transformation. If, say, 90% of the area is natural vegetation, the landscape may be most appropriately maintained as a natural landscape, and removing development rather than vegetation may be the priority. If, however, 60% is vegetated, the identification of first removals may be especially useful. If 30% is natural vegetation, almost any loss probably causes significant ecological damage. Identifying first removals is still useful but removal should only occur after serious societal consideration. If only 10% remains natural, probably only last stands exist.

How could a land planner, conservationist or manager use these guidelines? We suggest that, in addition to the normal plan for a landscape required by the client or user, five long- term plans be provided in skeletal form. These will be alternative scenarios assuming a 20% increase, no change, 20% loss, 50% loss and 80% loss in the area covered by natural ecosystems. Such plans will provide alternatives for informed decision-making and at least one plan for comparison with the actual landscape change. The plans will also highlight removal targets and last stands, as a graphic demonstration of long-term land values and societal priorities. Pinpointing last stands (Figure 15.5) long before they actually are last stands is an early warning system. Thus their planning, conservation and management can begin early and be accomplished intelligently.

Although removal of natural vegetation is expected in most landscapes, land planners, conservationists and managers are needed to show how to reverse this process in some landscapes. The 'aggregate-with-outliers' guideline described in section 15.2.2 for meshing land uses within a landscape emphasizes the ecological importance of aggregating land uses. This produces large natural-vegetation patches, agricultural areas and built areas (in addition to outliers). The same concept applies for whole landscapes within a region. Aggregating people in certain landscapes, natural vegetation in others and pastureland in others makes ecological sense. At present the natural-vegetation landscapes are frequently being degraded by other land uses; in contrast, increasing this natural vegetation is important. Therefore identifying and protecting the highest priority sites within natural-vegetation landscapes is just as important as pinpointing the target removals and last stands in other landscapes.

15.7 ECOLOGICAL PLANNING OF REGIONS, EVEN CONTINENTS

The preceding section implies that whole landscapes of natural vegetation are critical components of long-term planning in the face of human population growth. Spatially meshing these landscapes with others in a region, and indeed in a continent, is also important. Regional planning is often primarily focused on issues of transportation, infrastructure and residential development. Regional planning focused on ecological issues, or on a balance between ecological and human issues, is rare. Landscape and regional ecology is the foundation of such approaches (Forman, 1995).

This final section briefly considers ten examples or approaches where landscape-sized areas of natural vegetation are protected. The examples differ partly in their protection strategy or history and partly in the arrangement of nature on the land. The first eight apply to the landscape or regional scale, the ninth at the continental scale and the last at any scale.

(a)　Scattered spots

The objective is to protect the most important details (e.g. rare species or habitats) of a landscape, generally located in numerous, relatively small scattered spots. This has been the major thrust of the Nature Conservancy in the USA and is illustrated by the Sites of Special Scientific Interest in Britain.

A major ecological advantage of this strategy is to rescue species on the brink of extinction, thus providing significant short-term protection of biological diversity in a landscape. The scattered small patches provide the values of bits of nature across a matrix discussed above. Disadvantages include the exceedingly high edge-to-interior ratio and long perimeter length subject to disturbances from the matrix. Small patches generally with small populations are problematic for long-term protection against changing environmental conditions (Quinn and Karr, 1993).

(b)　Large pristine area

The protection of a large, relatively pristine patch has been a priority of conservationists. In some important cases this has been or become a whole landscape, or even a region, surrounded by other landscapes. Yellowstone National Park and its surroundings is a familiar example in the USA. Other examples are the several huge Costa Rican conservation areas, the Pine Barrens biosphere reserve (New Jersey, USA), the Nature Conservancy's Virginia Coast Reserve (USA) and Kakadu National Park (Australia).

The ecological advantages include an extensive interior area, numerous protected habitats in proximity, and other characteristics described above for large patches. Ecological disadvantages include a lack of risk spreading and somewhat reduced species richness compared with a few large patches. Some large patches are surrounded by a tightening noose of development, such as the White Mountain National Forest (USA). Some have a low density of mainly indigenous people and human impacts, as illustrated by Uluru National Park (Australia) and La Amistad Park (Costa Rica). In some cases the regional economy does not permit adequate protection of the large patch, so people colonize, fragment and degrade portions, as at the Tortuguero Conservation Area (Costa Rica).

(c) Large degraded area

Instead of protecting a large nearly pristine area, an alternative approach is to protect a large degraded area and then wait a few generations for ecological succession to repair the degradation and produce a valuable natural vegetation area. The key here is a large area. A familiar example is the 1625 km^2 (625 square miles) Okefenokee National Wildlife Refuge of Georgia, USA (Figure 15.6). During the period of about 1915–1929, this area was intensively and extensively logged. Consequently the cost of the land was very low and a large area was acquired. Today Okefenokee is the largest national wildlife refuge in the eastern United States and, after several decades of natural regrowth, is one of the biologically richest refuges in North America.

The ecological advantages of a large degraded area are the options available in a massive restoration project in addition to protecting a large patch. Disadvantages include dealing with residual effects of the preceding degradation process and, in early stages, the effects that the large degraded patch has on surrounding areas.

(d) Large fragmented area

Effectively an intermediate condition of land transformation between the two preceding cases, here prominent human-developed patches and corridors are enmeshed in a large patch or matrix of natural vegetation. Rather than being simply perforated or dissected, the natural vegetation is fragmented in pieces (see land transformation model in section 15.4). Examples include in the United States the Adirondacks of New York, Acadia National Park in Maine and the Pine Barrens landscape of New Jersey, and in Germany the Black Forest. Roads, villages, agriculture, tourist facilities and so on are typically within the large patch.

The ecological advantages of a large patch plus some risk spreading are present. However, human disturbances throughout the patch are especially prevalent. The edge-to-interior ratio is very high, and extinctions and recolonizations in a metapopulation may be high.

(e) Overlapping targets

An alternative approach for protection of a region is to address a number of targets, each requiring a large, yet somewhat different, area. For example, the proposal to restore the North Woods from Minnesota to Nova Scotia includes reintroduction of wolves (*Canis lupus*) to remote areas, designation of the Atlantic salmon (*Salmo salar*) as an endangered species for legal protection, establishment of a large Maine Woods National Park, and limiting the locations and levels of logging in national forests to maintain forever healthy water quality, rare animal populations and wood production (M.K. Kellett, personal communication). Meshing the areas required for these targets, plus detailed infilling, would provide protection of a whole region.

Figure 15.6 Nature conservation from seven decades of protection in a large patch at Okefenokee National Wildlife Refuge, Georgia, USA. (a) Improved water quality and extensive regrowth of cypress swamp (*Taxodium distichum*) and pine forest (*Pinus*) has occurred. (b) Alligator (*Alligator mississippiensis*) is one of the previously uncommon species that has re-established viable populations. (Photographs courtesy of US Fish and Wildlife Service.)

The ecological advantages include a few large patches, discussed in earlier sections. Also the very large total area enhances natural processes, including natural disturbance regimes in which species evolved and which are required by some species for persistence. Disadvantages depend on how much the widespread human disturbances associated with separated patches are eliminated.

(f) Dumbbells

Two large patches connected by a wide corridor is a logical design in some areas where animals migrate seasonally between the patches. Also the corridor provides a dispersal route that may reduce local extinction and enhance recolonization (Brown and Kodric-Brown, 1977; LaPolla and Barrett, 1993). An ecological disadvantage is that the corridor acts as a funnel that may channel animals to predators, hunters or disturbances. Probably examples of the dumbbell design exist.

(g) Corridors and nodes

Elaborating on the previous design, a series of patches can be connected by corridors. An example is a recent project to identify a biological corridor that links large patches of native forest throughout Central America (Carr *et al.*, 1994). The plan connects national parks and other protected areas in the region with the goals of maintaining regional biological diversity, reducing rates of soil erosion and sedimentation and improving water quality and quantity. The criteria for locating corridors include the incorporation of large natural areas already protected, areas of low human population density and forested rather than deforested areas.

(h) Network with nodes

A further development is a network of corridors that connect patches (Harris, 1984; Recher *et al.*, 1987; Saunders and Hobbs, 1991; Forman, 1991). For instance, in northern Florida the Suwannee River corridor and others in the area interconnect dozens of parks, nature reserves, refuges and forests of diverse ownership (Noss and Harris, 1986).

In addition to the advantages of the preceding design, the network provides optional routes for animals departing from a patch, as well as moving across the landscape. An ecological disadvantage is the relatively large proportion of edge habitat in the network.

(i) Continental network

This approach is a modification of the preceding to focus on large patches and wide corridors surrounded by buffers at the continental scale. An example is a Wildlands Project proposed for North America that focuses on protecting representatives of all native ecosystem types, as well as particular requirements of large-home-range mammals such as grizzly bear (*Ursus horribilis*), wolf (*Canis lupus*) and mountain lion (*Felis concolor*) (Noss, 1993b; Mann and Plummer, 1993). This is a set of core reserves large enough to maintain viable populations of all native species, connected by corridors and surrounded by buffer or multiple-use zones. The spatial pattern builds on the existing areas of natural vegetation but requires extensive relocation of people and human activities.

The major advantages are the significant protection given to the threatened large-home-range species and to biodiversity in general. The major ecological disadvantages are the omissions of, or lack of emphasis on, the numerous non-biodiversity components related to water, wind, soil and production. In addition, success in ecological protection probably depends at least as much on planning most of the continent that lies outside the network and buffers.

(j)   The spatial solution

This approach, described in the lead-off section, is based on spatial patterns that make ecological sense in any landscape or region, even though detailed inventories are unavailable or human imprints are widespread. The indispensable patterns, the aggregate-with-outliers arrangement and the strategic points highlight the keystones in an area and mesh different land uses to address flexibly a range of ecological objectives. This solution appears applicable at landscape, regional and continental scales.

The probability of successfully protecting nature and natural processes over human generations (i.e. in a sustainable environment), in view of expected population growth and activities, has not been evaluated for these ten alternatives. Yet, in addition to ecological advantages and disadvantages, the probability of success must be considered before choosing a strategy. For ecological, economic or political reasons, most of these options appear to have only a low to fair probability of successfully sustaining the environment. However, we believe that the emerging spatial solution offers room for serious optimism in protecting nature and natural processes, along with humans, in land mosaics worldwide.

15.8   CONCLUSION

A realistic view of land degradation around us, its monotonic increase and the woefully inadequate solutions available, help to pinpoint and clarify the spatial solution. Evaluations, critiques, alterations and additions will show whether a new-found optimism is well founded.

All of the preceding principles and concepts are spatially focused at landscape and regional scales. Major research frontiers are evident and important additional components will be added to the spatial solution. Nevertheless, science and humankind have crossed a threshold. After a decade of rapid-fire discoveries and associated controversies in landscape ecology, conservation biology, forestry, wildlife biology and related fields, the discoveries have coalesced to provide a remarkably clear picture. The spatial solution represents planning, conservation, design, management and policy of the future. More to the point, it is now on the table for solving today's crucial, vexing environmental and human issues of land use.

15.9 SUMMARY

The chapter outlines a 'spatial solution' that meshes biodiversity and natural processes with other major land uses in any changing landscape or region. Certain patterns of natural vegetation such as major stream corridors and a few large patches are 'indispensable', i.e. no known or feasible alternatives exist for their benefits, while others are especially effective in meshing land uses. Disturbance is a key creator of landscape pattern, and faunas respond to and depend on it. Habitat fragmentation, as one of five common spatial processes, is ecologically most important in the middle phase of land transformation. In changing land from a more to a less suitable type, an ecologically optimum sequence of mosaics is pinpointed. In a world of shrinking natural vegetation, several types of location are recommended for 'first removal' and several are identified to remain as 'last stands'. Since whole landscapes of natural vegetation are, and will be, especially significant, a range of approaches is presented for their planning and conservation. We conclude that science and humankind have crossed a threshold, whereby a spatial solution for conserving biodiversity in landscapes and regions has emerged. Working out its complete form is an exciting frontier, because this solution will rapidly become central in future land-use planning, conservation, design, management and policy.

15.10 REFERENCES

Arno, S.F. (1980) Forest fire history in the northern Rockies. *Journal of Forestry* 78:460–465.

Arnold, T.W. and Fritzell, E.K. (1990) Habitat use by male mink in relation to wetland characteristics and avian prey abundances. *Canadian Journal of Zoology* 68:2205–2208.

Arnold, G.W., Weeldenburg, J.R. and Steven, D.E. (1991) Distribution and abundance of two species of kangaroo in remnants of native vegetation in the central wheatbelt of Western Australia and the role of native vegetation along road verges and fencelines as linkages, in *Nature Conservation 2: The Role of Corridors*, (eds D.A. Saunders and R.J. Hobbs), pp. 273–280. Surrey Beatty & Sons, Chipping Norton, Australia.

Ayres, J.M. and Clutton-Brock, T.H. (1992) River boundaries and species range size in Amazonian primates. *American Naturalist* 140:531–537.

Baker, W.L. (1993) Spatially heterogeneous multi-scale response of landscapes to fire suppression. *Oikos* **66**:66–71.

Barton, D.R., Taylor, W.D. and Biette, R.M. (1985) Dimensions of riparian buffer strips required to maintain trout habitat in southern Ontario streams. *North American Journal of Fisheries Management* 5:364–378.

Beier, P. (1993) Determining minimum habitat areas and habitat corridors for cougars. *Conservation Biology* 7:94–108.

Best, L.B., Whitmore, R.C. and Booth, G.M. (1990) Use of cornfields by birds during the breeding season: the importance of edge habitat. *American Midland Naturalist* **123**:84–99.

Binford, M.W. and Buchenau, M. (1993) Riparian greenways and water resources, in *Ecology of Greenways: Design and Function of Linear Conservation Areas*, (eds D.S. Smith and P.C. Hellmund), pp. 69–104. University of Minnesota Press, Minneapolis, Minnesota.

Brocke, R.H., O'Pezio, J.P. and Gustafson, K.A. (1990) A forest managment scheme mitigating impact of road networks on sensitive wildlife species, in *Is Forest Fragmentation a Management Issue in the Northeast?*, pp. 13–17. General Technical Report NE-140, US Forest Service, Radnor, Pennsylvania.

Brown, J.H. and Kodric-Brown, A. (1977) Turnover rates in insular biogeography: effect of immigration on extinction. *Ecology* 58:445-449.

Buechner, M. (1989) Are small-scale landscape features important factors for field studies of small mammal dispersal sinks? *Landscape Ecology* 2:191–199.

Carr, M.H., Lambert, J.D. and Zwick, P.D. (1994) *Mapping of continuous biological corridor potential in Central America. Final Report for Paseo Pantera*. University of Florida, Gainesville, Florida.

Collinge, S.K. and Forman, R.T.T. (in preparation) *Consequences of four spatial configurations of landscape transformation for insects in native mixed-grass prairie*.

DeMers, M.N. (1993) Roadside ditches as corridors for range expansion of the western harvester ant (*Pogonomyrmex occidentalis* Cresson). *Landscape Ecology* 8:93–102.

Dunne, T. and Leopold, L.B. (1978) *Water in Environmental Planning*. W.H. Freeman, San Francisco.

Dunn, C.P., Sharpe, D.M., Guntenspergen, G.R. *et al.* (1991) Methods for analyzing temporal changes in landscape pattern, in *Quantitative Methods in Landscape Ecology*, (eds M.G. Turner and R.H. Gardner), pp. 173–198. Springer-Verlag, New York.

Ehrlich, P.R. (1983) Genetics and the extinction of butterfly populations, in *Genetics and Conservation: A Reference for Managing Wild Animals and Plant Populations*, (eds C.M. Schonewald-Cox, S.M. Chambers, B. MacBryde and L. Thomas), pp. 152–163. Benjamin/Cummings, Menlo Park, California.

Ehrlich, P.R. (1988) The loss of diversity, in *Biodiversity*, (ed. E.O. Wilson), pp. 21–27. National Academy Press, Washington, DC.

Ehrlich, P.R., White, R.R., Singer, M.C. *et al.* (1975) Checkerspot butterflies: a historical perspective. *Science* 188:221–228.

Foppen, R. and Reijnen, R. (1994) The effects of car traffic on breeding bird populations in woodland. II. Breeding dispersal of male willow warblers (*Phylloscopus trochilus*) in relation to the proximity of a highway. *Journal of Applied Ecology* 31:95–101.

Forman, R.T.T. (1987) The ethics of isolation, the spread of disturbance, and landscape ecology, in *Landscape Heterogeneity and Disturbance*, (ed. M.G. Turner), pp. 213–219. Springer-Verlag, New York.

Forman, R.T.T. (1991) Landscape corridors: from theoretical foundations to public policy, in *Nature Conservation 2: The Role of Corridors*, (eds D.A. Saunders and R.J. Hobbs), pp. 71–84. Surrey Beatty, Chipping Norton, Australia.

Forman, R.T.T. (1995) *Land Mosaics: The Ecology of Landscapes and Regions.* Cambridge University Press, Cambridge.

Forman, R.T.T. and Boerner, R.E.J. (1981) Fire frequency and the Pine Barrens of New Jersey. *Bulletin of the Torrey Botanical Club* 108:34–50.

Forman, R.T.T. and Godron, M. (1986) *Landscape Ecology.* John Wiley, New York.

Forman, R.T.T., Galli, A.E. and Leck, C.F. (1976) Forest size and avian diversity in New Jersey woodlots with some land use implications. *Oecologia* 26:1–8.

Foster, D.R. (1983) The history and pattern of fire in the boreal forest of southeastern Labrador. *Canadian Journal of Botany* 61:2459–2471.

Foster, D.R. (1988a) Disturbance history, community organization and vegetation dynamics of the old-growth Pisgah Forest, south-western New Hampshire, USA. *Journal of Ecology* 76:105–134.

Foster, D.R. (1988b) Species and stand response to catastrophic wind in central New England, USA. *Journal of Ecology* 76:135–151.

Foster, D.R. and Boose, E. (1992) Patterns of forest damage resulting from catastrophic wind in central New England, USA. *Journal of Ecology* 80:79–99.

Franklin, J.F. (1993) Preserving biodiversity: species, ecosystems, or landscapes? *Ecological Applications* 3:202–205.

Franklin, J.F. and Forman, R.T.T. (1987) Creating landscape patterns by forest cutting: ecological consequences and principles. *Landscape Ecology* 1:5–18.

Game, M. and Peterken, G.F. (1984) Nature reserve selection strategies in the woodlands of Central Lincolnshire, England. *Biological Conservation* 29:157–181.

Garshelis, D.L. and Pelton, M.R. (1981) Movements of black bears in the Great Smoky Mountains National Park. *Journal of Wildlife Management* 45:912–925.

Getz, L.L., Cole, F.R. and Gates, D.L. (1978) Interstate roadsides as dispersal routes for *Microtus pennsylvanicus*. *Journal of Mammalogy* 59:208–212.

Gustafson, E.J., Parker, G.R. and Backs, S.E. (1994) Evaluating spatial pattern of wildlife habitat: A case study of the Wild Turkey (*Meleagris gallopavo*). *American Midland Naturalist* 131:24–33.

Hansen, A.J., Urban, D.L. and Marks, B. (1992) Avian community dynamics: the interplay of landscape trajectories and species life histories, in *Landscape Boundaries: Consequences for Biotic Diversity and Ecological Flows*, (eds A.J. Hansen and F. di Castri), pp. 170–195. Springer-Verlag, New York.

Harris, L.D. (1984) *The Fragmented Forest: Island Biogeography Theory and the Preservation of Biotic Diversity.* University of Chicago Press, Chicago.

Harris, L.D. and Gallagher, P.B. (1989) New initiatives for wildlife conservation: the need for movement corridors, in *Preserving Communities and Corridors*, (ed. G. Mackintosh), pp. 11–34. Defenders of Wildlife, Washington, DC.

Henderson, M.T., Merriam, G. and Wegner, J. (1985) Patchy environments and species survival: chipmunks in an agricultural mosaic. *Biological Conservation* 31:95–105.

566   The spatial solution

Hole, F.D. and Campbell, J.B. (1985) *Soil Landscape Analysis*. Routledge & Kegan Paul, London.

Hunter, M.L., Jr (1993) Natural fire regimes as spatial models for managing boreal forests. *Biological Conservation* 65:115–120.

Johnson, W.C. and Adkisson, C.S. (1985) Dispersal of beech nuts by Blue Jays in fragmented landscapes. *American Midland Naturalist* 113:319–323.

Karr, J.R. (1991) Biological integrity: a long-neglected aspect of water resource management. *Ecological Applications* 1:66–84.

Knaapen, J.P., Scheffer, M. and Harms, B. (1992) Estimating habitat isolation in landscape planning. *Landscape and Urban Planning* 23:1–16.

Kozakiewicz, M., Kozakiewicz, A., Lukowski, A. and Gortat, T. (1993) Use of space by bank voles (*Clethrionomys glareolus*) in a Polish farm landscape. *Landscape Ecology* 8:19–24.

Kremsater, L.L. and Bunnell, F.L. (1992) Testing responses to forest edges: the example of black-tailed deer. *Canadian Journal of Zoology* 70:2426–2435.

Kruess, A. and Tscharntke, T. (1994) Habitat fragmentation, species loss, and biological control. *Science* 264:1581–1584.

Landers, J.L., Hamilton, R.J., Johnson, A.S. and Marchinton, R.L. (1979) Foods and habitat of black bears in southeastern North Carolina. *Journal of Wildlife Management* 43:143–153.

La Polla, V.N. and Barrett, G.W. (1993) Effects of corridor width and presence on the population dynamics of the meadow vole (*Microtus pennsylvanicus*). *Landscape Ecology* 8:25–37.

Li, H., Franklin, J.F., Swanson, F.J. and Spies, T.A. (1993) Developing alternative forest cutting patterns: a simulation approach. *Landscape Ecology* 8:63–75.

Likens, G.E. (ed.) (1985) *An Ecosystem Approach to Aquatic Ecology*. Springer-Verlag, New York.

Liu, J., Cubbage, F.W. and Pulliam, H.R. (1994) Ecological and economic effects of forest landscape structure and rotation length: simulation studies using ECOLECON. *Ecological Economics* 10:249–263.

Lovejoy, T.E., Rankin, J.M., Bierregaard, R.O., Jr *et al.* (1984) Ecosystem decay of Amazon forest remnants, in *Extinctions*, (ed. M.H. Niteki), pp. 295–325. University of Chicago Press, Chicago.

Malanson, G.P. (1993) *Riparian Landscapes*. Cambridge University Press, Cambridge.

Mann, C.C. and Plummer, M.L. (1993) The high cost of biodiversity. *Science* 260:1868–1871.

McLellan, B.N. and Shackleton, D.M. (1988) Grizzly bears and resource-extraction industries: effects of roads on behaviour, habitat use and demography. *Journal of Applied Ecology* 25:451–460.

Mech, L.D., Fritts, S.H., Raddle, G.L. and Paul, W.J. (1988) Wolf distribution and road density in Minnesota. *Wildlife Society Bulletin* 16:85–87.

Meffe, G.K. and Carroll, C.R. and contributors (1994) Principles of Conservation Biology. Sinauer Associates, Sunderland, Massachusetts.

Merriam, G. and Lanoue, A. (1990) Corridor use by small mammals: field measurement for three experimental types of *Peromyscus leucopus*. *Landscape Ecology* 4:123–131.

Noss, R. (1993a) Wildlife corridors, in *Ecology of Greenways: Design and Function* of *Linear Conservation Areas*, (eds D.S. Smith and P.C. Hellmund), pp. 43–68. University of Minnesota Press, Minneapolis, Minnesota.

Noss, R. (1993b) A conservation plan for the Oregon Coast Range: some preliminary suggestions. *Natural Areas Journal* 13:276–290.

Noss, R.F. and Harris, L.D. (1986) Nodes, networks and MUMS: preserving diversity at all scales. *Environmental Management* 10:299–309.

Ohsawa, M. and Liang-Jun, D. (1987) Urbanization and landscape dynamics in a watershed of the Miyako River, Chiba, Japan, in *Integrated Studies in Urban Ecosystems as the Basis of Urban Planning (II)*, (ed. H. Obara), pp. 187–197. Special Research Project on Environmental Science B334-Rl5-3, Ministry of Education, Culture and Science, Tokyo, Japan.

Pickett, S.T.A. and Thompson, J.N. (1978) Patch dynamics and the design of nature reserves. *Biological Conservation* 13:27–37.

Pickett, S.T.A. and White, P.S. (eds) (1985) *The Ecology of Natural Disturbance and Patch Dynamics*. Academic Press, New York.

Quinn, J.F. and Karr, J.R. (1993) Habitat fragmentation and global change, in *Biotic Interactions and Global Change*, (eds P.M. Karieva, J.G. Kingsolver and R.B. Huey), pp. 451–463. Sinauer Associates, Sunderland, Massachusetts.

Rackham, O. (1986) *The History of the Countryside*. J.M. Dent, London.

Raven, P. (1988) Our diminishing tropical forests, in *Biodiversity*, (ed. E.O. Wilson), pp. 119–122. National Academy Press, Washington, DC.

Recher, H.F., Shields, J., Kavanagh, R. and Webb, G. (1987) Retaining remnant mature forest for nature conservation at Eden, New South Wales: a review of theory and practice, in *Nature Conservation: The Role of Remnants of Native Vegetation*, (eds D.A. Saunders, G.W. Arnold, A.A. Burbidge and A.J.M. Hopkins), pp. 177–194. Surrey Beatty, Chipping Norton, Australia.

Romme, W. H. (1982) Fire and landscape diversity in subalpine forests of Yellowstone National Park. *Ecological Monographs* 52:199–221.

Romme, W.H. and Knight, D.H. (1981) Fire frequency and subalpine forest succession along a topographic gradient in Wyoming. *Ecology* 62:319–326.

Rost, G.R. and Bailey, J.A. (1979) Distribution of mule deer and elk in relation to roads. *Journal of Wildlife Management* 43:634–641.

Saunders, D.A. and Hobbs, R.J. (eds) (1991) *Nature Conservation 2: The Role of Corridors*. Surrey Beatty, Chipping Norton, Australia.

Saunders, D.A. and Ingram, J.A. (1987) Factors affecting survival of breeding populations of Carnaby's cockatoo *Calyptorhynchus funereus latirostris* in remnants of native vegetation, in *Nature Conservation: The Role of Remnants of Native Vegetation*, (eds D.A. Saunders, G.W. Arnold, A.A. Burbridge and A.J.M. Hopkins), pp. 249–258. Surrey Beatty, Chipping Norton, Australia.

Saunders, D.A., Arnold, G.W., Burbidge, A.A. and Hopkins, A.J.M. (eds) (1987) *Nature Conservation: The Role of Remnants of Native Vegetation*. Surrey Beatty, Chipping Norton, Australia.

Sharpe, D.M., Guntenspergen, G.R., Dunn, C.P. *et al.* (1987) Vegetation dynamics in a southern Wisconsin agricultural landscape, in *Landscape Heterogeneity and Disturbance*, (ed. M.G. Turner), pp. 137–155. Springer-Verlag, New York.

Soule, M.E. (ed) (1987) *Viable Populations for Conservation*. Cambridge University Press, Cambridge.

Suffling, R., Lihou, C. and Morand, Y. (1988) Control of landscape diversity by catastrophic disturbance: a theory and a case study of fire in a Canadian boreal forest. *Environmental Management* 12:73–78.

Turner, M.G. (ed.) (1987) *Landscape Heterogeneity and Disturbance*. Springer-Verlag, New York.

Turner, M.G. and Bratton, S.P. (1987) Fire, grazing and the landscape heterogeneity of a Georgia barrier island, in *Landscape Heterogeneity and Disturbance*, (ed. M.G. Turner), pp. 85–101. Springer-Verlag, New York.

Turner, M.G. and Gardner, R.H. (eds) (1991) *Quantitative Methods in Landscape Ecology: The Analysis and interpretation of Landscape Heterogeneity*. Springer-Verlag, New York.

Turner, M.G., Wu, Y., Romme, W.H. and Wallace, L.L. (1993) A landscape simulation model of winter foraging by large ungulates. *Ecological Modelling* 69:163–184.

Wauters, L., Casale, P. and Dhondt, A.A. (1994) Space use and dispersal of red squirrels in fragmented habitats. *Oikos* 69:140–146.

Wegner, J.F. and Merriam, G. (1979) Movements by birds and small mammals between a wood and adjoining farmland habitats. *Journal of Applied Ecology* 16:349–358.

Wiens, J.A. and Milne, B.T. (1989) Scaling of 'landscapes' in landscape ecology, or, landscape ecology from a beetle's perspective. *Landscape Ecology* 3:87–96.

Wilcove, D.S., McLellan, C.H. and Dobson, A.P. (1986) Habitat fragmentation in the temperate zone, in *Conservation Biology*, (ed. M.E. Soule), pp. 879–887. Sinauer Associates, Sunderland, Massachusetts, USA.

Wilcox, B.A. and Murphy, D.D. (1985) Conservation strategy: the effects of fragmentation on extinction. *American Naturalist* 125:879–887.

Wilson, E.O. (ed.) (1988) *Biodiversity*. National Academy Press, Washington, DC.

Wilson, E.O. (1992) *The Diversity of Life*. Harvard University Press, Cambridge, Massachusetts.

With, K.A. (1994) Using fractal analysis to assess how species perceive landscape structure. *Landscape Ecology* 9:25–36.

Worldwatch Institute (1992) *State of the World*. W.W. Norton, New York.

16

Ecosearch: a new paradigm for evaluating the utility of wildlife habitat

Henry L. Short, Jay B. Hestbeck and Ralph W. Tiner

16.1 INFORMATION NEEDS OF NATURAL RESOURCE MANAGERS

Resource managers must respond to legislative and societal demands to manage wildlife habitats in a cost-effective manner. To be successful, managers need a methodology:

1. to predict the composition or the biodiversity of the wildlife community that can occur on a unit of landscape at a given time;
2. to plan management strategies to enhance the future status of the wildlife community or of selected species of special interest;
3. to evaluate the impacts that proposed developments might have on the wildlife community;
4. to provide a means to predict trends in the status of the wildlife community over time.

This chapter describes such a methodology, Ecosearch, which is based on the assumption that many wildlife species are closely associated with certain habitat attributes that can be identified on aerial photographs. Using this methodology, we produce maps featuring the location of those habitat attributes and maps predicting areas that are apparently suitable as habitat for individual wildlife species. Our methodology is driven by basic natural history information that associates species with specific habitat attributes, uses photointerpretations of recent aerial photography to assess and physically locate those habitat attributes, and uses Geographic Information System (GIS) technologies to manipulate and integrate data sets to predict species presence and richness across a landscape. Our methodology operates

Conservation of Faunal Diversity in Forested Landscapes. Edited by
R.M. DeGraaf and R.I. Miller. Published in 1996 by Chapman & Hall. ISBN 0 412 61890 7.

under the paradigm that habitat for a given species can be estimated from a hypothesized species–habitat model and from point intercept and other data organized in geographic-information matrices.

16.2 BUILDING MAPS TO REPRESENT WILDLIFE HABITATS

Assessments of wildlife habitat can occur at different levels of resolution. Constraints of time and money have favored attempts to evaluate habitats using some form of remotely sensed data. Such efforts strive to attain predictability with affordability. These efforts build on the ecological literature but pragmatically accept the limitations of scale in the predictions they seek to achieve. For example, ecological studies may indicate that the diversity of breeding bird species in deciduous forests depends on foliage profiles (MacArthur *et al.*, 1962) and not upon plant species composition. While it may not be practical to measure foliage profiles in remotely sensed data, it is possible to consider structural features, such as components of habitat layers (a surrogate for a foliage profile within a tree canopy), and to de-emphasize cover types (as a surrogate for plant species composition) in the development of predictive methodologies. Likewise, many forest wildlife species are fossorial. While remotely sensed data often cannot predict the friability of soils (the exceptions include visibly rocky terrains, agricultural fields and sandy beaches), a surrogate measure of the suitability of the sub-surface for fossorial species can be based on the estimate of soil water saturation as judged by photographic signatures. It is assumed that vegetated areas with signatures characteristic of water-saturated soils may be estimated as suitable habitat for crayfish (e.g. *Cambarus carolinus*) but not for woodchucks (*Marmota monax*). The landscape geometry of habitats, which is important in animal habitation patterns, can be evaluated especially well with the interpretation of aerial photography. Patches, corridors, boundaries, habitat juxtaposition, edge, fragmentation, etc. (Chapter 15) can be statistically assessed if structural features discernible on aerial photographs are interpreted as point intercept data.

Thus the problem is to express wildlife–habitat associations in a format that can be remotely perceived, and that will allow reliable predictors about wildlife habitats to be made for resource managers.

The Ecosearch methodology provides a way to predict the utility of habitats depicted in maps prepared from aerial photography. Maps prepared from thematic imagery or aerial photography provide a framework for bounding, associating and locating features, but often are not very useful predictors of the utility of landscapes to wildlife species. These maps generally fail as wildlife predictors because the maps are based on vegetation classification systems that were developed for reasons other than the assessment of wildlife habitats. Classification systems based on concepts of floristics and dominance, general appearance and dominant life form, potential vegetation related to site constraints, etc. (Wenger, 1984) are generally weak predictors of habitat quality for vertebrate wildlife species, because many wildlife species are not associated as closely with vegetative cover types as they are with components of habitat structure. Maps need to represent the presence or absence of necessary components of structure if they are to have much predictive significance in describing the utility of an area as habitat to wildlife species. The challenge is to develop a mapping convention that describes habitat structure in a way that is predictive of the way many wildlife species utilize habitats.

Many studies have associated wildlife species such as birds (Karr, 1971; Rabenold, 1978; Geibert, 1979), mammals (Maser *et al.*, 1981) and herpetofauna (Heatwole, 1982) with the vertical structure of habitats. Such studies often suggest that wildlife species partition the vertical structure of vegetation as if that vegetation comprised a series of habitat layers, although the studies do not agree on the numbers of habitat layers that may exist in tropical and temperate habitats nor on the physical dimensions of such habitat layers. Several authors (e.g. Short, 1983, 1989; Thomas, 1979; DeGraaf *et al.*, 1992) have sought to associate wildlife species with habitat layers or other structural components of habitats.

One such effort clumps wildlife species into species–habitat groups that utilize the same habitat structure for feeding and breeding activities. Species–habitat groups are established by organizing information within a species–habitat matrix (like that in Figure 16.1) and then sorting the species into groups having similar habitat layer dependencies (Table 16.1).

All species in a group are expected to be similarly impacted by changes that might occur to the required habitat layers. Species within a species–habitat group can be differentiated using attributes like nest selection strategies based on vegetation cover types or soil textures that occur within a single layer of habitat (Short, 1989), or feeding strategies. MacArthur (1958), for example, determined that five species of warbler, which are sometimes found together in the breeding season in relatively mature boreal forests, could be distinguished because they fed in different positions, indulged in hawking and hovering to different extents, moved in different directions through the trees, and may have had differing foraging needs at different times corresponding to different nesting dates. Root (1967) described a guild as a group of species that exploit a resource in a similar fashion, and determined that each species in his study guild had a unique foraging strategy that reduced competition and allowed coexistence.

The concept of habitat layers has been an important one in bird-habitat studies for over 30 years. MacArthur *et al.* (1962) determined that a fairly accurate census of breeding birds could be predicted from measures of the amounts of foliage in three horizontal layers. Short (1988) reported the development of a Habitat Layers Index (HLI) model, which had values scaled between 0 and 1.0 and which was intended to describe the relative

Habitat layer where feeding occurs

		A	WTC	WSC	WU	WS	WC	BWC	UTC	USC	UU	SS
	WTC											
	WSC											
	WU											
	WS											
	WC											
	BWC											
	UTC											
	USC											
	UU											
	SS											
	BE											

(left axis label: Habitat layer where breeding occurs)

Figure 16.1 Format for organizing information about wildlife species. A = air; WTC = wetland tree canopy; WSC = wetland shrub canopy; WU = wetland understory; WS = water surface; WC = water column; BWC = bottom of water column; UTC = upland tree canopy; USC = upland shrub canopy; UU = upland understory; SS = subsurface; BE = breeds elsewhere.

complexity of the habitat structure in a study area. The Index, which compared the product of the number of habitat layers and the area covered by each layer found present on an area to the maximum potential layer × area product for the area, sought to provide a general estimate of species richness for a study area. A field validation of the HLI in southern Colorado found a significant correlation with bird species richness (r = 0.94) (Short, 1992).

Short (1988) described habitat layers not only as a predictor of bird species richness but also as a common denominator for expressing the interrelationships between a broad group of land-use activities such as agricultural, grazing, mineral, recreational and scenic, timber, wildlife resources, etc. He hypothesized that one approach to a multiple-use management problem could be to determine the mix of habitat layers that will maximize product values over a diverse group of desired activities (Short, 1985). Managers could then concentrate on how to produce combinations of habitat layers after an optimization analysis provided an estimate of the quantity of each habitat layer that needed to be produced to satisfy their management goal.

Efforts to map habitats in terms of structure thus seem justified because many wildlife species seem to be associated with structural components of

Table 16.1 Habitat layers used for feeding and breeding by some groups of primary consumers that occur in cottonwood–willow riparian habitats in west-central Arizona (after Short, 1983)

Group No.	Group members	Feeding												Breeding									
		A	UTC	USC	UU	SS	WTC	WSC	WU	WS	WC	BWC	BE	UTC	USC	UU	SS	WTC	WSC	WU	WS	WC	BWC
1	American coot (Fulica americana)				×				×	×	×	×									×		
2	Mallard (Anas platyrhynchos)				×				×	×	×	×								×			
3	Canada goose (Branta canadensis)				×				×	×			×			×							
	Gadwall (Anas strepera)									×	×	×	×										
4	Longfin dace (Agosia chrysogaster)									×	×	×									×	×	×
	Red shiner (Notropis lutrensis)									×	×	×									×	×	×
5	Gila sucker (Catostomus insignis)									×	×	×									×		×
6	Carp (Cyprinus carpio)										×	×										×	×
	Speckled dace (Rhinichthys osculus)										×	×										×	×

Table 16.1 Continued

Group No.	Group members	Feeding											Breeding										
		A	UTC	USC	UU	SS	WTC	WSC	WU	WS	WS	BWC	BE	UTC	USC	UU	SS	WTC	WSC	WU	WS	WC	BWC
7	Black bullhead (*Ictalurus melas*)											×											×
	Yellow bullhead (*Ictalurus natalis*)											×											×
	Gila mountain-sucker (*Pantosteus clarki*)											×											×
8	Roundtail chub (*Gila robusta*)									×											×	×	
9	Green sunfish (*Chaenobryttus cyanellus*)										×									×	×	
10	Soft-shelled turtle (*Trionyx spiniferus*)				×						×					×							
11	Porcupine (*Erethizon dorsatum*)		×	×	×		×	×	×							×	×						
12	Beaver (*Castor canadensis*)				×			×	×								×	×					

Codes for columns are identified in Figure 16.1.

habitat. What is to be mapped needs to be both evident on photography and relevant to the way that wildlife species use habitats. One study mapped a portion of the Big Stone National Wildlife Refuge in Minnesota in terms of areas where different layers of habitat occurred (Short and Williamson, 1986). Color infra-red aerial photography was interpreted so that combinations of habitat layers could be represented in a cellular format. Wildlife species–habitat groups were developed to indicate groups of species that required specific combinations of habitat layers for usual breeding and feeding activities. Areas within the refuge that provided specific structural configurations were then presented as potentially providing habitat for individuals within a species–habitat group requiring those habitat layer combinations.

Our present effort describes a map convention that incorporates attributes of habitat structure, soils, topography and water regimes to estimate habitat conditions, and a modeling convention that estimates species presence based on estimated habitat conditions and specific habitat requirements of vertebrate wildlife species. The modeling convention and the map convention use the same variables, allowing predictions of areas of potential habitat to be made for particular vertebrate wildlife species for specific land units. The number of species estimated to be present on a land unit can be considered a rough estimate of biodiversity.

16.3 ECOSEARCH: A WAY TO DESCRIBE AND MAP WILDLIFE HABITATS

Ecosearch produces a description of available habitat using data about the horizontal and vertical patterns of vegetative cover, the location of wetlands and deepwater habitats and the location of soil attributes and topographic features on a study area. Patterns of vegetative cover are estimated from a point intercept assessment of the vegetative canopy (whether overstory, midstory or understory) or surface structure determined from systematic sampling. Inferences from this estimation include a representation of the vertical structure of the vegetation community throughout wetland and upland areas, and descriptions of horizontal patterns within vertical habitat layers. The location of wetlands and deepwater habitats are determined from National Wetland Inventory (NWI) maps or from aerial photographs interpreted according to NWI procedures (National Wetlands Inventory, 1990). The location of major soil attributes is based on published US Soil Conservation Service maps and the location of topographic descriptors is based on published US Geological Survey maps.

The description of the habitat requirements of individual species is based on the apparent dependency of the species on particular wetland regimes, soil conditions, topographic conditions and/or some attribute of habitat structure like canopy height, layering and closure of canopies, tree diameter, understory height and volume, ground cover and horizontal patchiness (Thomas, 1979; Verner and Boss, 1980; Cody, 1985; DeGraaf *et al.*, 1992). Vertebrate non-fish wildlife species are modeled using the same variables or surrogates that are used to characterize the habitat. Areas of habitat that seem to satisfy the habitat requirements of species are then identified, and the number of species within the community that can utilize a common habitat locale provides estimate of species richness for that locale.

An example of the Ecosearch assessment is provided below to describe the procedure, to illustrate some of the products and to describe the apparent utility of the products and procedure.

16.3.1   Ecosearch: an example

The example cited here describes the Ecosearch methodology used to provide a wildlife habitat assessment for a 36 km^2 portion of Hampshire County, Massachusetts, within the towns of Amherst, Belchertown, Granby and Pelham. The central feature of this land area is Lawrence Swamp, with portions of the Holyoke Range, the Pelham Hills and the agricultural lands around South Amherst also prominent and providing ecological diversity. The study area encompasses two US Geological Survey topographic quadrangles: Belchertown and Mount Holyoke.

16.3.2   Characterization of wetland habitats

The production of a wetland data base is essential to Ecosearch because many wildlife species have wetland dependencies, because some wetland designations indicate the probable location of water-saturated soils which limit the distribution of many fossorial species, and because the structure of wetland and upland understories may be very different and represent habitats to different fauna.

New data on wetland and deepwater habitats for the western Massachusetts study area were acquired in lieu of using existing National Wetlands Inventory (NWI) maps, due to the date of the source photographs for the existing maps (*c.* 1977). Current mapping techniques used by the NWI program provide more accurate and detailed information than do the original NWI maps. A new photointerpretation was performed so that the more detailed information would be available for this Ecosearch assessment. Data on the structure of habitats and on the location of wetland and deepwater habitats were obtained from the photointerpretation of 1:40 000 scale color infra-red aerial photographs of the study area, acquired from the National Aerial Photography Program (NAPP). Photointerpretation was conducted using a 6-power mirror stereoscope. Wetland and deepwater habitats were interpreted and classified according to the US Fish and Wildlife Service's official wetland classification system (Cowardin *et al.*,

1979), following standard NWI mapping conventions (National Wetlands Inventory, 1990). Wetland delineations produced on a clear mylar overlay were converted to digital form using equipment compatible with the NWI operational version of the Wetland Analytical Mapping System software (National Wetlands Inventory, 1992). The digital data were later formated in a cellular form using GRID programs in ARC/INFO (ESRI, 1992).

The operational significance of using NWI maps in Ecosearch is that a massive, significant, high-resolution data base, already developed for about 75% of the land area in the conterminous 48 States and already digitized for about 35% of the same land area, is available for Ecosearch assessments. Ecosearch thus adds a significant assessment capability to this important natural resource data base.

16.3.3 Characterization of habitat structure

The characterization of habitat structure using point intercept data provides a way to locate important vertical structures and to describe the horizontal patterns of those structures.

The point intercept appraisal was accomplished with intercept points describing canopy structure, located at 50 m intervals across the study area, on a 120 row × 120 column sampling grid. The sampling grid was imprinted on a 36 sq in clear mylar overlay which was pin registered to the aerial photographs, allowing repeatable subsampling. Each of the 14 400 intercept points in the Massachusetts study was classified according to a structural classification system developed for this study. As indicated below, an intercept point might be classified as an upland tree but its interpretation for modeling was as an upland point with overstory, midstory and understory layers present. The classified point data was recorded in tabular form in Word Perfect 5.2 and later transformed into the GRID programs of ARC/INFO (ESRI, 1992) with 50 m × 50 m cells.

The point intercept assessment is a feasible, technique. Quality control of the point intercept data was achieved by checking row counts and randomly resampling selected rows. Fewer than 25 hours were required to classify the 14 400 intercept points because the average time to classify an individual point required only 3–4 seconds. Certain potential sources of error occurred: the relatively large intercept point occasionally intercepted two cover types; points that fell on shadowed areas were sometimes difficult to interpret; points identified as a tree crown within a wooded landscape could have occasionally fallen between tree crowns; palustrine aquatic bed wetlands generally do not appear on the aerial photos because they were photographed before leaf-out occurred, so this category may be under-represented. The palustrine unconsolidated bottom and deepwater habitat data into which the aquatic bed points were clumped may therefore be over-represented in the data. Difficulties were also encountered in distinguishing pastures from tilled row crops. A small number of intercept points fell on cover types that are not recognized as wildlife habitat and which are not represented in any of the species–habitat models developed for this study.

In addition, systematic sampling may provide a biased estimate of certain cover types under certain circumstances. For example, if a row or column of our sample intercepts fell along a road, that road signature could be over-represented in our samples for a neighborhood. Likewise, if our intercept samples ran parallel to the road, the road signature could be under-represented in the neighborhood.

The great descriptive detail in the relevant US Soil Conservation Service maps was reconfigured to emphasize soil particle size and drainage classes. The soils data and the topographic gradients from the US Geological Survey maps were also digitized and represented the third and fourth data layers in the habitat descriptions.

16.3.4   Species–habitat models

Species–habitat models transform natural history information into a template with dimensions described by wetland modifiers, soil and topographic conditions, vertical structure and horizontal patterns. Specific required micro-habitats that would not be recognizable in aerial photographs are located within macro-habitats that are recognizable in aerial photographs, so those landscapes can be searched to estimate the location of habitats suitable for the species. This section describes the logic used in describing the 'habitat templates' for species.

Thirty-nine species of amphibians and reptiles, 47 mammals and 130 species of birds were identified as possibly breeding on the Massachusetts study area. Species–habitat models were developed for each of the 216 species, and a small subset of these species was arbitrarily selected to illustrate the Ecosearch methodology. The species and their habitat criteria are listed in Table 16.2 and are described below. DeGraaf and Rudis (1992) and DeGraaf (personal communication) were extensively used as a natural history data base for developing the species–habitat models.

The logic used in creating the species–habitat models is listed in Figure 16.2. It is assumed that habitat, to be useful to a particular species, must provide those structural layers used by the species. Thus, an important first step is to identify, for each species, the habitat layers where the species' usual feeding and breeding activities occur. This is also the information used to develop species–habitat groups (Table 16.1). The model then strives to define the habitat template for the species by determining if the species has a dependency on the presence of an aquatic habitat layer and more specifically a particular water regime to complete its breeding cycle or to obtain essential foodstuffs. The model recognizes that many aquatic depen-

Table 16.2 Habitat information coded into Ecosearch models used to estimate habitat distribution areas for three species in the Western Massachusetts study area

Habitat criteria	Spotted salamander reproduction habitats	Meadow vole	Wood thrush
Habitat area (ha)	0.4	0.4	2
No. of neighborhood cells	9	9	9
Topography	valley floor	valley floor midslope	valley floor midslope
Soil types	poorly drained very poorly drained	excessively drained well drained–fine poorly drained	no specific requirement
Water regime	seasonally flooded semi-permanently flooded	no specific requirement	no specific requirement
Vertical structure	surface water in swamps, marshes, bogs, ponds and vernal ponds	herbaceous understory in: fields, pastures, fresh and salt water marshes, stream and lake borders, and open and wooded swamps	overstory and midstory in deciduous or mixed upland and wetland forests with abundant sapling growth bordering streams, lakes and swamps
Proportion of neighborhood cells that need provide the stated habitat condition	30%	30%	70%
	Non-reproduction habitat		
Habitat area (ha)	2		
No. of neighborhood cells	9		
Topography	valley floor midslope		
Soil types	well drained poorly drained		
Water regime	no specific requirement		
Vertical structure	A mesic leaf littered surface under a deciduous tree overstory and/or shrub midstory		
Other considerations	This surface structure needs to occur within 200 m of the reproduction wetland and to be contiguous to that wetland		
Proportion of neighborhood cells that need provide the stated habitat condition	30%		

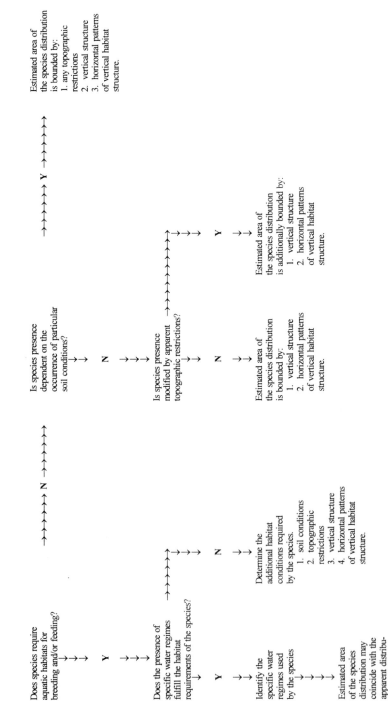

Figure 16.2 Logic used in Ecosearch.

dent species also utilize non-aquatic habitats for feeding and/or breeding activities at some time during their life cycle, and it considers the soil conditions, topographic restrictions and vertical habitat structures that describe those habitat conditions. Soils, topography and vertical structure are also emphasized in the models of species that do not have a water regime dependency. The specific habitat criteria used by wildlife species (moist deciduous leaf litter, tree cavities, peripheral foliage in the tree canopy, burrows in friable soils, etc.) are identified, and the vertical structures likely to provide these habitat characteristics are identified. After water regime, soil conditions, topographic restrictions and vertical structure requirements have been identified, the model strives to further define the habitat template by incorporating geometric features like block size, distance to water, degree of canopy closure, length and width of corridors, etc. Information within the four data layers for each cell on the study area provides the data to assess where the conditions described by the species–habitat models and represented in the habitat template might be satisfied on a study area.

16.3.5 Performing the Ecosearch assessment

A popular word game challenges participants with a series of words and a matrix with a jumble of letters. The goal is to find the target words which are arrayed within the matrix in a variety of patterns. Ecosearch, in some ways, is similar to this word game. The matrix represents data layers that describe an area in a cellular format. One layer describes the location of aquatic resources; a second layer describes soil conditions; a third layer describes topography and a fourth describes vertical habitat structures. The hypothesized species–habitat models represent target words insofar as they describe individualized arrangements of the same characters that are found listed in the data matrices. Ecosearch strives to predict where the 'species–habitat words' represented in habitat templates occur in the landscape matrices.

Ecosearch uses a user-friendly FORTRAN program to conduct 'word' searches through geographic-information matrices. The program first queries the user for information pertaining to the habitat template developed from the hypothesized species–habitat model. This information includes the following:

1. The habitat area required by the species.
2. The presumed size of the habitat block represented by the number of cells to be considered in the assessment. Although the analysis predicts the suitability of an individual cell as habitat for the species, a 'neighborhood' of cells is actually used to describe habitat patterns around the individual cell. The sampling neighborhood (a group like 3×3, 5×5, 7×7, 9×9, etc.) is defined by the habitat template.

3. Topographic restrictions.
4. Required soil conditions.
5. Water regime requirements.
6. Vertical structure types apparently required by the species.
7. The proportion of cells in the neighborhood that need to provide the required habitat conditions.

The appropriateness of each cell in the point intercept sampling grid is estimated, based on the hypothesized requirements in the habitat template developed from the species–habitat model. The same logic (Figure 16.2) used to build species–habitat models and to develop habitat templates is generally followed in performing an Ecosearch assessment. To estimate whether a given grid cell is habitat for a given vertebrate wildlife species, the model first determines whether the species has a breeding or feeding dependency on aquatic habitats. If an aquatic dependency exists, the model determines if the cell comprises an appropriate wetland. A given habitat condition is predicted to be appropriate if the value for the condition estimated from the data matrices is consistent with the value defined in the hypothesized species–habitat model. For appropriate wetlands, the cell is predicted as habitat if the topography, soil conditions and vegetative structure and composition within the sampling neighborhood are also appropriate. It is otherwise considered as non-habitat. If the cell is other than an appropriate wetland, then the model determines if the cell has appropriate topography, soil conditions and vegetative structure and composition. If a non-appropriate wetland cell also has inappropriate topography, soil conditions or vegetative cover, it is then predicted to be non-habitat. If the intercept cell is found to have an appropriate cover type but to be an inappropriate wetland type, then the model checks to determine if the cell represents an appropriate corridor to an appropriate wetland. If it does not, it is then predicted to be non-habitat.

The model uses topography, soil conditions and point intercept data about habitat structure if the species is not dependent on aquatic habitats for its breeding and feeding activities. A cell that does not fulfil the criteria specified in the habitat template derived from the species–habitat model is considered to be non-habitat for the species.

16.4 THE DEVELOPMENT OF MAPS PREDICTING HABITATS FOR WILDLIFE SPECIES

This study describes a way to predict the utility (as wildlife habitat) of land areas that are assessed using remotely-sensed data. The three maps presented here describe areas of a western Massachusetts study area that are apparently suitable habitat for the meadow vole (*Microtus pennsylvanicus*), the wood thrush (*Hylocichla mustelina*) and the spotted salamander

(*Ambystoma maculatum*). The maps describe areas that satisfy conditions described in habitat templates that have been developed from species–habitat models associating the feeding and breeding activities of wildlife species with identifiable components of landscapes.

16.4.1 Predicting habitat for the meadow vole

The meadow vole inhabits fields, pastures, orchards, freshwater marshes and meadows, borders of streams and lakes, open and wooded swamps and, less commonly, open woods and clearcuts (DeGraaf and Rudis, 1992). The species has a special habitat requirement for herbaceous vegetation and loose organic soils, and has a home range of up to 0.2 acres (DeGraaf and Rudis, 1992). The model for the meadow vole (Table 16.2) indicates that the species breeds and feeds in the upland understory and the wetland understory layers in habitats that were identified at intercept points as upland herbaceous, upland herbaceous mowed, and palustrine emergent persistent which includes sedge meadows, shallow marshes, deep marshes and seep meadows. The assignment of breeding and feeding activities to the understory layers correctly reflects the species' tunneling habits at or near the terrestrial surface, and the assignment of feeding and breeding activities to both upland and wetland understories reflects the species' presence in both upland fields and wetland meadows and shallow marshes. Ecosearch considered each individual cell within the study area as potential habitat for the meadow vole because each cell represented an area (0.6 acres) that exceeded the estimated home range of the vole (0.2 acres). The habitat template produced from the hypothesized species–habitat model considered a sampling neighborhood of nine cells (3 × 3) so that an isolated remnant of grassland or a single isolated grassy backyard would not be identified as vole habitat. The vole model emphasized that a habitat patch needed to be covered predominantly by herbaceous vegetation. Ecosearch predicted a given cell to be vole habitat when at least 30% of the neighborhood of nine cells provided the required habitat conditions. Figure 16.3 represents the map of apparent habitat for the meadow vole on the western Massachusetts study area.

16.4.2 Predicting habitat for the wood thrush

The wood thrush typically places its nest in a fork or saddled on a horizontal limb of a sapling or tree between 1.5 and 15 m (5–50 ft) above the ground or hides its nest in dense shrubbery within the forest (DeGraaf and Rudis, 1992). The wood thrush is thus considered to use the upland tree canopy, the upland shrub canopy, the wetland tree canopy and the wetland shrub canopy for nest sites within forested habitats, and its foraging habits of ground gleaning, scratching and leaf turning on the forest floor clearly

584   *Ecosearch*

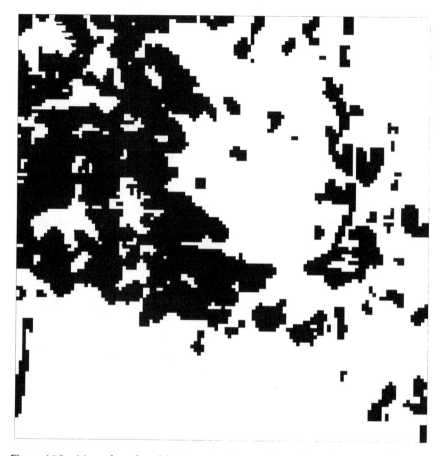

Figure 16.3 Map of predicted habitats for the meadow vole within a 36 km² study area in western Massachusetts.

occur in the upland understory and the wetland understory layers (Table 16.2). The wood thrush requires mature lowland forests (mainly deciduous or mixed); shady, cool, mature upland forests, often near a swamp, pond, stream or lake; and abundant undergrowth in forested habitats (DeGraaf and Rudis, 1992). The species has a special habitat requirement for deciduous or mixed forest with tall trees and abundant sapling growth, and a territory of about 2 ha (DeGraaf and Rudis, 1992). The model for the wood thrush assumes that habitat conditions can best be described at intercept points identified as upland deciduous trees, and wetland deciduous trees especially in areas bordering swamps, ponds, lakes or streams. The model also assumes that habitat for the wood thrush includes a 70% closed canopy and that large (i.e. non-fragmented) forest tracts comprise preferred

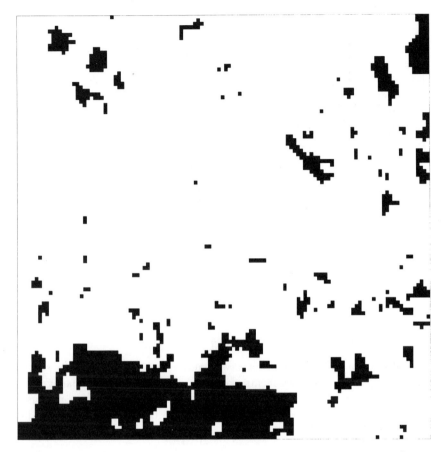

Figure 16.4 Map of predicted habitats for the woodthrush within a 36 km² study area in western Massachusetts.

habitats. Figure 16.4 represents the map of predicted habitat for the wood thrush in the western Massachusetts study area.

16.4.3 Predicting habitat for the spotted salamander

The spotted salamander breeds in the water surface layer and at various stages during its life cycle it feeds in the upland understory layer, the wetland understory layer and the water surface layer (Table 16.2). The animal seems to prefer deciduous or mixed woods on rocky hillsides and shallow woodland ponds or marshy pools that retain water suitable for reproduction. The major habitat requirements for reproduction are mesic woods and semi-permanent water (DeGraaf and Rudis, 1992). The spotted salamander is modeled as a wetland dependent species requiring seasonally flooded or semipermanently flooded wetlands – habitats readily identified on our digitized NWI map. Surface water, in seasonally flooded habitats, is present for extended periods, especially early in the growing season, but is absent by the end of the growing season in most years; it persists throughout the growing season in most years in semi-permanently flooded habitats (Cowardin *et al.*, 1979). Water needs to be retained for a sufficiently long time for eggs to hatch and for larval growth and transformation to occur. Non-permanent water is desirable so that fish predation on eggs and larvae will not occur. The spotted salamander will also utilize a habitat block about 2 ha in area for non-reproductive activities. The habitat model for the spotted salamander indicates that at least 30% of nine cells, each 0.25 ha in area, need to provide the necessary topographic, soil and vertical structure condition identified in Table 16.2 and need to occur within 200 m (within four cells) of seasonally flooded or semi-permanently flooded wetland. Habitat could be a block or a linear corridor with the stipulation that a forested continuity exists from the grid cell to an acceptable water source. Figure 16.5 represents the map of apparent habitat for the spotted salamander on our study area.

16.5 THE UTILITY OF ECOSEARCH

The evaluation of the importance of wildlife habitat to wildlife species drives habitat impact assessments, land-use planning, habitat management and wildlife population management. Such evaluations require a substantial understanding of animal–habitat relationships so that estimates and predictions of animal response to habitat change will be meaningful.

16.5.1 Assessment of habitats

Many efforts have been made to associate wildlife species with habitats in order to predict the usefulness of habitats to a species or the vertebrate wildlife community. These efforts can be arrayed along a gradient which ranges from causal models identifying limiting factors for a species to weak associations of wildlife species with vegetative cover types (Table 16.3).

Species–habitat, cause–effect models strive to determine limiting factors for species or to describe strong inferences that may lead to the rejection of null hypotheses. They can provide strong predictors of habitat requirements for individual species but because of their design they usually have limited capabilities for predicting vegetative cover types, habitat structure or biodiversity. They are detailed studies that are expensive to conduct, but such studies frequently provide basic data used in subsequent modeling efforts.

Ecosearch predicts areas of habitat for an individual wildlife species, determined from a hypothesized species–habitat model based on topo-

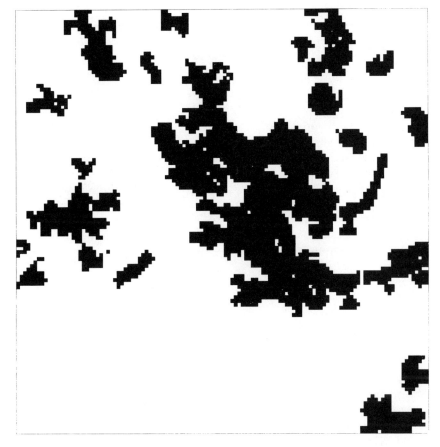

Figure 16.5 Map of predicted habitats for the spotted salamander within a 36 km^2 study area in western Massachusetts.

graphy, soil conditions, water regime and habitat structure. Ecosearch provides a moderate to strong predictor of habitat structure on a study area, a moderate to strong predictor of the suitability of habitat on a study area for individual species, and a moderate to strong predictor of possible biodiversity on a study area. Ecosearch strives to estimate habitat quality by distinguishing useful habitat from non-useful habitat for individual species, but does not try to rank the quality of two or more areas as habitat for that species. We do not rank habitats because Ecosearch is accomplished at a scale that is inappropriate for distinguishing many limiting factors for wildlife species and because it is very difficult to assess and verify relative habitat quality. Scale limits Ecosearch as it does all habitat evaluation assessments. For example, Ecosearch can estimate that a habitat seems

Table 16.3   An array of habitat assessment technologies

Factor	Species habitat, cause–effect models	Eco-search	HEP	Habitat layer models	GAP
Predictive capabilities					
Cover types	L	L	L	L	M–H
Habitat structure	L	M–H	L	M	L
Suitability for species	M–H	M–H	M–H	L	L
Estimate biodiversity	L	M–H	L	M	L
Cost	H	L–M	M–H	L	L–M
Use GIS technologies	N	Y	N	Y	Y
Use statistical assessments	Y	Y	N	N	N

L = Low
M = Moderate
H = High
N = No
Y = Yes

suitable for a cavity-dependent species on the basis of the structure of the tree canopy layer observed in the sampling neighborhood and the likelihood that cavities will exist under those conditions, but field verification for the presence of tree cavities will be needed to verify the accuracy of the predictive model as well as the actual presence of the species. Ecosearch can probably be used to predict the impacts from local or large-scale changes on individual species or the total vertebrate non-fish wildlife community. In the examples cited in this chapter, Ecosearch extensively used the descriptive narratives, habitat relationship matrices and status matrices developed for the management of wildlife habitat on national, state and private forests in New England (DeGraaf and Rudis, 1992; DeGraaf *et al.*, 1992). That extensive data base associated amphibians, reptiles, birds and mammals with general features such as forest and non-forest cover types, within-stand features such as degree of canopy closure, the presence of large-cavity tree holes, the presence of midstory and understory layers, and the presence of dead and down materials on the forest floor, caves, seeps, gravel pits, cliffs and ledges, cut banks, slash piles, etc. Ecosearch also uses current GIS and computer capabilities, and statistical assessments to develop its predictive models.

HSI models have been developed by the US Fish and Wildlife Service and use information about the physical and biological attributes of a particular habitat to yield an index of habitat suitability that is assumed to be proportional to the habitat's carrying capacity for a species (Anderson and Gutz-willer, 1994). HSI models provide moderate to strong predictions of habitat quality for individual species, but are not designed to predict the habitat structure required by the species, and usually they do not try to predict bio-diversity. They are developed from descriptive data and are intended to provide an index representing the value of habitat conditions on a study area compared with the optimum habitat conditions that could exist for that species. The goal of the index – to attain a quantified relationship to carrying capacity – is difficult to achieve. Difficulties occur, and many HSI-models were awkward to validate for several reasons: because frequently it is not evident how physical and biological information about species requirements can be integrated into acceptable and useful predictive models; because the information base describing habitat-limiting factors of indivi-dual species is often inadequate; and because agreement does not exist about what to measure to depict relative habitat quality. For example, population density does not adequately reflect habitat quality because den-sities may mirror short-term changes in environmental conditions, and dominant individuals of some populations may inhabit the most favorable habitats, forcing high-density concentrations of sub-dominant individuals into lower quality habitats (Anderson and Gutzwiller, 1994).

The HSI models were intended to be used in the Habitat Evaluation Pro-cedures to provide a habitat analysis that could assess the biological costs of proposed habitat modifications. A finite number of species are usually selected for an HEP analysis and HSI values are determined for each selected species in the study area. The product of HSI values for each species and the areas of available habitat for each species in the study area, when combined for all the species in the HEP analysis, comprise Habitat Units (HU) which act as currency in assessment or planning studies. HEP analyses, unfortunately, can produce a variety of assessments for the same area in the same time period because the species selected for appraisal can significantly affect the HU values determined present on an area.

The Habitat Layers Index (HLI) Model (Short, 1988) emphasized the association of wildlife species with habitat structure and provided HEP with one way to estimate the utility of a habitat for the total wildlife commu-nity. This model provided a moderately robust predictor of habitat struc-ture on a study area and a moderately robust predictor of biodiversity for a study area but it was not predictive of cover types or the suitability of habitats for individual species. It was based on the assumption that structu-rally complex habitats tend to provide more niche spaces and consequently to accommodate more wildlife species than do structurally simple habitats. The HLI model considered cover types as combinations of habitat layers which are volumes of habitat at prescribed distances above or below the air–terrestrial or air–aquatic surface interfaces. Habitats are compared and evaluated by calculating a summed product of the number of habitat layers present and the area of each of those habitat layers on a study area. The summed product is compared with a maximum measure of layers of habitat area, to provide an HSI measure. It was believed that the resulting index value was related to total species richness on the study area. A field study of the Layers of Habitat model, accomplished in Colorado, provided an index value that was significantly correlated with bird species richness ($r =$ 0.94) (Short, 1992). The Layers of Habitat Model contributed to the development of Ecosearch but does not provide a similar level of predictability or detail.

The simplest and perhaps least predictive habitat assessment model is the association of wildlife species with vegetative cover types. This association forms the basis of the GAP analysis process which maps cover types, and attempts to identify areas of high vegetative diversity, with a high degree of species range overlap, as candidate areas for acquisition as new sanctuaries or as regions where management can enhance biodiversity (Scott *et al.*, 1993). The present GAP project seems to provide a moderate to strong predictor of the location of vegetation cover types but represents a weak predictor of species presence and of biodiversity. Short and Hestbeck (1995) suggest that the hypothesis that species can be associated with vegetative cover types so that the mapping of cover types will predict the presence of species may not be false. However, predictive capabilities may be weak with large error terms. This occurs because species are presumably associated with habitat structure, a variety of structural conditions can occur within a polygon mapped as a specific cover type, and few vertebrate species are restricted to individual vegetation cover types. Only 5% (18 species) of the 338 non-fish vertebrate species in New England have forest-only dependencies and only 17% (58 species) are dependent on a single habitat type, even when a habitat type has an inclusive definition like forest or non-forest (designations that include several vegetation cover types) or water (DeGraaf *et al.*, 1992). Vegetative-cover maps, using GIS capabilities, clearly have utility for estimating the apparent distribution of cover types and for providing a substrate for overlays depicting the distribution of a variety of geological, physical, social or economic attributes. However, the estimation of candidate areas for biological sanctuaries or the definition of areas of high biodiversity are probably among the least precise products that can be developed from these extensive mapping efforts.

16.5.2   Inventory and monitoring of wildlife habitats

The estimate of wildlife resources on an area that is of limited size is a substantial task. Sampling protocols are inadequately determined and standardized for many taxa groups and few agencies and individuals are dedicated to the routine accomplishment of inventories and archiving and analyzing data from such inventories. In addition, areas where wildlife inventories can occur may diminish in number because it may become increasingly difficult to gain access periodically to many private lands for accomplishing on-site surveys. Local inventories describe present status and, if accomplished over time, monitor changes that may be attributable to land-use change or other causal agents. The problems associated with local inventories are daunting. They pale, however, when information needs become regional or national in scope. What is difficult on a limited scale is frequently impossible in a practical sense in a regional or national context.

Ecosearch can advantageously be used in regional or national inventory and monitoring efforts. This is a relevant capability. The mission of the National Biological Service (NBS) of the US Department of the Interior is to describe and understand the status and trends of the nation's biota and to provide that understanding to decision-makers in a form that allows them to assess and predict the biological consequences of various policies and management practices. Ecosearch can become part of a sampling protocol. Thematic mapping can be used to stratify landscapes for sampling. Images with similar signatures comprise strata and the location and areas of these strata can be estimated. Areas within each strata that are in the public domain can be identified and notated. For example, one of the most appropriate places for NBS to conduct inventory and monitoring studies may be in units of the National Wildlife Refuge System of the US Fish and Wildlife Service. Ecosearch provides the mechanism for estimating the taxa that may be expected to occur on samples of the strata determined as being present within a region. Aerial photographs of the selected sites are interpreted to describe water regimes and habitat structure and these data are supplemented with soils and topographic data and are evaluated to predict habitats for wildlife species. On-site assessments accomplished within a stratum on public lands (or elsewhere when permission has been granted) provide positive, negative or inconclusive verification of predictions developed with Ecosearch. The inventory and monitoring products in regional and national inventories are estimates of the quantity of habitat presumed suitable for individual species within the region.

16.5.3 Status and trend information

Ecosearch provides a variety of measures that can be used to estimate status and trends of habitat quantity, habitat quality and biodiversity. The estimate of the area of different wetland regimes on a common area as determined from the interpretation of registered and standardized aerial photographs taken at times t_0 and t_1 is routinely accomplished by the National Wetland Inventory (e.g. Foulis and Tiner, 1994a,b). Such studies identify the acreage of individual wetland types at different time periods, describe the changes and conversions that have occurred to different types of wetlands, and describe the causes of wetland loss or gain.

The summary of point intercept data accomplished on a common area with the interpretation of registered and standardized aerial photographs taken at times t_0 and t_1 can be used to estimate changes in vegetative and landscape cover either within a total test area or within individually described subunits of the test area. Summary data indicating changes from shrub to tree cover can indicate changes in successional state; tree maturation can be suggested by changes in the type of crowns encountered in the assessment; habitat fragmentation can be indicated by changes in the pattern of intercept data within a study area; changes to the wetland regime can be hypothesized on the basis of the vegetative structure encountered by point intercept data; and land-use changes can be substantially described on the basis of the composition of the point intercept data.

Ecosearch can also be used to estimate status and trends of apparent habitat available for individual wildlife species. Clearly the quantity of predicted habitat available for a given species at times t_0 or t_1 on a common area represents two measures of status and the comparison of the two status measurements represents a trend assessment. The Ecosearch analysis performed on a common area identified on registered and standardized aerial photographs taken at times t_0 and t_1 will describe areas of potential habitat for individual species at the two points in time. The presence of predicted habitat for an individual species within a prescribed block at t_0 can represent a data point for that block at t_0. If a comparable assessment is accomplished for all species that can occur in that habitat block then a data set can be compiled which can be used to estimate the potential biodiversity in the habitat block at t_0. If the exercise is repeated for the same block at t_1, then a trend estimate of predicted biodiversity can be accomplished.

16.6   SUMMARY

Maps that predict areas of habitat for wildlife species can be an important tool for enhancing the management of those species. Such maps are especially meaningful if they reflect attributes important to species and can be driven by remotely sensed information. Ecosearch is a methodology for determining landscape areas important as habitats for wildlife species. Ecosearch uses natural history information to build models that describe the 'niche space' of a species and then transforms that niche space description into a habitat template that can be recognized on interpreted aerial photography. Ecosearch also provides a convention for interpreting aerial photographs so that the habitat template is recognizable, and a software program to affect the recognition of habitat templates and to map areas of habitat templates as predicted areas of habitat for an individual species. We describe a variety of potential uses of Ecosearch on blocks of land up to 3000–5000 ha in area.

16.7 ACKNOWLEDGEMENTS

Several persons contributed significantly to this study. David B. Foulis interpreted aerial photographs of the western Massachusetts study area, to provide both a new NWI map of the study area and a point intercept interpretation of vertical structures across the study area. John Eaton digitized the NWI wetland delineations. Diane Pupek compiled and organized data and provided assistance in manuscript preparation. David W. Goodwin, Tamia C. Rudnicky and R. P. Schauffler managed the data through the GRID programs within ARC/INFO.

16.8 REFERENCES

Anderson, S.H. and Gutzwiller, K.J. (1994) Habitat evaluation methods. *Research and Management Techniques for Wildlife and Habitats*, (ed. T.A. Bookout), pp. 592–606. Wildlife Society, Bethesda, MD.

Cody, M.L. (ed.) (1985) *Habitat Selection in Birds*. Academic Press, Orlando, FL. 558 pp.

Cowardin, L.M., Carter, V., Golet, F.C. and LaRoe, E.T. (1979) *Classification of wetlands and deepwater habitats of the United States*. USDI Fish and Wildlife Service, Washington, DC. 131 pp.

DeGraaf, R.M. and Rudis, D.D. (1992) *New England Wildlife: Habitat, Natural History, and Distribution*. USDA Forest Service. Northeastern For. Exp. Stat. Gen. Tech. Rept. NE-108. 491 pp.

DeGraaf, R.M., Yamasaki, M., Leak, W.B. and Lanier, J.W. (1992) *New England Wildlife: Management of Forested Habitats*. USDA Forest Service. Northeastern For. Exp. Stat. Gen. Tech. Rept. NE-144. 271 pp.

ESRI (1992) *Cell-based Modeling with GRID*^{TM}, Version 6, 2nd edn. Environmental Systems Research Institute, Redlands, CA.

Foulis, D.B., and Tiner, R.W. (1994a) *Wetland trends for selected areas of the Casco Bay Estuary of the Gulf of Maine (1974–77 to 1984–87)*. USDI Fish and Wildlife Service, Hadley, MA. Ecological Services report R5-94/1. 15 pp.

Foulis, D.B. and Tiner, R.W. (1994b) *Wetland status and trends in St Mary's County, Maryland (1981–82 to 1988–89)*. USDI Fish and Wildlife Service, Hadley, MA. Ecological Services report R5-93/20. 13 pp.

Geibert, E.H. (1979) Songbird diversity along a powerline right-of-way in an urbanizing Rhode Island environment. *Trans. NE. Sect. Wildl. Soc.* 36:32–44.

Heatwole, H. (1982) A review of structuring in herpetofaunal assemblages, in *Herpetological Communities*, (ed. N.J. Scott), pp. 1–19. USDI Fish and Wildlife Service. Wildl. Res. Rept. 13.

Karr, J.R. (1971) Structure of avian communities in selected Panama and Illinois habitats. *Ecol. Monogr.* 41:207–233.

MacArthur, R.H. (1958) Population ecology of some warblers of northeastern coniferous forests. *Ecology* 39(4):599–619.

MacArthur, R.H., MacArthur, J.W. and Preer, J. (1962) On bird species diversity: II. Prediction of bird census from habitat measurements. *American Naturalist* 96(888):167–174.

Maser, C., Mate, B.R., Franklin, J.F. and Dryness, C.T. (1981) *Natural history of Oregon coast mammals.* USDA Forest Service, Pacific Northwest For. and Range Exp. Stat. Gen. Tech. Rept. PNW-133.

National Wetlands Inventory (1990) *Photointerpretation conventions for the National Wetlands Inventory.* USDI Fish and Wildlife Service, St. Petersburg, FL. 45 pp + Appendices.

National Wetlands Inventory (1992) *Digitizing conventions for the National Wetlands Inventory.* USDI Fish and Wildlife Service, St. Petersburg, FL. 22 pp + attachments.

Rabenold, K.N. (1978) Foraging strategies, diversity, and seasonality in bird communities of Appalachian spruce–fir forests. *Ecol. Monogr.* **48**:397–424.

Root, R.B. (1967) The niche exploitation pattern of the Blue-gray Gnatcatcher. *Ecol. Monogr.* **37**(4):317–350.

Scott, J.M., Davis, F., Csuti, B. *et al.* (1993) GAP analysis: A geographic approach to protection of biological diversity. *Wildl. Monogr.* **123**:1–41.

Short, H.L. (1983) *Wildlife Guilds in Arizona Desert Habitats.* USDI, Bur. Land Manage. Tech. Note 362. 258 pp.

Short, H.L. (1985) Management goals and habitat structure, in *Riparian Ecosystems and their Management: Reconciling Conflicting Uses,* (tech. coords R.R. Johnson, C.D. Zieball, D.R. Patton *et al.*), pp. 257–262. First North American Riparian Conference. USDA For. Serv. Gen. Tech. Rept. RM 120, 523 pp.

Short, H.L. (1988) A habitat structure model for natural resource management. *Journal of Environmental Management* **27**:289–305.

Short, H.L. (1989) A wildlife habitat model for predicting effects of human activities on nesting birds, in *Freshwater Wetlands and Wildlife,* (eds R.R. Sharitz and J.W. Gibbons), pp. 957–973. USDOE Office of Scientific and Technical Information, Oak Ridge, TN.

Short, H.L. (1992) Use of the habitat linear appraisal system to inventory and monitor the structure of habitats, in *Ecological Indicators*, Vol. 2, (eds D.H. McKenzie, D.E. Hyatt and V.J. McDonald), pp. 961–974. Elsevier Applied Science, London and New York.

Short, H.L. and Hestbeck, J.B. (1995) National biotic resource inventories and GAP analysis. *BioScience* **45**(8):535–539.

Short, H.L. and Williamson, S.C. (1986) Evaluating the structure of habitat for wildlife, in *Wildlife 2000: Modeling Habitat Relationships of Terrestrial Vertebrates,* (eds J. Verner, M.L. Morrison and C.J. Ralph), pp. 97–104. Univ. Wisconsin Press, Madison, WI. 470 pp.

Thomas, J.W. (ed.) (1979) *Wildlife Habitats in Managed Forests: The Blue Mountains of Oregon and Washington.* USDA Forest Service. Agriculture Handbook no. 553. 512 pp.

Verner, J. and Boss, A.S. (eds) (1980) *California Wildlife and their Habitats: Western Sierra Nevada.* USDA Forest Service. General Technical Rep. PSW-37. 439 pp.

Wenger, K.F. (ed.) (1984) *Forestry Handbook*, 2nd edn. SAF Publ. No. 84-01. John Wiley and Sons, NY. 1335 pp.

-17

Modern approaches to monitoring changes in forests using maps

Ronald I. Miller

17.1 INTRODUCTION

This chapter is about mapping for planning the wise use of the forest. In many instances, this will involve the integration of timber harvesting together with the protection of forest wildlife. This chapter does not therefore concentrate upon maps of either scientific research plots or timber harvest parcels. It focuses upon modern mapping techniques that can be used for the integrated stewardship of the forests.

Both practical mapping and data collection methods used to monitor forested areas have changed dramatically during the past 20 years. As a result of these innovations, the capability of monitoring the impacts of change on forest ecosystems has substantially improved.

Today policy-making and planning for the use and conservation of forest lands is often conducted at a scale that requires consideration of relatively large tracts of forested landscape. Maps produced at these scales are especially effective for influencing policy-making and planning related to forest change. 'Change' in this chapter refers to changes that occur in the forested landscape from both natural and human disturbance phenomena.) This chapter covers recent developments in the uses of maps for monitoring forest landscapes.

Research in the natural sciences has traditionally focused upon areas that could be mapped at large 'close to the ground' scales that require little reduction for map representation. In contrast, as a result of modern planning demands and developments in landscape ecology, many forest landscape projects focus upon the effects of landscape changes over extensive areas, i.e. the small map scale. This focus upon extensive areas has also

Conservation of Faunal Diversity in Forested Landscapes. Edited by
R.M. DeGraaf and R.I. Miller. Published in 1996 by Chapman & Hall. ISBN 0 412 61890 7.

been promoted with the development of new technologies for data collection (e.g. satellite remote sensing, Global Positioning Systems – GPS) and for map production (i.e. Geographical Information Systems – GIS). Some contemporary issues related to the interpretation and mapping of forest biodiversity data across extensive areas are presented here.

Significantly more information is generated by traditional ecological research at site and local scales. Therefore current information about the ecology of forests is primarily available for mapping at the large scale. Modern landscape change maps, however, often focus upon small map scales.

Today biodiversity mapping projects developed at a local–regional perspective (e.g. Hollander *et al.*, 1994) are rapidly being expanded to include much larger areal extents. In many instances, this will involve the integration of timber harvesting with the protection of forest wildlife. This chapter demonstrates the potentials offered by modern technologies for monitoring the effects of change (e.g. climatic, disturbance) on the ecology of the forest.

17.2   ISSUES OF SCALE AND RESOLUTION THAT PERTAIN TO MODERN FOREST MAPS

Today the monitoring of the distributions of forest vegetation and fauna for conservation purposes requires the integration of many different perspectives (Franklin, 1995) that have developed in the past as parts of separate scientific disciplines. For example, the ecological sciences and the geographical sciences have primarily developed independently from one another. With the advent of Geographical Information Systems, there is a need for a common framework for the overlapping concerns shared by these two sciences.

An analogous situation exists in the domain of computer technology where the scope in the past 20 years has evolved at an exponential rate. For example, a variety of modeling approaches has been used to model forest ecosystems (e.g. Botkin, 1992) and this has produced the need for a greater integration between rapid developments in computer technology and developments in the ecological sciences and in conservation biology (e.g. Haines-Young *et al.*, 1993; Miller, 1994a). These disciplines similarly require a common framework within which to operate (ibid). Ecologists, geographers and computer scientists need to act together today to create this common framework.

17.2.1   Ecological spatial analysis

The answers to ecological questions must be pursued on many different spatial and temporal scales (May, 1994; Miller, 1994b). These scales vary as a result of the very diverse functional and structural aspects that affect ecological systems. Historically studies of ecological phenomena have frequently focused upon short time periods and spatial scales between 1 and 100 m^2. Therefore our understanding of the ecological processes that transpire at the long-term temporal perspective and at the broad spatial scale is limited.

Pattern analysis in ecology often relates the coordinates of species data to the spatial coordinates of environmental factors (Ludwig and Reynolds, 1988; Miller *et al.*, 1989). In this regard, there are generally two basic types of data arrays in landscape ecology. One is used to represent temporal distribution patterns and the other is used to represent spatial distribution patterns (Ludwig and Reynolds, 1988). The focus of this chapter is the representation of spatial distribution patterns.

(a) Ecosystems

In recent years a hierarchical approach to ecosystem modeling is becoming more widely accepted (Klijn and Udo de Haes, 1994; King, 1993; O'Neill *et al.*, 1986; Urban *et al.*, 1987). This approach is exemplified by the different levels of ecosystem homogeneity that can be distinguished today in landscape modeling. The approach can be used as a tool for relating ecosystem processes to spatial scales and thus it provides an ecological framework for environmental policy (Klijn and Udo de Haes, 1994). The use of this hierarchical approach helps to mitigate many problems encountered repeatedly during the past 25 years in attempts at landscape classification.

Using all possible field techniques, patterns and mosaics can be repeatedly subdivided into finer patterns and mosaics. For a single area, given enough time and money, field surveys conducted by *n* individuals will produce *n* different maps (Lowell, 1994) and perhaps *n* different forest landscape classifications. The perception of patterns and mosaics, therefore, is only limited by the tools that are used to collect the data. In addition, the scale of the available data significantly influences our ability to portray the forest landscape. Today at the small map scale our perception of patterns and mosaics is limited and influenced by the available tools. This is particularly true since the techniques that we require to employ these tools most effectively are only now being developed.

(b) Species and habitats

A wide spectrum of interrelated biological and environmental factors influences the distribution of species across the surface of the earth. The long-term conservation of species is therefore dependent upon the consideration of both environmental conditions and biological distributions (Kirkpatrick and Brown, 1994).

Each factor considered in the analysis of a species distribution is usually represented by an extensive range of values. A compilation of the value ranges for the factors that influence the distribution of some rare vertebrate species in Venezuela (e.g. elevation, climate, vegetation types, etc.) illustrates this point (Miller, 1994b:10, Table 1.1). A map needs to consider these value ranges if it is to portray accurately the distribution of a forest wildlife species. Often a species distribution is spatially analyzed using a grid overlay. The accuracy of such an analysis is determined by the minimum value represented in a grid cell. Therefore the ranges of each of the most important factors (e.g. 0–1500 m elevation; 300–800 mm/yr rainfall) need to be considered for an accurate analysis and mapping of forest wildlife species.

The current status of a landscape is dependent upon the interaction of a multitude of biological factors that interface with a multitude of environmental and disturbance factors (Reice, 1994). The mosaic pattern of species, populations and communities that comprise forested landscapes emerges from the interaction between ecological and biological parameters, the distribution of physical features, and the disturbance patterns that dominate in an area (Davis *et al.*, 1994). Techniques are being developed to model spatial distribution patterns of faunal species from assessments of variations in environmental conditions (Aspinall, 1992; Lavers and Haines-Young, 1993).

Modern mapping technologies and spatial information systems are particularly useful for analyzing spatial patterns in the habitats of forested temperate landscapes (e.g. Pastor and Broschart, 1990). For example, in the United States intensive land use in the corn belt and in the lower Mississippi valley has greatly simplified vertical habitat complexity (Flather *et al.*, 1992). Conversely, the landscapes in New England, the southern Appalachians and the northern portions of Michigan and Wisconsin have retained a greater proportion of their vertical habitat complexity (ibid). Modern technologies provide a common system for analyzing the relation between the disparate factors which produce these patterns.

17.2.2   National and international planning

Today, demand is great for ecological data that can be used to model landscapes at the macroscale (Brown, 1995; Flather *et al.*, 1992). The data used for scientific analysis occurs across a wide spectrum of resolution. The resolution of data required to produce maps for national and international policy and planning is usually coarse in comparison.

Presently coarse resolution data have a great value in regional studies conducted within the global climatic change context. Maps at the small scale serve as reference documents for environmental planning and for nature protection. Most importantly, these maps promote the consideration of environmental issues in development planning. These types of map are currently proving useful for environmental assessment work by the World Bank, the Inter-American Development Bank and many other international institutions.

The use of maps at the small scale provides a variety of advantages for conservation/development planning in environmentally sensitive areas. The visual presentation of important environmental factors makes the available data more accessible to project officers under time pressure during the formation of a project. For example, a consultant hired by a country proposing a loan to the World Bank could identify, compile and map the critical factors related to the biodiversity in the region of the project. A data base of environmental information produced in coordination with a GIS system then becomes a useful product to a country. The small-scale map and its associated data base can then be used to track changes in the elements of biodiversity and in the impacts of development.

17.3   CONTEMPORARY CLASSIFICATION OF THE LANDSCAPE

The categories of polygons that comprise any landcover classification system are created to suit the needs and the perspectives of both the architects and the users of the system. The architects' conceptual framework and the needs of the map users serve as the guideposts in the creation of each classification system. However, these perspectives often diverge. For example, the category types useful to scientists are not necessarily the category types most useful to planners and policy makers. Today the classification of satellite imagery requires the use of many different aspects of an integrated GIS (IGIS – Davis and Simonett, in press).

Ecologists and biogeographers have used a variety of mapping approaches over the last 30 years (Plate L) to portray the vegetation cover on the surface of the globe (WCMC, 1992:248). Biogeographers commonly classify landcover types based upon the environmental factors that influence the habitat. A biogeographer classifying habitat types defines categories based upon heat factors, water factors, various chemical factors and various mechanical factors (e.g. Walter, 1973). Development and environmental planners may define categories of landcover based upon the use that is being made of each landscape patch (i.e. land use). For example, a recent World Bank study of land-use changes in Uganda (Jaggannathan *et al.*, 1990) defines landcover categories based upon the factors regulating the population growth, the agricultural system and the livelihood system being used in an area. These categories of land use are the most useful classifiers of landscapes for planners. It is most important that any landcover classification be designed in relation to the uses that will be required of it.

17.3.1   Boundary issues in the definition of landcover categories

Boundaries between landscape classification units in nature are usually not sharp: they are vague and constitute broad transition zones (Hengvelde, 1990). (Hengvelde refers to the boundaries of 'sampling units' in biogeography while this discussion focuses upon the boundaries of polygons on maps that display landscape classifications.) For forests, boundaries exist as transition zones rather than as the limits represented by exact polygon peripheries. Therefore on most depictions of forested areas a somewhat continuous phenomenon is being represented in a discreet fashion (Lowell, 1994).

Many factors determine the sharpness of boundaries between units in nature. The factors that define the sharpness of boundaries between landscape classes produced from an interpretation of satellite imagery (e.g. Thematic Mapper (TM) imagery) are not the same factors that establish landscape boundaries in nature. The validity of boundaries in nature is difficult to assess since boundary definition is based upon the choice of classification criteria and the numerical procedures employed (Hengvelde, 1990). The landcover boundaries derived from satellite imagery interpretation are dependent upon the reflectance characteristics of infra-red light wavelengths. Therefore the future successful use of satellite imagery to map natural landcover types on the earth's surface is dependent upon resolving these different varieties of boundary definition difficulties.

Recent studies suggest that the boundaries of land-cover categories derived from satellite imagery are sharper, more distinct and more accurate using a dynamic learning algorithm in conjunction with a neural network approach (Chen *et al.*, 1995). The use of neural networks presents the capability to resolve highly non-linear and complex boundary problems commonly encountered during the interpretation of remotely sensed data (Chen *et al.*, 1995; Foody *et al.*, 1995).

17.3.2   Accuracy assessment

Scientists have recently focused upon accuracy assessment to define the placement of boundaries for vegetation landcover types on maps (Congalton, 1994). One issue in measuring map accuracy involves testing mapped vegetation categories in relation to known point locations in the field. Another issue involves the accuracy of classification within mapped polygons of a single vegetation category.

Different map scales and data resolutions are employed to produce vegetation classifications. As a result, vegetation landcover categories that exist at one scale of observation may not exist at another. For example, many vegetation 'types' represented at the 1:1 000 000 map scale become mosaics of distinct patches when represented at the 1:50 000 map scale. Therefore the vegetation detail necessary to be represented on a map is defined by the ultimate applications of the map and the scale requirements.

The use of 'fuzzy surfaces' was recently proposed as a methodology that may provide a more accurate representation of forest boundaries on maps (Lowell, 1994). The use of fuzzy surfaces with a GIS permits the representation of forest boundaries as continuous map elements rather than as discreet polygons. This modeling approach is currently being tested with values derived from fuzzy surfaces, conventional maps and field observations (ibid). Further study is required before we know if fuzzy surfaces can replace polygons on conventional maps.

Very similar accuracy assessment issues arise in the placement of mapped polygon boundaries on animal species distribution patterns (Miller, 1994a). Significant further research is necessary before boundaries at small map scales for both vegetation and animal distribution patterns can be prudently and accurately defined.

17.3.3 Classification of forests on the landscape

Today the common framework that is used to monitor forest landcover changes is based upon a 'polygon' perspective. The surface of the globe is viewed as a collection of polygons in two dimensions. The availability of imagery data collected from satellites (i.e. Landsat 1972) has provided a strong impetus for this perspective. This conceptual framework has spawned a multitude of efforts, at every scale, to classify the earth's surface into landcover categories. Forests are presented as a tesselation of polygons that represent different forest types differentiated by factors such as species composition, succession and disturbance.

Compatibility between different landcover classifications is a difficult problem to resolve. Traditionally a unique classification system is developed for the local area identified with a specific project. The shift between local areas, each with a unique landcover classification system, becomes difficult because of the different criteria used to construct each system. The differences between landcover classification systems are based upon both different design approaches and differing ecological conditions. This problem was recently encountered in the northeastern United States during efforts to map the vegetation of New England. For example, in New Hampshire and Connecticut different definitions are used for palustrine categories because of the different hydrologic conditions found in these two states. Therefore swampy areas in each of these states are distinguishable from the consideration of different ecological factors. Since ecological conditions vary considerably, for regionally consistent forest landcover classifications it will always be necessary to somehow conform classification categories.

The estimation of landcover areal extents based upon satellite imagery data is very dependent upon factors such as the spatial scale and the specific design of the landcover classes (Cherrill and McClean, 1995; Moody and Woodcock, 1994). For example, vegetation classes defined from satellite imagery are dominated by differences in spectral reflectance which may represent distinctions in biomass. Many coarse modern estimates of percentage forest cover for single countries are derived from estimates of the mass of carbon in live vegetation per unit area (Olson *et al.*, 1983). These approximations are generally derived using vegetation indices such as the Normalized Difference Vegetation Index (NDVI) or the Least Area Index (LAI) and these approaches permit vegetation characterization at continental scales. These estimates are summarized in a recent global biodiversity status report (WCMC, 1992:Table 18.2).

The design of landcover classes is also strongly regulated by the data used to formulate them. The identification of forest landcover classes on maps is confounded by the different data types used to define the classes. This is because the characteristics of each data type will usually define the configuration of the final forest landcover classes. For example, infra-red imagery is related to vegetation pigments, leaf cell structure and leaf water content. These characteristics regulate the landcover classes that are produced from satellite imagery data. Therefore the reflectance characteristics that chlorophyll lends to vegetation define the vegetation landcover classes that are produced from satellite imagery.

Vegetation classification systems today often include some floristic elements (e.g. areas dominated by deciduous or coniferous species). However, forest landcover classifications built upon satellite imagery usually do not incorporate considerations such as species composition and structural configuration of the forest. In contrast the forest landcover classes formulated by lumber companies are often dominated by timber age-class distinctions that permit estimation of wood harvest rates. Hence the criteria used to define forest landcover classes that appear on regional thematic maps are usually best adapted to the uses for which these maps are designed. In the future, some consistency will need to be brought to these maps so that they can most effectively be used to guide environmental planning and policy-making.

17.4   IMPORTANT SCALE AND RESOLUTION CONCERNS REGARDING FOREST DATA

Field data that document factors such as climate, topography and soils are the data most useful for discriminating among forest habitat types. In contrast, satellite imagery data are most widely used today to define forest landcover types. This contrast between habitat and landcover information underlies some of the current problems related to two of the major categories of information for mapping biodiversity distribution patterns in the forest.

The availability of these different data types varies considerably. For example, the relevant data are often either unavailable or inaccessible for many regions of the world. The following sections document some of the issues encountered in handling one predominant category of forest data: current forest landcover information.

17.4.1 Map scales

A map frequently presents a model of the forest within a temporal and spatial context. Two general categories of map scales are distinguished in the spatial context. A large map scale refers to a map with a reduction ratio of 1:50 000 or less, while a small map scale refers to a map with a reduction ratio of greater than 1:50 000 (Miller, 1994a). Estes and Mooneyhan (1994) have recently presented a useful elaboration of these scale categories that corresponds well with how these categories are used today by both planners and scientists (Table 17.1).

Scientists involved in ecological research generally focus upon the site or local scale. Currently a more comprehensive coverage of the earth by maps that display biodiversity information at site and local scales is needed to improve our understanding of the dynamics between ecological phenomena and the landscape (e.g. Jones and Lawton, 1995).

In recent years, ecological research has more frequently focused upon local and national/regional scales than in the past (e.g. Miller, 1986; Miller *et al.*, 1989). A recent monitoring program in Canada distinguishes between intensive monitoring sites and extensive monitoring sites (Shackell *et al.*, 1993). This distinction provides a setting for the identification of the most appropriate scale for handling each ecological factor under consideration in a monitoring program.

The comprehensive data required to enhance our understanding at the continental or global scales is still not available (Estes and Mooneyhan, 1994). An improved understanding of the earth's integrated ecological systems will only come from future research that considers the interrelationships between ecological factors at different scales of observation.

17.4.2 An example of global forest data currently derived from small-scale maps

Efforts to compile summary information about natural resources across the globe are vital for environmentally effective international planning and

Table 17.1 Scale categories (after Estes and Mooneyhan, 1994)

Scale	Reduction rate
Site	1:10 000 or larger
Local	1:10 000 to 1:50 000
National/regional	1:50 000 to 1:250 000
Continental	1:250 000 to 1:1 000 000
Global	1:1 000 000 or smaller

decision-making. Modern decision-making necessitates utilization and integration of all available data (Cherrill and McClean, 1995). This is especially problematic where the definitions of land-cover categories are set by organizations with different objectives. The World Resources Institute (WRI) is a leader in the quest for summarized environmental data that can be portrayed at the small map scale. WRI compiles summary information about natural resources on a country by country basis. This generalized information is often useful for broad planning purposes. The summary tables are based upon data at a scale that fits the national and international perspective.

The most recent WRI report includes up-to-date information regarding forest change, protected areas and levels of human disturbance for each nation (World Resources Institute, 1994:318–319). This summary information was in general extracted from pre-existing maps. The map used to classify the areas of low, medium and high human disturbance in this presentation was at a relatively coarse resolution (it was produced with a minimum mapping unit of 40 000 ha). Another table from the WRI report presents information about the extent and loss of forest landcover in the 1980s (WRI, 1994:320, Table 20.3). This table represents a fusion of many different data sources that yields a summary of the best current forest landcover information available at the national scale. It contains coarse estimates of changes that are occurring in the extents of forests. Most importantly, the WRI report includes explicit documentation about the maps and data sources from which the summary information is derived. The documentation allows users of this forest landcover information to validate any subsequent applications of the information.

A true understanding of forest change at either the local or the national/regional scale requires integration of a historical perspective together with current field data (e.g. Chapter 1; also Phillips and Gentry, 1994). Many of the processes that contribute to forest change are phenomena that can only be understood at the local and site scales. At the present time, large-scale air photos are effective for generating authentic maps of forested areas in a region. The airborne application of various developing technologies (section 17.5.4) provides the potential to produce authentic, detailed maps of forested areas at the regional scale. The successful application of these approaches at this scale will invoke a reduced cost from the traditional aerial photography approach. However, this technology still requires a significant investment of time and money for its successful implementation in a specific region.

The data presented in the tables of the WRI report are very coarse by scientific standards and therefore would not usually be used for scientific analysis purposes, particularly since disturbance is inherently scale-dependent (Reice, 1994). However, these are probably the best national data available that can be used effectively for planning purposes and for pointing out broad trends in human disturbance in specific countries. The status of this summary information points out the crux of a gulf which exists between policy-making organizations and scientists. Scientists have historically worked with data at relatively fine detail, often with minimum mapping units of 1–2 ha or less. In contrast, national and international organizations find this summary information adequate for their present needs – particularly since this is the best information available that relates to the national regional scale.

17.5   CONTEMPORARY MONITORING OF FOREST BIODIVERSITY AT THE BROAD SCALE

17.5.1   Some modern digital landcover maps

Digital forest landcover maps produced from AVHRR (Advanced Very High Resolution Radiometry) imagery are now available for the United States. (Zhu and Evans, 1994) and for South America (Stone *et al.*, 1994). The forested landscape for the US maps was classified into 23 landcover types and the minimum mapping unit (= map resolution) was 1 km². The forest landcover on these US maps compares well with the Forest Inventory and Analysis (FIA) maps produced by the US Forest Service (Evans, 1994). However, significant complications affect the adequate statistical field validation of maps produced using AVHRR data (Evans, 1994). Consequently these maps are useful only to observe trends at the national scale (Zhu and Evans, 1994).

For the United Kingdom, landcover maps are now available that are produced from Landsat TM (Thematic Mapper) imagery data (Fuller *et al.*, 1994). These maps classify the landscape into 25 classes with a minimum mapping unit of 25 m. Therefore the map of the United Kingdom provides greater detail due to its use of Thematic Mapper imagery than the US map produced with AVHRR imagery. A greater accuracy in landcover classification was also achieved on these UK maps with the use of satellite imagery from several seasons. Yet several difficulties were encountered during the production of the maps and cartographic ambiguities persist (Fuller *et al.*, 1994; R. Haines-Young, personal communication).

17.5.2   Current approaches to mapping the forest

The development of GIS technology in recent years has far outpaced the collection of information related to the ecology of the landscape. Errors in regard to the identification of landcover types and their areal extents currently far exceed errors that involve cartographic precision (Estes and Mooneyhan, 1994; Moody and Woodcock, 1994). One result of this dichotomy between the technology and the data is that modern attempts at landcover classification on the global scale vary considerably (Figure 17.1). In this regard the landcover heterogeneity and the species diversity of forest cover in many regions of the world provide serious difficulties for mapping vegetation over broad areas.

Problems that involve the accurate representation of forests at the small map scale involve definition problems that are due to the heterogeneity of forest habitats (e.g. the northeastern United States). Similarly the mapping of ecosystem changes at the small map scale is difficult as a result of the spectrum of ecosystem conditions and ecosystem change factors (WCMC, 1992:250). In contrast, boundaries of forested areas are easier to define and identify than boundaries of many other types of vegetation. Both naturally occurring and disturbance-induced successional changes are the dominant factors that sometimes make the identification of forest boundaries difficult. In forested vegetation the ability to map accurately different forest types on the regional scale is primarily dependent upon the resolution capabilities of the data collection techniques.

The TM sensors aboard Landsats 4 and 5, launched in 1982 and 1984, respectively, provide enhanced radiometric resolution superior to the MSS sensor. However, considerable difficulties are encountered today during the interpretation of all types of satellite imagery for the production of forest landcover maps. The many categories of problems include issues related to validation, classification and image processing. These problems may involve issues such as:

- mapping the distribution of tree species in the forest ecosystem, which is a signal detection problem (Lees, 1995); it is therefore not possible to distinguish the characteristics that differentiate between tree species with widely used satellite imagery technologies (i.e. AVHRR, TM, SPOT);
- the use of differing field survey data for the validation of landcover classes;
- attempts to 'cross-walk' (interrelate) landcover classifications produced by different sources within a region;
- software difficulties that involve the 'edge-matching' of contiguous digital map quadrangles.

Some of these difficulties were discussed at a recent symposium which focused upon the range of problems currently encountered in the use of spatial data (Congalton, 1994). These problems create inconsistencies in the landcover maps and make it currently difficult to place confidence in the reliability of some of the final map products.

The approaches that employ aerial photography, aerial video, the GPS, field surveys and satellite imagery interpretation are currently the most effective approaches for mapping the forest. The production of reliable landcover maps for regional, national, continental and global scales is becoming increasingly dependent upon the integration of the capabilities of these many different developing technologies for data collection.

17.5.3   Monitoring faunal changes in forests

Many causes of current forest decline are hypothesized. Changes in forest habitat, perhaps brought about by processes such as global climate change, are a result of the interaction between ecological and anthropogenic phenomena. Many current studies map forest ecosystems to measure the potential impact of global change (e.g. Baker *et al.*, 1995). Future approaches to the effects of forest change will integrate the impacts of ecosystem, species and human phenomena (Jones and Lawton, 1995).

The impact of climate change is currently an important consideration for environmental policy-makers and planners. However, spatial-analytical methods regarding species–climate relationships have developed slowly because of data inavailability and the incompatibilities between the data types (Miller *et al.*, 1989). Yet the importance of microhabitats as useful predictors of species abundance patterns at the landscape scale is well known (e.g. Urban and Smith, 1989). Recently a modeling method was introduced to measure the effects of climate change in the abundance and distribution of animal species (Aspinall and Matthews, 1994). This modeling approach was tested with a butterfly species occupying semi-natural deciduous woodland in Scotland. Data resolution problems still existed, i.e. the species data originated at a resolution of 10 km^2 while the climate data was at a resolution of 1 km^2. However, this approach provides a potential practical analytical method for analyzing climate–species relationships.

The influence of forest habitat change on birds is being appraised frequently today. Mapping techniques that configure the spatial effects of processes influencing the vulnerability of forest birds are currently being developed (Gustafson *et al.*, 1993). Recent studies of boreal forest versus clearcut habitats in Sweden found that the forest bird species showed great flexibility in habitat selection (Hansson, 1983, 1994). These studies suggest that human impact resulting in patchiness of forests has actually increased the adaptability of many of the forest vertebrates in temperate and boreal forests. In general, forest fragmentation is considered one of the most serious threats to birds in forested landscapes today (Chapter 7; Kattan *et al.*, 1994; Willson *et al.*, 1994).

For the effective monitoring of faunal changes in forests of a region, the elements underlying the faunal changes need to be accurately represented on small map scales. This will only be accomplished conclusively from an understanding of the common functional elements and processes that underlie the current decline of forests and forest biota. Spatially explicit models have the potential to facilitate the future management of bird species populations in forested landscapes (Liu *et al.*, 1995). In the future, accurate map representations of predicted faunal changes in forests will permit policy decision-makers and planners to initiate programs which will reduce the loss of forest fauna.

17.5.4   Some promising modern approaches for monitoring the forest at the broad scale

Since the introduction of the Landsat technology in the 1970s, a major focus in the advancement of this technology has been the development of its use in vegetation mapping (Davis and Simonett, in press). Some of the major areas that have required development have included:

- the registration of the imagery and associated field data to geographic coordinates;
- the correct processing of these data and their conversion to a valid and usable format;
- the accumulation of sufficient field data for the verification of a landscape classification.

Successful developments in these areas would permit the production of relatively detailed small-scale vegetation maps for large areas (e.g. regions, nations, continents, etc.).

The complexity of the New England forest landscape is well documented. Forests in New England are made up of heterogeneous assemblages of plant species that have been acutely influenced by significant historical land-use patterns (Chapter 1). It is therefore difficult to identify individual New England forest patches in the field with specific landscape classes on a map. The principal difficulty when using satellite imagery in this region is that the forest classes are identified by such a variation of spectral values. A relatively accurate classification of the forested landscape (Plate M) has now been produced by using imagery from several dates together with hyperclustering – a recent technique that generates a large number of classes (categories) that serve to form the polygons in a preliminary unsupervised classification of satellite imagery. These polygons are then characterized using interactive techniques (Slaymaker *et al.*, in press). Combined with airborne video techniques, which collect data over wide regions for validation of the forest maps, this approach is currently being used to map the forested landscape in the New England region (Plate N).

The airborne collection of field data is a very promising development which is leading to the accurate verification of maps that are needed for land areas of extents above the site and local scales. The use of airborne technologies permits the collection of field data that covers a wide enough extent of the landscape for the production of small-scale forest landscape maps. Some of the current technological possibilities include: traditional aerial photography (Chapter 16); 35 mm photography stored as digital data on CD-ROM; and airborne video stored as digital data. The digital nature of these techniques provides the capability of map production and analysis using a GIS. Most importantly, these techniques provide us with the ability to verify and refine vegetation features identified using satellite imagery.

The cost of available and developing airborne technologies is the factor that currently limits their use. At this time, the airborne video technique seems to provide the most cost-effective means for data collection on the broad scale (Bossler *et al.*, 1994; Evans, 1994; Graham, 1993; Slaymaker *et al.*, in press). In the future the costs will most probably determine which of these technologies becomes most widely used for the verification of small-scale forest maps.

17.5.5   Monitoring forests for planning and policy-making

Management strategies for protecting biodiversity on forested landscapes are still being developed (Kuusipalo and Kangas, 1994). Nevertheless, assessing the landcover change in forested areas is becoming more exact using comparisons of satellite imagery from different dates (Green *et al.*, 1994; Wolter *et al.*, 1995) together with the development of ground-truth data collection techniques. An increase in the accuracy of the classification of temperate forest has been achieved using satellite imagery from different seasons of the year (Wolter *et al.*, 1995). Identification of phenological changes enhances the potential for more accurate classification of tree genera and species (Wolter *et al.*, 1995). However, relating land-use change to landcover change is still difficult and requires further study (e.g. Green *et al.*, 1994). Future planning in both conservation and development requires a greater understanding of this relationship between land use and landcover changes.

The new concept of sustainable forestry proposes a more integrated approach to the use and conservation of forest ecosystems. However, research to date is insufficient to indicate how sustainable forest ecosystem management will evolve from conventional sustained yield timber practices (Aplet *et al.*, 1993).

Many hypotheses are being proposed today to explain the possible effects of climate change on the forest landscape and its associated biodiversity. A case in point is a recent hypothesis that modern forest decline is caused partially by simplification of forest structure and by the onslaught of more frequent and intensified climatic perturbations (Mueller-Dombois, 1993). Hypotheses such as this require substantial testing and analysis at the national/regional, continental and global scales. The eventual verification of such hypotheses will permit realistic modeling of the anthropogenic and ecological factors that produce forest decline.

The past 75 years have seen the formation and alliance of national and international organizations at a rapid pace. This has resulted in international agreements which have incorporated some commonality of purpose. (The United Nations and the World Bank originated from just this impetus.) In recent years environmental issues have become a focus of international concern and the need for environmental information has grown precipitously. These two trends have now converged so that the national and international organizations now require extensive natural resources information for planning purposes. Maps produced today using components from Integrated Geographical Information Systems (IGISs) (Davis and Simonett, in press) are the most effective means for representing this information and communicating its current status on the small map scales most useful to these organizations.

17.6   CONCLUSIONS

Individuals and institutions across the spectrum of current environmental concern require a mass of information at many different scales concerning the status of the forests. Yet comprehensive data at scales above the site level are often not available. The strategy presented in Chapter 1 can be effectively used to portray present change patterns. This approach provides a template upon which regional change concerns can be measured. Some of the techniques presented in this chapter provide capabilities for the collection of more comprehensive information about the forest at scales above the local level.

Some promising new spatial modeling techniques may provide useful tools for more accurately portraying the forest landscape. For example, a greater understanding of mapped transition zones between forest types is still required. Evidence suggests that fuzzy surfaces may provide a more realistic depiction of forest landscape maps than is currently possible with conventional thematic maps (Lowell, 1994).

In science the limitations of available technologies have often regulated the answers that could be provided. Answers to the questions being asked today regarding forested landscapes are governed by the data collection capabilities of the available technologies. The answers to questions about the effects of change on forest communities need to be more clearly presented within the context of historical change patterns. In the future this will permit clearer and more effective design and implementation of policy-making in regard to forested landscapes.

17.7   SUMMARY

This chapter is a review of recent applications of mapping technologies and spatial information systems that can be used for the monitoring of forested landscapes. This encompasses aspects of species, habitat and ecosystem conservation and also applications that can be used for the execution of conservation/development policy-making and planning.

Some important issues are discussed in this chapter in regard to mapping and the delineation of patch boundaries on the forested landscape. The process for classifying forest types and for assessing the accuracy of mapped boundaries of forest patches are considered. Also some important issues in regard to the resolution of forest data and the scale of forest maps are discussed.

The chapter examines some issues that concern the use of global forest information derived from small-scale maps. Some ideas regarding the functionality of this information are discussed. In regard to small-scale maps, some current map coverages at the national, continental and global scales are described.

A multistage approach to landcover mapping is considered which employs satellite imagery interpretation in conjunction with video techniques for ground-truth. A practical application of inexpensive airborne video techniques for enhancing regional forest maps is considered.

17.8 ACKNOWLEDGEMENTS

I thank Dana Slaymaker at the University of Massachusetts for providing me with a valuable account of the video procedures being developed to produce the vegetation maps for the New England GAP analysis project. I would also like to thank Frank Davis at the University of California in Santa Barbara, California, and Douglas Muchoney, formerly at the Smithsonian Conservation and Research Center in Front Royale, Virginia, for their valuable, thorough reviews of this material.

17.9 REFERENCES

Aplet, G., Johnson, N., Olson, J. and Sample, V. (eds) (1993) *Defining Sustainable Forestry*, Island Press, Washington DC.

Aspinall, R. (1992) An inductive modelling procedure based on Bayes' theorem for analysis of pattern in spatial data. *Int. J. Geographical Information Systems* 6(2):105–121.

Aspinall, R. and Matthews, K. (1994) Climate change impact on distribution and abundance of wildlife species: An analytical approach using GIS. *Environmental Pollution* 86:217–223.

Baker, W.L., Honaker, J.J. and Weisberg, P.J. (1995) Using aerial photography and GIS to map the forest-tundra ecotone in Rocky Mountain National Park, Colorado, for global change research. *Photogrammetric Engineering and Remote Sensing* 61(3):313–320.

Bossler, J.D., Novak, K. and Johnson, P.C. (1994) Digital mapping on the ground and from the air. *Geographical Information Systems*:44–48.

Botkin, D.B. (1992) *Forest Dynamics: An Ecological Model.* Oxford University Press, Oxford, UK.

Brown, J.H. (1995) *Macroecology.* University of Chicago Press, Chicago, USA. 269 pp.

Chen, K.S., Tzeng, Y.Z., Chen, C.F. and Kao, W.L. (1995) Land-cover classification of multispectral imagery using a dynamic learning neural network. *Photogrammetric Engineering and Remote Sensing* 61(4):403–408.

Cherrill, A. and McClean, C. (1995) An investigation of uncertainty in field habitat mapping and the implications for detecting land cover change. *Landscape Ecology* 10(1):5–22.

Congalton, R.G. (ed.) (1994) *International Symposium on the Spatial Accuracy of Natural Resource Databases, 16–20 May, Williamsburg, Virginia.* American Society for Photogrammetry and Remote Sensing, Bethesda, Maryland USA.

Davis, F.W. and Simonett, G.S. (in press) GIS and remote sensing.

Davis, M.B., Sugita, S., Calcote, R.R. *et al.* (1994) Historical development of alternate communities in a hemlock–hardwood forest in northern Michigan, in *Large-scale Ecology and Conservation Biology*, (eds P.J. Edwards, R.M. May and N.R. Webb), pp. 19–40. Blackwell Scientific, London, England.

Estes, J.E. and Mooneyhan, D.W. (1994) Of maps and myths. *Photogrammetric Engineering and Remote Sensing* 60(5):517–524.

Evans, D.L. (1994) Some considerations for AVHRR forest classification accuracy assessments, in *International Symposium on the Spatial Accuracy of Natural Resource Data Bases*, (ed. R.G. Congalton), pp. 161–167. American Society for Photogrammetry and Remote Sensing, Bethesda, Maryland.

Flather, C.H., Brady, S.J. and Inkley, D.B. (1992) Regional habitat appraisals of wildlife communities: A landscape level evaluation of a resource planning model using avian distribution data. *Landscape Ecology* 7(2):137–147.

Foody, G.M., McCulloch, M.B. and Yates, W.B. (1995) Classification of remotely sensed data by an artificial neural network: issues related to training data characteristics. *Photogrammetric Engineering and Remote Sensing* 61(4):391–401.

Franklin, J.F. (1995) Why link species conservation, environmental protection, and resources management? in *Linking Species and Ecosystems*, (eds G. Jones and J.H. Lawton), pp. 326–335. Chapman & Hall, New York and London, 383 pp.

Fuller, R.M., Groom, G.B. and Jones, A.R. (1994) The land cover map of Great Britain: an automated classification of Landsat Thematic Mapper data. *Photogrammetric Engineering and Remote Sensing* 60(5):553–562.

Graham, L. (1993) Airborne video for near-real-time natural resource applications. *Journal of Forestry* 91:28–32.

Green, K., Kempka, D. and Lackey, L. (1994) Using remote sensing to detect and monitor land-cover and land-use change. *Photogrammetric Engineering and Remote Sensing* 60(3):331–337.

Gustafson, E.J., Crow, T.R. and Thompson, F.R., III (1993) Assessing the impact of forest management strategies on the vulnerability of forest birds to cowbird brood parasitism with a spatially explicit GIS model. *Second International Conference/Workshop on Integrating GIS and Environmental Modeling, September 26–30, 1993*, Breckenridge, Colorado, USA.

Haines-Young, R., Green, D.R. and Cousins, S.H. (eds) (1993) *Landscape Ecology and GIS.* Taylor & Francis, London.

Hansson, L. (1983) Bird numbers across edges between mature conifer forests and clearcuts in central Sweden. *Ornis Scandinavia* 14:97–103.

Hansson, L. (1994) Vertebrate distributions relative to clear-cut edges in a boreal forest landscape. *Landscape Ecology* 9(2):105–115.

Hengvelde, R. (1990) *Dynamic Biogeography.* Cambridge University Press, Cambridge, England.

Hollander, A.D., Davis, F.W. and Stoms, D.M. (1994) Hierarchical representations of species distributions using maps, images, and sighting data, in *Mapping the Diversity of Nature*, (ed. R.I. Miller), pp. 71–88. Chapman & Hall, London.

Jaggannathan, N.V., Mori, H. and Hassan, H.M. (1990) *Applications of Geographical Information Systems in Economic Analysis: A case study of Uganda.* The World Bank, Environment Working Paper No 27, April 1990.

Jones, C.G. and Lawton, J.H. (1995) *Linking Species and Ecosystems.* Chapman & Hall, New York and London, 383 pp.

Kattan, G.H., Alvarez-Lopez, H. and Giraldo, M. (1994) Forest fragmentation and bird extinctions: San Antonio eighty years later. *Conservation Biology* 8(1):138–146.

King, A.W. (1993) Considerations of scale and hierarchy, in *Ecological Integrity and the Management of Ecosystems*, (eds S. Woodley, J. Kay and G. Francis), pp. 19–45. St Lucie Press, 220 pp.

Klijn, F., and Udo de Haes, H.A. (1994) A hierarchical approach to ecosystems and the implications for ecological land classification. *Landscape Ecology* 9(2):89–104.

Kirkpatrick, J.B. and Brown, M.J. (1994) A comparison of direct and environmental domain approaches to planning reservation of forest higher plant communities and species in Tasmania. *Conservation Biology* 8(1):217–224.

Kuusipalo, J. and Kangas, J. (1994) Managing biodiversity in a forestry environment. *Conservation Biology* 8(2):450–460.

Lavers, C. and Haines-Young, R. (1993) The use of landscape models for the prediction of the environmental impact of forestry, in *Landscape Ecology and GIS*, Taylor & Francis, London, pp. 273–281.

Lees, B. (1995) Species mapping in forest ecosystems. Paper presented at the International Union of Forestry Research Organizations (IUFRO) XX World Congress, *Caring for the Forest: Research in a Changing World*, August 6–12, 1995, Tampere, Finland.

Liu, L., Dunning, J.B.J. and Pulliam, H.R. (1995) Potential effects of a forest management plan on Bachman's Sparrows (*Aimophila aestivalis*): linking a spatially explicit model with GIS. *Conservation Biology* 9(1):62–75.

Lowell, L. (1994) Initial studies in fuzzy surface-based cartographic representations of forests, in *International Symposium on the Spatial Accuracy of Natural Resource Data Bases*, (ed. R.G. Congalton), pp. 168–177. American Society for Photogrammetry and Remote Sensing, Bethesda, Maryland.

Ludwig, J.A. and Reynolds, J.F. (1988) *Statistical Ecology: A Primer on Methods and Computing.* John Wiley & Sons, New York.

May, R.M. (1994) The effects of spatial scale on ecological questions and answers, in *Large-scale Ecology and Conservation Biology*, (eds P.J. Edwards, R.M. May and N.R. Webb), pp. 1–18. Blackwell Scientific, London.

Miller, R.I. (1986) Predicting distribution patterns of vascular plants in the southern Appalachians of the southeastern United States. *Journal of Biogeography* 13:193–311.

Miller, R.I. (1994a) *Mapping the Diversity of Nature.* Chapman & Hall, London.

Miller, R.I. (1994b) Setting the scene, in *Mapping the Diversity of Nature*, (ed. R.I. Miller), pp. 3–17. Chapman & Hall, London.

Miller, R.I., Stuart, S.N. and Howell, K.N. (1989) A methodology for analyzing rare species distribution patterns utilizing GIS technology: The rare birds of Tanzania. *Journal of Landscape Ecology* 2(3):173–189.

Moody, A. and Woodcock, C.E. (1994) Scale-dependent errors in the estimation of land-cover proportions: Implications of global land-cover datasets. *Photogrammetric Engineering and Remote Sensing* 60(5):585–594.

Muchoney, D.M. (1994) Relationships and divergence of vegetation and mapping classifications. *Proceedings of the 5th ASIS SIG/CR Classification Research Workshop, 57th ASIS Annual Meeting, October 16–20, 1994*, Alexandria, Virginia.

Muchoney, D.M. and Haack, B.N. (1994) Change detection for monitoring forest defoliation. *Photogrammetric Engineering and Remote Sensing* 60(10):1243–1251.

Mueller-Dombois, D. (1993) Biotic impoverishment and climatic change: Global causes of forest decline, in *Forest Decline in the Atlantic and the Pacific Region*, (ed. R.F. Huettl). Springer Verlag, New York.

Olson, J.S., Watts, J.A. and Allison, L.J. (1983) *Carbon in Live Vegetation of Major World Ecosystems*. Oak Ridge National Laboratory, Oak Ridge, Tennessee, USA.

O'Neill, R.V., DeAngelis, D.L., Waide, J.B. and Allen, T.F.H. (1986) *A Hierarchical Concept of Ecosystems*. Princeton University Press, Princeton, New Jersey, USA.

Pastor, J. and Broschart, M. (1990) The spatial pattern of a northern conifer–hardwood landscape. *Landscape Ecology* 4(1):55–68.

Phillips, O.L. and Gentry, A.H. (1994) Increasing turnover through time in tropical forests. *Science* 263:954–958.

Reice, S.R. (1994) Nonequilibrium determinants of biological community structure. *American Scientist* 82:424–435.

Shackell, N.L., Freedman, B. and Staicer, C. (1993) National environmental monitoring: a case study of the Atlantic Maritime Region, in *Ecological Integrity and the Management of Ecosystems*, (eds S. Woodley, J. Kay and G. Francis), pp. 131–153. St Lucie Press, 220 pp.

Slaymaker, D.M., Jones, K.M.L., Griffin, C.R. and Finn, J.T. (in press) Mapping deciduous forests in New England using aerial videography and multi-temporal Landsat TM imagery.

Stone, T.A., Schlesinger, P., Houghton, R.A. and Woodwell, G.M. (1994) A map of the vegetation of South America based on satellite imagery. *Photogrammetric Engineering and Remote Sensing* 60(5):541–551.

Urban, D.L. and Smith, T.M. (1989) Microhabitat pattern and the structure of forest bird communities. *American Naturalist* 133(6):811–829.

Urban, D.L., O'Neill, R.V. and Shugart, H.H. Jr (1987) A hierarchical perspective can help scientists understand spatial patterns. *Bioscience* 37(2):119–127.

Walter, H. (1973) *Vegetation of the Earth*, Vol. 15. Springer-Verlag, New York.

Willson, M.F., De Santo, T.L., Sabag, C. and Armesto, J.J. (1994) Avian communities of fragmented south-temperature rainforests in Chile. *Conservation Biology* 8(2):508–520.

Wolter, P.T., Mladenoff, D.J., Host, G.E. and Crow, T.R. (1995) Improved forest classification in the northern Lake States using multi-temporal Landsat imagery. *Photogrammetric Engineering and Remote Sensing* LXI(9):1129–1143.

World Conservation Monitoring Centre (WCMC) (1992) *Global Biodiversity: Status of the Earth's Living Resources*. Chapman & Hall, London.

World Resources Institute (1994) *World Resources 1994–95*. Oxford University Press, New York.

Zhu, Z. and Evans, D.L. (1994) US forest types and predicted percent forest cover from AVHRR data. *Photogrammetric Engineering and Remote Sensing* 60(5):525–531.

Some conclusions

The presentations in this volume demonstrate convincingly that the policies and practices that regulate the conservation and management of forest landscapes are notably interdependent with the dynamics of forest wildlife populations. This fact is emphasized in the chapters on waterbirds, capercaillie, raptors, boreal forest birds, carnivores and ungulates, especially moose.

Wildlife species that depend on habitats in forested ecosystems for critical resources in their life cycles (e.g. breeding sites, food resources, etc.) are seriously being affected by the encroachment of human civilization and by intensive forest management. For example, presently forest wildlife are particularly affected by the conversion of forested landscapes to human activities such as cropland, pasture and highway development as well as exploitative deforestation, especially in the tropical regions of the world.

The concept of change is complex and it is comprised of different dimensions. Each of these dimensions can be characterized by many different axes of variation. In most regions of the globe, it is difficult to separate clearly the forest changes that are due to human impact from those that are due to variations in natural phenomena such as climate. For instance, in the northeastern region of the United States, changes in the forest landscape have followed cyclical patterns formed as a result of both naturally occurring phenomena and from the changing use of forest resources by human inhabitants.

The chapters on tools in this volume indicate that, during the implementation of environmental policy and planning in regard to forest vegetation and wildlife, a diverse spectrum of technological and data resources is available and should be brought to bear on the implementation of effective forest conservation. In this regard it is noted that monitoring forested ecosystems is as much influenced by information that is collected using the most mundane and down-to-earth field techniques as by data that are collected from technologies in airplanes and satellites. Advances in technologies are currently benefiting forest monitoring. However, the context of planning and policy-making in regard to forested landscapes should not be set by the limits of technologies. The context should be set by known historical and modern ecological information about the involved landscapes and their natural disturbance regimes. Research in reference landscapes is vital to our understanding of the roles of natural disturbances on the distributions of different taxa on different sites and in different regions of the world. Such work requires international cooperation between managers and researchers in reference areas and in landscapes in need of restoration or improved management. Current technological capabilities should then be used to enhance the depiction of landscapes and to assess their dynamic characteristics.

Species index

Page numbers in **bold** refer to figures, those in *italic* refer to tables

Subject index

Page numbers in **bold** refer to figures, those in *italic* refer to tables